Nucleic Acids and Molecular Biology

14

Series Editor
H. J. Gross

Springer

Berlin
Heidelberg
New York
Hong Kong
London
Milan
Paris
Tokyo

Alfred M. Pingoud (Ed.)

Restriction Endonucleases

With 100 Figures, 27 of Them in Color

 Springer

Professor Dr. ALFRED M. PINGOUD
Institute for Biochemistry
Justus-Liebig-University
Heinrich-Buff-Ring 58
35392 Giessen
Germany

ISSN 0933-1891

ISBN 3-540-20502-0 Springer-Verlag Berlin Heidelberg New York

Library of Congress Cataloging-in-Publication Data

Restriction endonucleases / Alfred M. Pingoud.
    p. cm. -- (Nucleic acids and molecular biology)
  Includes bibliographical references and index.
  ISBN 3-540-20502-0 (alk. paper)
    1. Restriction enzymes, DNA. 2. Endonucleases. 3. Gene amplification. I. Pingoud, A.
(Alfred) II. Series.

QP609.R44R4732003
572'.785--dc22

2003064952

Springer-Verlag is a part of Springer Science+Business Media

springeronline.com

© Springer-Verlag Berlin Heidelberg 2004

Printed in Germany

The use of general descriptive names, registered names, trademarks, etc. in this publication does not imply, even in the absence of a specific statement, that such names are exempt from the relevant protective laws and regulations and therefore free for general use.

Production and typesetting: Friedmut Kröner, 69115 Heidelberg, Germany
Cover design: *design & production* GmbH, 69126 Heidelberg, Germany

31/3150 YK – 5 4 3 2 1 0 – Printed on acid free paper

Preface

Approximately 50 years ago the phenomenon of restriction was first described. This led later to the discovery of restriction (and modification) enzymes by Werner Arber, Daniel Nathans and Hamilton Smith who received the Nobel Prize in Physiology or Medicine 1978 "for the discovery of restriction enzymes and their application to problems of molecular genetics". In retrospect, it is clear that the impact of this discovery was much greater: there would be no recombinant DNA revolution and no gene technology without restriction endonucleases. However, beyond their tremendous importance as tools for the analysis and recombination of DNA, restriction enzymes have provided outstanding model systems to study many aspects of protein-nucleic acid interactions, e.g., how proteins find their target sites within a great excess of non-specific sites, how short DNA sequences are recognized with such extreme accuracy, and how recognition is so efficiently coupled to catalysis.

Restriction enzymes are usually part of restriction-modification (R-M) systems that serve to protect bacteria (Prokarya as well as Archaea) against phage infections and uptake of foreign DNA. There are three main types (I, II and III) of restriction enzymes, which differ in subunit composition, cofactor requirement and mode of action. The best-studied (and most simple) ones are the Type II restriction endonucleases of which more than 3500 are known. It was not recognized initially that restriction endonucleases are evolutionarily related, because – with few exceptions – they have little if any sequence homology. Only after the first crystal structures were determined did it become clear that most Type II restriction endonucleases are related in evolution and that they share a similar active site with Type I and III restriction endonucleases. The diversity of restriction enzymes makes them fascinating objects for studying the evolution of a family of enzymes with a common basic function – highly specific DNA recognition and cleavage. Surprisingly, with probably only very few exceptions, they do not seem to be related to another family of site-specific endonucleases, the homing endonucleases.

In the last few years, considerable progress has been made regarding the mechanism of action of restriction enzymes. Still, many open questions remain, even very basic ones. For example, there is not yet a consensus on the mechanism of target site location or the details of the mechanism of DNA cleavage, and only little is known about the conformational changes following initial DNA binding by restriction endonucleases and leading to the activation of their catalytic centers. Major challenges remain, the most ambitious one being to use the knowledge that we have regarding structure-function relationships of these enzymes to engineer variants with new specificities.

I am very pleased that so many of the top scientists in the field could be convinced to provide chapters and I am grateful to them for taking the time to do so. The result of their efforts is a book that represents the most detailed account of restriction endonucleases currently available.

Giessen, September 2003   *Alfred Pingoud*

Contents

R.J. Roberts, M. Belfort, T. Bestor, A.S. Bhagwat, T.A. Bickle,
J. Bitinaite, R.M. Blumenthal, S.K. Degtyarev, D.T.F. Dryden,
K. Dybvig, K. Firman, E.S. Gromova, R.I. Gumport, S.E. Halford,
S. Hattman, J. Heitman, D.P. Hornby, A. Janulaitis, A. Jeltsch,
J. Josephsen, A. Kiss, T.R. Klaenhammer, I. Kobayashi, H. Kong,
D.H. Krüger, S. Lacks, M.G. Marinus, M. Miyahara, R.D. Morgan,
N.E. Murray, V. Nagaraja, A. Piekarowicz, A. Pingoud, E. Raleigh,
D.N. Rao, N. Reich, V.E. Repin, E.U. Selker, P.-C. Shaw, D.C. Stein,
B.L. Stoddard, W. Szybalski, T.A. Trautner, J.L. Van Etten,
J.M.B. Vitor, G.G. Wilson, S.-Y. Xu

Restriction-Modification Systems as Minimal Forms of Life   19

I. Kobayashi

The Type I and III Restriction Endonucleases: Structural
Elements in Molecular Motors that Process DNA   111

S.E. McCLELLAND, M.D. SZCZELKUN

The Integration of Recognition and Cleavage : X-Ray Structures
of Pre-Transition State Complex, Post-Reactive Complex
and the DNA-Free Endonuclease   137

A. GRIGORESCU, M. HORVATH, P.A. WILKOSZ, K. CHANDRASEKHAR,
J.M. ROSENBERG

Structure and Function of EcoRV Endonuclease 179

F.K. WINKLER, A.E. PROTA

Two of a Kind: BamHI and BglII   215

É. Scheuring Vanamee, H. Viadiu, C.M. Lukacs, A.K. Aggarwal

Analysis of Type II Restriction Endonucleases That Interact with Two Recognition Sites

A.J. Welsh, S.E. Halford, D.J. Scott

The Role of Water in the EcoRI–DNA Binding

N. Sidorova, D.C. Rau

Role of Metal Ions in Promoting DNA Binding and Cleavage by Restriction Endonucleases

J.A. Cowan

Restriction Endonucleases: Structure of the Conserved Catalytic Core and the Role of Metal Ions in DNA Cleavage

J.R. Horton, R.M. Blumenthal, X. Cheng

Protein Engineering of Restriction Enzymes

J. Alves, P. Vennekohl

Contributors

AGGARWAL, ANEEL K.

Structural Biology Program, Department of Physiology and Biophysics, Mount Sinai School of Medicine, 1425 Madison Ave, New York, New York 10029, USA

ALVES, JÜRGEN

Department of Biophysical Chemistry, Medical School Hanover, Carl-Neuberg-Str. 1, 30625 Hanover, Germany

BELFORT, MARLENE

Molecular Genetics Program, New York State Department of Health, Albany, New York 12201–2002, USA

BESTOR, TIMOTHY

Genetics and Development, Columbia University, New York, New York 10032, USA

BHAGWAT, ASHOK S.

Department of Chemistry, Wayne State University, Detroit, Michigan 48202, USA

BICKLE, THOMAS A.

Department of Microbiology, Biozentrum, Universitat Basel, 4056 Basel, Switzerland

BITINAITE, JURATE

New England Biolabs, 32 Tozer Road, Beverly, Massachusetts 01915, USA

BLUMENTHAL, ROBERT M.

Department of Microbiology & Immunology, and Program in
Bioinformatics & Proteomics/Genomics, Medical College of Ohio, Toledo,
Ohio 43614–5806, USA

BUJNICKI, JANUSZ M.

Bioinformatics Laboratory, International Institute of Molecular and Cell
Biology in Warsaw, Trojdena 4, 02-109 Warsaw, Poland

CHANDRASEGARAN, SRINIVASAN

Department of Environmental Health Sciences, The Johns Hopkins
University Bloomberg School of Public Health, 615 North Wolfe Street,
Baltimore, Maryland 21205–2179, USA

CHANDRASEKHAR, K.

Department of Biological Sciences, University of Pittsburgh; Pittsburgh,
Pennsylvania 15260, USA

CHENG, XIAODONG

Department of Biochemistry, Emory University School of Medicine, 1510
Clifton, Atlanta, Georgia 30322, USA

COWAN, J.A.

Evans Laboratory of Chemistry, The Ohio State University, 100 West 18th
Avenue, Columbus, Ohio 43210, USA

DEGTYAREV, SERGEY K.

SibEnzyme, 630090 Novosibirsk, Russia

DRYDEN, DAVID T.F.

School of Chemistry, University of Edinburgh, The King's Buildings,
Edinburgh EH9 3JJ, UK

DURAI, SUNDAR

Department of Environmental Health Sciences, The Johns Hopkins
University Bloomberg School of Public Health, 615 North Wolfe Street,
Baltimore, Maryland 21205–2179, USA

DYBVIG, KEVIN

Department of Genetics, University of Alabama at Birmingham,
Birmingham, Alabama 35294, USA

FIRMAN, KEITH

Biophysics Laboratories, School of Biological Sciences, University of Portsmouth, Portsmouth PO1 2DT, UK

GRAZULIS, SAULIUS

Institute of Biotechnology, Graiciuno 8, Vilnius 2028, Lithuania

GRIGORESCU, ARABELA

Department of Biological Sciences, University of Pittsburgh; Pittsburgh, Pennsylvania 15260, USA

GROMOVA, ELIZAVETA S.

A.N. Belozersky Institute of Physico-Chemical Biology, Moscow State University, 119992 Moscow, Russia

GUMPORT, RICHARD I.

The University of Illinois College of Medicine, Urbana, Illinois 61801–3602, USA

HALFORD, STEPHEN E.

Department of Biochemistry, School of Medical Sciences, University Walk, University of Bristol, Bristol BS8 1TD, UK

HATTMAN, STANLEY

Department of Biology, University of Rochester, Rochester, New York 14627–0211, USA

HEITMAN, JOSEPH

Howard Hughes Medical Institute, Duke University Medical Center, Durham, North Carolina 27710, USA

HORNBY, DAVID P.

Department of Molecular Biology and Biotechnology, University of Sheffield, Firth Court, Western Bank, Sheffield S10 2TN, UK

HORTON, JOHN R.

Department of Biochemistry, Emory University School of Medicine, 1510 Clifton, Atlanta, Georgia 30322, USA

HORVATH, MONICA

Department of Biological Sciences, University of Pittsburgh; Pittsburgh,
Pennsylvania 15260, USA

HUBER, ROBERT

Max Planck Institute of Biochemistry, Am Klopferspitz 18a, 8215
Martinsried, Germany

JANULAITIS, ARVYDAS

Institute of Biotechnology, 2028 Vilnius, Lithuania

JELTSCH, ALBERT

Institut für Biochemie, FB 8, Heinrich-Buff-Ring 58, Justus-Liebig-
Universität, 35392 Giessen, Germany

JOSEPHSEN, JYTTE

Department of Dairy and Food Science, Royal Veterinary and Agricultural
University, 1958 Frederiksberg C, Denmark

KANDAVELOU, KARTHIKEYAN

Department of Environmental Health Sciences, The Johns Hopkins
University Bloomberg School of Public Health, 615 North Wolfe Street,
Baltimore, Maryland 21205–2179, USA

KISS, ANTAL

Institute of Biochemistry, BRC, 6701 Szeged, Hungary
Klaenhammer

KOBAYASHI, ICHIZO

Institute of Medical Science, University of Tokyo, 4-6-1, Shirokanedai,
Minato-ku, Tokyo 108–8639, Japan

KONG, HUIMIN

New England Biolabs, 32 Tozer Road, Beverly, Massachusetts 01915, USA

KRÜGER, DETLEV H.

Institut für Virologie, Universitätsklinikum Charite,
Humboldt-Universität, Schumannstr. 20/21, 10098 Berlin, Germany

LACKS, SANFORD

Brookhaven National Laboratory, Upton, New York 11973–5000, USA

LUKACS, CHRISTINE M.

Structural Biology Program, Department of Physiology and Biophysics, Mount Sinai School of Medicine, 1425 Madison Ave, New York, New York 10029, USA

MANI, MALA

Department of Environmental Health Sciences, The Johns Hopkins University Bloomberg School of Public Health, 615 North Wolfe Street, Baltimore, Maryland 21205–2179, USA

MARINUS, MARTIN G.

Department of Pharmacology, University of Massachusetts Medical School, Worcester, Massachusetts 01655, USA

McCLELLAND, SARAH E.

DNA-Protein Interactions Group, Department of Biochemistry, School of Medical Sciences, University of Bristol, Bristol, BS8 1TD, UK

MIYAHARA, MICHIKO

National Institute of Health Sciences, 1–18–1, Kamiyoga, Setagaya-ku, Tokyo 158–8501, Japan

MORGAN, RICHARD D.

New England Biolabs, 32 Tozer Road, Beverly, Massachusetts 01915, USA

MÜCKE, MERLIND

Institut für Virologie, Universitätsklinikum Charité, Humboldt-Universität, Schumannstr. 20/21, 10098 Berlin, Germany

MURRAY, NOREEN E.

Institute of Cell and Molecular Biology, University of Edinburgh, The King's Buildings, Edinburgh EH9 3JR, UK

NAGARAJA, VALAKUNJA

Department of Microbiology and Cell Biology, Indian Institute of Science, 560012 Bangalore, India

PIEKAROWICZ, ANDRZEJ

Institute of Microbiology, Warsaw University, Miecznikowa 1, 02–096 Warsaw, Poland

PINGOUD, ALFRED

Institut für Biochemie, Justus-Liebig-Universität, 35392 Giessen, Germany

PROTA, ANDREA E.

Biomolecular Research, Paul Scherrer Institut, 5232 Villigen, Switzerland

RALEIGH, ELISABETH

New England Biolabs, 32 Tozer Road, Beverly, Massachusetts 01915, USA

RAO, DESIRAZU N.

Department of Microbiology and Cell Biology, Indian Institute of Science, 560012 Bangalore, India

RAU, DONALD C.

Laboratory of Physical and Structural Biology, National Institute of Child Health and Human Development, National Institutes of Health, Bld. 9, Room 1E108, Bethesda, Maryland 20892, USA

REICH, NORBERT

University of California, Santa Barbara, Santa Barbara, California 93106–0001, USA

REPIN, VLADIMIR E.

State Research Center of Virology and Biotechnology 'Vektor', Koltsovo, Novosibirsk Region 633059, Russia

REUTER, MONIKA

Institut für Virologie, Universitätsklinikum Charite, Humboldt-Universität, Schumannstr. 20/21, 10098 Berlin, Germany

ROBERTS, RICHARD J.

New England Biolabs, 32 Tozer Road, Beverly, Massachusetts 01915, USA

ROSENBERG, JOHN M.

Department of Biological Sciences, University of Pittsburgh; Pittsburgh, Pennsylvania 15260, USA

SCHEURING, VANAMEE ÉVA

Structural Biology Program, Department of Physiology and Biophysics, Mount Sinai School of Medicine, 1425 Madison Ave, New York, New York 10029, USA

Scott, David J.

Department of Biochemistry, School of Medical Sciences, University Walk, University of Bristol, Bristol BS8 1TD, UK

Selker, Eric U.

Institute of Molecular Biology, University of Oregon, Eugene, Oregon 97403, USA

Shaw, Pang-Chui

Department of Biochemistry, The Chinese University of Hong Kong, Hong Kong

Sidorova, Nina

Laboratory of Physical and Structural Biology, National Institute of Child Health and Human Development, National Institutes of Health, Bld. 9, Room 1E108, Bethesda, Maryland 20892, USA

Siksnys, Virginijus

Institute of Biotechnology, Graiciuno 8, Vilnius 2028, Lithuania

Stein Daniel D.

Department of Microbiology, University of Maryland, College Park, Maryland 20742, USA

Stoddard, Barry L.

Fred Hutchinson Cancer Research Center, Seattle, Washington 98109, USA

Szczelkun, Mark D.

DNA-Protein Interactions Group, Department of Biochemistry, School of Medical Sciences, University of Bristol, Bristol, BS8 1TD, UK

Szybalski, Waclaw

McArdle Laboratory, University of Wisconsin, Madison, Wisconsin 53706, USA

Todd R.

Departments of Food Science and Microbiology, North Carolina State University, Raleigh, North Carolina 27695–7624, USA

TRAUTNER, THOMAS A.

MPI für Molekulare Genetik, Ihnestrasse 73, 14195 Berlin, Germany

URBANKE, CLAUS

Zentrale Einrichtung Biophysikalisch-Biochemische Verfahren, Medizinische Hochschule, 30625 Hannover, Germany

VAN ETTEN, JAMES L.

Department of Plant Pathology, University of Nebraska-Lincoln, Lincoln, Nebraska 68583, USA

VENNEKOHL, PETRA

Department of Biophysical Chemistry, Medical School Hanover, Carl-Neuberg-Str. 1, 30625 Hanover, Germany

VIADIU, HECTOR

Structural Biology Program, Department of Physiology and Biophysics, Mount Sinai School of Medicine, 1425 Madison Ave, New York, New York 10029, USA

VITOR, JORGE M.B.

Faculdade de Farmacia de Lisboa, 1600–083 Lisbon, Portugal

WELSH, ABIGAIL J.

Department of Biochemistry, School of Medical Sciences, University Walk, University of Bristol, Bristol BS8 1TD, UK

WILKOSZ, PATRICIA A.

Department of Biological Sciences, University of Pittsburgh; Pittsburgh, Pennsylvania 15260, USA

WILSON, GEOFFREY G.

New England Biolabs, 32 Tozer Road, Beverly, Massachusetts 01915, USA

WINKLER, FRITZ K.

Biomolecular Research, Paul Scherrer Institut, 5232 Villigen, Switzerland

XU, SHUANG-YONG

New England Biolabs, 32 Tozer Road, Beverly, Massachusetts 01915, USA

A Nomenclature for Restriction Enzymes, DNA Methyltransferases, Homing Endonucleases, and Their Genes[1]

R.J. Roberts, M. Belfort, T. Bestor, A.S. Bhagwat, T.A. Bickle,
J. Bitinaite, R.M. Blumenthal, S.K. Degtyarev, D.T.F. Dryden,
K. Dybvig, K. Firman, E.S. Gromova, R.I. Gumport, S.E. Halford,
S. Hattman, J. Heitman, D.P. Hornby, A. Janulaitis, A. Jeltsch,
J. Josephsen, A. Kiss, T.R. Klaenhammer, I. Kobayashi, H. Kong,
D.H. Krüger, S. Lacks, M.G. Marinus, M. Miyahara, R.D. Morgan,
N.E. Murray, V. Nagaraja, A. Piekarowicz, A. Pingoud, E. Raleigh,
D.N. Rao, N. Reich, V.E. Repin, E.U. Selker, P.-C. Shaw, D.C. Stein,
B.L. Stoddard, W. Szybalski, T.A. Trautner, J.L. Van Etten,
J.M.B. Vitor, G.G. Wilson, S.-Y. Xu

R.J. Roberts (✉), J. Bitinaite, H. Kong, R.D. Morgan, E. Raleigh, G.G. Wilson, S.-Y. Xu
New England Biolabs, Beverly, Massachusetts 01915, USA
M. Belfort, Molecular Genetics Program, New York State Department of Health, Albany, New York 12201–2002, USA
T. Bestor. Genetics and Development, Columbia University, New York, New York 10032, USA
A.S. Bhagwat, Department of Chemistry, Wayne State University, Detroit, Michigan 48202, USA
T.A. Bickle, Department of Microbiology, Biozentrum, Universitat Basel, 4056 Basel, Switzerland
R.M. Blumenthal, Program in Bioinformatics and Proteomics/Genomics, Medical College of Ohio, Toledo, Ohio 43699–0008, USA
S.K. Degtyarev, SibEnzyme, 630090 Novosibirsk, Russia
D.T.F. Dryden, School of Chemistry, University of Edinburgh, The King's Buildings, Edinburgh EH9 3JJ, UK
K. Dybvig, Department of Genetics, University of Alabama at Birmingham, Birmingham, Alabama 35294, USA
K. Firman, Biophysics Laboratories, School of Biological Sciences, University of Portsmouth, Portsmouth PO1 2DT, UK
E.S. Gromova, A.N. Belozersky Institute of Physico-Chemical Biology, Moscow State University, 119992 Moscow, Russia
R.I. Gumport, The University of Illinois College of Medicine, Urbana, Illinois 61801–3602, USA

---

[1] Originally published in Nucleic Acids Research, Vol. 31 No. 7, 1805–1812 by Oxford University Press 2003. Reprinted with permission of Oxford University Press.

S.E. Halford, Department of Biochemistry, University of Bristol Medical School, Bristol BS8 1TD, UK

S. Hattman, Department of Biology, University of Rochester, Rochester, New York 14627–0211, USA

J. Heitman, Howard Hughes Medical Institute, Duke University Medical Center, Durham, North Carolina 27710, USA

D.P. Hornby, Department of Molecular Biology and Biotechnology, University of Sheffield, Firth Court, Western Bank, Sheffield S10 2TN, UK

A. Janulaitis, Institute of Biotechnology, LT-2028 Vilnius, Lithuania

A. Jeltsch, A. Pingoud, Institut für Biochemie, Justus-Liebig-Universität, 35392 Giessen, Germany

J. Josephsen, Department of Dairy and Food Science, Royal Veterinary and Agricultural University, 1958 Frederiksberg C, Denmark

A. Kiss, Institute of Biochemistry, BRC, 6701 Szeged, Hungary

T.R. Klaenhammer, Departments of Food Science and Microbiology, North Carolina State University, Raleigh, North Carolina 27695-7624, USA

I. Kobayashi, Department of Molecular Biology, Institute of Medical Science, University of Tokyo, 4-6-1, Shirokanedai, Minato-ku, Tokyo 108–8639, Japan

D.H. Krüger, Institut für Virologie-Charite, Humboldt Universitat, 10098 Berlin, Germany

S. Lacks, Brookhaven National Laboratory, Upton, New York 11973–5000, USA

M.G. Marinus, Department of Pharmacology, University of Massachusetts Medical School, Worcester, Massachusetts 01655, USA

M. Miyahara, National Institute of Health Sciences, 1-18-1, Kamiyoga, Setagaya-ku, Tokyo 158-8501, Japan

N.E. Murray, Institute of Cell and Molecular Biology, University of Edinburgh, The King's Buildings, Edinburgh EH9 3JR, UK

V. Nagaraja, D.N. Rao, Department of Microbiology and Cell Biology, Indian Institute of Science, IN-560012 Bangalore, India

A. Piekarowicz, Institute of Microbiology, Warsaw University, Miecznikowa 1, 02–096 Warsaw, Poland

N. Reich, University of California, Santa Barbara, Santa Barbara, California 93106–0001, USA

V.E. Repin, State Research Center of Virology and Biotechnology 'Vektor', Koltsovo, Novosibirsk region 633059, Russia

E.U. Selker, Institute of Molecular Biology, University of Oregon, Eugene, Oregon 97403, USA

P.-C. Shaw, Department of Biochemistry, The Chinese University of Hong Kong, Hong Kong

D.C. Stein, Department of Microbiology, University of Maryland, College Park, Maryland 20742, USA

B.L. Stoddard, Fred Hutchinson Cancer Research Center, Seattle, Washington 98109, USA

W. Szybalski, McArdle Laboratory, University of Wisconsin, Madison, Wisconsin 53706, USA

T.A. Trautner, MPI für Molekulare Genetik, Ihnestrasse 73, 14195 Berlin, Germany

J.L. Van Etten, Department of Plant Pathology, University of Nebraska-Lincoln, Lincoln, Nebraska 68583, USA

J.M.B. Vitor, Faculdade de Farmacia de Lisboa, 1600–083 Lisbon, Portugal

1 Introduction

There are three main groups of restriction endonucleases (REases) called Types I, II, and III (1,2). Since 1973, REases and DNA methyltransferases (MTases) have been named based on an original suggestion by Smith and Nathans (3). They proposed that the enzyme names should begin with a three-letter acronym in which the first letter was the first letter of the genus from which the enzyme was isolated and the next two letters were the first two letters of the species name. Extra letters or numbers could be added to indicate individual strains or serotypes. Thus, the enzyme HindII was one of four enzymes isolated from *Haemophilus influenzae* serotype d. The first three letters of the name were italicized. Later, a formal proposition for naming the genes encoding REases and MTases was adopted (4). When there were only a handful of enzymes known, these schemes were very useful, but as more enzymes have been found, often from different genera and species with names whose three-letter acronyms would be identical, considerable laxity in naming conventions has appeared. In addition, we now know that each major type of enzyme can contain subtypes. This especially applies to the Type II enzymes, of which more than 3500 have been characterized (5). In this paper we revisit the naming conventions and outline an updated scheme that incorporates current knowledge about the complexities of these enzymes. We describe a set of naming conventions for REases and their associated MTases. Since the homing endonucleases (6) have been named in an analogous fashion, we propose that similar guidelines be applied to that group of enzymes. Finally, it is important to realize that the aim of this document is to provide a nomenclature for these enzymes, not to provide a rigorous classification.

2 General Rules

First, we introduce a number of general changes, standard abbreviations and definitions that are recommended for use.

1. 'Restriction enzyme' and 'restriction endonuclease' should be regarded as synonymous and the abbreviation REase (or in some cases, R) is preferred. However, the abbreviation ENase, which has been used extensively, may also be used. Alternative names such as restrictases should be avoided. The abbreviation R-M should be used for restriction modification. Homing endonuclease should be abbreviated HEase.
2. Methyltransferase is the preferred name, since it correctly describes the activity. Methylase, while in common use, is not strictly accurate and should be avoided in print. The abbreviation MTase (or in some cases, M) should be the standard.
3. Italics will no longer be used for the first three-letter acronym of the REase or MTase name. Many journals already avoid italics and retaining

the italic convention is not easily translated to computers and serves no essential purpose. The convention of naming different enzymes from the same isolate of the same organism with increasing Roman numerals will continue.

4. Restriction enzyme names should not include a space between the main acronym and the Roman numeral. This practice, which has been employed to avoid the inelegant look caused when characters in italic fonts are juxtaposed next to characters in a regular font, is incorrect. Now that italics will no longer be used in names there is no reason to continue this practice. The previous scheme of using a raised dot after the prefix will be abandoned and a normal dot (period) should be used. Furthermore, except for the single period or hyphen (in homing endonucleases) that is used to separate the prefix from the main part of the name, no punctuation marks, such as parentheses, periods, commas or slashes, should be used in REase or MTase names. Only alphanumeric characters should appear. Already the enzymes from *Nostoc* species C have been changed from their original Nsp(7524)I to NspI and many others have also changed. The most recent is Bst4.4I, which has changed to Bst44I.

5. The designation of the three main types of REases as Type I, Type II, and Type III will continue, with the capital 'T' preferred. However, they will be divided into subtypes as indicated below. One new type of REase will be added. This is Type IV, which will include those systems that cleave only methylated DNA as their substrate and show only weak specificity, such as the McrA, McrBC and Mrr systems of *Escherichia coli*.

6. The sequence databases contain many genes that are excellent candidates to encode DNA MTases and REases, based on sequence similarity. These will be named according to the same guidelines as are used for biochemically characterized enzymes, but will carry the suffix P to indicate their putative nature. Once they have been characterized biochemically and shown to be active, the P will be dropped and their names will be changed to a regular name with the next Roman numeral that is appropriate.

7. The current convention of naming R-M enzymes with a prefix M, R, etc. will be expanded to include the protein products of related genes such as the controlling proteins (e.g., C.BamHI) and the nicking enzymes that cleave G/T mismatches (e.g., V.HpaII for the vsr-like enzyme associated with the HpaII system) and N.BstNBI for the regular nicking enzymes. In addition, up to two characters will be allowed in the prefix. This will enable enzymes, such as Eco57I, with both REase activity and MTase activity fused in a single protein to be designated RM.Eco57I. Its accompanying MTase would remain as M.Eco57I. Note that the current convention of permitting the REase to be named either with or without the 'R' prefix will be continued. Thus, R.EcoRI and EcoRI will be considered synonymous as will RM.Eco57I and Eco57I. For certain nicking enzymes that have been obtained from the Type IIT enzymes, where one of the two heterodimeric

subunits has been inactivated the resultant mutant nicking enzymes should be called Nt.BbvCI or Nb.BbvCI, where the 't' and the 'b' indicate cleavage of the top or bottom strand of the normal recognition sequence.

8. When two REase or MTase genes are present and associated with a single R-M system, they should be referred to with the second character of the prefix being an Arabic 1 or 2. Thus, the two M gene products of the HphI R-M system would be M1.HphI and M2.HphI.

9. The standard abbreviations for methylated bases should be 5-methylcytosine (m5C), N4-methylcytosine (m4C) and N6-methyladenine (m6A). It is not necessary to use a superscript for the number.

10. Isoschizomers are REases that recognize the same sequence. The first example discovered is called a prototype and all subsequent enzymes that recognize the same sequence are isoschizomers of the prototype. Neoschizomers are that subset of isoschizomers that recognize the same sequence, but cleave at different positions from the prototype. Thus, AatII (recognition sequence: GACGT↓C) and ZraI (recognition sequence: GAC↓GTC) are neoschizomers of one another, while HpaII (recognition sequence: C↓CGG) and MspI (recognition sequence: C↓CGG) are isoschizomers, but not neoschizomers. Analogous designations are not appropriate for MTases, where the differences between enzymes are not so easily defined and usually have not been well characterized.

11. The solitary MTases (i.e., not associated with an REase) such as the Dam and Dcm MTases of E.coli and the eukaryotic MTases such as Dnmt1 and Dnmt3a will be named systematically in accordance with the general rules established for the prokaryotic enzymes. Thus, the systematic name for the Dam MTase of *E.coli* K12 will be M.EcoKDam and the murine maintenance MTase will be M.MmuDnmt1. However, it will be acceptable to refer to them by their more commonly used trivial names, Dam, Dcm, Dnmt1, etc., but it will simplify automated searching and cross-referencing of the literature if the systematic name, including the M prefix, also appears at least initially in a publication. Solitary MTases that are phage or virus borne are also named with the prefix M and the name of the phage or virus that carries them. Optionally, the host name may be included. Thus, the MTase encoded by phage SPR of Bacillus subtilis is named M.SPRI (7) and the MTase encoded by the archaeal virus φCh1 of *Natrialba magadii* is named M.NmaPhiCh1I (8).

12. The rules for naming genes of Type II R-M systems should adhere to the proposals of Szybalski et al. (4) with the obvious extensions to accommodate C, V and N genes. Thus, the entire name should be italicized, the first letter will be lower case and the capital letter(s) used as a prefix for the protein will become the suffix for the gene. The gene for EcoRI thus becomes *ecoRIR* and that for its MTase is *ecoRIM*. In the case of genes with two prefixes in the protein name the gene name would incorporate both

letters of the prefix. Thus, the gene for RM.Eco57I would become *eco57IRM*. In the case of Type I enzymes, an acronym for the source organism should be followed by the traditional gene designations, hsdS, hsdR and hsdM. Thus, the three genes of the EcoKI restriction system would be *ecoKIhsdM*, *ecoKIhsdR,* and *ecoKIhsdS*. However, it will be acceptable to omit the *ecoKI* where appropriate.

13. It sometimes happens that two genes are required for a single enzyme activity, effectively encoding two subunits. In these situations the two genes and their products should carry a suffix A and B. For example, BbvCI is a heterodimeric REase. The two gene products would be called R.BbvCIA (or just BbvCIA) and R.BbvCIB (or just BbvCIB), and the active holoenzyme would be BbvCI. Note that the two separate MTases of this system would be M1.BbvCI and M2.BbvCI. For enzymes like Eco57I, which have both endonuclease and MTase activity in the same polypeptide chain, the endonuclease would be referred to as RM.Eco57I, but the second MTase activity associated with this system would be called M.Eco57I. For MTases, an example is M.AquI, which has one gene encoding the N-terminal region up to the middle of the variable region of this m5C MTase and a second gene encoding the remaining C-terminal region (9). In this case, the two parts of this protein should be referred to as M.AquIA and M.AquIB and the genes as *aquIAM* and *aquIBM*.

3 Details of Types and Subtypes

3.1 Types I, II, III, and IV

The original subdivision of Types I, II, and III will be maintained and a new Type IV added to accommodate a class of methyl-dependent restriction enzymes. The previously proposed candidates for new types, such as Eco57I and GsuI, will be incorporated as subtypes of existing Type II enzymes.

3.2 Type I

The key characteristics of the Type I R-M systems are that these enzymes are multisubunit proteins that function as a single protein complex and usually contain two R subunits, two M subunits and one S subunit (10). The symbol for Type I systems is *hsd*, thus the genes are *hsdR, hsdM,* and *hsdS,* and their protein products are HsdR, HsdM, and HsdS, respectively. The protein products can be abbreviated by omitting Hsd. The S subunit is the specificity subunit that determines which DNA sequence is recognized. The R subunit is essential for cleavage (restriction) and the M subunit catalyzes the methylation reaction: in all known cases the methylated base formed is m6A. When

Type I enzymes act on unmethylated substrates, they function mainly as REases (they may also methylate unmodified sites with a low probability) and have an absolute requirement for ATP during cleavage. They cleave the DNA at variable positions away from their recognition sequence. The location of the cleavage sites is determined by either the collision and stalling of two such complexes during translocation along a DNA chain, or the stalling of a single enzyme on a single-site circular substrate following DNA translocation. The biochemical nature of the termini produced upon cleavage is unknown and the enzymes do not turn over in the cleavage reaction. In contrast, when these complexes encounter a hemimethylated substrate, in which one strand of the recognition sequence is methylated, as would occur immediately after DNA replication of a fully methylated substrate, then the complex functions as a DNA MTase, using S-adenosylmethionine (AdoMet) as the donor of the methyl group. A complex of two M subunits and one S subunit is fully functional as an MTase. Probably the best known Type I enzyme is EcoKI (11). The REase is referred to as R.EcoKI or EcoKI, but it is important to remember that it is also an MTase. The MTase complex of two HsdM and one HsdS is referred to as M.EcoKI. When referring to phenotypes the preferred convention is r_{KI}^+ m_{KI}^+ etc.

Four subcategories of Type I enzymes (A, B, C, and D) are in common use (12). These are based on genetic complementation and their use will be continued. If experimental evidence defines new subtypes, then additional letters may be used as suffixes to describe them. A number of artificially created hybrid enzymes have been described (13), which often include those with new specificities. These should be named as deemed appropriate, but without a Roman numeral at the end.

3.3 Type II

The Type II REases recognize specific DNA sequences and cleave at constant positions at or close to that sequence to produce 5′-phosphates and 3′-hydroxyls. Usually they require Mg^{2+} ions as a cofactor, although some have more exotic requirements (see below). They may act as monomers, dimers or even tetramers and usually act independently of their companion MTase. The MTases usually act as monomers and transfer a methyl group from the donor S-adenosyl-L-methionine directly to double-stranded DNA and form m4C, m5C, or m6A. Because of the interest in these Type II REases for recombinant DNA technology, more than 3500 have been characterized (5). Given the assay that is used to find them, which detects any activity yielding a consistent DNA fragmentation pattern, it is no surprise that they come in a large variety of 'flavors'. Early on it was recognized that while then-normal Type II enzymes recognized palindromic sequences and cleaved symmetrically within them, the Type IIS enzymes cut outside their normally asymmetric sequences and differed in other interesting ways (14). We now know of additional enzymes that cleave on both sides of their recognition sequence (e.g., BcgI), are activated by AdoMet (e.g., Eco57I), interact with two copies of their recognition sequence (e.g., EcoRII) or have unusual subunit structures (e.g., BbvCI).

These additional kinds of enzymes will be considered subdivisions of Type II. It should be recognized that for the purposes of nomenclature some enzymes would fall into more than one subdivision. Specifically, some of the criteria are based on the sequence cleaved and others on the structure of the enzymes themselves, so not all subdivisions are mutually exclusive, e.g., BcgI is both Type IIB and IIH. Type IIS enzymes, originally designated as enzymes with cleavage sites shifted away from their recognition sequence (4), will be retained, but a new Type IIA will be defined that includes all Type II REases that recognize asymmetric sequences. A new Type IIP will be used to designate the enzymes that recognize symmetric sequences (palindromes).

The overriding criterion for inclusion as a Type II enzyme would be that it yields a defined fragmentation pattern and cleaves either within or close to its recognition sequence at a fixed site or with known and limited variability. In general, the Type II REases and their associated MTases are separate, independent enzymes, but in several classes (e.g., IIB, IIG, and IIH) the R and M genes are fused into a single composite gene. The nomenclature for the subtypes of the Type II enzymes currently known is shown below. It should be noted that these designations are not intended to be exclusive, but rather to permit enzymes with common characteristics to be referred to as a group. Conservation of structural domains with associated enzymatic activities is observed between different classes of Type II enzymes and also between other types of R-M enzymes.

The Type II subdivisions are summarized in Table 1 and described in more detail below.

3.4 Type IIP

This would be used as a generic description for all enzymes that recognize symmetric sequences, often termed palindromes, and cleave at fixed symmetrical locations either within the sequence or immediately adjacent to it. The recognition sequences and cleavage sites of these enzymes should be represented as in the following example: EcoRI: G↓AATTC. In full double-stranded form this corresponds to:

5′ G↓AATTC
3′CTTAA↑G

Note that enzymes such as SinI (recognition sequence: GGWCC), BglI (recognition sequence: GCCNNNN↓NGGC) and HindII (recognition sequence: GTYRAC) belong to Type IIP because the recognition mechanism still involves a symmetric homodimer.

Table 1. Subtypes of Type II REases

Subtype[a]	Defining feature	Examples	Recognition sequence
A	Asymmetric recognition sequence	FokI	GGATG (9/13)
		AciI	CCGC (−3/−1)
B	Cleaves both sides of target on both strands	BcgI	(10/12) CGANNNNNNTGC (12/10)
		GsuI	CTGGAG (16/14)
		HaeIV	(7/13) GAYNNNNNRTC (14/9)
C	Symmetric or asymmetric target. R and M functions in one polypeptide	BcgI	(10/12) CGANNNNNNTGC (12/10)
		EcoRII	↓CCWGG
E	Two targets; one cleaved, one an effector	NaeI	GCC↓GGC
F	Two targets, both cleaved coordinately	SfiI	GGCCNNNN↓NGGCC
		SgrAI	CR↓CCGGYG
G	Symmetric or asymmetric target. Affected by AdoMet	BsgI	GTGCAG (16/14)
		Eco57I	CTGAAG (16/14)
H	Symmetric or asymmetric target. Similar to Type I gene structure	BcgI	(10/12) CGANNNNNNTGC (12/10)
		AhdI	GACNNN↓NNGTC
M	Subtype IIP or IIA. Require methylated target	DpnI	Gm6 A↓TC
P	Symmetric target and cleavage sites	EcoRI	G↓AATTC
		PpuMI	RG↓GWCCY
		BslI	CCNNNNN↓NNGG
S	Asymmetric target and cleavage sites	FokI	GGATG (9/13)
		MmeI	TCCRAC (20/18)
T	Symmetric or asymmetric target. R genes are heterodimers	Bpu10I	CCTNAGC (−5/−2)[b]
		BslI	CCNNNNN↓NNGG

[a] Note that not all subtypes are mutually exclusive. For example, BslI is of subtype P and T.
[b] The abbreviation indicates double-strand cleavage as shown below:

```
5′CC↓TNAGC
3′GGANT↑CG
```

3.5 Type IIA

This would be used as a generic designation for any Type II enzymes that recognize asymmetric sequences irrespective of whether they cleave away from the sequence or within the sequence. Typically these systems have one REase gene and two MTase genes, one to modify each strand of the asymmetric recognition sequence. However, occasionally two R genes are found as with Bpu10I (15), or both M genes are fused as with M.FokI (16). When more than one R or M gene is present the genes and their protein products should be named with either an Arabic 1 or 2 in the prefix of the name. Thus, the two MTases of the SapI system would be named M1.SapI and M2.SapI if the proteins are being referred to, or *sapIM1* and *sapIM2* for the genes. However, the two subunits of the Bpu10I REase would be designated R.Bpu10IA and R.Bpu10IB and their genes *bpu10IAR* and *bpu10IBR*. The recognition sequences and cleavage sites of the Type IIS REases should be represented as in the following example:

HphI: GGTGA(8/7) where the first numeral in the parentheses indicates the position of cleavage on the strand written and the second numeral indicates the cleavage position on the complementary strand. In full double-stranded form this corresponds to:

5′GGTGANNNNNNNN↓
3′CCACTNNNNNNN↑

Note that when recognition sequences are assigned, the convention is to write the single-stranded sequence such that cleavage lies downstream of the sequence. If cleavage takes place within the sequence, then the single-strand designation is always written so that the sequence of the strand is first alphabetically.

3.6 Type IIB

This would be used for enzymes that cleave on both sides of the recognition sequence. At present there are many well defined members of this class (AloI, BplI, Bst44I, BaeI, BcgI, BsaXI, Bsp24I, CjeI, CjePI, HaeIV, Hin4I, and PpiI). In this case the recognition sequence and cleavage sites should be represented as exemplified for BcgI:

BcgI– recognition sequence: (10/12)CGANNNNNNTGC(12/10)

Here, the (10/12) preceding the recognition sequence indicates that cleavage occurs 10 bases in front of the sequence on the strand written and 12 bases before the sequence on the complementary strand. The (12/10) following the recognition sequence indicates cleavage 12 bases after the recognition sequence on the strand written and 10 bases after the sequence on the complementary strand. In double-stranded form this would be written:

↓NNNNNNNNNNNCGANNNNNNTGCNNNNNNNNNNNNN↓
↑NNNNNNNNNNNNNGCTNNNNNNACGNNNNNNNNNNN↑

3.7 Type IIC

This would be used as a generic term for all enzymes that have a hybrid structure containing both cleavage and modification domains within a single polypeptide. Examples include all of the Type IIB, IIG and some Type IIH enzymes.

3.8 Type IIE

This would be used for enzymes that interact with two copies of the recognition sequence, one being the actual target of cleavage, the other being the allosteric effector. The best studied examples are EcoRII (17) and NaeI (18). FokI, MboII, and Sau3AI were found to exhibit similar properties. Other enzymes such as Acc36I, AtuBI, BsgI, BpmI, Cfr9I, Eco57I, HpaII, Ksp632I, NarI, SacII, and SauBMKI are likely to be members of this group because they are reported to be stimulated by oligonucleotide duplexes containing the specific recognition site.

3.9 Type IIF

This would be used for enzymes that interact with, and cleave coordinately, two copies of their recognition sequence. Examples include BspMI, Cfr10I, NgoMIV, SfiI and SgrAI.

3.10 Type IIG

This would be used for enzymes that have both R and M domains fused to form single polypeptides and that may be stimulated or inhibited by AdoMet, but otherwise resemble Type II enzymes. These include Bce83I, BseMII, BseRI, BsgI, BspLU11III, Eco57I, GsuI, MmeI, and Tth111II. The recognition sequences may or may not be asymmetric. Thus, both Type IIA and Type IIP enzymes may be of Type IIG.

3.11 Type IIH

This would be used for enzymes that contain genetic features resembling Type I enzymes, but biochemically behave as Type II enzymes. At present three examples have been characterized: AhdI and PshAI, both of which comprise a three gene system akin to that of a typical Type I enzyme (G.G.Wilson, unpublished results), and BcgI, which is a two gene system. Several hypothetical systems have gene organizations that resemble that of BcgI.

3.12 Type IIM

This would be used for DpnI and similar enzymes that recognize a specific methylated sequence in DNA and cleave at a fixed site. Note that the methyl-dependent enzymes such as McrA, McrBC are not considered members of this subclass, because they do not have well defined recognition sequences and cleavage sites. They are included within the Type IV enzymes.

3.13 Type IIS

This would be used for Type IIA enzymes that cleave at least one strand of the DNA duplex outside of the recognition sequence (i.e. cleavage is shifted relative to the recognition sequence). Note that for some enzymes, such as BsmI (recognition sequence: GAATGC), cleavage of the strand written takes place outside of the recognition sequence, whereas cleavage of the complementary strand takes place within the recognition sequence. This is still considered a Type IIS enzyme. However, in most cases both strands are cleaved away from the recognition sequence, which therefore remains intact. This was the earliest sub-class of the Type II restriction enzymes to be recognized (14).

3.14 Type IIT

This would be used for enzymes that are composed of heterodimeric sub-units. This subtype includes enzymes like BbvCI, Bpu10I and BslI.

3.15 Nicking Enzymes

Two types of nicking enzymes are known. One type includes those that behave functionally like REases, but cleave only one strand of the DNA substrate. These enzymes should be named with the prefix N and their recognition sequences should be written such that the strand displayed is the strand nicked. Thus, N.BstSEI has the recognition sequence: GAGTCNNNN↓ which is abbreviated to GAGTC (4). Similarly, the mutants of AlwI and MlyI that have interrupted the dimerization function, and which have become nicking enzymes, are named N.AlwI (19) and N.MlyI (20). For enzymes such as Bpu10I, where the wild-type REase has two subunits, each of which nicks a different strand, the mutant nicking enzymes made by inactivating one or the other subunit should be named Nt.Bpu10I for the enzyme that nicks the top strand of the normal recognition sequence and Nb.Bpu10I for the enzyme that nicks the bottom strand.

   In full double-stranded format Nt.Bpu10I would recognize
5′CC↓TNAGC
3′GGANTCG
while Nb.Bpu10I would recognize
5′CCTNAGC
3′GGANT↑CG
Alternatively this may be written
5′GC↓TNAGG
3′CGANTCC
A single-stranded representation of their recognition sites would be Nt.Bpu10I (recognition sequence: CC↓TNAGC) and Nb.Bpu10I (recognition sequence: GC↓TNAGG or CCTNA↑GC). Note that the use of ↑ always denotes cleavage of the lower strand.

   A second type of nicking enzyme is found exclusively in association with m5C-MTases, where it serves to nick the G/T mismatches that can result from deamination of m5C within the recognition sequence of the MTase. The best studied of these is the Vsr protein that accompanies the Dcm MTase of *E.coli* K-12, M.EcoKDcm. Vsr recognizes the specific G/T mismatch that occurs if there is deamination of the methylated cytosine residue within the context of the CCWGG recognition sequence of M.EcoKDcm (21). These kinds of mismatch nicking enzymes are named with the prefix V and should be given the acronym of the MTase gene with which they are associated. Thus, Vsr, the product of the V gene that overlaps with the gene for M.EcoKDcm, is systematically named V.EcoKDcm. However, the trivial name Vsr, which was originally designated for this protein, is an acceptable synonym. For other V genes and their products the systematic names are preferred. Thus, V.HpaII is the preferred name for the mismatch nicking endonuclease that accompanies M.HpaII.

3.16 Control Proteins

Some R-M systems are found to have an additional gene that encodes a protein involved in the control of expression of the R gene. The best studied examples are the PvuII and BamHI systems, where the products of the C genes, C.PvuII (22) and C.BamHI (23), serve as transcriptional activators; this prevents the expression of the R genes following transfer of the systems into naive hosts, until such time as C protein has accumulated and methylation is sufficient to provide protection against what would otherwise be the deleterious action of the REase.

3.17 Type III

These systems are composed of two genes (*mod* and *res*) encoding protein subunits that function either in DNA recognition and modification (Mod) or restriction (Res) (10, 24, 25). Both subunits are required for restriction, which also has an absolute requirement for ATP hydrolysis. For DNA cleavage, the enzyme must interact with two copies of a non-palindromic recognition sequence and the sites must be in an inverse orientation in the substrate DNA molecule. Cleavage is preceded by ATP-dependent DNA translocation as with the Type I REases. The enzymes cleave at a specific distance away from one of the two copies of their recognition sequence. The Mod subunit can function independently of the Res subunit to methylate DNA: in all known cases the methylated base formed is m6A and full modification is actually hemimethylation. This is not deleterious because of the requirement for two unmodified sites in inverse repeat orientation for cleavage. DNA replication puts all of the unmodified sites in the same orientation. The best-known examples of Type III enzymes are EcoP1I and EcoP15I. Putative Type III R-M systems are easily recognized because of their similarity at the sequence level. When naming the genes for these enzymes the mod gene of EcoP1I would be systematically named *ecoP1Imod*, but the abbreviation mod is acceptable when it does not result in confusion.

3.18 Type IV

These systems are composed of one or two genes encoding proteins that cleave only modified DNA, including methylated, hydroxymethylated and glucosyl-hydroxymethylated bases. Their recognition sequences have usually not been well defined except for EcoKMcrBC, which recognizes two dinucleotides of the general form RmC (a purine followed by a methylated cytosine – either m4C or m5C) and which are separated by anywhere from 40 to 3000 bases. Cleavage takes place ~30 bp away from one of the sites. The best studied example at both the genetic and biochemical level is EcoKMcrBC of *E.coli* (26, 27), but on the basis of sequence similarity it is likely that there are many such systems in other bacteria and archaea. As with the genes of the Type I and Type III systems, the abbreviations McrBC for the enzyme and *mcrBC* for the gene are acceptable.

3.19 Hypothetical Enzymes

Hypothetical REases and DNA MTases can often be found by similarity searching in DNA sequences or their presence may be inferred when specific sequences in plasmid or bacterial DNAs are found to be methylated. It is convenient and useful to be able to refer to such hypothetical enzymes by name. The following convention for naming these enzymes is proposed. They should be named as though they were normal R-M systems, but should carry the suffix 'P' to indicate their putative nature. Once biochemical or unequivocal genetic activity, such as phage restriction, is demonstrated the suffix 'P' and any open reading frame (ORF) designations can be dropped allowing the main element of the name to be retained. Furthermore a Roman numeral should be included to indicate whether it is the first, second, third, etc. enzyme to be found in that organism. Note that the P extension should remain with the gene until such time as a gene product has been demonstrated to be functional.

This 'P' convention is illustrated with genes from *H.influenzae* serotype d. Two Type II REases, HindII and HindIII, and their associated MTases had been characterized biochemically (28–31). One Type I system had been demonstrated genetically (32) and the MTase, presumably associated with this system, had been partially characterized biochemically (30, 31). In the genome there are two putative Type I systems, although only one has a complete set of intact genes (33). The intact system therefore carries the designation HindI. In addition to these three systems, there was also known to be a Dam-like MTase, now called M.HindDam. However, also in the genome are putative m5C-MTase and REase genes (genes HI1040 and HI1041) that show high similarity to the known R-M system, HgiDI (34). The MTase encoded by HI1041 leads to a functional protein with specificity identical to that of M.HgiDI (R.D. Morgan, J. Patti, and R.J. Roberts, unpubl. results). It is therefore named M.HindV. However, the adjacent gene for the putative endonuclease is inactive and so it is named HindVP. One other R-M system can also be seen in the genome, this time encoding a Type III system. Neither the R nor the M genes have yet been demonstrated to be active and so these are named HindORF1056P and M.HindORF1056P. If they are shown to be active they would be renamed HindVI and M.HindVI. The convention here is to name the system after the ORF encoding the MTase gene. This is to ensure that the two genes are given names that indicate they are part of the same R-M system.

3.20 Homing Endonucleases

Homing endonucleases have been classified into four families according to conserved sequence motifs. These are the LAGLIDADG, GIY-YIG, H-N-H and His-Cys box families (35). Nomenclature of the homing endonucleases is patterned after that of REases, with a three-letter genus-species designation, followed by a Roman numeral (6). Whereas intron endonucleases are characterized by the prefix I- (for intron), the intein endonucleases are characterized by the prefix PI(for protein insert), and where the endonuclease is not intron- or intein-encoded, the prefix is F- (for freestanding). The systematic nomenclature does not preclude maintaining historic names. Counter to the original conventions proposed (6), the above nomenclature will extend to putative homing endonucleases without demonstrated catalytic activity. As with hypothetical REases, the suffix P will be used to denote the putative nature of the assignment, and the P will be dropped once nuclease activity has been confirmed. Hybrid homing endonucleases will be preceded by the prefix H-, followed by the authors' designation, e.g., an I-DmoI/I-CreI chimera could be H-DreI, or an I-TevI/I-BmoI hybrid could be H-TevBmo. Those homing endonucleases that have been characterized biochemically will continue to be listed within REBASE (5).

3.21 Adherence to These Conventions and Updates

The authors of this proposal have all agreed to follow these recommendations and it is hoped that other authors and journals will also adhere to these conventions. If further changes become appropriate, then REBASE (5) should be consulted for the latest modifications and practices.

References

1. Boyer HW (1971) DNA restriction and modification mechanisms in bacteria. Annu Rev Microbiol 25:153–176
2. Yuan R (1981) Structure and mechanism of multifunctional restriction endonucleases. Annu Rev Biochem 50:285–315
3. Smith H O, Nathans D (1973) A suggested nomenclature for bacterial host modification and restriction systems and their enzymes. J Mol Biol 81:419–423
4. Szybalski W , Blumenthal RM, Brooks JE, Hattman S, Raleigh EA (1988) Nomenclature for bacterial genes coding for class-II restriction endonucleases and modification methyltransferases. Gene 74:279–280
5. Roberts RJ, Macelis D (2003) REBASE – restriction enzymes and methylases. Nucleic Acids Res 31:418–420
6. Belfort M, Roberts RJ (1997) Homing endonucleases: keeping the house in order. Nucleic Acids Res 25:3379–3388
7. Noyer-Weidner M, Jentsch S, Pawlek B, Gunthert U, Trautner TA (1983) Restriction and modification in *Bacillus subtilis*: DNA methylation potential of the related bacteriophages Z, SPR, SPb, f3T, and r11. J Virol 46:446–453
8. Klein R, Baranyi U, Rossler N, Greineder B, Scholz H, Witte A (2002) *Natrialba magadii* virus fCh1: first complete nucleotide sequence and functional organization of a virus infecting a haloalkaliphilic archaeon. Mol Microbiol 45:851–863
9. Karreman C, de Waard A (1990) *Agmenellum quadruplicatum* M.AquI, a novel modification methylase. J Bacteriol 172:266–272

10. Dryden DT, Murray NE, Rao DN (2001) Nucleoside triphosphate-dependent restriction enzymes. Nucleic Acids Res 29:3728–3741

11. Murray NE (2000) Type I restriction systems: sophisticated molecular machines. Microbiol Mol Biol Rev 64:412–434

12. Titheradge AJ, King J, Ryu J, Murray NE (2001) Families of restriction enzymes: an analysis prompted by molecular and genetic data for type ID restriction and modification systems. Nucleic Acids Res 29:4195–4205

13. Gubler M, Braguglia D, Meyer J, Piekarowicz A, Bickle TA (1992) Recombination of constant and variable modules alters DNA sequence recognition by type IC restriction-modification enzymes. EMBO J 11:233–240

14. Szybalski W, Kim SC, Hasan N, Podhajska AJ (1991) Class-IIS restriction enzymes – a review. Gene 100:13–26

15. Stankevicius K, Lubys A, Timinskas A, Vaitkevicius D, Janulaitis A (1998) Cloning and analysis of the four genes coding for Bpu10I restriction-modification enzymes. Nucleic Acids Res 26:1084–1091

16. Looney MC, Moran LS, Jack WE, Feehery GR, Benner JS, Slatko BE, Wilson GG (1989) Nucleotide sequence of the FokI restriction-modification system: separate strand-specificity domains in the methyltransferase. Gene 80:193–208

17. Reuter M, Kupper D, Meisel A, Schroeder C, Kruger DH (1998) Cooperative binding properties of restriction endonuclease EcoRII with DNA recognition sites. J Biol Chem 273:8294–8300

18. Huai Q, Colandene JD, Topal MD, Ke H (2001) Structure of NaeI-DNA complex reveals dual-mode DNA recognition and complete dimer rearrangement. Nat Struct Biol 8:665–669

19. Xu Y, Lunnen KD, Kong H (2001) Engineering a nicking endonuclease N.AlwI by domain swapping. Proc Natl Acad Sci USA 98:12990–12995

20. Besnier CE, Kong H (2001) Converting MlyI endonuclease into a nicking enzyme by changing its oligomerization state. EMBO Rep 2:782–786

21. Hennecke F , Kolmar H, Brundl K, Fritz HJ (1991) The vsr gene product of E. coli K-12 is a strand- and sequence-specific DNA mismatch endonuclease. Nature 253:776–778

22. Tao T, Bourne JC, Blumenthal RM (1991) A family of regulatory genes associated with Type II restriction-modification systems. J Bacteriol 173:1367–1375

23. Sohail A, Ives CL, Brooks JE (1995) Purification and characterization of C.BamHI, a regulator of the BamHI restrictionmodification system. Gene 157:227–228

24. Mucke M, Reich S, Moncke-Buchner E, Reuter M, Kruger DH (2001) DNA cleavage by type III restriction-modification enzyme EcoP15I is independent of spacer distance between two head to head oriented recognition sites. J Mol Biol 312:687–698

25. Janscak P, Sandmeier U, Szczelkun MD, Bickle TA (2001) Subunit assembly and mode of DNA cleavage of the type III restriction endonucleases EcoP1I and EcoP15I. J Mol Biol 306:417–431

26. Raleigh EA, Wilson G (1986) Escherichia coli K-12 restricts DNA containing 5-methylcytosine. Proc Natl Acad Sci USA 83:9070–9074

27. Stewart FJ, Panne D, Bickle TA, Raleigh EA (2000) Methylspecific DNA binding by McrBC, a modification-dependent restriction enzyme. J Mol Biol 298:611–622

28. Smith HO, Wilcox KW (1970) A restriction enzyme from Hemophilus influenzae. I. Purification and general properties. J Mol Biol 51:379–391

29. Kelly TJ Jr, Smith HO (1970) A restriction enzyme from Hemophilus influenzae. II. Base sequence of the recognition site. J Mol Biol 51:393–409

30. Roy PH, Smith HO (1973) DNA methylases of Hemophilus influenzae Rd. I. Purification and properties. J Mol Biol 81:427–444

31. Roy PH, Smith HO (1973) DNA methylases of Haemophilus influenzae Rd. II. Partial recognition site base sequences. J Mol Biol 81:445–459

32. Gromkova R, Bendler J, Goodgal S (1973) Restriction and modification of bacterio-phage S2 in *Haemophilus influenzae*. J Bacteriol 114:1151–1157
33. FleischmannRD, Adams MD, White O, Clayton RA, Kirkness EF, Kerlavage AR, Bult CJ, Tomb J, Dougherty BA, Merrick JM et al. (1995) Whole-genome random sequenc-ing and assembly of *Haemophilus influenzae* Rd. Science 269:496–512
34. Dusterhoft A, Erdmann D, and Kroger M (1991) Stepwise cloning and molecular characterization of the HgiDI restriction-modification system from *Herpetosiphon giganteus* Hpa2. Nucleic Acids Res 19:1049–1056
35. Belfort M, Derbyshire V, Cousineau B, Lambowitz A (2002) Mobile introns: pathways and proteins. In: Craig N, Craigie R, Gellert M, Lambowitz A (eds) Mobile DNA II. ASM Press, Washington, DC, pp 761–783

Restriction-Modification Systems as Minimal Forms of Life

I. Kobayashi

1 Introduction

A restriction (R) endonuclease recognizes a specific DNA sequence and introduces a double-strand break (Fig. 1A). A cognate modification (M) enzyme methylates the same sequence and thereby protects it from cleavage. Together, these two enzymes form a restriction-modification system. The genes encoding the restriction endonuclease and the cognate modification enzyme are often tightly linked and can be termed a restriction-modification gene complex. Restriction enzymes will cleave incoming DNA if it has not been modified by a cognate or another appropriate methyltransferase (Fig. 1B). Consequently, it is widely believed that restriction-modification systems have been maintained by bacteria because they serve to defend the cells from infection by viral, plasmid, and other foreign DNAs (*cellular defense hypothesis*).

An alternative hypothesis for the maintenance of restriction-modification systems is based on the observation that several restriction-modification gene complexes in bacteria are not easily replaced by competitor genetic elements because their loss leads to cell death (*post-segregational killing*; Naito et al. 1995; Handa et al. 2001; Sadykov et al. 2003; Figs. 1C, 2B). This finding led to the proposal that these complexes may actually represent one of the simplest forms of life, similar to viruses, transposons, and homing endonucleases. This *selfish gene hypothesis* (Naito et al. 1995; Kusano et al. 1995; Kobayashi 1996, 1998, 2001) is now supported by many lines of evidence from genome analysis and experimentation.

A third type of hypothesis that explains why restriction-modification systems are present assumes that they aid the generation of diversity (Arber 1993; Price and Bickle 1986; *variation hypothesis*). Supporting this notion is that these systems are indeed associated with genome variation in a number of different ways (Sect. 2). However, such restriction-modification-associated

I. Kobayashi
ikobaya@ims.u-tokyo.ac.jp, Institute of Medical Science,
University of Tokyo, 4-6-1 Shirokanedai, Minato-ku, Tokyo 108-8639, Japan

Nucleic Acids and Molecular Biology, Vol. 14
Alfred Pingoud (Ed.)
Restriction Endonucleases
© Springer-Verlag Berlin Heidelberg 2004

A. Enzymes and reactions

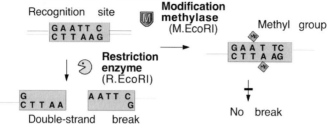

B. Attack on unmethylated incoming DNA

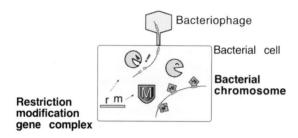

C. Post-segregational host killing

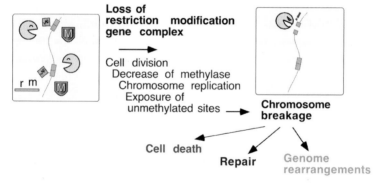

Fig. 1. Action of a restriction-modification gene complex. **A** Restriction enzyme (toxin) and modification methyltransferase (antitoxin). The antitoxin (modification enzyme) protects the targets of the toxin (restriction enzyme) by methylation. **B** Attack on incoming DNA. An attack on invading DNA that is not appropriately methylated is likely to be beneficial to the restriction-modification gene complex and to its host. **C** A simple dilution model for post-segregational killing. After loss of the restriction-modification gene complex, the toxin (restriction enzyme) and antitoxin (modification enzyme) will become increasingly diluted through cell division. Finally, too few modification enzyme molecules remain to defend all the recognition sites present on the newly replicated chromosomes. Any one of the remaining molecules of the restriction enzyme can attack these exposed sites. The chromosome breakage then leads to extensive chromosome degradation, and the cell dies unless the breakage is somehow repaired. The chromosome breakage may stimulate recombination and generate a variety of rearranged genomes, some of which might survive. *rm* Restriction-modification gene complex. Reproduced from *Nucleic Acids Research* (Kobayashi 2001)

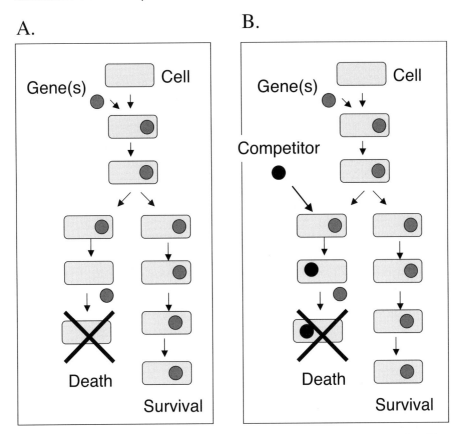

Fig. 2. The principle of post-segregational killing. **A** Once established in a cell, the addiction gene complex is difficult to eliminate because its loss, or some sort of threat to its persistence, leads to cell death. Intact copies of the gene complex survive in the other cells of the clone. **B** Advantage in competitive exclusion. A specific case of post-segregational cell killing showing the fight the gene complex raises against an incoming competing genetic element

genome variation can also be explained by the selfish gene hypothesis, as will be outlined (Sect. 6).

In this chapter, I will first review the evidence supporting the notion that some restriction-modification gene complexes behave as mobile genetic elements that may induce genome variability (Sect. 2). Next, I will describe their attacks on the genome and their consequences and then present the selfish gene hypothesis (Sect. 3). This is followed by a review of the gene organization of these complexes and how they are regulated in relation to their life cycle (Sect. 4). The competition that exists between restriction-modification gene complexes is then described along with the other types of intragenomic interactions involving these complexes (Sect. 5). The effect of the parasitic selfish behavior of restriction-modification complexes on the genome, in particular, their ability to induce mutagenesis and recombination, will then be discussed. This will illustrate how the host-parasite-type interactions between restriction-modification complexes and the genome contribute to genomic evolution (Sect. 6). How the selfish gene point of view can aid the classification of these complexes is described in Section 7. The next section (Sect. 8) discusses how these systems can be utilized in practical terms. The penultimate section (Sect. 9) proposes that the attack on the host by restriction-modification systems upon their disturbance reflects a general feature of genes that are assembled in a chromosome. The last section (Sect. 10) draws some conclusions.

This work owes much to other publications and databases. Particularly helpful were a brief but insightful review on programmed cell death in bacteria (Yarmolinsky 1995), my own reviews on restriction-modification systems (Kobayashi 2001) and on post-segregational killing systems (Kobayashi 2003b), and an extensive database on restriction enzymes and their genes, namely, REBASE (Roberts et al. 2003b) [http://rebase.neb.com]. To minimize the number of citations, only a few of the possible references have been cited. I welcome feedback on the novel view of restriction enzymes that is detailed in this chapter.

2 Genomics and Mobility of Restriction-Modification Systems

The decoding of several bacterial genomes has provided ample evidence of the variability and potential mobility of restriction-modification systems. Here I will review these lines of evidence. The question of how this variability/mobility is generated will be addressed again in Section 3.4.

2.1 Genomics

The restriction-modification gene homologues that have been identified in completely sequenced bacterial genomes are listed in REBASE. Some of these genomes – for example, those of *Haemophilus influenzae*, *Methanococcus jannaschii*, *Helicobacter pylori*, *Neisseria meningitidis*, *Neisseria gonorrhoeae*, and *Xylella fastidiosa* – have impressive numbers of restriction-modification gene homologues. Many restriction-modification gene homologues are specific to one strain within a given species, as has been noted for *Escherichia coli* [REBASE], *N. gonorrhoeae* [REBASE] and *H. pylori* (Alm et al. 1999; Nobusato et al. 2000a) [REBASE]. For example, comparison of the modification enzyme homologues in two completely sequenced strains of *H. pylori* revealed that while many pairs were very similar to each other, some homologues occurred in only one strain (Alm et al. 1999; Nobusato et al. 2000a).

2.2 Horizontal Gene Transfer Inferred from Evolutionary Analyses

Various types of evolutionary analyses suggest that restriction-modification genes have undergone extensive horizontal transfer between different groups of microorganisms (Table 1 (4); Kobayashi et al. 1999; Kobayashi 2001). Early studies found that close homologues occur in distantly related organisms such as Eubacteria and Archaea (archaebacteria) (e.g. Nolling and de Vos 1992). Extensive sequence alignment and phylogenetic tree construction now provide strong support for this point (Nobusato et al. 2000a; Bujnicki 2001). The incongruence in the same species of the phylogenetic tree of the methyltransferases with the tree of ribosomal RNA genes is additional evidence of the extensive horizontal transfer that the restriction-modification genes appear to have experienced (Nobusato et al. 2000a). Moreover, the GC content and/or codon usage of restriction-modification genes often differ from those of the majority of the genes in the genome (Jeltsch and Pingoud 1996; Alm et al. 1999; Nobusato et al. 2000a; Chinen et al. 2000b). This indicates that some restriction-modification genes may have joined the genome relatively recently by horizontal transfer from distantly related bacteria.

2.3 Presence on Mobile Genetic Elements

Sometimes there are hints for the molecular basis of the variability and horizontal transfer of restriction-modification gene complexes. One of these hints is that these complexes are often found on a variety of mobile genetic elements [Table 1 (2)]. For example, many restriction-modification gene complexes reside on plasmids (Table 2, B). Many of the cases of strain-specific restriction-modification systems in *E. coli* can be explained by their presence on plasmids [REBASE]. Moreover, some restriction-modification gene homologues have been found in a prophage in the chromosome [Table 1 (2)]. Others are on transposons, conjugative transposons (or integrative conjugative elements), genomic islands, and integrons [Table 1 (2)]. Restriction-modification gene homologues are also sometimes found to be linked with mobility-related genes, although the significance of this linkage is less clear than with the above cases (Xu et al. 1998; Vaisvila et al. 1995).

2.4 Genomic Contexts and Genome Comparison

Close examination of the genomic neighborhood of restriction-modification gene homologues and its comparison with a closely related genome also sometimes provide hints as to how restriction-modification gene complexes can enter a genome and be associated with genome rearrangements [Table 1 (3), Fig. 3].

Table 1. Evidence of the mobility of restriction-modification gene complexes

Types	Examples
(1) Strain-specific presence within a species	*Escherichia coli* [REBASE]; *Helicobacter pylori* [REBASE] (Akopyants et al. 1998; Alm et al. 1999; Nobusato et al. 2000a; Xu et al. 2000; Lin et al. 2001)
(2) Presence on mobile genetic elements	
(2–1) Plasmid	Many. See Table 2, B
(2–2) Bacteriophage (prophage)	*hindIII* on Phi-flu in *Haemophilus influenzae* (Hendrix et al. 1999) *sau42I* on Phi-42 in *Staphylococcus aureus* (GenBank X94423; Kobayashi et al. 1999) *ecoO109I* on P4 in *Escherichia coli* (Kita et al. 1999) *ecoT38I* on P2 in *Escherichia coli* (Kita et al. 2003) *bsuMI* on prophage 3 of *Bacillus subtilis* (Ohshima et al. 2002) *dam* on Bacteriophage T2 of *Escherichia coli* (Miner and Hattman 1988)
(2–3) Integrative conjugative element/ Genomic island	*sth368I* on integrative conjugative element ICESt1 in *Streptococcus thermophilus* (Burrus et al. 2001) A solitary DNA methyltransferase gene in conjugative transposon *Tn5252* (Sampath and Vijayakumar 1998) *hsdMS* homologues on genomic islands in *Staphylococcus aureus* (Kuroda et al. 2001)
(2–4) Transposon	*rle39BI* flanked by IS (insertion sequence) as inverted repeats (Rochepeau et al. 1997)
(2–5) Integron	*xbaI* in *Xanthomonas campestris* pv. *badrii* (Kobayashi et al. 1999; Rowe-Magnus et al. 2001) Gene for M.VchAORF447P (= M.Vch01I) on mega-integron in *Vibrio cholerae* (Kobayashi et al. 1999; Heidelberg et al. 2000) *hphI* homologue in a cassette in superintegron in *Vibrio metschnikovii* (Rowe-Magnus et al. 2003) A homologues of methyltransferase CAA68045 of *Lactococcus lactis* in a cassette in super-integron in *Vibrio metschnikovii* (Rowe-Magnus et al. 2003)

Table 1. (*Continued*)

Types	Examples
(3) Genomic context and/or genome comparison	
(3–1) Insertion into an operon-like gene cluster (Fig. 3A)	*haeII* between *mucE* and *mucF* in *Haemophilus aegyptius* (Stein et al. 1998)
	nmeBI or *nmeDI* between *pheS* and *pheT* in *Neisseria meningitidis* (Claus et al. 2000; Saunders and Snyder 2002)
	ssuDAT1I between *purH* and *purD* in *Streptococcus suis* (Sekizaki et al. 2001)
(3–2) Insertion with a long target duplication (Fig. 3B)	HP1366/ HP1367/ HP1368 (Type IIS restriction-modification gene homologues) into jhp1284/ jhp1285 (Type III restriction-modification system in *Heliobacter pylori* (Nobusato et al. 2000b)
(3–3) Simple substitution (Fig. 3C)	Restriction-modification homologues in *Helicobacter pylori* strains (Alm et al. 1999; Nobusato et al. 2000b) and in *Pyrococcus* species (Chinen et al. 2000b)
(3–4) Substitution adjacent to large inversion (Fig. 3D)	Restriction-modification homologues in *Helicobacter pylori* strains (Alm et al. 1999) and in *Pyrococcus* species (Chinen et al. 2000b)
(3–5) Apparent transposition (Fig. 3E)	Restriction-modification homologues in *Helicobacter pylori* strains (Alm et al. 1999; Kobayashi 2001) and in *Pyrococcus* species (Chinen et al. 2000b)
(4) Evolutionary/ informatic analyses	
(4–1) Phylogenetic analysis through sequence alignments	Close homology between archaeal and eubacterial restriction-modification genes (Nolling and de Vos 1992) Modification genes (Jeltsch et al. 1995; Nobusato et al. 2000a; Bujnicki 2001); Restriction genes (Bujnicki 2001)
(4–2) GC3 (GC content of the third letter of a codon; difference from bulk of the genome)	Restriction modification genes in many eubacteria and Archaea, e.g. *Helicobacter pylori* (Alm et al. 1999; Nobusato et al. 2000a)
(4–3) Codon usage (difference from bulk of the genome)	Restriction modification genes in many eubacteria and archaea, e.g. (Jeltsch and Pingoud 1996; Karlin et al. 1998b)

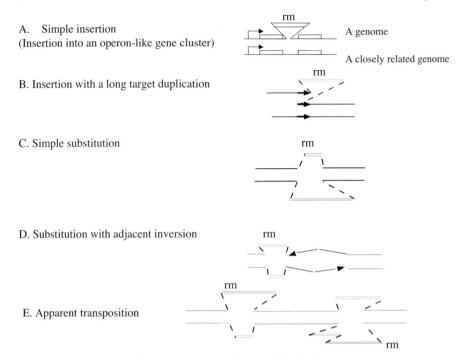

A. Simple insertion
(Insertion into an operon-like gene cluster)

B. Insertion with a long target duplication

C. Simple substitution

D. Substitution with adjacent inversion

E. Apparent transposition

Fig. 3. Restriction-modification gene homologues linked with large genome polymor-
phism. *rm* A restriction-modification gene complex. **A** Insertion into an operon-like
gene cluster. The *boxes* indicate open reading frames constituting an operon-like gene
cluster. An *arrow* indicates transcription.
B Insertion of a restriction-modification complex with long target duplication. The *thick
arrow* indicates a duplicated sequence that is in the order of 100 bp in length. **C** Simple
substitution of a restriction-modification complex. **D** Substitution of a restriction-modification complex adjacent to a large inversion. The *bent arrows* indicate a segment of the
length of the order of 10 kbp that appears inverted when two genomes are compared. **E**
Apparent transposition. The *double lines* marked as *rm* indicate two regions that are
highly homologous with each other and carry a restriction-modification gene complex

2.4.1 Insertion into an Operon-Like Gene Cluster

Comparison of the sequence of *Haemophilus aegyptius* with that of *Haemophilus influenzae* suggests that the HaeII restriction-modification gene complex was inserted into an operon-like gene cluster in an ancestor of *H. aegyptius* and that this replaced a short intergenic sequence [Fig. 3A, Table 1 (3–1)].
A similar insertion was found between *phS* and *phT* in *Neisseria*. There are
several nonhomologous alleles, not necessarily restriction-modification gene
homologues, that can occupy this region (Saunders and Snyder 2002). In the
pur cluster of *Streptococcus suis*, the insertion of the *ssuDAT1I* genes carries a
3-bp terminal duplication (Sekizaki et al. 2001).

2.4.2 Insertion with Long Target Duplication

Some restriction-modification gene homologues are found to be flanked by long (in the order of 100 bp) direct repeats (Lubys et al. 1996; Gunn and Stein 1997; Nobusato et al. 2000b; Aras et al. 2001). The comparison of two genomes of *H. pylori* has suggested that this duplication is generated when a restriction-modification gene complex inserts itself into the genome (Fig. 3B: Nobusato et al. 2000b).

2.4.3 Substitution Adjacent to a Large Inversion

When closely related bacterial genome sequences are compared, it appears that restriction-modification gene complexes often substitute other sequences [Fig. 3C, Table 1 (3–3)]. Sometimes the substitution lies adjacent to a large inversion [Fig. 3D, Table 1 (3–4)].

2.4.4 Apparent Transposition

Two closely homologous DNA segments carrying a restriction-modification gene complex homologue are sometimes found at different locations in two closely related genomes [Fig. 3E, Table 1 (3–5)]. This provides evidence for the *transposition* of a restriction-modification gene complex in the formal sense of the word.

2.4.5 Linkage of a Restriction-Modification Gene Complex with Another Restriction-Modification Gene Complex or a Cell Death-Related Gene

Some subgenomic regions in *H. pylori* are full of restriction-modification gene homologues (Alm et al. 1999; Nobusato et al. 2000b; Kobayashi 2001). Indeed, restriction-modification gene complexes are often found to be linked with another restriction-modification gene complex [REBASE] or with other types of cell death-related genes (Kobayashi 2003). They are similar to subgenomic regions full of mobile elements in some prokaryotic genomes and many eukaryotic genomes.

2.5 Defective Restriction-Modification Gene Complexes

Many restriction-modification gene homologues in the sequenced genomes appear to be defective due to insertions, deletions and point mutations [REBASE]. For example, many of the restriction-modification homologues present in the fully sequenced *H. pylori* genomes are defective (Nobusato et al. 2000b; Lin et al. 2001). When two *H. pylori* strains were analyzed, while all the strain-specific restriction-modification gene homologues were found to be active, most of the shared restriction-modification genes turned out to be inactive in both strains (Lin et al. 2001). This supports the notion that the strain-specific restriction-modification genes were acquired more recently through horizontal transfer from other bacteria and were selected for function. This is reminiscent of the defective transposons in many genomes. Sometimes a restriction-modification gene complex appears to have been inactivated by the insertion of another restriction-modification gene complex [Table 1 (3–2); Nobusato et al. 2000b]. This is similar to the insertion of transposons into other transposons, which is frequently observed in plant and other genomes. These defective restriction-modification gene complex homologues may have some activity, as do some defective viruses and transposons (see Sect. 5.2). Such mutational inactivation of restriction-modification genes can be distinguished from the phase variation of restriction-modification systems (Dybvig et al. 1998; Saunders et al. 1998; Alm et al. 1999; Kobayashi 2001).

3 Attack on the Host Genome and the Selfish Gene Hypothesis

The above observations are in harmony with the hypothesis that restriction-modification gene complexes represent a class of mobile genes that move between genomes and cause genome rearrangements of various types. Viruses, transposons and other mobile genetic elements all employ unique strategies for their own survival. The strategy used by restriction-modification gene complexes is straightforward, namely, they destroy nonself DNA that is marked by the absence of proper methylation. Their attack on invading DNAs represents one simple manifestation of this strategy.

3.1 Post-Segregational Host Cell Killing

The resistance of restriction-modification gene complexes to their loss from a cell through *post-segregational host killing* (the programmed death of cells that have been freed of a restriction-modification gene complex; Fig. 2A) was observed when a restriction-modification gene complex was challenged by a competitor genetic element (Fig. 2B). A plasmid carrying a restriction-modification gene complex cannot be readily displaced by an incompatible plasmid in transformation (Naito et al. 1995). Similarly, a restriction-modification gene complex on the bacterial chromosome cannot be easily replaced by a homologous stretch of DNA through homologous recombination in bacteriophage-mediated transduction (Handa et al. 2001) and in natural transformation (Sadykov et al. 2003). It turned out that the replacement does occur but that those cells experiencing the replacement fail to survive.

A plausible reason for this phenomenon is illustrated in Fig. 1C. If a restriction-modification gene complex is lost, the cell's descendants will contain fewer and fewer molecules of the modification enzyme because of dilution. Eventually, the modification enzyme's capacity to protect the many recognition sites on newly replicated chromosomes from the remaining pool of restriction enzymes becomes inadequate. Chromosomal DNA will then be cleaved at these exposed sites, leading to cell death (unless the break is repaired; see Sect. 6 below). Naturally, the restriction enzyme will also be diluted by cell division following the loss of the gene complex. However, there is asymmetry between the roles of the methyltransferase and the restriction enzyme – the asymmetry between life and death. For the restriction enzyme to kill the cell, a single break on the chromosome may well be sufficient. In contrast, for the modification enzyme to keep the host alive, all (or sufficiently many) of the hundreds of recognition sites along the chromosome need to be methylated.

Evidence that supports this simple dilution model was obtained by experiments in which a restriction-modification gene complex on a plasmid with a temperature-sensitive replication machinery was lost from *E. coli* cells after a temperature shift (Naito et al. 1995; Kusano et al. 1995; Handa and Kobayashi 1999; Handa et al. 2000; Chinen et al. 2000a). The blockage of plasmid replication stopped the increase in viable cell counts and resulted in loss of cell viability. This led to induction of the SOS response. Many cells formed long filaments, some of which were multinucleated and others anucleated. The accumulation of very long noncircular forms of chromosome followed by extensive DNA degradation was observed.

The author's group has been successful in demonstrating that all the Type II restriction-modification systems examined so far (PaeR7I, EcoRI, EcoRV, EcoRII, SsoII and BamHI) show post-segregational killing activity. Consequently, I am inclined to believe that this may be a general property of Type II restriction-modification systems. The strength of the activity, however, varies widely among these Type II restriction-modification systems and among host genetic backgrounds (see Sects. 5 and 6). Post-segregational cell killing by other types of restriction-modification systems has not yet been reported, although host attack by a Type I restriction-modification gene complex under special environmental and/or genetic conditions has been detected (O'Neill et al. 1997; Makovets et al. 1999; Cromie and Leach 2001; Murray 2002).

3.2 Comparison with Other Post-Segregational Cell Killing Systems

Post-segregational killing has been long recognized as a mechanism by which plasmids are stably maintained (Table 2; Yarmolinsky 1995; Gerdes et al. 1997; Gerdes 2000; Kobayashi 2003b). Such plasmid post-segregational killing systems consist of a pair of linked genes that encode a *toxin* protein (also called

Table 2. Post-segregational-killing gene complexes on plasmids

Locus (plasmid)	Bacteria	Killer	Target	Anti-killer	References on addiction activity
A. Classical proteic systems					
ccd (=*lyn* =*let*) (F)	*Escherichia coli*	CcdB (=LetD)	DNA gyrase	CcdA (= LetA)	Jaffe et al. (1985)
segB (pSM19035)	*Streptococcus pyogenes* (broad host range in gram-positive bacteria)	Zeta (phospho-transferase with ATP/GTP)	Unknown	Epsilon (inhibitor of binding of ATP/GTP)	Meinhart et al. (2003)
B. Restriction-modification systems					
paeR7I (pMG7)	*Pseudomonas aeruginosa*	PaeR7I	5' CTCGAG in the chromosome	M.PaeR7I	Naito et al. (1995)
ecoRI (RTF-1)	*Escherichia coli*	EcoRI	5' GAATTC in the chromosome	M.EcoRI	Naito et al. (1995); Kulakauskas et al. (1995); Kusano et al. (1995)
ecoRII (Fig. 6; RTF-2; N-3)	*Escherichia coli*	EcoRII	5' CCWGG in the chromosome	M.EcoRII	Chinen et al. (2000a); Takahashi et al. (2002)
ecoRV (pLG13)	*Escherichia coli*	EcoRV	5' GATATC in the chromosome	M.EcoRV	Nakayama and Kobayashi (1998)
ssoII (Fig. 6; P4)	*Shigella sonnei*	SsoII	5' CCNGG in the chromosome	M.SsoII	Chinen et al. (2000a)
bsp6I (pXH13)	*Bacillus* sp. strain RFL6	Bsp6I	5' GCNGC in the chromosome	M.Bsp6I	(Kulakauskas et al. 1995b)
C. Antisense-RNA-regulated systems					
hok/sok (R1)	*Escherichia coli*	Hok (Holin)	Membrane	Sok-RNA (anti-sense RNA)	Gerdes et al. (1997); Moller-Jensen et al. (2001)

Some parts were modified from (Gerdes 2000) and from REBASE (http://rebase.neb.com; Roberts et al. 2003b)

killer or poison) and an *antitoxin* (also called anti-killer or antidote). The antitoxin blocks the action of the toxin by interacting with it either as a protein (classical proteic toxin/antitoxin systems; Gerdes 2000) or as an antisense RNA that inhibits the expression of the toxin (anti-sense-RNA-regulated systems; Gerdes et al. 1997). (The word *classical* is used because the restriction-modification systems are also made up of a protein toxin and a protein antitoxin.) Loss of the gene pair from a cell brings about the lethal action of the toxin because the antitoxin gene product is metabolically unstable. Since the cells that have lost the plasmid die, the plasmid is maintained at a high frequency in the population of viable cells (Fig. 2A). The targets and mechanisms of action of several post-segregational killing systems are now being understood at the level of protein structure (Table 2). The toxin protein may be a holin (which makes a hole in the cytoplasmic membrane), a phosphotransferase, or a codon-specific mRNA nuclease (Gerdes et al. 1997; Kamada et al. 2003; Meinhart et al. 2003; Pedersen et al. 2003).

As expected, the insertion of several Type II restriction-modification gene complexes into a plasmid increases the stability of its maintenance in pure culture (Naito et al. 1995; Kulakauskas et al. 1995a; Kusano et al. 1995; Handa and Kobayashi 1999; Handa et al. 2000; Nakayama and Kobayashi 1998; Chinen et al. 2000a; Takahashi et al. 2002). Figure 4 shows the plasmid stabilization that occurs when the EcoRII restriction-modification gene complex is inserted (Chinen et al. 2000a; Takahashi et al. 2002). This is similar to the maintenance of a plasmid that carries an antibiotic-resistance gene (such as *amp*) in the presence of the antibiotic (such as ampicillin). The loss of the plasmid and hence the resistance gene leads to cell death, so that all the surviving cells carry the resistance gene (*amp*) and hence the plasmid. If the selection by the antibiotic can be called external, then the selection imposed by the restriction-modification gene complex (and other post-segregational killing systems) can be called internal.

The concept that a genetic element is maintained because it is toxic to the genome rather than because it is useful for the genome is not as paradoxical as it appears. First of all, the genomes emerging from the decoding efforts do not look like a well-designed and optimized blueprint for a machine. Rather, they look like a relatively disorganized and ever-changing community of genes that can have potentially different interests to that of serving the community. These genes can move into the genome and can also disappear. Bacteriophage genomes, for example, stay as a prophage in the genome and are induced to replicate under some stressful conditions. Second, the toxic effect of post-segregational killing systems is tightly repressed under most conditions (see Sect. 4). Third, it is likely that pure cultures of a bacterial line carrying a plasmid occur only rarely in the natural environment. Normally the plasmid (as with other pieces of DNA) would regularly encounter competitors. Consequently, its post-segregational killing activity provides it with a competitive advantage as discussed (Sect. 3.1; Fig. 2B), since the killing of the

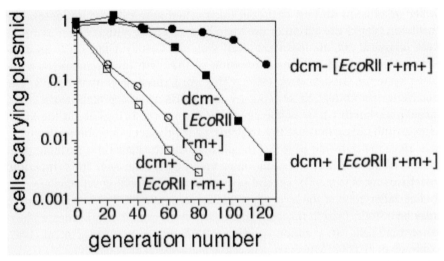

Fig. 4. Carriage of the EcoRII restriction-modification gene complex stabilizes the maintenance of a plasmid, but this is suppressed when *dcm* for a solitary methyltransferase is present on the chromosome. The plasmid carries either an intact restriction gene or its mutant form. The chromosome of the *Escherichia coli* host carries either an intact *dcm* gene or its mutant form. The cells with one of the plasmids were grown in liquid medium after removing the agent that selects the plasmid. The culture was continued with appropriate dilutions, after which the cells were spread on agar that is either selective or not selective for the plasmid, followed by colony counting. The number of plasmid-carrying cells (colony formers on agar selective for the plasmid) divided by the number of viable cells (colony formers on unselective agar) was plotted on the vertical axis. The number of viable cells was used to calculate generation numbers on the horizontal axis. Modified from *Journal of Bacteriology* (Takahashi et al. 2002)

host that has been infected with the competitor DNA allows copies of the plasmid to survive in the neighboring clonal cells. The killing that is induced by post-segregational killing genes resembles altruistic suicide despite the fact that it is induced by a self-interested piece of DNA.

3.3 Selfish Gene Hypothesis

The resistance of chromosomal restriction-modification gene complexes to being replaced by a homologous stretch of DNA as described above (Handa et al. 2001; Sadykov et al. 2003) demonstrates that the restriction-modification gene complex, rather than the plasmid, is the unit of post-segregational killing and hence is likely the unit of selection. The above arguments (Sect. 3.2) that point out that plasmids carrying a post-segregational killing gene complex have a competitive advantage can also be applied to restriction-modification gene complexes. The killing provides an advantage to the restriction-modification gene complex and any linked genes as well. It has been hypothesized that the competitive advantage of post-segregational killing has, at least in part, assured the maintenance of these restriction-modification systems (Naito et al. 1995; Kusano et al. 1995; Kobayashi 1996; 1998b). This suggests that restriction-modification gene complexes could be regarded as a form of life. More specifically, these restriction-modification elements deserve to be called selfish genes in the sense that this term is used in genetics, evolutionary biology and behavioral ecology (Kobayashi 1998b; Haig 1997; Hurst and Werren 2001). This mode of cell death – apparently altruistic cell death programmed by a resident genetic element upon invasion of its competitor genetic element (Fig. 2B) – is also seen in phage exclusion (Snyder 1995).

The term *genetic addiction* has also been applied to post-segregational killing systems (Lehnherr et al. 1993; Kobayashi 2003). This term explains how these gene sets can work symbiotically with a host genome. The host also could gain an advantage when a post-segregational killing complex is present on its genome. Its cell killing activity would defend the host genome against infection (Fig. 2B). More generally speaking, the deaths induced by these post-segregational killing genes could be useful for bacterial survival just as programmed cell deaths are useful for survival of multi-cellular organisms (Yarmolinsky 1995). Another argument for using "addiction" instead of "post-segregational killing" is that it is a better term in cases where these elements' action can only result in inhibition of growth, as opposed to outright killing, of its host. See Section 6 for more discussion about this point. A thorough discussion of genetic addiction can be found elsewhere (Kobayashi 2003).

3.4 Genomics as Explained by the Selfish Gene Hypothesis

The primary property of restriction-modification gene complexes, namely, post-segregational cell killing, may explain their mobility and other genomic aspects (see Sect. 2).

The restriction-modification gene complex would provide a competitive advantage to not just plasmids but also to any DNA that is linked with it (Fig. 2B). Any type of genetic element that is relatively independent from the remainder of the genome (e.g. plasmids, bacteriophages, proviruses) is essentially unstable in its maintenance, and therefore it would benefit from carrying a restriction-modification gene complex. Thus, the selfish behavior of the restriction-modification gene complex would increase the stability of the associated DNA. In return, the restriction-modification gene complex would acquire mobility. This explains the carriage of a restriction-modification gene complex by the many different mobile elements (and, of course, by plasmids) discussed above [Sect. 2.3; Tables 1 (2), 2, B].

A feature that is common to some of those genomes that bear many restriction-modification homologues (see Sect. 2.1) is their marked capacity for natural transformation. This mechanism allows all the genes in the genome to move between genomes of a population by means of homologous recombination. Thus, a restriction-modification gene complex does not have to depend on a mobile element for its mobility. This susceptibility to natural transformation also means that chromosomal genes would be frequently replaced by incoming homologous stretches of DNAs. The restriction-modification gene complexes will resist their loss by host killing as has been demonstrated (Sadykov et al. 2003). Moreover, the restriction-modification gene complexes would also defend themselves and their host bacterial cells against invading DNAs by cleaving this foreign DNA.

The features of the restriction-modification genes complexes that were revealed by the analyses of genome contexts and the comparison of closely related genomes [Sect. 2.4; Table 1 (3)] may be explained by the selfish nature of these complexes. *Insertion into an operon* (Sect. 2.4.1) would be beneficial for the complexes as it allows the restriction-modification genes to spread horizontally by homologous recombination by taking advantage of the flanking sequences within a species (Saunders and Snyders 2002). Homologous recombination may lead to an increase of their frequency in the population because of their resistance to loss by post-segregational killing. The operon would also acquire stability in the maintenance of its persistence and expression, especially when it is in competition with an operon of a similar function.

Insertion with long target duplication (Sect. 2.4.2) benefits the restriction-modification genes similarly to the *insertion into an operon* described above in several points. A disadvantage of this structure for the inserted gene complex is that later host-mediated homologous recombination between the duplicated regions could result in the deletion of the restriction-modification gene complex (Aras et al. 2001). However, the restriction-modification gene complex would resist such loss through post-segregational killing. The long flanking direct repeats could also benefit restriction-modification gene complexes as it would promote their multiplication, as discussed later (see Sect. 4.3).

The *linkage of restriction-modification homologues with a large inversion* (Sect. 2.4.3) would provide another means of resistance to loss, as it would inhibit its replacement by its allele through homologous recombination, which needs homology in both sides.

4 Gene Regulation in the Life Cycle of Restriction-Modification Systems

Restriction-modification systems are regulated by a variety of mechanisms. These different mechanisms may be interpreted in terms of their behavior as post-segregational killing gene complexes, namely, that they have to establish themselves, maintain themselves, and engage in post-segregational killing

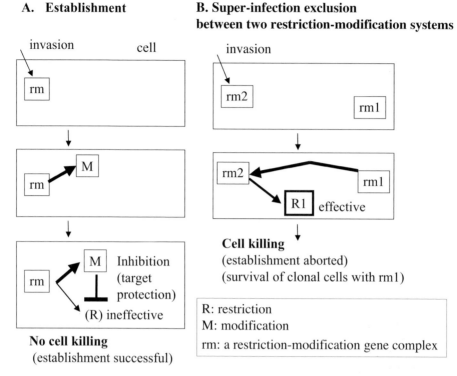

A. Establishment

B. Super-infection exclusion between two restriction-modification systems

Cell killing
(establishment aborted)
(survival of clonal cells with rm1)

No cell killing
(establishment successful)

R: restriction
M: modification
rm: a restriction-modification gene complex

Fig. 5. Establishment in a new host and super-infection exclusion. **A** Invasion and establishment of a restriction-modification gene complex in a new host. The regulatory system of the gene complex allows the modification enzyme to be expressed first to prevent chromosome breakage and host cell killing by the restriction enzyme. This gene complex thus successfully establishes itself in its new host cell. **B** Super-infection exclusion upon invasion of a restriction-modification gene complex into a cell that harbors another restriction-modification gene complex with a similar specificity in its regulatory system. The regulatory system of the resident complex forces the incoming complex to express its restriction enzyme first. The cell is killed and the establishment of the incoming restriction-modification gene complex is aborted. The resident restriction-modification gene complex survives in the neighboring clonal cells. This represents another example of defense by cell death (see Fig. 2B)

when they are threatened. When a restriction-modification gene complex attempts to establish itself in a new host cell, it must avoid cell killing by expressing its modification enzyme (antitoxin) first, after which its restriction enzyme (toxin) can be expressed (Fig. 5A). In the maintenance phase, the amounts of its products should be tightly auto-regulated so that the restriction attack on the host is minimized until a signal to attack has been received. The strategies used to establish, auto-regulate and kill the cell appear to vary among the restriction-modification systems and depend on their gene organization (Fig. 6).

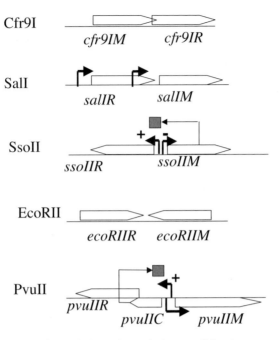

Fig. 6. Organization and regulation of restriction-modification gene complexes. The *pointed box* indicates a gene together with its direction. The *thick arrow* indicates transcription. The *two gray boxes* indicate M.SsoII and C.PvuII proteins. The *plus sign* indicates a positive effect of the protein on gene expression, while the *minus sign* indicates a negative effect. See Table 2 and text for references

4.1 Gene Organization

The toxin and antitoxin genes in other (non-restriction-modification) post-segregational killing systems are compactly organized (Gerdes 2000; Kobayashi 2003b). Similarly, the constituent genes of most of the restriction-modification systems identified so far are also tightly linked with each other. Indeed, sometimes the genes actually overlap (Lubys et al. 1994; Sekizaki et al. 2001). Such compact organization would promote the simultaneous loss of restriction and modification genes that would then induce post-segregational killing. Thus, such clusters represent units of mobility and selection. In several cases, however, restriction-modification and/or specificity gene homologues are separated in the genome. These cases may represent, in principle, either (1) the decay of a restriction-modification gene cluster (Sect. 2.5), (2) a unique function as a solitary gene (Sect. 5.2), and/or (3) a restriction-modification system that is composed of unlinked genes (Schouler et al. 1998; Dandekar et al. 2000; Kuroda et al. 2001). These cases may represent the loss of virulence ('selfishness') of restriction-modification gene complexes towards their hosts as discussed earlier (Kobayashi 2001).

In many of the classical proteic toxin/antitoxin systems characterized so far, the genes form an operon with the antitoxin gene followed by the toxin gene (Kobayashi 2003). In another system, however, this order is reversed (Tian et al. 1996). Such co-linear arrangements are also found in most restriction-modification systems (Wilson and Murray 1991) [REBASE] (Fig. 6). In the PaeR7I and Cfr9I systems, the modification gene is the first gene, but in the SalI and EcoRI systems, the restriction gene is the first. Several restriction-modification systems, such as SsoII, show divergent organization, but several others, such as EcoRII, show convergent organization. Several addiction gene complexes such as *segB* (Table 2) carry another gene whose product is involved in regulating gene expression. Likewise, several Type II restriction-modification systems also encode a control protein. Many restriction-modification gene complexes show a more complex organization [REBASE] (Roberts et al. 2003a).

4.2 Gene Regulation

4.2.1 Restriction Gene Downstream of Modification Gene

In the Cfr9I operon (Fig. 6), the last codon of the modification gene overlaps with the start codon for the restriction gene (ATGA). Moreover, a nucleotide sequence that is complementary to a predicted Shine-Dalgarno sequence precedes *cfr9IR* and lies within the *cfr9IM* gene. These features may relate to some regulatory mechanism that controls the expression of the restriction gene (Lubys et al. 1994).

4.2.2 Restriction Gene Upstream of Modification Gene

In the SalI operon (Fig. 6), the toxin (restriction enzyme) gene lies upstream of the antitoxin (modification enzyme) gene, similar to what was observed for one of the classical proteic post-segregational killing systems (Tian et al. 1996). These two operons are regulated by a similar mechanism (Alvarez et al. 1993). The operon is mainly transcribed from a promoter located immediately upstream of *salIR*. However, there is another promoter within the 3′ end of the *salIR* coding region that allows the modification gene to be expressed in the absence of the former promoter. The latter promoter might be involved in the establishment of modification activity prior to restriction endonuclease activity within a new host.

4.2.3 Modification Enzyme as a Regulator

In many of the classical proteic post-segregational killing systems, the antitoxin protein serves as a regulator (Kobayashi 2003). Likewise, in some restriction-modification gene complexes, the modification enzyme is involved in the gene regulation. For example, EcoRII methyltransferase (M.EcoRII) represses its own expression at the transcriptional level (Som and Friedman 1993). Likewise, M.MspI represses, *in trans*, the expression of the *mspIM-lacZ* fusion by binding to the intergenic region between *mspIM* and *mspIR* (Som and Friedman 1997). Moreover, in *lacZ* fusion experiments, M.SsoII was shown to repress its own synthesis but to stimulate expression of the cognate restriction endonuclease. These mechanisms would help in the establishment of the complex and would tightly regulate the amounts of the gene products. The N-terminus of M.SsoII, which is predicted to form a helix-turn-helix, was shown to be responsible for its regulatory functions as well as its specific DNA-binding activity (Karyagina et al. 1997). The methyltransferase M.ScrFIA from the ScrFI system also carries a helix-turn-helix motif at its N-terminus and serves as a regulator (Butler and Fitzgerald 2001).

Transcription of the restriction-modification operons can also be regulated by the methylation of cognate recognition sites. An example of this is the CfrBI system. Here M.CfrBI decreases the transcription from the *cfrBIM* promoter but increases the transcription from the *cfrBIR* promoter. These effects depend on the methylation at the CfrBI site in the promoter region by M.CfrBI (Beletskaya et al. 2000). Similar apparently strategically placed cognate recognition sites in the regulatory region are found with other restriction-modification gene complexes (Wilson and Murray 1991) [REBASE].

4.2.4 Regulatory Proteins

The control (C) protein present in several restriction-modification gene complexes stimulates the expression of the restriction gene and/or represses the expression of the modification gene (Tao et al. 1991; Lubys et al. 1999; Kroger et al. 1995; Anton et al. 1997; Vijesurier et al. 2000). Some C proteins also regulate their own synthesis (Kita et al. 2002; Cesnaviciene et al. 2003). Thus, these proteins may play a role in the establishment and auto-regulation of restriction-modification systems. As with the regulatory methyltransferases described above (see Sect. 4.2.3), these proteins also have a helix-turn-helix domain that is probably responsible for DNA binding. Their modes of action and their recognition sites have been analyzed (Rimseliene et al. 1995; O'Sullivan and Klaenhammer 1998; Bart et al. 1999; Kita et al. 2002; Cesnaviciene et al. 2003). That C regulatory proteins actually participate in post-segregational killing has been demonstrated for EcoRV (Nakayama and Kobayashi 1998).

4.2.5 Type I Restriction-Modification Systems

Following the conjugal transfer of the *hsdK* genes (*hsdR*, *hsdM*, and *hsdS*) of *E. coli* K-12, restriction activity was first detected only after approximately 15 generations, whereas modification activity was observed immediately. This sequential expression may play a role in the establishment of the *hsdK* genes in an unmodified host and suggests that restriction activity is regulated after conjugal transfer (Prakash-Cheng and Ryu 1993).

Another regulatory protein is the ClpXP protein, which regulates the endonuclease activity of EcoKI by degrading the subunit that is essential for restriction but not for modification. ClpXP protects the bacterial chromosome but few such restriction-alleviating effects were detected with unmodified foreign DNA within the cytoplasm of a restriction-proficient cell (Doronina and Murray 2001).

4.3 Restriction-Modification Gene Complexes May Be Able to Multiply Themselves

Favoring the selfish gene point of view is the observation that a restriction-modification gene complex engages in virus genome-like multiplication by itself (Sadykov et al. 2003). In this case, when a *Bacillus subtilis* clone carrying a copy of the BamHI gene complex flanked by long direct repeats was propagated, extensive tandem amplification of *bamHI* was observed in some clones. Mutational analysis suggested that restriction cutting of the genome partici-

*Bam*HI rm single copy

*Bam*HI rm amplified

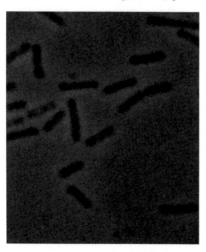
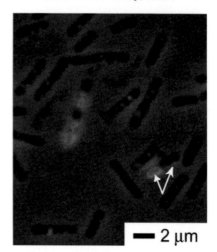

Fig. 7. Multiplication of the BamHI restriction-modification gene complex. *Left* A line of *Bacillus subtilis* carrying a single copy of *bamHI* on the chromosome. *Right* A line carrying, on the average, 11 copies of *bamHI*. The *arrows* indicate a pair of daughter cells, only one of which shows the strong signal that is indicative of burst-like amplification. The fixed cells were treated with Cy3-labeled probe for FISH (fluorescent in situ hybridization). Reproduced from *Molecular Microbiology* (Sadykov et al. 2003)

pates in the amplification. Visualization by fluorescent *in situ* hybridization (FISH) revealed that the amplification occurred in single cells in a burst-like fashion, which is reminiscent of the induction of provirus replication (Fig. 7). The long direct repeats flanking several restriction-modification gene complexes in several naturally competent bacteria (see Fig. 3B; Sect. 2.4.2) would aid their amplification. The ability of a restriction-modification gene complex to amplify itself may help its maintenance, multiplication and spreading within a population of naturally competent bacteria. These observations suggest that a restriction-modification gene complex might be able to increase its frequency in the cell population through a life cycle that is similar to that of a DNA virus. More precisely, they may be regarded as DNA viroids, which are DNA viruses that lack a capsid (Kobayashi 2002).

5 Intra-Genomic Competition Involving Restriction-Modification Gene Complexes

Just as with other mobile elements, such as viruses, there are various forms of competition between restriction-modification systems during several phases of their life cycle. For example, when one restriction-modification system tries to kill the host, another restriction-modification system may prevent the killing. Conversely, when one restriction-modification system tries not to kill the host, the other restriction-modification system may force it to kill the host. Such interference takes place when there is some similarity in the two restriction-modification systems and may help in classifying restriction-modification systems (Sect. 7). Insertional inactivation of one restriction-modification gene complex by another (Sect. 2.5) represents another form of their competition.

5.1 Two Restriction-Modification Systems with the Same Recognition Sequence can Block the Post-Segregational Killing Potential of Each Other

When two addiction systems that have antitoxins with the same activity are present in the same cell, interference of the post-segregation killing activity of the two systems would result. In other words, the loss of one addiction gene complex would not lead to killing because the antitoxin of the other addiction gene complex would prevent the action of the toxin (Fig. 8). In restriction-modification systems, this type of competition can be clearly described as the protection of the recognition sequences on the chromosome by another methyltransferase (Kusano et al. 1995). For example, EcoRI cannot cause post-segregational killing when RsrI is present in the same host because the methyltransferases of both systems protect the same recog-

A. Post-segregational cell killing by a restriction-modification gene complex

B. Inhibition of the post-segregational killing by another restriction-modification gene complex of the same recognition sequence

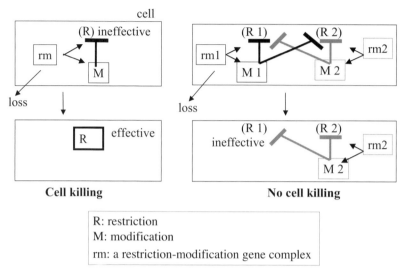

Cell killing   No cell killing

R: restriction
M: modification
rm: a restriction-modification gene complex

Fig. 8. Inhibition of post-segregational killing by another restriction-modification gene complex of the same recognition sequence. **A** Post-segregational killing programmed by one restriction-modification gene complex. This restriction-modification gene complex can force its maintenance on the host. **B** Inhibition of this post-segregational killing by a second restriction-modification gene complex. The second modification enzyme (anti-toxin) can protect, through methylation, the common recognition sequence along the genome from lethal attack by the first restriction enzyme (toxin) after loss of the first restriction-modification gene complex. The first restriction-modification gene complex cannot force its maintenance on the host

nition sequences (5′ GAATTC). Such within-host competition for recognition sequences may explain the evolution of the individual specificity and collective diversity in target sequence recognition by restriction-modification systems. The evolution of this feature of restriction-modification systems has been explained previously as frequency-dependent selection by invaders (Levin 1988).

The competition for recognition sequences can be one-sided when the targets of one post-segregational killing system are included in the targets of the other system. For example, SsoII recognizes 5′ CCNGG (where N=A, T, G or C) while EcoRII recognizes 5′ CCWGG (where W=A or T). The SsoII system can prevent the post-segregational killing of the EcoRII system because it protects all of the 5′ CCWGG sequences. In contrast, the EcoRII system cannot prevent the post-segregational killing by the SsoII system because it cannot protect the 5′ CCGGG and CCCGG sequences recognized by SsoII (Chinen et al. 2000a). This one-sided incompatibility implies that there is selective pressure on restriction-modification systems to recognize less specific recognition sequences.

5.2 Solitary Methyltransferases Can Attenuate the Post-Segregational Killing Activity of Restriction-Modification Systems

There is wide variation in the strength of host killing (virulence) even between closely related addiction systems (for example, see (Chinen et al. 2000a)). Restriction-modification gene complexes or their restriction genes often suffer from a mutation that reduces their virulence, as discussed above (Sect. 2.5), probably as a result of selection against their virulence (Sect. 6.1). The above type of competition between restriction-modification systems (Sect. 5.1) could help a less virulent system to attenuate the attack of a virulent system. This benefit may promote the acquisition by hosts of less virulent systems. We may call this an immunization or vaccination effect.

An extreme case of this is where a solitary antitoxin (methyltransferase) that is not paired with a toxin (restriction enzyme; Sect. 2.5) attenuates the activity of the toxin (restriction enzyme) from another system. An example is the solitary methyltransferase Dcm, which occurs in the chromosome of *E. coli* and related bacteria. Dcm methylates 5′ CCWGG sites (where W=A or T) and thus defends its host's genome from the toxin of the EcoRII restriction-modification system, which recognizes exactly the same sequence, as demonstrated in Fig. 4B (Takahashi et al. 2002). Another example is a plasmid that carries a homologue of the SsoII methyltransferase together with a truncated form of the SsoII restriction enzyme gene (Ibanez et al. 1997). This solitary methyltransferase could conceivably protect cells from the post-segregational activity of the intact SsoII system. A solitary DNA methyltransferase homologue that could play a similar role has been found in an integron cassette in *Vibrio cholerae* (Table 1). In addition, some bacteriophages are known to carry a solitary DNA methyltransferase (Hattman et al. 1985; Trautner and Noyer-Weidner 1993) that may serve to defend their genomes from restriction attack in the new host cells that they have entered.

5.3 Resident Restriction-Modification Systems Can Abort the Establishment of a Similar Incoming Restriction-Modification System

A resident post-segregational killing system can abort the establishment of another post-segregational killing system if both systems regulate their establishment in a new host in a similar fashion. As discussed above (Sect. 4), when restriction-modification systems establish themselves in a new host, they first express the modification gene and then the restriction gene to prevent cell killing. Here, the accumulation of a regulatory protein – the modification enzyme itself or a C regulatory protein – leads to the expression of the restriction enzyme (Fig. 5A). When a resident restriction-modification system has the same specificity in this establishment-regulating mechanism, its *trans*-activation may lead to the cell killing (Fig. 6B). In other words, the regulatory proteins of the resident restriction-modification system that induces restriction enzyme expression may act on the incoming restriction-modification gene complex, forcing the premature expression of its restriction enzyme in the absence of the prior expression of its modification enzyme. This would kill the host, thereby aborting the establishment of the incoming gene complex. This type of trans-activation has been found between restriction-modification systems that carry C regulatory proteins of the same specificity and has been designated as super-infection exclusion, mutual exclusion or apoptotic mutual exclusion (Nakayama and Kobayashi 1998). Such mutual competition between restriction-modification systems may have driven the evolution of specificity in the mechanisms that regulate establishment. This feature may be useful in classifying restriction-modification systems (Sect. 7).

5.4 Suicidal Defense Against Restriction-Modification Gene Complexes

From the host's point of view, the above apoptotic mutual exclusion mechanism resembles a suicidal strategy to block infection (Fig. 2B). The same type of mechanism could also be employed when the cell bears a solitary toxin that can sense the invasion of a restriction-modification system. Examples of this are methylated DNA-specific endonucleases (such as McrBC in *E. coli*) that recognize the genome methylation by an incoming restriction-modification system and then cut the chromosome (Fig. 9A; Kobayashi 1996).

5.5 Defense Against Invaders by Restriction-Modification Systems

Post-segregational killing by a restriction-modification system after the invasion of its competitor (Fig. 2B) is beneficial to the restriction-modification system and, in most cases, to the host as well, as discussed above (Sect. 3.3). From the host's point of view, this mechanism represents a suicidal strategy to block infection. Direct attack on invading DNAs by a resident restriction-modification system (Fig. 1B) is also beneficial to the restriction-modification system and, in most cases, to the host.

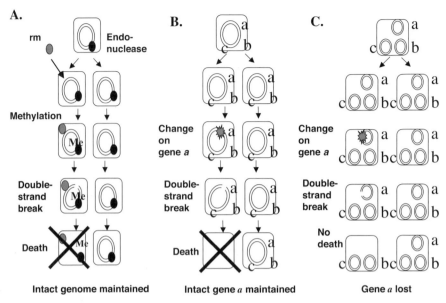

Fig. 9. The chromosome as a vehicle for mutual addiction among genes. **A** Chromosome breakage by a methylated-DNA-specific endonuclease in defense against an invading restriction-modification gene complex. A restriction-modification gene complex enters a cell and starts methylating its target, namely, chromosomal recognition sites. A methylated-DNA-specific endonuclease (a solitary toxin) senses these changes and triggers cell suicide by chromosomal cleavage and degradation. The uninfected genome survives in the neighboring clonal cells. **B** Chromosome breakage by an endonuclease in defense against an altered state of a gene. Some alteration, such as a DNA damage, takes place on gene *a* in a chromosome. An endonuclease recognizes this alteration and makes a DNA breakage there. This suicide of gene *a* can lead to the loss of all the remaining genes in the chromosome and cell death. The unaltered gene *a* and genome survive in the neighboring clonal cells. Each gene on the chromosome can thus force its maintenance on the genome by post-disturbance cell killing through chromosome breakage. **C** The case when each gene is on an independent replication unit. Suicide of one gene by DNA breakage cannot lead to loss of all the remaining genes and cell death. A gene thus cannot force its maintenance on the genome. *rm* Restriction-modification gene complex. *Me* Methyl group on the chromosome. *a, b, c* gene a, gene b, gene c

6 Genome Dynamics and Genome Co-Evolution with Restriction-Modification Gene Complexes

The restriction-modification systems and the host bacteria thus show a wide variation with respect to virulence and susceptibility. This variability in both sides may be a central property of their interaction and likely reflect their past interaction. The restriction-modification gene complexes have a long-term as well as a short-term effect on their hosts. By killing off hosts that lose the restriction-modification gene complexes, they select for hosts that have a reduced tendency to lose them. The host genome will also select the restriction-modification gene complexes with reduced virulence as discussed above (Sect. 5), although the host might benefit from the restriction-modification gene complexes under some conditions (Sect. 3.3). There could be a fine balance between the costs and benefits of the restriction-modification-induced death for the restriction-modification system (symbiont), the host, and the host-symbiont complex. There are signs of host-symbiont co-evolution in the genomes and genome dynamics that may reflect such a balance. The choice between life and death will depend on the various forms of genome dynamic, namely mutagenesis and recombination, induced by the action of restriction-modification systems on the genome. These processes, in turn, will directly affect evolution of the genome. Table 3 tries to incorporate these topics and previously discussed ones and to evaluate the possible benefits/costs to the restriction-modification gene complex, to the genome and to the third party.

6.1 Some Restriction-Modification Gene Complexes and Restriction Sites are Eliminated from the Genome

Signs of strong selection against restriction sites that are the targets of attack by restriction-modification systems are seen in the genomes. Such systematic avoidance of potential restriction sites (palindromes) in bacterial genomes is called *restriction avoidance* and has been well characterized from the informatic point of view (Gelfand and Koonin 1997; Rocha et al. 1998; 2001). For example, EcoRI sites are rare in the genome of its natural host, *E. coli* (Gelfand and Koonin 1997). This is likely to have resulted from selection due to host attack by restriction-modification systems. (However, it is difficult to evaluate the contribution, if any, of selection through restriction attacks on transferring genomic DNAs within a group.) In many genomes, restriction avoidance is more pronounced in the genome proper than in its prophages. This is probably because the genome proper has long experienced selection by restriction-modification systems, while the prophages, as newcomers, may have not yet experienced this selection (Rocha et al. 1998; 2001). Such restriction avoidance would lead to a decrease in the virulence of the particular restriction-modification system, which will almost always be beneficial for the host and can be beneficial to the restriction-modification system itself under some conditions (Table 3).

The presence of mutationally inactivated restriction-modification gene complexes in sequenced genomes such as *H. pylori* (Nobusato et al. 2000b; Lin et al. 2001) [REBASE] (see Sect. 2.5) may have resulted from similar selection against toxic restriction-modification systems and possibly from some mutagenesis processes (see Sect. 6.2 below).

Table 3. Intra-genomic interactions involving restriction-modification systems and their short-term benefits for the participants

Types	Examples/comments
[1] Competition between restriction-modification systems	See Sect. 4, Figs. 5B, 8B
[2] Vaccination with a solitary methyltransferase	See Sect. 5.2
[3] Death, upon invasion of a restriction-modification system, programmed by a methylated-DNA-specific endonuclease	See Sect. 5.4, Fig. 9A
[4] Defense against invasion of competitor through post-segregational killing	Fig. 2B
[5] Defense against infection through direct attack	Fig. 1B
[6] Prevention/repair of DNA breakage by bacteriophages	Various restriction alleviation mechanisms of bacteriophages (Bickle and Kruger 1993); DNA double-strand break-repair through homologous recombination by bacteriophage system (Takahashi and Kobayashi 1990; see Sect. 6.4; Fig. 10, left)
[7] Host exonuclease/recombinase action at a restriction break	Destruction of incoming nonself DNA and recombination repair of chromosomal self-DNA by RecBCD enzyme (Heitman et al. 1999; Handa et al. 2000; see Sect. 6.5; Fig. 10, right)
[8] SOS responses to DNA breakage by restriction enzymes and methylated-DNA-specific endonucleases	(Heitman and Model 1987, 1991; Handa et al. 2000)
[9] Mutagenesis	SOS mutagenesis; elevated mutation at 5-methylcytosine (Lieb 1991; Friedberg et al. 1995; see Sect. 6.2)
[10] Anti-mutagenesis	Very-short-patch mismatch repair (Lieb 1991); V.MthTI (Nolling et al. 1992; Mol et al. 2002; see Sect. 6.2)
[11] Selection against the target sequences resulting in restriction site avoidance in bacterial genomes	See Sect. 6.1 (Gelfand and Koonin 1997; Karlin et al. 1998a; Rocha et al. 1998, 2001)
[12] Mutational inactivation of restriction-modification gene complex (or of restriction gene only) and its selection	See Sect. 2.5. Restriction-modification gene homologues in *Helicobacter pylori* (Nobusato et al. 2000b; Lin et al. 2001)

[a] Negative

Benefit to the restriction-modification system	Benefit to the host # (genome)	Benefit to the third party
+/–[a]	+/–	–/+ (the other restriction-modification system)
–	+	+/–(invaders)
–	+	
+	+	– (invader)
+	+	– (invader)
–	–	+ (bacteriophage)
–	+	– (invader)
–	+	
–	+/–	
+	–/+	
–	+	
–	+	+ (invader)

6.2 Mutagenesis and Anti-Mutagenesis

The presence of a restriction-modification gene complex (Handa et al. 2000) and the action of a restriction enzyme (Heitman and Model 1991) induce the SOS response (Table 3 [10]). The action of a methylated DNA-specific endonuclease also induces the SOS response (Heitman and Model 1987). The global mutagenesis would generate heterogeneity in the cell population as it does in other stressful conditions (Higgins 1992) and may thereby help the survival of the genome. These mechanisms could contribute to the evolution of restriction avoidance and to the inactivation of a restriction-modification gene complex.

DNA methylation may locally increase mutation. 5-methylcytosine shows a higher rate of deamination than cytosine. Its deamination at the C/G pair in duplex DNA results in the formation of thymine and hence the generation of a T/G mispair (Friedberg et al. 1995; Lieb 1991). Very short patch mismatch correction by a *vsr* gene linked to *dcm*, a solitary cytosine methyltransferase in *E. coli* (see Sect. 5.2), can repair this mispair and restore the mC/G pair (Lieb 1991). A homologue of the *vsr* gene is linked with several restriction-modification gene complexes that produce 5-methylcytosine [REBASE], some of which have been shown to be active (Table 3 [11]). The local mutagenesis induced by the action of the restriction-modification systems leads to loss of the restriction sites along the genome, which can be beneficial to the host. The anti-mutagenesis of *Vsr* homologues linked with restriction-modification would prevent such loss and would be immediately beneficial to the restriction-modification system.

6.3 End Joining

A restriction break on the chromosome may be repaired through precise end-joining by DNA ligase (Heitman et al. 1989). Such repair by ligation may be associated with the insertion of a DNA fragment generated by the restriction enzyme (Schiestl and Petes 1991). It is not clear whether these REMI (restriction-enzyme-mediated integration) events do actually take place under natural conditions.

Joining at nonhomologous ends and other forms of illegitimate (nonhomologous) recombination might restore a viable but rearranged genome. A nonhomologous end-joining mechanism associated with restriction breakage and with homologous DNA interaction has been observed (Kusano et al. 1997).

6.4 Homologous Recombination by Bacteriophages

Of the many antirestriction strategies used by bacteriophages and plasmids (Bickle and Kruger 1993), the homologous recombination machinery carried by bacteriophages appears to be particularly well adapted to counteracting attacks by a variety of restriction-modification systems (Kobayashi 1998). Lambdoid (lambda-related) bacteriophages may repair the restriction break by a double-strand break repair mechanism (Fig. 10, left), in which a double-strand break is repaired by copying homologous DNA with or without associated crossing-over of the flanking sequences (Takahashi and Kobayashi 1990).

Recombination stimulated by a restriction break may also take place between co-infecting sister genomes in a clone. In addition, recombination may take place with a partner – possibly present as a prophage – from another phage clone possessing a divergent genome. Such out-crossing could confer several advantages in addition to the primary advantage of the immediate restoration of the cleaved phage chromosome. If the template DNA were to lack the recognition site, recombination might result in a DNA region that lacks this particular restriction site. The crossing-over (or half crossing-over) of flanking sequences could also confer a third kind of advantage. Here, alleles at different locations that are either sensitive or resistant to a particular restriction enzyme could be recombined to generate rare combinatory genotypes. Some would be more resistant to attack by the present set of restriction-modification systems than the current major combinations, and they would increase in number. As a phage population encounters bacterial populations possessing various combinations of restriction-modification systems with diverse sequence specificities, these processes of breakage, repair, gene conversion and crossing-over will take place. Such host-parasite-type arms races could continue forever.

6.5 Cellular Homologous Recombination in Conflict and Collaboration with Restriction-Modification Gene Complexes

The chromosomal breakage generated following the loss of a restriction-modification gene complex leads to extensive chromosome degradation by RecBCD exonuclease in *E. coli* (Fig. 10, right; Handa et al. 2000). This exonuclease switches to recombination repair when it encounters a specific sequence, called chi, on the genome (Handa et al. 2000). The specific sequence varies among bacterial groups (el Karoui et al. 1998) and may serve as an identification marker of the genome of a group. The specificity in this sequence recognition can be altered by a mutation in this exonuclease/recombinase (Handa et al. 1997; Arnold et al. 2000).

This exonuclease-based system may represent another mechanism that allows the genome to distinguish between 'self' and 'nonself', similar to the

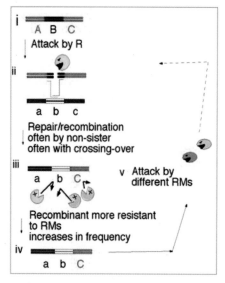

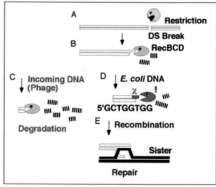

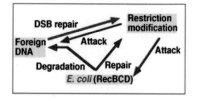

Fig. 10. Homologous recombination machinery and restriction-modification systems. *Left* Recombination repair of a restriction break by bacteriophage and its consequences. A restriction enzyme cuts an invading bacteriophage chromosome (*A–B–C*) at its recognition sequence (*B; i, ii*). The break stimulates (the double-strand break-repair type) homologous recombination of this chromosome with a homologous uncut chromosome, occasionally one without the recognition sequence (*a–b–c; ii*). Break repair through gene conversion regenerates an intact duplex DNA, which might lack the sequence (*B*) attacked by this restriction modification system (*iii*). The flanking crossing-over as well as the gene conversion will generate rare recombinant genotypes (*a–b–C*) that are more resistant to the attacks by the present set of restriction modifications systems (*iv*). These genotypes will relatively increase in frequency. The phage population will encounter bacterial populations with different combination of restriction-modification systems with a variety of recognition sequences (*v*). The above genotypes may not be most resistant to these. Their attack would trigger another round of repair and recombination. *R* Restriction enzyme; *RM* restriction-modification system; *A* and *a* allelic sequences; *B* and *b* allelic sequences, the former being sensitive to the first restriction enzyme and the latter insensitive; *C* and *c* allelic sequences. *Right* Destruction/ repair by bacterial homologous recombination machinery at a restriction break. From a restriction break on any DNA within the cell (*A*), the bacterial (*Escherichia coli*) RecBCD enzyme starts exonucleolytic degradation (*B*). It will completely destroy the DNA (*C*) unless a chi sequence on the chromosome attenuates degradation and switches the enzyme action to homologous recombination (*D, E*). *Bottom* Conflict and collaboration among an invading DNA, a restriction-modification system, and RecBCD system within a genome. The RecBCD system will collaborate with a restriction-modification gene complex in destroying invading DNAs, but it will repair its own chi-marked chromosome after attack by some restriction-modification gene complex. The foreign DNA may carry a DSB (double-strand break) repair machinery to repair the breaks

endonuclease-based restriction-modification systems. The host RecBCD system is known to destroy invading bacteriophage DNAs (nonself DNA) after restriction cleavage (Fig. 10, bottom). If incoming DNA carries hi, the exonucleolytic degradation would be attenuated and parts of the fragment may be incorporated into the chromosome, if not rejected by the mismatch repair system. This would result in the mosaic polymorphisms of the chromosome seen within a bacterial group (McKane and Milkman 1995).

6.6 Selfish Genome Rearrangement Model

The restriction-modification genes are often strain-specific and are associated with genome rearrangements as described above (Sect. 2). Some of the rearrangements may be ascribed to the activity of their vehicles – mobile ele-

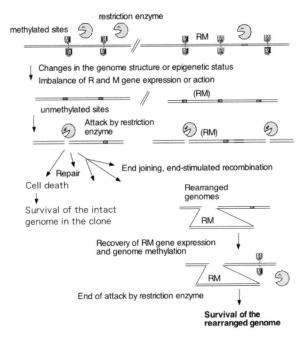

Fig. 11. Selfish genome rearrangement model. When the expression of a restriction-modification gene complex is somehow disturbed, the unmethylated sites are exposed to attack by the restriction enzyme. This leads to death of the cell and survival of the intact genome in the other cells in the clone. The restriction breaks will stimulate various forms of recombination and create numerous rearranged genomes. If a particular rearranged genome allows the expression of the restriction-modification gene complex, the methylation of the recognition sites along the genome will resume and the restriction attack will cease. *R* restriction; *M* modification; *RM* a restriction-modification gene complex. (*RM*) The restriction-modification gene complex that is not expressed. Reproduced from *Nucleic Acids Research* (Kobayashi 2001)

ments such as prophages and transposons. However, some forms of the rearrangements, such as *operon insertion* and *insertion with long target duplication* (Fig. 3), suggest direct participation of the restriction-modification gene complexes in the rearrangement processes (Kobayashi et al. 1999; Kobayashi 2001).

Supporting this is that a threat to a restriction-modification gene complex under laboratory conditions leads to the recovery of variously rearranged genomes (Handa et al. 2001; Sadykov et al. 2003). The restriction-modification gene complex might both generate various rearranged genomes through attack, and select those genomes that have allowed their genes to be expressed, as illustrated in Fig. 11. There is evidence for both the aspects of this 'selfish genome rearrangement' model (Kobayashi 2001).

7 Towards Natural Classification of Restriction Enzymes

The decoding of bacterial genome sequences revealed the presence of a wide variety of restriction-modification genes and restriction enzymes. This diversity requires a new type of classification of restriction enzymes and their genes. A new nomenclature, as opposed to classification, for restriction enzymes and their genes has been proposed (Roberts et al. 2003a). Given that the diversity of restriction-modification systems is equivalent to the diversity of life, these systems may be viewed as a form of life, just as viruses, transposons and other mobile elements are. This view might further aid the classification of these systems. Such natural classification based on their evolution would help us to understand these systems and ultimately would promote their use.

There are already extensive works on the phylogeny of modification enzymes that are similar to one another (for example, (Bujnicki 2001; Nobusato et al. 2000a)). Structural genomic analyses will also soon reveal the hidden evolutionary relationships between the restriction enzymes (Bujnicki 2000), and the construction of satisfactory phylogenetic trees or networks for restriction enzymes will then be possible. There are many clearly dead restriction and/or modification genes in the sequenced genomes (see above). These are likely to be particularly informative about the biology and evolution of restriction-modification systems. These dead or inactivated genes could be somehow linked with the functional genes as in the case of defective transposons or defective viruses.

One natural way of classifying organisms is by their mutual competition. Plasmids are classified according to their incompatibility: when two plasmids are similar in the regulatory system in replication, they fail to co-exist in the same host. Restriction-modification systems compete with each other for recognition sequences (Sect. 5.1; Kusano et al. 1995). The recognition sequences define the ecological niches for their adaptation and hence provide one clear criterion of their classification.

Another type of incompatibility is mutual exclusion (Sect. 5.3; Nakayama and Kobayashi 1998). This interference takes place because their specificity in the regulatory system is the same and because their recognition sequences are not identical. Therefore, mutual exclusion provides another aspect for the elementary classification of restriction-modification systems. The restriction-modification systems have been classified into the following exclusion groups to date: BamHI, PvuII, EcoRV, EcoRII, SsoII (Nakayama and Kobayashi 1998; Chinen et al. 2000a).

Some Type II restriction enzymes show a unique property that distinguishes them from the prototype Type II restriction enzymes (Roberts et al. 2003a). For example, some need two copies of their recognition sequence. Others show atypical subunit composition and/or gene fusion. These properties are expected to have resulted from adaptation through natural selection acting on the restriction-modification systems (or from a mere accident in their evolution). Unfortunately, we do not know yet the underlying natural selection or the relevant ecological niche in the genome that drove the acquisition of most of these properties. Understanding *why* will be a major goal of the study of restriction enzymes in coming years.

8 Application of the Behavior of Restriction-Modification Gene Complexes as Selfish Elements

The re-discovery of restriction-modification gene complexes as a form of life makes it possible to utilize them in science, biotechnology and medicine in several ways. One way is to use restriction enzyme genes for positive selection in gene cloning (O'Connor and Humphreys 1982; Kuhn et al. 1986; Jamsai et al. 2003). Those clones that have lost the ability to attack the host genome will be recovered. The restriction-modification gene complexes can also be used to control survival and death of bacteria (Torres et al. 2000; Takahashi et al. 2002). A plasmid carrying a toxin gene (for EcoRI) cannot escape a host cell carrying a cognate antitoxin gene (for M.EcoRI) on its chromosome (Torres et al. 2000). Restriction-modification gene complexes have already been used for the stable maintenance of plasmids (see Table 2B) and should be useful for the stable maintenance of other types of mobile genetic elements. Restriction-modification gene complexes have also been shown to help the maintenance of chromosomal linked genes (Handa et al. 2001; Sadykov et al. 2003). In addition, the ability of a restriction-modification gene complex to amplify itself together with a linked gene and with an allelic gene (Sect. 4.3; Sadykov et al. 2003) could help the amplification of useful genes, which in turn would help to elevate their expression and efficient spread. The hypothetical virus-like life cycle of restriction-modification gene complexes, if proven, may also make it possible to use them as a new type of vector. Furthermore, the ability of restriction-modification gene complexes to cause various types of genome rearrangements and/or to select some of these would provide a novel method of inducing artificial genome evolution (Handa et al. 2001). Artificial homing of *ecoRI* flanked by two EcoRI sites into an unmodified EcoRI site (Sect. 6.6; Eddy and Gold 1992) may provide another way of moving genes.

9 A Hypothesis on the Attack by Restriction-Modification Gene Complexes on the Chromosomes

When the destruction of one gene leads to the death of an organism under a standard condition, this gene is defined as being *essential* to this organism. The assumption implicit here is that the function of this gene is important to the activities of this organism. This dogma in current biology is challenged by the observation of post-segregational killing by, for example, a restriction-modification gene complex. This is because the inactivation of a modification gene of a restriction-modification gene complex leads to cell death, and therefore this antitoxin gene should be classified as an essential gene.

This argument implies that the addiction capacity, or the ability to kill a host or inhibit its growth when their persistence is threatened, can be a general property of gene sets controlling any function. A genome may gradually become dependent on or "addicted" to a set of initially "dispensable" genes that have joined it (Fig. 2A; Yarmolinsky 1995). In other words, these gene sets may co-evolve with the host genome since their first appearance to the point that they can program cell death upon being eliminated from the genome (Kobayashi 2003).

Attempts to resolve the conflicts between potentially selfish genes in a genome through addiction-related mechanisms may be a basis for the evolution and maintenance of several genetic systems (Kobayashi 2003). Use of the chromosome, a one-dimensional continuous structure where genes reside, as genetic material can be also understood from the point of view of post-segregational killing as discussed below (Kobayashi 2003).

Recent works revealed that chromosomes in bacteria and vertebrates suffer double-strand breaks at a high frequency (Michel et al. 1997; Handa et al. 2000; Handa and Kobayashi 2003; Sonoda et al. 1998) even in the absence of external agents. In eukaryote cells, the double-strand breakage turns on checkpoint control mechanisms leading to repair or cell death. The endonuclease that presumably introduces a break at a stalled replication fork in the eukaryotes and in the archaea shares structural similarities with restriction enzymes (Nishino et al. 2003).

In a chromosome, all the genes align hand in hand in the continuity of duplex DNA (Fig. 9B). The suicide of one gene (gene *a*) by DNA breakage at its resident locus in the chromosome would lead, through chromosome degradation, to death of all the other genes residing on the same chromosome and, eventually, to the death of the cell (Fig. 9B). As a result, the intact copies of the gene (gene *a*) would survive in the neighboring clonal cells. Every gene making up a community of thousands of genes can force its maintenance in the genome in this way. This would be impossible if each gene resides on an independent replication unit (as in Fig. 9C). (This situation is not entirely hypothetical because there are such mini-chromosomes in the macronuclei of some unicellular eukaryotes.) This may explain why a plasmid benefits from carriage of a special addiction gene set, such as a restriction-modification gene complex, which also forces its maintenance by chromosome breakage (although *in trans*). The selfishness of the genes and the social order as a genome are unified in the structure called a chromosome. This combination of the chromosomal double-strand breakage and degradation may play the role of a social contract that enforces order onto the society of genes (the genome) by assuring the healthy status of each gene. In other words, a chromosome represents a genetic system that is held together by the mutual addiction of the genes. Spontaneous chromosomal breakage becomes evident when subsequent steps of the recombination repair of the breaks are blocked. The breakage will be repaired depending on the conditions (see Sect. 6.5; Fig. 10, right). The restriction-modification systems could represent parasites/symbionts that take advantage of this programmed cell death/revival process.

10 Conclusions

In conclusion, there is now convincing evidence that several restriction-modification systems may be regarded as representing a form of life. One strategy that promotes their survival is genetic addiction, where the host genome is attacked when their persistence is threatened. This results from their elementary activity of cutting nonself DNA marked by the absence of proper methylation. They may move within a genome and between genomes and multiply their own DNA. Their dynamic interactions with the host genome are likely to have affected genome evolution. The structure and function of restriction enzymes may have resulted from adaptation, through selection, to their symbiotic life cycle.

Despite the more than 50,000 publications that are related to restriction enzymes and, presumably, the millions of experiments that involve them, we are far from understanding these proteins. Indeed, we rediscovered them only recently. The study of these simple and yet fascinating forms of life will provide insights into life and death at the DNA level.

Acknowledgements. I am grateful to Alfred Pingoud for encouragement, patience, and generosity. Michael Yarmolinsky and Dhruba Chattoraj provided helpful discussions. Miki Watanabe, Ken Ishikawa, Marat Sadykov, and Keiko Kita provided useful comments on the manuscript. My work during this writing was supported by grants from MEXT (Genome Homeostasis, Evolutionary Biology, Protein 3000).

References

Akopyants NS, Fradkov A, Diatchenko L, Hill JE, Siebert PD, Lukyanov SA, Sverdlov ED, Berg DE (1998) PCR-based subtractive hybridization and differences in gene content among strains of *Helicobacter pylori*. Proc Natl Acad Sci USA 95:13108–13113

Alm RA, Ling LS, Moir DT, King BL, Brown ED, Doig PC, Smith DR, Noonan B, Guild BC, deJonge BL, Carmel G, Tummino PJ, Caruso A, Uria-Nickelsen M, Mills DM, Ives C, Gibson R, Merberg D, Mills SD, Jiang Q, Taylor DE, Vovis GF, Trust TJ (1999) Genomic-sequence comparison of two unrelated isolates of the human gastric pathogen *Helicobacter pylori*. Nature 397:176–180

Alvarez MA, Chater KF, Rodicio MR (1993) Complex transcription of an operon encoding the SalI restriction-modification system of *Streptomyces albus* G. Mol Microbiol 8:243–252

Anton BP, Heiter DF, Benner JS, Hess EJ, Greenough L, Moran LS, Slatko BE, Brooks JE (1997) Cloning and characterization of the BglII restriction-modification system reveals a possible evolutionary footprint. Gene 187:19–27

Aras RA, Takata T, Ando T, van der Ende A, Blaser MJ (2001) Regulation of the HpyII restriction-modification system of *Helicobacter pylori* by gene deletion and horizontal reconstitution. Mol Microbiol 42:369–382

Arber W (1993) Evolution of prokaryotic genomes. Gene 135:49–56

Arnold DA, Handa N, Kobayashi I, Kowalczykowski SC (2000) A novel, 11-nucleotide variant of chi, chi*: one of a class of sequences defining the *E. coli* recombination hotspot, chi. J Mol Biol 300:469–479

Bart A, Dankert J, van der Ende A (1999) Operator sequences for the regulatory proteins of restriction modification systems. Mol Microbiol 31:1277–1278

Beletskaya IV, Zakharova MV, Shlyapnikov MG, Semenova LM, Solonin AS (2000) DNA methylation at the CfrBI site is involved in expression control in the CfrBI restriction-modification system. Nucleic Acids Res 28:3817–3822

Bickle TA, Kruger DH (1993) Biology of DNA restriction. Microbiol Rev 57:434–450

Bujnicki JM (2000) Phylogeny of the restriction endonuclease-like superfamily inferred from comparison of protein structures. J Mol Evol 50:39–44

Bujnicki JM (2001) Understanding the evolution of restriction-modification systems: clues from sequence and structure comparisons. Acta Biochim Pol 48:935–967

Burrus V, Bontemps C, Decaris B, Guedon G (2001) Characterization of a novel type II restriction-modification system, Sth368I, encoded by the integrative element ICESt1 of *Streptococcus thermophilus* CNRZ368. Appl Environ Microbiol 67:1522–1528

Butler D, Fitzgerald GF (2001) Transcriptional analysis and regulation of expression of the ScrFI restriction-modification system of *Lactococcus lactis* subsp. *cremoris* UC503. J Bacteriol 183:4668–4673

Cesnaviciene E, Mitkaite G, Stankevicius K, Janulaitis A, Lubys A (2003) Esp1396I restriction-modification system: structural organization and mode of regulation. Nucleic Acids Res 31:743–749

Chinen A, Naito Y, Handa N, Kobayashi I (2000a) Evolution of sequence recognition by restriction-modification enzymes: selective pressure for specificity decrease. Mol Biol Evol 17:1610–1619

Chinen A, Uchiyama I, Kobayashi I (2000b) Comparison between *Pyrococcus horikoshii* and *Pyrococcus abyssi* genome sequences reveals linkage of restriction-modification genes with large genome polymorphisms. Gene 259:109–121

Claus H, Friedrich A, Frosch M, Vogel U (2000) Differential distribution of novel restriction-modification systems in clonal lineages of *Neisseria meningitidis*. J Bacteriol 182:1296–1303

Cromie GA, Leach DR (2001) Recombinational repair of chromosomal DNA double-strand breaks generated by a restriction endonuclease. Mol Microbiol 41:873–883

Dandekar T, Huynen M, Regula JT, Ueberle B, Zimmermann CU, Andrade MA, Doerks T, Sanchez-Pulido L, Snel B, Suyama M, Yuan YP, Herrmann R, Bork P (2000) Re-annotating the *Mycoplasma pneumoniae* genome sequence: adding value, function and reading frames. Nucleic Acids Res 28:3278–3288

Doronina VA, Murray NE (2001) The proteolytic control of restriction activity in *Escherichia coli* K-12. Mol Microbiol 39:416–428

Dybvig K, Sitaraman R, French CT (1998) A family of phase-variable restriction enzymes with differing specificities generated by high-frequency gene rearrangements. Proc Natl Acad Sci USA 95:13923–13928

Eddy SR, Gold L (1992) Artificial mobile DNA element constructed from the EcoRI endonuclease gene. Proc Natl Acad Sci USA 89:1544–1547

el Karoui M, Ehrlich D, Gruss A (1998) Identification of the lactococcal exonuclease/recombinase and its modulation by the putative Chi sequence. Proc Natl Acad Sci USA 95:626–631

Friedberg EC, Walker GC, Siede W (1995) DNA repair and mutagenesis. ASM Press, Washington, D.C.

Gelfand MS, Koonin EV (1997) Avoidance of palindromic words in bacterial and archaeal genomes: a close connection with restriction enzymes. Nucleic Acids Res 25:2430–2439

Gerdes K (2000) Toxin-antitoxin modules may regulate synthesis of macromolecules during nutritional stress. J Bacteriol 182:561–572

Gerdes K, Gultyaev AP, Franch T, Pedersen K, Mikkelsen ND (1997) Antisense RNA-regulated programmed cell death. Annu Rev Genet 31:1–31

Gunn JS, Stein DC (1997) The *Neisseria gonorrhoeae* S.NgoVIII restriction/modification system: a type IIs system homologous to the *Haemophilus parahaemolyticus* HphI restriction/modification system. Nucleic Acids Res 25:4147–4152

Haig D (1997) The social gene. In: Krebs J, Davies NB (eds) Behavioural ecology: an evolutionary approach, 4th edn. Blackwell Science, Oxford, pp 284–304

Handa N, Ichige A, Kusano K, Kobayashi I (2000) Cellular responses to postsegregational killing by restriction-modification genes. J Bacteriol 182:2218–2229

Handa N, Kobayashi I (1999) Post-segregational killing by restriction modification gene complexes: observations of individual cell deaths. Biochimie 81:931–938

Handa N, Kobayashi I (2003) Accumulation of large non-circular forms of the chromosome in recombination-defective mutants of *Escherichia coli*. BMC Mol Biol 4:5

Handa N, Nakayama Y, Sadykov M, Kobayashi I (2001) Experimental genome evolution: large-scale genome rearrangements associated with resistance to replacement of a chromosomal restriction modificaiton gene complex. Mol Microbiol 40:932–940

Handa N, Ohashi S, Kusano K, Kobayashi I (1997) Chi*, a chi-related 11-mer partially active in an *E. coli* recC* strain. Genes Cells 2:525–536

Hattman S, Wilkinson J, Swinton D, Schlagman S, Macdonald PM, Mosig G (1985) Common evolutionary origin of the phage T4 dam and host *Escherichia coli* dam DNA-adenine methyltransferase genes. J Bacteriol 164:932–937

Heidelberg JF, Eisen JA, Nelson WC, Clayton RA, Gwinn ML, Dodson RJ, Haft DH, Hickey EK, Peterson JD, Umayam L, Gill SR, Nelson KE, Read TD, Tettelin H, Richardson D, Ermolaeva MD, Vamathevan J, Bass S, Qin H, Dragoi I, Sellers P, McDonald L, Utterback T, Fleischmann RD, Nierman WC, White O (2000) DNA sequence of both chromosomes of the cholera pathogen *Vibrio cholerae*. Nature 406:477–483

Heitman J, Ivanenko T, Kiss A (1999) DNA nicks inflicted by restriction endonucleases are repaired by a RecA- and RecB-dependent pathway in *Escherichia coli*. Mol Microbiol 33:1141–1151

Heitman J, Model P (1987) Site-Specific Methylases Induce the SOS DNA Repair Response in *Escherichia coli*. J Bacteriol 169:3243–3250

Heitman J, Model P (1991) SOS induction as an in vivo assay of enzyme-DNA interactions. Gene 103:1–9

Heitman J, Zinder ND, Model P (1989) Repair of the *Escherichia coli* chromosome after in vivo scission the *Eco*RI endonuclease. Proc Natl Acad Sci USA 86:2281–2285

Hendrix RW, Smith MC, Burns RN, Ford ME, Hatfull GF (1999) Evolutionary relationships among diverse bacteriophages and prophages: all the world's a phage. Proc Natl Acad Sci USA 96:2192–2197

Higgins NP (1992) Death and transfiguration among bacteria. Trends Biochem Sci 17:207–211

Hurst GD, Werren JH (2001) The role of selfish genetic elements in eukaryotic evolution. Nat Rev Genet 2:597–606

Ibanez M, Alvarez I, Rodriguez-Pena JM, Rotger R (1997) A ColE1-type plasmid from *Salmonella enteritidis* encodes a DNA cytosine methyltransferase. Gene 196:145–158

Jaffe A, Ogura T, Hiraga S (1985) Effects of the ccd function of the F plasmid on bacterial growth. J Bacteriol 163:841–849

Jamsai D, Nefedov M, Narayanan K, Orford M, Fucharoen S, Williamson R, Ioannou PA (2003) Insertion of common mutations into the human beta-globin locus using GET Recombination and an EcoRI endonuclease counterselection cassette. J Biotechnol 101:1–9

Jeltsch A, Kroger M, Pingoud A (1995) Evidence for an evolutionary relationship among type-II restriction endonucleases. Gene 160:7–16

Jeltsch A, Pingoud A (1996) Horizontal gene transfer contributes to the wide distribution and evolution of type II restriction-modification systems. J Mol Evol 42:91–96

Kamada K, Hanaoka F, Burley SK (2003) Crystal structure of the MazE/MazF complex. Molecular bases of antidote-toxin recognition. Mol Cell 11:875–884

Karlin S, Campbell AM, Mrazek J (1998a) Comparative DNA analysis across diverse genomes. Annu Rev Genet 32:185–225

Karlin S, Mrazek J, Campbell AM (1998b) Codon usages in different gene classes of the *Escherichia coli* genome. Mol Microbiol 29:1341–1355

Karyagina A, Shilov I, Tashlitskii V, Khodoun M, Vasil'ev S, Lau PC, Nikolskaya I (1997) Specific binding of sso II DNA methyltransferase to its promoter region provides the regulation of sso II restriction-modification gene expression. Nucleic Acids Res 25:2114–2120

Kita K, J. Tsuda, K. Okamoto, H. Yanase, Tanaka M (1999) Evidence of horizontal transfer of the EcoO109I restriction-modification gene to *Escherichia coli* chromosomal DNA. J Bacteriol 181:6822–6827

Kita K, Kawakami H, Tanaka H (2003) Evidence for horizontal transfer of the EcoT38I restriction-modification gene to chromosomal DNA by the P2 phage and diversity of defective P2 prophages in *Escherichia coli* TH38 strains. J Bacteriol 185:2296–2305

Kita K, Tsuda J, Nakai SY (2002) C.EcoO109I, a regulatory protein for production of EcoO109I restriction endonuclease, specifically binds to and bends DNA upstream of its translational start site. Nucleic Acids Res 30:3558–3565

Kobayashi I (1996) DNA modification and restriction: Selfish behavior of an epigenetic system. In: Russo V, Martienssen R, Riggs A (eds) Epigenetic mechanisms of gene regulation. Cold Spring Harbor Laboratory Press, New York, pp 155–172

Kobayashi I (1998) Selfishness and death: raison d'être of restriction, recombination and mitochondria. Trends Genet 14:368–374

Kobayashi I (2001) Behavior of restriction-modification systems as selfish mobile elements and their impact on genome evolution. Nucleic Acids Res 29:3742–3756

Kobayashi I (2002) Life cycle of restriction-modification gene complexes, powers in genome evolution. In: Yoshikawa H, Ogasawara N, Satoh N (eds) Genome science: towards a new paradigm? Elsevier, Amsterdam, pp 191–200

Kobayashi I Addiction as a principle of symbiosis of genetic elements in the genome – restriction enzymes, chromosome and mitochondria. In: Sugiura M (ed) Symbiosis and cellular organelles. Logos, Berlin (in press)

Kobayashi I Genetic addiction: a principle of gene symbiosis in a genome. In: Phillips G, Funnell B, (eds) Plasmid Biology. ASM Press, Washington, DC (in press)

Kobayashi I, Nobusato A, Kobayashi-Takahashi N, Uchiyama I (1999) Shaping the genome – restriction-modification systems as mobile genetic elements. Curr Opin Genet Dev 9:649–656

Kroger M, Blum E, Deppe E, Dusterhoft A, Erdmann D, Kilz S, Meyer-Rogge S, Mostl D (1995) Organization and gene expression within restriction-modification systems of *Herpetosiphon giganteus*. Gene 157:43–47

Kuhn I, Stephenson FH, Boyer HW, Greene PJ (1986) Positive-selection vectors utilizing lethality of the EcoRI endonuclease. Gene 42:253–263

Kulakauskas S, Lubys A, Ehrlich SD (1995) DNA restriction-modification systems mediate plasmid maintenance. J Bacteriol 177:3451–3454

Kuroda M, Ohta T, Uchiyama I, Baba T, Yuzawa H, Kobayashi I, Cui L, Oguchi A, Aoki K, Nagai Y, Lian J, Ito T, Kanamori M, Matsumaru H, Maruyama A, Murakami H, Hosoyama A, Mizutani-Ui Y, Takahashi NK, Sawano T, Inoue R, Kaito C, Sekimizu K, Hirakawa H, Kuhara S, Goto S, Yabuzaki J, Kanehisa M, Yamashita A, Oshima K, Furuya K, Yoshino C, Shiba T, Hattori M, Ogasawara N, Hayashi H, Hiramatsu K (2001) Whole genome sequencing of meticillin-resistant *Staphylococcus aureus*. Lancet 357:1225–1240

Kusano K, Naito T, Handa N, Kobayashi I (1995) Restriction-modification systems as genomic parasites in competition for specific sequences. Proc Natl Acad Sci USA 92:11095–11099

Kusano K, Sakagami K, Yokochi T, Naito T, Tokinaga Y, Ueda E, Kobayashi I (1997) A new type of illegitimate recombination is dependent on restriction and homologous interaction. J Bacteriol 179:5380–5390

Lehnherr H, Maguin E, Jafri S, Yarmolinsky MB (1993) Plasmid addiction genes of bacteriophage P1: doc, which causes cell death on curing of prophage, and phd, which prevents host death when prophage is retained. J Mol Biol 233:414–428

Levin BR (1988) Frequency-dependent selection in bacterial populations. Philos Trans R Soc Lond (B)319:459–472

Lieb M (1991) Spontaneous mutation at a 5-methylcytosine hotspot is prevented by very short patch (VSP) mismatch repair. Genetics 128:23–27

Lin LF, Posfai J, Roberts RJ, Kong H (2001) Comparative genomics of the restriction-modification systems in *Helicobacter pylori*. Proc Natl Acad Sci USA 98:2740–2745

Lubys A, Jurenaite S, Janulaitis A (1999) Structural organization and regulation of the plasmid-borne type II restriction-modification system Kpn2I from *Klebsiella pneumoniae* RFL2. Nucleic Acids Res 27:4228–4234

Lubys A, Lubiene J, Kulakauskas S, Stankevicius K, Timinskas A, Janulaitis A (1996) Cloning and analysis of the genes encoding the type IIS restriction- modification system HphI from *Haemophilus parahaemolyticus*. Nucleic Acids Res 24:2760–6

Lubys A, Menkevicius S, Timinskas A, Butkus V, Janulaitis A (1994) Cloning and analysis of translational control for genes encoding the Cfr9I restriction-modification system. Gene 141:85–89

Makovets S, Doronina VA, Murray NE (1999) Regulation of endonuclease activity by proteolysis prevents breakage of unmodified bacterial chromosomes by type I restriction enzymes. Proc Natl Acad Sci USA 96:9757–9762

McKane M, Milkman R (1995) Transduction, restriction and recombination patterns in *Escherichia coli*. Genetics 139:35–43

Meinhart A, Alonso JC, Strater N, Saenger W (2003) Crystal structure of the plasmid maintenance system epsilon/zeta: functional mechanism of toxin zeta and inactivation by epsilon 2 zeta 2 complex formation. Proc Natl Acad Sci USA 100:1661–1666

Michel B, Ehrlich SD, Uzest M (1997) DNA double-strand breaks caused by replication arrest. EMBO J 16:430–438

Miner Z, Hattman S (1988) Molecular cloning, sequencing, and mapping of the bacteriophage T2 dam gene. J Bacteriol 170:5177–184

Mol CD, Arvai AS, Begley TJ, Cunningham RP, Tainer JA (2002) Structure and activity of a thermostable thymine-DNA glycosylase: evidence for base twisting to remove mismatched normal DNA bases. J Mol Biol 315:373–384

Murray NE (2002) Immigration control of DNA in bacteria: self versus non-self. Microbiology-Sgm 148:3–20

Naito T, Kusano K, Kobayashi I (1995) Selfish behavior of restriction-modification systems. Science 267:897–899

Nakayama Y, Kobayashi I (1998) Restriction-modification gene complexes as selfish gene entities: Roles of a regulatory system in their establishment, maintenance, and apoptotic mutual exclusion. Proc Nat Acad Sci USA 95:6442–6447

Nishino T, Komori K, Ishino Y, Morikawa K (2003) X-ray and biochemical anatomy of an archaeal XPF/Rad1/Mus81 family nuclease. Similarity between Its endonuclease domain and restriction enzymes. Structure (Cambridge) 11:445–457

Nobusato A, Uchiyama I, Kobayashi I (2000a) Diversity of restriction-modification gene homologues in *Helicobacter pylori*. Gene 259:89–98

Nobusato A, Uchiyama I, Ohashi S, Kobayashi I (2000b) Insertion with long target duplication: a mechanism for gene mobility suggested from comparison of two related bacterial genomes. Gene 259:99–108

Nolling J, de Vos WM (1992) Characterization of the archaeal, plasmid-encoded type II restriction- modification system MthTI from *Methanobacterium thermoformicicum* THF: homology to the bacterial NgoPII system from *Neisseria gonorrhoeae*. J Bacteriol 174:5719–5726

Nolling J, van Eeden FJ, Eggen RI, de Vos WM (1992) Modular organization of related Archaeal plasmids encoding different restriction-modification systems in *Methanobacterium thermoformicicum*. Nucleic Acids Res 20:6501–6507

O'Connor CD, Humphreys GO (1982) Expression of the Eco RI restriction-modification system and the construction of positive-selection cloning vectors. Gene 20:219–229

O'Neill M, Chen A, Murray NE (1997) The restriction-modification genes of *Escherichia coli* K-12 may not be selfish: they do not resist loss and are readily replaced by alleles conferring different specificities. Proc Natl Acad Sci USA 94:14596–14601

O'Sullivan DJ, Klaenhammer TR (1998) Control of expression of LlaI restriction in *Lactococcus lactis*. Mol Microbiol 27:1009–1020

Ohshima H, Matsuoka S, Asai K, Sadaie Y (2002) Molecular organization of intrinsic restriction and modification genes BsuM of *Bacillus subtilis* Marburg. J Bacteriol 184:381–389

Pedersen K, Zavialov AV, Pavlov MY, Elf J, Gerdes K, Ehrenberg M (2003) The bacterial toxin RelE displays codon-specific cleavage of mRNAs in the ribosomal A site. Cell 112:131–140

Prakash-Cheng A, Ryu J (1993) Delayed expression of in vivo restriction activity following conjugal transfer of *Escherichia coli* hsdK (restriction-modification) genes. J Bacteriol 175:4905–4906

Price C, Bickle TA (1986) A possible role for DNA restriction in bacterial evolution. Microbiol Sci 3:296–299

Rimseliene R, Vaisvila R, Janulaitis A (1995) The eco72IC gene specifies a trans-acting factor which influences expression of both DNA methyltransferase and endonuclease from the Eco72I restriction-modification system. Gene 157:217–219

Roberts RJ, Belfort M, Bestor T, Bhagwat AS, Bickle TA, Bitinaite J, Blumenthal RM, Degt-yarev S, Dryden DT, Dybvig K, Firman K, Gromova ES, Gumport RI, Halford SE, Hattman S, Heitman J, Hornby DP, Janulaitis A, Jeltsch A, Josephsen J, Kiss A, Klaen-hammer TR, Kobayashi I, Kong H, Kruger DH, Lacks S, Marinus MG, Miyahara M, Morgan RD, Murray NE, Nagaraja V, Piekarowicz A, Pingoud A, Raleigh E, Rao DN, Reich N, Repin VE, Selker EU, Shaw PC, Stein DC, Stoddard BL, Szybalski W, Trautner TA, Van Etten JL, Vitor JM, Wilson GG, Xu SY (2003a) A nomenclature for restriction enzymes, DNA methyltransferases, homing endonucleases and their genes. Nucleic Acids Res 31:1805–1812

Roberts RJ, Vincze T, Posfai J, Macelis D (2003b) REBASE: restriction enzymes and methyltransferases. Nucleic Acids Res 31:418–420

Rocha EP, Danchin A, Viari A (2001) Evolutionary role of restriction/modification sys-tems as revealed by comparative genome analysis. Genome Res 11:946–958

Rocha EP, Viari A, Danchin A (1998) Oligonucleotide bias in *Bacillus subtilis*: general trends and taxonomic comparisons. Nucleic Acids Res 26:2971–2980

Rochepeau P, Selinger LB, Hynes MF (1997) Transposon-like structure of a new plasmid-encoded restriction- modification system in *Rhizobium leguminosarum* VF39SM. Mol Gen Genet 256:387–396

Rowe-Magnus DA, Guerout A-M, Ploncard P, Dychinco B, Davies J, Mazel D (2001) The evolutionary history of chromosomal super-integrons provides an ancestry for mul-tiresistant integrons. Proc Natl Acad Sci USA 98:652–657

Rowe-Magnus DA, Guerout A-M, Biskri L, Bouige P, Mazel D (2003) Comparative analy-sis of superintegrons: engineering extensive genetic diversity in the Vibrionaceae. Genome Res 13:428–442

Sadykov M, Asami Y, Niki H, Handa N, Itaya M, Tanokura M, Kobayashi I (2003) Multi-plication of a restriction-modification gene complex. Mol Microbiol 48:417–427

Sampath J, Vijayakumar MN (1998) Identification of a DNA cytosine methyltransferase gene in conjugative transposon Tn5252. Plasmid 39:63–76

Saunders NJ, Peden JF, Hood DW, Moxon ER (1998) Simple sequence repeats in the *Heli-cobacter pylori* genome. Mol Microbiol 27:1091–1098

Saunders NJ, Snyder LAS (2002) The minimal mobile element. Microbiology 148:3756–3760

Schiestl RH, Petes TD (1991) Integration of DNA fragments by illegitimate recombina-tion in *Saccharomyces cerevisiae*. Proc Natl Acad Sci USA 88:7585–7589

Schouler C, Gautier M, Ehrlich SD, Chopin MC (1998) Combinational variation of restriction modification specificities in *Lactococcus lactis*. Mol Microbiol 28:169–178

Sekizaki T, Otani Y, Osaki M, Takamatsu D, Shimoji Y (2001) Evidence for horizontal transfer of SsuDAT1I restriction-modification genes to the *Streptococcus suis* genome. J Bacteriol 183:500–511

Snyder L (1995) Phage-exclusion enzymes: a bonanza of biochemical and cell biology reagents? Mol Microbiol 15:415–420

Som S, Friedman S (1993) Autogenous regulation of the EcoRII methylase gene at the transcriptional level: effect of 5-azacytidine. EMBO J 12:4297–4303

Som S, Friedman S (1997) Characterization of the intergenic region which regulates the MspI restriction-modification system. J Bacteriol 179:964–967

Sonoda E, Sasaki MS, Buerstedde J-M, Bezzubova O, Shinohara A, Ogawa H, Takata M, Yamaguchi-Iwai Y, Takeda S (1998) Rad51-deficient vertebrate cells accumulate chro-mosomal breaks prior to cell death. EMBO J 17:598–608

Stein DC, Gunn JS, Piekarowicz A (1998) Sequence similarities between the genes encod-ing the S.NgoI and HaeII restriction/modification systems. Biol Chem 379:575–578

Takahashi N, Kobayashi I (1990) Evidence for the double-strand break repair model of bacteriophage l recombination. Proc Natl Acad Sci USA 87:2790–2794

Takahashi N, Naito Y, Handa N, Kobayashi I (2002) A DNA methyltransferase can protect the genome from postdisturbance attack by a restriction-modification gene complex. J Bacteriol 184:6100–6108

Tao T, Bourne JC, Blumenthal RM (1991) A family of regulatory genes associated with type II restriction- modification systems. J Bacteriol 173:1367–1375

Tian QB, Hayashi T, Murata T, Terawaki Y (1996) Gene product identification and promoter analysis of hig locus of plasmid Rts1. Biochem Biophys Res Commun 225:679–684

Torres B, Jaenecke S, Timmis KN, Garcia JL, Diaz E (2000) A gene containment strategy based on a restriction-modification system. Env Microbiol 2:555–563

Trautner TA, Noyer-Weidner M (1993) Restriction/modification and methylase systems in *Bacillus subtilis*, related species, and their phages. In: Sonenshein AL, Hoch JA, Losick R (eds) *Bacillus subtilis* and other gram-positive bacteria: Biochemistry, physiology, and molecular genetics. ASM Press, Washington, DC, pp 539–552

Vaisvila R, Vilkaitis G, Janulaitis A (1995) Identification of a gene encoding a DNA invertase-like enzyme adjacent to the PaeR7I restriction-modification system. Gene 157:81–84

Vijesurier RM, Carlock L, Blumenthal RM, Dunbar JC (2000) Role and mechanism of action of C. PvuII, a regulatory protein conserved among restriction-modification systems. J Bacteriol 182:477–487

Wilson GG, Murray NE (1991) Restriction and modification systems. Annu Rev Genet 25:585–627

Xu Q, Morgan RD, Roberts RJ, Blaser MJ (2000) Identification of type II restriction and modification systems in *Helicobacter pylori* reveals their substantial diversity among strains. Proc Natl Acad Sci USA 97:9671–9676

Xu SY, Xiao JP, Ettwiller L, Holden M, Aliotta J, Poh CL, Dalton M, Robinson DP, Petronzio TR, Moran L, Ganatra M, Ware J, Slatko B, Benner J (1998) Cloning and expression of the ApaLI, NspI, NspHI, SacI, ScaI, and SapI restriction-modification systems in *Escherichia coli*. Mol Gen Genet 260:226–231

Yarmolinsky MB (1995) Programmed cell death in bacterial populations. Science 267:836–837

Molecular Phylogenetics of Restriction Endonucleases

J.M. Bujnicki

1 Discovery and Classification of Restriction Enzymes

The phenomenon of restriction and modification (R-M) was first discovered in the early 1950s. It was observed that certain strains of bacteria inhibited ('restricted') the growth of bacteriophages previously propagated on a different strain. In the early 1960s, it was found that the restriction is due to the enzymatic cleavage of the phage DNA by sequence-specific endonucleases (REases), which are sensitive to covalent modification of bases in the target sequence. Some of the REases produced discrete DNA fragments upon cleavage. This property proved very useful for analyzing and rearranging DNA, which soon prompted the rapid development of genetic engineering techniques as well as the search for more REases with novel recognition sequences (early review: Arber and Linn 1969). It was in the mid-1970s when cloning of R-M enzymes themselves began (review by Lunnen et al. 1988). It was found that most of restriction enzymes are genetically linked with modification enzymes of cognate specificity, forming R-M systems, but a few solitary enzymes were also characterized (reviews: Wilson 1991; Wilson and Murray 1991).

Several kinds of R-M systems have been discovered. The differences concern enzyme composition and cofactors, the symmetry or asymmetry of the target sequence, and position of the cleavage site with respect to the target sequence. The classification and nomenclature of R-M systems have been recently revised and reviewed (Roberts et al. 2003a; see also this Volume). Briefly, restriction enzymes have been classified into four types: I (Enzyme Commission number: EC 3.1.21.3), II (EC 3.1.21.3), III (EC 3.1.21.4), and modification-directed (M-D, not classified by EC). Type II has been subdivided into numerous subtypes (Roberts et al. 2003a). Only selected relevant members of subtypes IIP, IIE, IIF, and IIS will be discussed in this article. Enzymes

J.M. Bujnicki
Bioinformatics Laboratory, International Institute of Molecular and Cell Biology in Warsaw, Trojdena 4, 02-109 Warsaw, Poland

Nucleic Acids and Molecular Biology, Vol. 14
Alfred Pingoud (Ed.)
Restriction Endonucleases
© Springer-Verlag Berlin Heidelberg 2004

of Types I, II, and III are parts of restriction-modification (R-M) systems, which comprise two opposing activities with similar specificities: a REase activity, and a methyltransferase (MTase) activity. While REase targets and cleaves specific sequences in the DNA, MTase modifies the same sequences by addition of a methyl group to cytosine or adenine, and thereby renders them invincible to cleavage by REase. Modification is required to protect the 'self' DNA against the cleavage. However, M-D restriction enzymes do not require the cognate MTase activity – conversely, they recognize and cleave sequences with modified bases, for instance those methylated by MTases from other R-M systems (Bickle and Kruger 1993). The subunit composition and sequence-structure-function relationships in different types of R-M systems have been recently reviewed (Bujnicki 2001b) and will not be discussed here in too much detail. Instead, this review will be focused on phylogenetics of restriction enzymes, their origin and evolutionary basis of the diversity and variability of their catalytic domains. For specialized reviews of classification of restriction enzymes (and R-M systems in general), their biochemistry and structural biology, see other chapters of this volume and other relatively recent reviews (Roberts et al. 2003a; Pingoud and Jeltsch 2001; Kobayashi 2001).

2 Genomic Context of R-M Systems

R-M systems occur ubiquitously among Eubacteria and Archaea (Roberts and Macelis 2001). Closely related systems are found quite commonly in phylogenetically distant organisms, the codon usage of their genes is often different from that of the host genome, and they are known to appear in distinct genomic context in different strains of the same organism. All these features suggest that R-M systems have been frequently transmitted by horizontal gene transfer (Jeltsch and Pingoud 1996; Nobusato et al. 2000a, b; Chinen et al. 2000).

It is commonly believed that the principal biological function of R-M systems is to protect the genome of the host organism against invasion of foreign genetic elements, such as phages or conjugative plasmids (Arber 1979). The DNA of the host is usually modified by the MTase, and hence resistant to cleavage, while the DNA of the invader is unmodified or exhibits 'foreign' patterns of methylation and can be shred into pieces upon its penetration into the recipient bacteria. Many phages have found it advantageous to circumvent restriction by various means, initiating an 'arms race' with the bacteria (review by Kruger and Bickle 1983). Some phages encode their own MTases with the same specificity as the host R-M systems or other DNA modification enzymes, such as hydroxymethyltransferases or glucosylases, which make them resistant to restriction by Type I, II, and III enzymes. It has been hypothesized that development of the ability to enzymatically introduce hydroxymethylcytosine (hmC) as well as 'typical' methylated bases (such as m⁶A) into the DNA by T-even phages was an evolutionary response to the presence of the 'classical' R-M systems in their hosts (Bickle and Kruger 1993). This in turn led to the development of M-D enzymes, which specifically targeted phage-like patterns of modification. As a response, the T-even phages developed the ability to glycosylate the hmC in their DNA, rendering them resistant to all known restriction enzymes of *E. coli*. Incidentally, the first observation of the restriction phenomenon was due to M-D enzymes (later described as McrA and McrBC), which were active against T-even phages containing nonglycosylated hmC in their DNA, but inactive against glycosylated hmC (Luria and Human 1952). The most recent stage in this intriguing evolutionary scenario is probably the development of a PvuRts1I nuclease from *Proteus vulgaris*, which seems to restrict glycosylated T4 DNA more effectively than the same DNA containing nonglycosylated hmC (Janosi et al. 1994).

In addition to their function of a simple "immune system" in prokaryotes, there is evidence of functional and structural similarity of restriction enzymes to recombinases and transposases (Carlson and Kosturko 1998; Hickman et al. 2000; Mucke et al. 2002). Recently, it has been proposed that the major function of R-M systems is to promote the frequency of genetic rearrangements, and thereby to ensure a rich diversity of genomes. Thereby, REases would act for the benefit of the biological evolution of the population rather than the benefit of individual organisms or the R-M systems themselves (Arber 2000). However, all these functions can be interpreted also in the light of the 'selfish gene' hypothesis, which requires less assumptions to be made. It states that restriction enzymes simply assure their own maintenance against all other genetic elements by cleaving 'non-self' DNA and that all other aspects of their function are of "second-order" (review: Kobayashi 2001). The REase activity not only includes destruction of incoming foreign DNA, but also contributes to a high frequency of horizontal transmission of R-M systems that allow them to invade new genomes (Jeltsch and Pingoud 1996; Kobayashi 2001). When an R-M system enters a new host on a DNA fragment and the REase is expressed, its cleavage activity creates recombinogenic free DNA ends within the host genome. This creates the opportunity of the entire R-M system to integrate into the cut site at increased frequency, for instance by illegitimate recombination. In this regard, R-M systems behave in a very similar way to the homing endonucleases (Gimble 2000). R-M systems can be regarded as intracellular parasites at the verge of mutualism in the same sense as lysogenic bacteriophages, which can confer advantage in certain contexts, such as bacterial resistance to antibiotics (Rocha et al. 2001). For a specialized review of "selfishness" of R-M systems and the evolutionary context of their interactions with the host genome, see Kobayashi (2001, and an article in this volume).

3 Historical Perspective of Comparative Analyses of Restriction Enzymes: Are They Products of Divergent or Convergent Evolution?

Early sequence comparisons of REases revealed a striking lack of common features between enzymes that recognize different sequences. Neither overall sequence similarity nor common motifs could be identified. However, amino acid sequences of some isoschizomers that cleave the same sequence at the same position were found to be evidently similar; e.g., EcoRI and RsrI (G/<u>AATT</u>C, where "/" denotes the site of cleavage and <u>underline</u> denotes the 'sticky end' produced after the cleavage)(Stephenson et al. 1989), MthTI and NgoPII (GG/CC)(Nolling and de Vos 1992), and XmaI and Cfr9I (C/<u>CCGG</u>G)(Withers et al. 1992). Typically, if statistically significant similarity between isoschizomers was observed, it was very high (often >90% identity), suggesting that they diverged very recently. Nonetheless, some isoschizomers were found to be dissimilar; for instance, HaeIII shares only 15% sequence identity with MthTI or NgoPII, despite these enzymes are functionally equivalent with respect to the recognition sequence and the cleavage site. Enzymes that cleave the same sequence at different positions (neoschizomers), are generally as dissimilar as typical pairs of non-izoschizomers, e.g., SmaI (CCC/GGG) and XmaI (C/<u>CCGG</u>G) share only 13% identity. The lack of any detectable common motifs or statistically significant sequence similarity between non-isoschizomeric REases suggested that they arose independently by convergence, and not by divergence from a common ancestor (Wilson 1991). Later, overall sequence similarity was detected in a few cases between enzymes that recognize sequences that differ by one or two base pairs, i.e., MunI (C/<u>AATT</u>G) and EcoRI (G/<u>AATT</u>C) (Siksnys et al. 1994) or SsoII (/<u>CCNGG</u>, where N=any base) and PspGI (/<u>CCWGG</u>, where W=A or T)(Morgan et al. 1998). Nonetheless, until the mid-1990s, the prevailing view was that REases evolved by convergence. The divergent evolution was recognized only within some small families that group enzymes with the same or nearly the same specificity and these families were believed to be unrelated to each other (review: Heitman 1993). This belief was soon to be shattered by the results of crystallographic analyses.

4 Crystallography of Type II REases: Exploration of the "Midnight Zone of Homology"

The first crystallographic analysis of a restriction enzyme (EcoRI complexed with its target DNA at 3-Å resolution) was published in 1986 (McClarin et al. 1986). However, the original model was found to be partly incorrect and a substantially revised structure has been released in 1990 (Kim et al. 1990). The second published REase structure was that of EcoRV, in 1993 (Winkler et al. 1993). In the years 1990–2002, structures of 14 type-II REases have been determined (Table 1). Comparisons of these structures revealed that all of them share a common structural core (four-stranded mixed β-sheet flanked by α-helices), which serves as a scaffold for the active site, typically comprising two or three acidic residues (Asp or Glu) and one Lys residue (Venclovas et al. 1994; Aggarwal 1995). In a few enzymes, Lys was found to be replaced by Glu or Gln (Newman et al. 1994; Lukacs et al. 2000). These residues are the

Table 1. Structurally characterized REases and related proteins with the PD-(D/E)xK fold

Name	PDB	Reference	Specificity
BamHI	2bam	Viadiu and Aggarwal (1998)	G/GATCC
BglI	1dmu	Newman et al. (1998)	GCCNNNN/NGGC
BglII	1d2i	Lukacs et al. (2000)	A/GATCT
Bse634I*	1knv	Grazulis et al. (2002)	R/CCGGY
BsoBI	1dc1	van der Woerd et al. (2001)	C/YCGRG
Cfr10I*	1cfr	Bozic et al. (1996)	R/CCGGY
EcoRI	1eri	Kim et al. (1990)	G/AATTC
EcoRV	1rvb	Kostrewa and Winkler (1995)	GAT/ATC
FokI	1fok	Wah et al. (1997)	GGATGN$_9$/NNNN[a]
HincII	1kc6	Horton et al. (2002)	GTY/RAC
MunI	1d02	Deibert et al. (1999)	C/AATTG
NaeI	1iaw	Huai et al. (2001)	GCC/GGC
NgoMIV	1fiu	Deibert et al. (2000)	G/CCGGC
PvuII	1pvi	Cheng et al. (1994)	CAG/CTG
λ exo*	1avq	Kovall and Matthews (1998)	5′-exonuclease
MutH*	1az0	Ban and Yang (1998)	/GATC[b]
Vsr	1cw0	Tsutakawa et al. (1999a)	C/(T:G)WGG[b]
TnsA*	1flz	Hickman et al. (2000)	Tn7 transposon end
T7 endo I*	1m0i	Hadden et al. (2001)	Holliday junction
Ss Hjc*	1hh1	Bond et al. (2001)	Holliday junction
Pf Hjc*	1gef	Nishino et al. (2001)	Holliday junction
Sc RPB5*	1dzf	Todone et al. (2000)	not a nuclease
Mj EndA*	1a79	Li et al. (1998)	tRNA introns
Af EndA*	NA	Li and Abelson (2000)	tRNA introns

The most representative entry from the Protein Data Bank (PDB) (http://www.rcsb.org) has been chosen for each enzyme, with the preference for protein-DNA complexes and structures solved at possibly highest resolution. * indicates the enzymes for which protein-DNA cocrystal structures are not available in PDB. *Ss, Sulfolobus solfataricus. Pf, Pyrococcus furiosus, Sc, Saccharomyces cerevisiae, Mj, Metanococcus jannaschii, Af, Archaeoglobus fulgidus, NA,* coordinates have not been deposited in PDB
[a] The isolated catalytic domain is nonspecific
[b] Nicks single strand

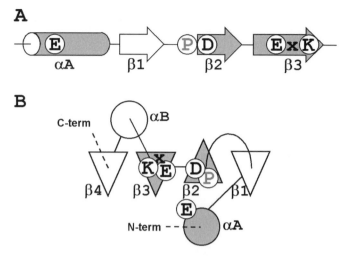

Fig. 1A, B. The conserved core and the most typical active site of enzymes from the PD-(D/E)xK superfamily. Structural elements that serve as a scaffold for the conserved residues are shown in *gray*. Elements that display little or no sequence conservation but are present in nearly all PD-(D/E)xK nucleases are shown in *white*. **A** Primary structure of the αβββ structural motif and the associated e-X_{10-30}-p-D-x_{10-30}-(D/E)-x-K sequence motif. The most typical residues are shown in *circles* (for simplicity, the 'D/E' carboxylate is represented by 'E'). Common secondary structural elements (α-helices and β-strands) are depicted as *cylinders* and *arrows*. **B** Topology diagram depicting the spatial orientation of the conserved elements (Bujnicki 2001b). *Circles*, *triangles*, and *connecting lines* represent α-helices, β-strands, and loops, respectively. Two alternative positions of the triangles correspond to the parallel/antiparallel orientation of the β-strands

only shared sequence feature, forming a very weakly conserved motif associated with a characteristic αβββ pattern of secondary structure elements (Fig. 1). The common sequence pattern is e-X_{10-30}-p-D-x_{10-30}-(D/E)-x-K (where lowercase indicates partial conservation, uppercase indicates nearly universal conservation and X is any amino acid) and hereafter will be referred to as "PD-(D/E)xK" for simplicity and in accord with the tradition. The structure-based predictions were supported by site-directed mutagenesis studies of several REases, which demonstrated that the conserved residues are indispensable for the DNA cleavage activity (Thielking et al. 1991; Lagunavicius and Siksnys 1997; Nastri et al. 1997). Nonetheless, the common secondary structures harboring the active site of different REases were found to be embedded in very different peripheral elements, which sometimes constituted the majority of the protein. Moreover, the dimerization modes of different enzymes were often dissimilar. Initially, it was not clear if the few common features represent sufficient evidence for the common ancestry of all REases or only of those most similar (e.g., EcoRI and BamHI vs. EcoRV and PvuII)(Aggarwal 1995).

Traditionally, homology[1] of genes/proteins was inferred from the similarity of DNA/amino acid sequences. As more and more protein structures were determined, it was found that conservation of structural and functional similarities despite extreme erosion of sequence similarity is common in the protein world (Murzin et al. 1995). Large-scale comparisons of protein sequences and structures revealed that in homologous superfamilies, protein sequences diverge most quickly, then functions, then structures (Rost 1997, 2002). In other words, homologous proteins usually retain the common three-dimensional fold regardless of the divergence at the sequence level. Hence, structural conservation should be regarded as the most robust evidence of protein homology. Nonetheless, evolutionary analyses of the proteins in the "midnight zone of homology" (where sequences of homologous proteins are saturated by amino acid substitutions to the level of less than 15–10 % identity and any genuine similarities disappear in the random noise) are complicated by the fact that protein structures also change in the course of evolution. These changes may be as drastic as insertions, deletions, swapping or substitutions of secondary structure elements or even entire domains, invasion or withdrawal of strands in the common β-sheet, or circular permutations of polypeptide segments (Murzin 1998; Kinch and Grishin 2002). At the level of the quaternary structure, typical evolutionary changes involve the development of new oligomerization modes and swapping of structural segments between the domains.

Because similar structures can arise independently in evolution (at least in theory) and truly homologous structures may become partly dissimilar due to divergence, structural similarity alone does not provide sufficient evidence of common ancestry. In the absence of obvious sequence similarity between the proteins of interest, their evolutionary relatedness is typically established based on the presence of a common structural scaffold (even if the peripheral structures are dissimilar) and in addition: the similarity of generic biological function (e.g., DNA binding) as well as molecular function (common residues in the binding site, common mechanism of reaction), retention of unusual structural features (unlikely to be reinvented independently), a common domain organization (conserved fusions with unquestionably homologous domains), a common genetic context (homologous neighbors on the chromosome) or, preferably, a combination of the above (Todd et al. 2001; Kinch and Grishin 2002).

As mentioned above, all REase structures determined to date (March 2003) share a geometrically and topologically equivalent core, which acts as a scaf-

---

[1] "Homology" is synonymous with "common origin", which is a qualitative feature, while "similarity" is a quantitative feature. These two terms should not be used interchangeably. Specifically, homology should not be expressed in percentiles – that two genes share "30 % homology" does not mean that their sequences are 30 % identical, but that in at least one of them 70 % of the original sequence has been replaced by a fragment from an unrelated gene. On the one hand, homologous proteins may share less than 10 % of identity. On the other hand, many short sequence fragments are 100 % identical, but it does not make them homologous.

fold for a similar metal-binding/catalytic site (the "PD-(D/E)xK" motif) located at the heart of the DNA-binding site (Fig. 1). Recent analysis of crystal structures of REases has suggested that the active site and the recognition sites are the major centers of structural stabilization preserved in the evolution (Fuxreiter and Simon 2002). Independently of the analysis of the crystal structures, Jeltsch et al. (1995) used a statistical Monte-Carlo procedure to find a highly significant correlation between the genotype (amino acid sequence) and the phenotype (target DNA sequence) of Type II REases and came up with a conclusion that they did not arise independently in evolution, but rather evolved from one or a few primordial nucleases. From the current perspective of protein evolution, the above-mentioned results are robust evidence for the common origin of those REases, whose structures are known, which rule out the lingering hypothesis of convergent evolution beyond any reasonable doubt (Venclovas et al. 1994; Huai et al. 2000; Pingoud and Jeltsch 2001). Since all structurally characterized REases turned out to be related despite they share little if any sequence similarity, it was hypothesized that all restriction enzymes (including those, whose structures remained unknown) may be related, regardless if their sequences exhibit significant similarity or not (Pingoud and Jeltsch 1997).

5 Homology Between Restriction Endonucleases and Other Enzymes Acting on Nucleic Acids

Crystallographic studies revealed that the PD-(D/E)xK superfamily comprises not only restriction enzymes, but also other nucleases, implicated in DNA recombination and repair, including: phage λ exonuclease, two bacterial enzymes exerting ssDNA nicking in the context of methyl-directed and very-short-patch DNA repair: MutH and Vsr, Tn7 transposase TnsA, two closely related Holliday junction resolvases (Hjc) from different species of Archaea, and a Holliday junction resolvase (endonuclease I) from phage T7 (see Table 1 for details). As with most of restriction enzymes, these nucleases did not exhibit easily identifiable sequence similarity to their remote homologues. That T7 endonuclease I and Hjc enzymes belong to the PD-(D/E)xK superfamily was predicted earlier based on combination of bioinformatics and mutagenesis (Bujnicki and Rychlewski 2001a; Daiyasu et al. 2000; Kvaratskhelia et al. 2000), but the relationship of other proteins to restriction enzymes came as a surprise.

Remarkably, crystallographic analyses revealed the PD-(D/E)xK fold in two proteins that do not function as deoxyribonucleases at all. The N-terminal domain (NTD) of the RPB5 subunit of RNA polymerase from yeast *S. cerevisiae* exhibits perfect conservation of the restriction enzyme-like structure, but lacks any catalytic residues – it is plausible that it functions as a nucleic acid binding domain devoid of any catalytic activity (Todone et al. 2000). The C-terminal catalytic domain (CTD) of tRNA splicing endonuclease EndA also bears striking resemblance to the minimal core of the PD-(D/E)xK fold (Bujnicki and Rychlewski 2001d). EndA-CTD also lacks the catalytic residues common to restriction enzymes and other related deoxyribonucleases. However, in the different part of the common fold, it developed an RNase active site, whose geometric configuration is very similar to that of a His-Tyr-Lys triad in structurally unrelated RNase A (Li et al. 1998). The structural similarities between these proteins and restriction enzymes have not been anticipated from sequence comparisons and were inferred only after the crystal structures were solved.

Since EndA and RPB5-NTD lack the PD-(D/E)xK active site, their evolutionary relationship to restriction enzymes is less certain compared to Holliday junction resolvases or other *bona fide* deoxyribonucleases. However, detailed analyses of sequence alignments generated by superposition of atomic coordinates, revealed common residues important for the formation of the hydrophobic core. The conservation of aromatic and hydrophobic residues in the critical positions of the PD-(D/E)xK fold is particularly striking between EndA-CTD, archaeal Hjc enzymes and λ exonuclease and between RPB5-NTD and the catalytic domain of FokI (data not shown). Moreover, PD-(D/E)xK nucleases are the only structures, to which EndA-CTD and RPB5-NTD exhibit any significant similarity. If EndA-CTD and RPB5-NTD acquired the PD-(D/E)xK fold 'by convergence', one could expect to find some lingering similarity between their structures and a non-PD-(D/E)xK ancestral fold. The present version of the protein structure database contains no such fold unrelated to the genuine PD-(D/E)xK nucleases, which could be credibly identified as a potential ancestor of EndA-CTD and/or RPB5-NTD. Hence, the most parsimonious hypothesis is that EndA-CTD and RPB5-NTD are true members of the PD-(D/E)xK superfamily. However the ultimate confirmation of this hypothesis (and other hypotheses based on structural similarities in the absence of sequence similarities) must await the identification of "evolutionary intermediates", whose sequences would provide links of statistically significant sequence similarity between now "isolated" members of the family. Alternatively, intermediates would be found that link EndA-CTD and RPB5-NTD with some other proteins.

6 Non-Homologous Active Sites in Homologous Structures

Solution of the crystal structure of Type II REase Cfr10I revealed the conservation of the classical PD-(D/E)xK fold, with unusual modification of the active site: the conserved "(D/E)" carboxylate has been replaced by Ser (Bozic et al. 1996). Strikingly, another carboxylate side chain has been identified in the active site, provided by a Glu residue located in a non-homologous structural element, namely in an α-helix that immediately follows the αβββ core (helix αB, see Figs. 1 2, and 3). That the nonclassical (version II) E204 residue

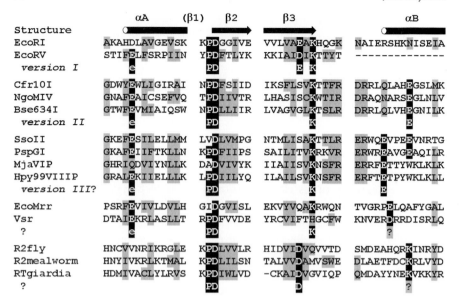

Fig. 2. Alignment of conserved sequence blocks depicting the variable architecture of the PD-(D/E)xK active site. Common secondary structures $\alpha(\beta)\beta\beta(+\alpha)$ are indicated ($\beta2$ is not shown because of poor sequence conservation) – in accordance with Fig. 1. The alignment is subdivided into five groups, based on different patterns of conservation of genuine and predicted catalytic residues. Conserved residues are shown on gray background, residues of the e-X_{10-30}-p-D-x_{10-30}-(D/E)-x-K motif are shown on black background. This figure represents a slightly modified version of an alignment in Bujnicki (2003)

is essential for catalysis, while the Ser residue is not, has been confirmed by mutagenesis (Skirgaila et al. 1998). Remarkably, the Ser-Glu swapped Cfr10I variant, in which the "classical" (version I) architecture of the active site has been restored, retained approximately 10 % activity of the wild-type enzyme, suggesting that the spatial conservation of the functional groups rather than sequence conservation plays a major role in the organization of the active sites of PD-(D/E)xK nucleases (Skirgaila et al. 1998). Crystallographic analyses of enzymes NgoMIV (Deibert et al. 2000) and Bse634I (Grazulis et al. 2002), revealed that they too exhibit version II of the active site. It is noteworthy that spatial migration of catalytic residues within the structural framework of the active site has been well-documented in many different enzymes, such as the thioredoxin superfamily (Martin 1995) and the enolase superfamily (Hasson et al. 1998) (for an excellent recent review see Todd et al. 2001).

Bioinformatics analyses suggested further tantalizing examples of potentially novel versions of the PD-(D/E)xK active site in restriction enzymes and related nuclease domains (Fig. 3). For instance, it has been found that SsoII and related nucleases possess an "alternative" carboxylate in the same posi-

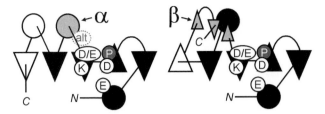

Fig. 3. Topology diagrams depicting major structural differences between the α (EcoRI-like) and β (EcoRV-like) subclasses of PD-(D/E)xK enzymes (Huai et al. 2000; Bujnicki 2001b). Common secondary structures (see Fig. 1) are shown in *black*. Key elements involved in DNA recognition are shown in *gray* (in α class it is a universally conserved α-helix B, in β class it is an additional small β-sheet). Other elements specific for α and β subclasses (including the topologically variable fifth β-strand) are shown in *white*. The alternative site in α-helix B, to which the D/E carboxylate migrated in some of the enzymes from the α subclass, is indicated as '*alt*'. This figure represents a slightly modified version of a diagram in Bujnicki (2003)

tion as Cfr10I and NgoMIV (version II), and an additional carboxylate, 4 residues towards the N-terminus (i.e., on the same face of the α-helix). Using site-directed mutagenesis, it has been demonstrated that either of the two (version II or version III) Glu residues in SsoII or PspGI is required to support catalysis, although the highest activity is displayed by the wild-type enzyme that retains both carboxylates (Pingoud et al. 2002). It remains to be determined experimentally, if putative restriction enzymes such as MjaVIP or Hpy99VIIIP, which possess only version III Glu and no version I, or II acidic residues (Fig. 2), represent yet another configuration of the PD-(D/E)xK active site.

A fourth alternative location of the essential carboxylate has been proposed for a DNA repair enzyme Vsr (Tsutakawa et al. 1999a, b) and an M-D restriction enzyme Mrr (Bujnicki and Rychlewski 2001b). In Vsr, divergence of the active site involved replacement of the "(D/E)xK" element by "FxH", and development of an additional, unique catalytic His residues in another part of the common fold (Tsutakawa et al. 1999a, b). Yet another version of the same active site was found in a PD-(D/E)xK domain found by sequence searches in R2 and other site-specific, non-long terminal repeat retrotransposable elements (Yang et al. 1999). Unlike for Vsr, whose crystal structures with and without the DNA are available, predictions of the unusual active site in Mrr and R2 have been supported only by limited mutagenesis studies, and the definite confirmation of these hypotheses must await determination of the crystal structures and detailed biochemical analyses.

Careful sequence database searches carried out after the Vsr structure was solved, provided a convincing evolutionary link between the Vsr family and other nucleases with "normal" (version I) active site, comprising typical carboxylates and Lys side chains (Aravind et al. 2000; Bujnicki and Rychlewski 2001a). These analyses proved that PD-(D/E)xK enzymes can lose the "PD-(D/E)xK" signature and develop novel active sites to catalyze the phosphodiester cleavage with different mechanisms. From the mechanistic point of view it is not clear "what was the purpose" of these conversions of the active site. From the evolutionary point of view, however, an enzyme with any version of the active site can be selected (and subsequently "fine-tuned") to catalyze a given reaction, as long as it is able to "do the job" and appears "at the right time and the right place" to provide a selective advantage for the host cell.

7 Cladistic Analysis of the PD-(D/E)xK Superfamily

Structural and functional differences prompted classification of Type II REases into two groups, represented by the archetypal enzymes EcoRI and EcoRV. On the one hand, enzymes that cleave to produce 5′-staggered ("sticky") ends usually recognize specific bases in the DNA mainly via residues from an α-helix and have been termed EcoRI-like or the α-class. On the other hand, REases that cleave to produce "blunt" ends usually recognize specific bases in the DNA mainly via residues from an additional β-sheet–they have been termed EcoRV-like, or the β-class (Aggarwal 1995; Huai et al. 2000; Bujnicki 2000; 2001b). Fig. 3 shows the topology diagrams of the common core of α and β-classes. It is evident that both α- and β-class REases retain the elements that serve as a scaffold for the active site, while the peripheral elements are quite distinct. It has been pointed out that even though all restriction enzymes comprise a 5-stranded β-sheet, the C-terminal (fifth) strand is topologically non-equivalent in the two classes – it is parallel or antiparallel with respect to the central β-sheet in α- and β-class REases, respectively (Venclovas et al. 1994; Huai et al. 2000; Bujnicki 2001b).

The observation of major structural differences between α- and β-class enzymes and relatively smaller variations within each of these two classes, suggested that the measure of structural differences could be used as an indicator of evolutionary divergence to infer the tree of REases. There is a number of computational methods which allow phylogenetic reconstructions to be made based on protein structure rather than amino acid sequence (review: May 1999). Some of these methods rely on inference of evolutionary distances using traditional superposition and calculation of the RMS distances for pairs of Cα atoms that are geometrically (and/or topologically) superimposable (Johnson et al. 1990; Grishin 1997). Others compare intrinsic features of individual protein structures, such as distribution of Cα-Cα distances within (not between) the structures of interest (Carugo and Pongor 2002). As with methods for sequence-based phylogenetic inference (review: Nei 1996), all methods for structure-based treeing have shortcomings. Moreover, it is not always clear if the measures of structural divergence implemented in these methods have been at all tuned up to correlate with the accepted measures of sequence divergence. Hence, the utility of these methods as tools for calculation of the absolute timing of evolutionary events is very limited. Nonetheless, the structure-based evolutionary inference can provide working models of phylogenies for such protein families, whose divergence definitely precludes 'classical' sequence-based treeing. The utility of such phylogenies can be assessed by their predictive power and the ability to explain the available experimental data.

Using the method of Johnson and co-workers (1990), I have inferred the molecular phylogeny of the PD-(D/E)xK superfamily using the nine crystal structures available in the Spring of 1999 (Bujnicki 2000). The results of that study were in excellent agreement with the function-based classification of REases into EcoRI-like (α-class) "5'-end cutters" and EcoRV-like (β-class) "blunt-end cutters": these two groups formed two monophyletic branches on the structure-based superfamily tree. One notable exception was the "3'-end cutter" BglI found to be related to the EcoRV-like enzymes. However, the functional difference between BglI and EcoRV could be easily rationalized by relatively minor modifications of the protein surface that resulted in formation of different dimers and hence – different distances between the two active sites in these two enzymes (Newman et al. 1998). The nicking enzyme MutH was found to be related to "blunt end cutters", in agreement with its overall similarity to the "blunt end cutter" PvuII reported earlier (Ban and Yang 1998), while the phage λ exonuclease was found to be relatively close to the root of the tree (i.e., close to the putative ancestor of all PD-(D/E)xK proteins).

Fig. 4 shows the revised and updated phylogenetic tree of the PD-(D/E)xK superfamily, inferred for all structures available to date (Table 1). It illustrates a hypothetical evolutionary scenario, in which contemporary members of the PD-(D/E)xK superfamily evolved by the gradual accumulation of various "decorations" of the common ancestral fold. The tree has been inferred based on the cladistic analysis, with different "structural peculiarities" used as informative characters (mainly the presence or geometry of certain non-universal β-strands and α-helices). The topology of this tree is in good agreement with the consensus of several different structure-based treeing methods (calculated using the STRUCLA server http://asia.genesilico.pl/strucla).

Analysis of the superimposed structures or, for simplicity, of the topological diagrams (Fig. 4), reveals the structural elements at the periphery of the common PD-(D/E)xK fold that occupy equivalent position with respect to the core, yet originate from non-homologous parts of the polypeptide. One example is the additional, sixth β-strand (present only in some nucleases): in BglII and BamHI it is encoded by the N-terminus and assumes antiparallel orientation with respect to the universally conserved third and fourth strands; in NgoMIV, FokI, PvuII, NaeI, and BglI it is parallel and in the C-terminus, while in Vsr it is antiparallel and in the C-terminus. An α-helix, which packs against strands $\beta4$, $\beta5$, (and $\beta6$, if it is present), also exhibits topological variability: in

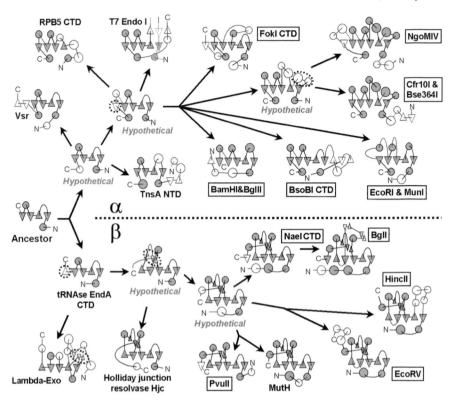

Fig. 4. Cladistic tree of known PD-(D/E)xK structures. Characters (features) that are either unique to one protein structure or universally conserved were ignored, because they do not provide any information to resolve inter-superfamily relationships. Evolutionary steps (acquisition and loss of structural elements) are indicated by *arrows*. Whenever a contemporary structure shares all major elements with the hypothetical ancestor, the contemporary structure is shown at the node instead of the ancestor, to illustrate its 'ancient' character. Elements that are conserved in a given lineage (apomorphies) are shown in *white* in this structure, in which they appeared for the first time. Elements inherited from ancestors older than the last common ancestor of the given lineage (symplesiomorphies) are shown in *gray*. The major features that allow distinction between different evolutionary lineages are depicted by *dotted circles*: (1) the directionality of the fifth β-strand (parallel in the α lineage and antiparallel in the β-lineage), (2) the additional small β-sheet in the β-lineage (note that the additional β-sheet of λ-exo is topologically different and has been acquired independently), (3) a unique tetramerization interface in the common ancestor of Cfr10I/Bse364I and NgoMIV (formed partly by a helical insertion in one of the loops). Other notable peculiarities are the unusual left-handed β-α-β elements in the C-terminal part of Vsr and FokI, as opposed to the typical right-handed structures in other proteins, and that T7 endonuclease I forms a swapped dimer and the core of one catalytic domain is made of fragments of two different monomers (one indicated by *dotted lines*)

EcoRV, HincII, NaeI, BglI, MutH, BglII, and BamHI it is encoded N-terminally to the common core and assumes antiparallel orientation with respect to the universally conserved helix αA, while in the Hjc enzymes (β-lineage) and all members of the α-class except BglII and BamHI, it is encoded C-terminally to the common core and assumes parallel orientation with respect to αA. In Vsr, this helix is formed by a polypeptide segment between strands β4 and β5, resulting in a left-handed β-α-β crossover, which is a very rare feature of protein structures in general. Remarkably, in FokI the geometrically equivalent element is formed by a connection between strands β5 and β6; it is actually broken into two helices connected by a kink, resulting in a geometrically similar, yet non-homologous left-handed β-α(kink)α-β crossover.

According to the most parsimonious evolutionary scenario shown in Fig. 4., these recurrent structural variants evolved by independent extension of the common fold in different evolutionary lineages (i.e., "by convergence"). Alternatively, some REase structures could be products of circular permutation or recombination between parts of catalytic domains derived from REases with different overall structures. Nonetheless, the latter scenario seems to me unlikely. Firstly, the peripheral elements of different enzymes share no detectable sequence similarity, (hence no convincing evidence for circular permutation or recombination). Secondly, to my knowledge, active PD-(D/E)xK REases with "chimeric" catalytic domains have been obtained only by recombination of segments derived from very closely related enzymes (Xu et al. 2001) and there have been no reports of active restriction enzymes produced by the spontaneous recombination of remote homologues in the nature. Thirdly, spatially equivalent "optional" structures in α-class and β-class REases often exhibit distinct pattern of contacts with the rest of the protein, while they tend to be more similar (and better superimposable) within each class (data not shown). In general, the natural development of functional hybrid domains is extremely rare compared to insertion or deletion of individual secondary structure elements at the periphery of the common fold (Grishin 2001; Kinch and Grishin 2002). Hence, it is reasonable to believe that different REases might have independently found ways to decorate the common core with a few similar secondary structures. Nonetheless, the structural plasticity of PD-(D/E)xK domains makes them attractive candidates for engineering protein variants with modified connectivity between the individual secondary structural elements. Studies on the activity and stability of such variants could serve as a test for the proposed evolutionary scheme of development of contemporary PD-(D/E)xK nucleases.

It is noteworthy that the structures of sequence-specific restriction enzymes are more elaborate and according to the presented scenario, occupy the terminal branches of the tree, while other nucleases, usually involved in generic metabolism of nucleic acids (DNA repair, recombination etc.), are typically located close to the root of the tree. Nucleases associated with the cellular DNA repair or recombination machinery evolve under the constant selective pressure stemming from their functionally critical interactions with conserved partners (Aravind et al. 2000). This pressure has apparently slowed down the divergence of generic nucleases, involved in recombination, and stabilized the ancestral state by limiting the development of major structural innovations. Restriction enzymes are neither essential to the cell nor involved in conserved interactions with common partners and hence are free to experiment with different extensions of the common fold. One notable exception from this rule is the MutH mismatch repair nuclease, whose ancestor could have been a type II restriction enzyme. The phylogenetic distribution of MutH is restricted to γ-Proteobacteria (Bujnicki 2001a), hence it seems that the MutH proteins have become adapted to function in cooperation with the MutL/MutS system relatively recently in the macroevolutionary time scale.

The extreme divergence between most of families of Type II enzymes (both at the sequence and structural level) paralleled by very close similarity between known isoschizomers can be explained as a consequence of fluctuations in the selection pressure. On the one hand, R-M systems have been identified in nearly all prokaryotic genomes, with the probable exception of a few intracellular pathogens (Roberts et al. 2003b). On the other hand, functional analyses have demonstrated that many of the REase genes are inactivated by mutations – some bacterial genomes are practically "graveyards" full of dead REases (Kong et al. 2000; Lin et al. 2001; Matveyev et al. 2001). It seems that periods of the strong selective pressure on both high activity and strict sequence specificity of Type II REases (such as during the attack of bacteriophages on the host) are separated by long "idle" interludes. Hence, restriction enzymes may enjoy long periods of relaxation, when they proliferate by horizontal gene transfer and at the same time accumulate random mutations that render most of the copies inactive, but also (with low probability) enable generation of variants with new interesting features. Attacks of phages on a bacterial population may kill those hosts, whose restriction enzymes relaxed too much. However, if any nuclease is able to "stand up and fight" with the invading DNA and allows it host to survive, its evolution will be directed towards optimization of both the activity and the specificity. An additional condition for survival is the presence of the functional MTase, which would protect the host against the active REase. Of course, under strong selective pressure, the specificity of the MTase may be concurrently tuned up to the specificity of the REase. Thereby, phage infections may serve as evolutionary bottlenecks for restriction enzymes, allowing selection of quite peculiar structural variants of the PD-(D/E)xK fold.

8 Identification of PD-(D/E)xK Domains in Other Nucleases and Prediction of Their Position on the Phylogenetic Tree

Using various bioinformatics methods, PD-(D/E)xK domains have been detected in nucleases, for which no crystal structure has been solved yet (including genuine REases, proteins implicated in DNA repair and/or recombination as well as many putative open reading frames). Most of these analyses failed to confidently identify the structures outside the common αββββ core, however in a few cases more extensive similarities to one REase structure or another have been found, suggesting a specific location where a sequence-based branch should be grown on the structure-based tree. Here, I will mention only these predictions that can be reliably integrated with the phylogeny of the PD-(D/E)xK superfamily, with the preference for those that have been validated by site directed mutagenesis. The readers are referred to previously published broad surveys for the extensive list of predicted REase domains, most of which still await experimental confirmation (refs. Aravind et al. 2000; Bujnicki and Rychlewski 2001a). For a recent review of bioinformatics methodology see (Bujnicki 2003). Figure 5 shows the phylogenetic tree of the PD-(D/E)xK superfamily, which combines the structural and bioinformatics data. Two notable cases of "secondary convergence", or independent invention of the same function in different evolutionary lineages are illustrated.

Multiple sequence alignment analysis, followed by structure prediction and identification of the characteristic αββββ pattern of secondary structures and conserved residues, were used to identify a PD-(D/E)xK domain fused to a tandem of P-loop ATPase domain in of the R subunits of type-I and type-III enzymes as well as in the RecB/RecE helicase family (Davies et al. 1999b; Aravind et al. 1999). In agreement with these predictions, alanine mutagenesis of the suspected catalytic Asp, Glu and Lys residues abolished the nuclease activity of these multifunctional enzymes, thereby uncoupling the cleavage and ATP-dependent DNA translocase activities (Davies et al. 1999a; Janscak et al. 1999, 2001; Wang et al. 2000; Chang and Julin 2001). Interestingly, in Types I and III REases the PD-(D/E)xK domain was found either in the N- or C-terminus of the ATPase module. It is not clear if these fusions were generated independently or one was derived by circular permutation from the other. Independent fusions of the PD-(D/E)xK domain with different helicase domains have been identified in many enzymes implicated in DNA repair and recombination (Aravind et al. 2000). According to secondary structure predictions, the nonspecific nuclease domain in most of these proteins represents a relatively compact version of the PD-(D/E)xK fold, which appeared quite early in the evolution of the superfamily.

A relatively compact PD-(D/E)xK domain was also identified in a functionally uncharacterized protein PrfA conserved in Gram-positive bacteria (Rigden et al. 2002). A statistically significant match between the PrfA sequence and the structure of REase PvuII was obtained using a fold-recog-

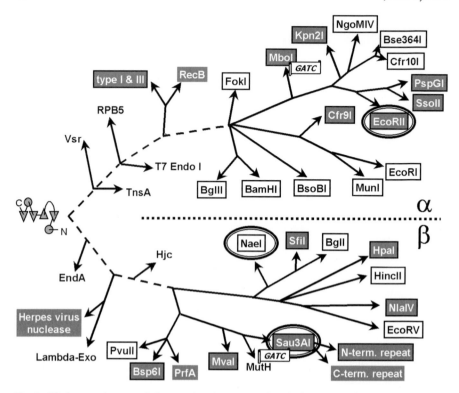

Fig. 5. Phylogenetic tree of the PD-(D/E)xK superfamily, based on the 'structural' tree (Fig. 4) and expanded to include additional members, identified by sequence analyses and protein-fold recognition (Siksnys et al. 1995; Aravind et al. 2000; Bujnicki and Rychlewski 2001a, c; Bujnicki 2001a; Friedhoff et al. 2001; Pingoud et al. 2002; Rigden et al. 2002). Restriction enzymes are shown in *black frames*. PD-(D/E)xK domains identified by bioinformatics and not by crystallography are shown in *white on gray background*. SubType IIE enzymes from three different lineages are indicated by *circles*. Isoschizomers MboI and Sau3AI that originated from two different lineages are indicated by a label with their recognition sequence GATC. Parts of the tree, which could not be confidently resolved based on either sequence or structural analyses, are shown in *broken lines*

nition method. Although the cellular role of PrfA remains to be determined, its predicted nuclease activity was confirmed by showing that it nicks supercoiled DNA. The structural model of PrfA was further supported by mutagenesis of one of the carboxylate residues (Rigden et al. 2002).

Iterative sequence database searches and secondary structure prediction were used to identify the relationship between two groups of PD-(D/E)xK REases: structurally characterized Type IIF enzymes that recognize different sequences with a common core X/**CCGG**X[2] and cleave to produce tetranu-

---

[2] "X" is used here to indicate a variable part of the DNA target sequence, i.e., a position, for which different enzymes may have different preference or no preference at all.

cleotide staggered ends (representative members: Cfr10I and NgoMIV), and Type IIP and IIE enzymes that recognize sequences with a common core /CCXGG and cleave to produce pentanucleotide staggered ends (representative members: SsoII, PspGI, and EcoRII) (Bujnicki and Rychlewski 2001a; Pingoud et al. 2002; Tamulaitis et al. 2002). This group of enzymes can be regarded as a model system for studying the evolution of three features: (1) development of novel versions of the PD-(D/E)xK active site (discussed earlier in this article); (2) development of additional protein-DNA contacts outside the common core that change the specificity; and (3) development of quaternary interactions characteristic for different functional subtypes (IIP, IIE, or IIF).[3]

Comparative analysis of the crystal structures of Cfr10I (recognition sequence: R/CCGGY (where R=G or A and Y=C or T), and NgoMIV (recognition sequence: G/CCGGC) revealed that they utilize similar residues and mechanism to recognize the central CCGG bases (Deibert et al. 2000). The "RxDR" motif present in both Cfr10I and NgoMIV was implicated in direct recognition of the "GG" segment. However, NgoMIV apparently developed a different N-terminal extension of the common core, which provided some of the key residues involved in recognition of the outer GC pair, including D34. Analysis of superimposed structures of Cfr10I and NgoMIV reveals that other residues that participate in direct recognition of the outer bases in NgoMIV are substituted in Cfr10I by different side chains (including R227 vs D240 and Q63 vs. K64). The availability of several isoschizomers of Cfr10I (including the crystallographic structure of Bse634I) as well as their T/CCGGA-recognizing homologues such as Kpn2I, suggests that careful examination of conservation patterns in this family of REases could help identify variable residues that control the specificity towards the outer base pair. Such "evolutionary hot spots" could be attractive targets for mutagenesis aiming at engineering of REases with non-native specificities. However, I am not aware of any attempts to engineer new specificities in Cfr10I or any of its homologues based on the evolutionary information.

The equivalent of the "RxDR" motif present in NgoMIV has been also identified in SsoII (recognition sequence /CCNGG), EcoRII, and PspGI (both recognizing /CCWGG, where W=A or T). This finding has been supported by the results of site-directed mutagenesis, suggesting that all these enzymes utilize similar mechanism to recognize the 'GG' module (Pingoud et al. 2002). Sequence analysis and theoretical modeling of the structures of SsoII and PspGI suggested that these enzymes share a common ancestor with NgoMIV

---

[3] The functional characteristics of subtypes of Type II REases are reviewed elsewhere in this volume. Briefly, Type IIP enzymes are "typical" dimers that bind and cleave a single DNA target, Type IIE are dimers that bind two identical sites and cleave only one of them, and Type IIF are tetramers that bind two identical sites and cleave them both in a concerted reaction.

(and Cfr10I, Bse634I etc.) and that their acquired the ability to produce pentanucleotide 5′-staggered ends by changing the respective orientation of the two monomers in the dimer, to increase the distance between the two catalytic sites compared to NgoMIV (Pingoud et al. 2002). A similar reorientation of monomers with respect to each other and to the DNA has been demonstrated by crystallographic analysis and used to explain the evolution of the BglI enzyme, which produces trinucleotide 3′-staggered ends, from EcoRV-like enzymes, which produce blunt ends (Newman et al. 1998). Thus, it appears that reorientation of monomers in a dimer is a common evolutionary mechanism for development of REases with new specificities.

Analyses of structural elements appended to the common PD-(D/E)xK fold as extensions or insertions in regions outside the common fold of SsoII, PspGI, NgoMIV, and EcoRII, revealed that they are responsible for development of features characteristic for different functional subtypes (IIP, IIF, or IIE). First, the catalytic domain of PspGI and SsoII retained the "orthodox" features of subType IIP. Second, the ancestor of the NgoMIV/Cfr10I lineage developed longer insertions in the loops between strands $\beta1/\beta2$ and $\beta2/\beta3$, which formed a novel tetramerization interface required to produce a subType IIF enzyme (Deibert et al. 2000). Third, the catalytic domain of EcoRII was N-terminally fused to a novel DNA-binding domain, which allowed the enzyme to simultaneously bind two identical sites, a feature typical for subType IIE enzymes (Reuter et al. 1999). Remarkably, it was shown that deletion of the N-terminal domain of EcoRII restores the "orthodox" characteristics of subType IIA (Mucke et al. 2002).

It is noteworthy, that the archetypal Type IIE enzyme, NaeI (Huai et al. 2000), belongs to the β-class of REases and shows no particular similarity to Type IIE enzyme EcoRII, the α class member. The subType IIE features of NaeI were also found to result from fusion of the PD-(D/E)xK domain with a DNA-binding domain, but in this case the additional domain was found C-terminally to the catalytic domain. Moreover, the auxiliary DNA-binding domains of EcoRII and NaeI show no relationship to each other (data not shown).

Another Type IIE enzyme, Sau3AI was also shown to be a member of the PD-(D/E)xK superfamily. It was found that Sau3AI comprises a tandem duplication of two PD-(D/E)xK domains, of which both exhibit statistically significant sequence similarity to the DNA repair enzyme MutH, but only the N-terminal copy retains the hallmark catalytic residues (Friedhoff et al. 2001; Bujnicki 2001a). The C-terminal copy of the PD-(D/E)xK domain in Sau3AI has apparently lost the active site, but retained the ability to bind the same sequence (GATC) as the N-terminal domain. Hence, both domains of Sau3AI and the sole catalytic domain of MutH share similar sequence specificity. It is striking that the catalytic domains of three subType IIE enzymes (NaeI, EcoRII, and Sau3AI) show different relationships with other PD-(D/E)xK nucleases and their auxiliary DNA-binding domains are unrelated. The only reasonable explanation is that different subType IIE enzymes have independently (i.e., "convergently") found a similar solution to the same problem of binding two identical DNA sites. Another evidence of convergent evolution of a similar function is presented by enzymes MboI and Sau3AI, which both recognize /GATC, yet apparently originated from two different lineages, i.e., α and β, respectively (Fig. 5). It will be of interest to find, if these isoschizomers use similar or different mechanisms for recognition of the same substrate. Since they are homologous, it could be expected that they interact with the DNA in a similar manner. Nonetheless, as mentioned earlier, α- and β-class REases utilize different structural elements as the scaffold for recognition of the target sequence, hence major differences should not be surprising.

Based on the results of protein fold-recognition analysis carried out via our "meta-server" (http://genesilico.pl/meta/), I have provisionally classified several Type II REases as members of the β-class, based on their extensive similarities to known structures: NlaIV to EcoRV, HpaI to EcoRV and HincII, SfiI to BglI, Bsp6I to PvuII, and MvaI to MutH. The 'new' members of the β-class recognize and cleave different sequences than their structural templates, yet share with them all key structural and functional elements. Theoretical models of NlaIV and Bsp6I were validated by site-directed mutagenesis of residues involved in dimerization, DNA-binding and catalysis (M.Radlinska and J.M.Bujnicki, manuscripts in preparation), other predictions remain to be confirmed experimentally. A large-scale comparative analysis of REases has been launched in my laboratory, aiming at detection of as many evolutionary links between different REases as possible, to use this information as a guide for experimental analyses.

9 In the End, Convergence Wins: Sequence Analyses Reveal Type II Enzymes Unrelated to the PD-(D/E)XK Superfamily

Despite the notion of 'secondary' convergence in the terminal branches of the PD-(D/E)xK superfamily, all crystallographic studies and sequence analyses of many nuclease families substantiated the hypothesis of primary common origin of all REase domains in restriction enzymes and many DNA repair and recombination enzymes. Ironically, at the turn of the millennium, soon after the compelling hypothesis of homology between all Type II enzymes started to be generally accepted as a general working model, new studies have demonstrated convincingly that some *bona fide* Type II REases are in fact evolutionarily unrelated and structurally dissimilar to the PD-(D/E)xK superfamily.

First, sequence analysis of a novel Type IIS restriction enzyme BfiI revealed that its N-terminal part exhibits low, statistically insignificant similarity to an EDTA-resistant nuclease (Nuc), a member of the phospholipase D superfamily (PLD) (Sapranauskas et al. 2000). The intriguing identity of several key residues between BfiI and Nuc prompted the Siksnys group to carry out further experiments. In accordance with the hypothesis of remote homology of the two enzymes, BfiI turned out to share the peculiar features of Nuc, namely to be active in the absence of Mg^{2+} as well as to hydrolyze an artificial substrate bis-(p-nitrophenyl) phosphate (Sapranauskas et al. 2000). Recently, site-directed mutagenesis and biochemical analyses demonstrated that BfiI, exactly like Nuc, forms a dimer with a single active site composed of residues from both subunits (Lagunavicius et al. 2003).

Second, bioinformatics analyses of the HNH endonuclease superfamily[4], revealed their relationship to many genuine REases, including M-D enzyme McrA, Type IIA restriction enzymes HpyI, NlaIII, SphI, SapI, NspHI, NspI and KpnI, as well as Type IIS enzyme MboII and its homologues from *Helicobacter pylori* (Bujnicki et al. 2000; 2001a; Aravind et al. 2000). To my knowledge, no crystallographic or mutagenesis studies have been published to support this prediction, but a few are currently in progress.

Third, closely related Type II enzymes Eco29kI, NgoMIII, and MraI were found to be remotely related to the GIY-YIG superfamily of nucleases (Bujnicki et al. 2001a). At the time of that analysis, no crystal structure of a GIY-YIG superfamily member was available. In order to provide a structural platform for experimental validation of our prediction, we attempted to predict the fold of an archetypal GIY-YIG nuclease, the N-terminal (NTD) catalytic domain of I-TevI, using ab initio simulations (Bujnicki et al. 2001b) based on sparse NMR restraints published by the Belfort group (Kowalski et al. 1999). Recently, a crystal structure of the I-TevI-NTD has been determined (Van Roey et al. 2002), confirming the three-dimensional fold predicted by our preliminary model. The refined model of Eco29kI was corroborated by the results of site-directed mutagenesis of the predicted catalytic residues (our unpubl. data).

10 Evolutionary Trajectories of Restriction Enzymes: Relationships to Other Polyphyletic Groups of Nuclease

Recent analyses reinforce the original suggestion that restriction enzymes as a whole are polyphyletic and that they have been re-invented multiple times. There are many protein folds that served as a scaffold for the development of various nuclease active sites. Most of them are α/β proteins, with the catalytic residues located at the C-termini of β-strands, but nucleases with all-α and all-β folds are also known (Fig. 6). It is remarkable that the majority of known

---

[4] Herein I use the name 'HNH superfamily' to refer to all proteins comprising the so-called '$\beta\beta\alpha$-Me' or 'His-Me finger' motif, including homologues of the orthodox HNH nucleases, His-Cys homing nucleases, T4 endonuclease VII, colicin E7 and E9 and a nonspecific *Serratia* nuclease among others.

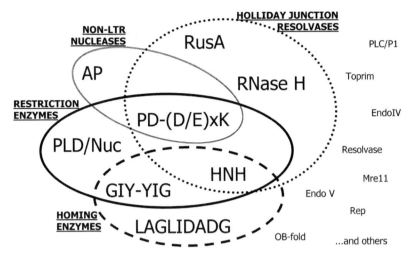

Fig. 6. Polyphyletic groups of nucleases. Restriction enzymes evolved from at least four different folds. Homing enzymes evolved from at least three different folds (review: Jurica and Stoddard 1999). Holliday-junction resolvases originated from at least four different folds (review: Aravind et al. 2000). The PD-(D/E)xK domain found in the non-LTR transposable is very often replaced by a nuclease domain related to a superfamily grouping DNase I and apurinic/apyrymidinic (AP) DNA repair endonucleases (Finnegan 1997). There are other protein folds, which comprise nuclease members. It remains to be seen, if any of them gave rise to restriction enzymes

REases originated from the PD-(D/E)xK superfamily and relatively few have been found to belong to other nuclease families. In this context it is noteworthy to mention the key difference between two large families of 'selfish' nucleases: restriction enzymes and homing endonucleases. Whereas the former typically recognize short, well-defined sequences and cleave them with remarkable specificity, the latter prefer very long and degenerate targets (Jurica and Stoddard 1999). Both homing enzymes and restriction enzymes evolved from several different folds: either types of nucleases were derived from the GIY-YIG and HNH superfamilies, while the PD-(D/E)xK and PLD superfamilies are so far unique to restriction enzymes, and the LAGLIDADG fold is unique to homing enzymes (Fig. 6). On the one hand, the LAGDLI-DADG fold is a saddle-shaped structure compatible with the major groove of DNA, which provides several β-turn structures that can be extended to form a very long binding site, as observed in known structures of homing enzymes (Jurica et al. 1998; Silva et al. 1999; Ichiyanagi et al. 2000). On the other hand, adaptation of the PD-(D/E)xK fold to cover larger fragments of the DNA requires adding new strands to the central beta sheet or fusions with additional domains. Nonetheless, the PD-(D/E)xK fold seems to be better suited for grasping the target DNA both from the major and the minor groove site, thereby increasing the number of possible protein-DNA contacts.

Incidentally, the EndA nuclease, which has been mentioned earlier in this article, is a fusion of PD-(D/E)xK-like and LAGLIDADG-like domains, of which neither exhibits the functional deoxyribonuclease active site (Bujnicki and Rychlewski 2001d). Other LAGLIDADG-like domains devoid of the REase active site have been found in various transcription factors (Iyer et al. 2002), suggesting that this fold may serve as a generic nucleic acid-binding module. The same is probably true for the PD-(D/E)xK fold. It remains to be seen if there are indeed no restriction enzymes related to the LAGLIDADG super-family or no homing enzymes from the PD-(D/E)xK superfamily.

There are many superfamilies of nucleases with distinct folds, which remain as potential structural templates for the majority of known restriction enzymes, whose sequences were too diverged to allow classifications, while bioinformatics analyses such as secondary structure prediction revealed no obvious link between their sequences and any of the four folds currently known to be the evolutionary sources of REases. Undoubtedly, the increasing size of the structural database will facilitate detection of remote relationships with the fold-recognition methods. However, some restriction enzymes may be representatives of extremely rare folds or entirely new folds and therefore not amenable to the protein-fold recognition analysis. In both cases, the avail-ability of homologous sequences may considerable improve the reliability of detection or dismissal of similarities to known structures. Hence, identifica-tion of a few sequences of homologous isoschizomers in order to improve the chance of success of bioinformatics analysis may be an alternative to costly crystallographic analyses. To date, many of the restriction enzymes were thought to represent distinct folds, until the structures were actually solved and "more of the same old PD-(D/E)xK" was identified. From the analyses of PD-(D/E)xK nucleases and many other proteins it is now clear that the crite-rion of sequence dissimilarity is worthless as an indicator of a lack of evolu-tionary and structural relationships. Guided by the working model of the REase evolution, bioinformatics analyses may help to focus the search of new, unusual restriction enzyme structures as well as the candidates that either seem homologous to other nuclease superfamilies or do not exhibit signifi-cant structural similarity to any protein in the database. As mentioned above, fold-recognition and homology modeling methods have been very useful in identification of potential active sites of REases. In the age of structural genomics more and more structures, including those of REases and related proteins, are expected to be solved by X-ray crystallography and NMR. It is anticipated that bioinformatics methods will serve as tools to fill in the gaps on the landscape of sequence-structure-function relationships, providing a useful platform for experimental studies on protein function in the structural and evolutionary context.

Acknowledgements. I am grateful to Aravind, Alfred Pingoud, Rich Roberts, and Virgis Siksnys for numerous stimulating discussions on various aspects of evolutionary and structural biology of restriction enzymes. My research on R-M enzymes is supported by EMBO, the Howard Hughes Medical Institute (Young Investigator Programme award), and by KBN (grant 6P04 B00519). The Fellowship for Young Scientists from the Foundation of Polish Science is also gratefully acknowledged.

References

Aggarwal AK (1995) Structure and function of restriction endonucleases. Curr Opin Struct Biol 5:11–19

Aravind L, Walker DR, Koonin EV (1999) Conserved domains in DNA repair proteins and evolution of repair systems. Nucleic Acids Res 27:1223–1242

Aravind L, Makarova KS, Koonin EV (2000) Holliday junction resolvases and related nucleases: identification of new families, phyletic distribution and evolutionary trajectories. Nucleic Acids Res 28:3417–3432

Arber W (1979) Promotion and limitation of genetic exchange. Science 205:361–365

Arber W (2000) Genetic variation: molecular mechanisms and impact on microbial evolution. FEMS Microbiol Rev 24:1–7

Arber W, Linn S (1969) DNA modification and restriction. Annu Rev Biochem 38:467–500

Ban C, Yang W (1998) Structural basis for MutH activation in *E. coli* mismatch repair and relationship of MutH to restriction endonucleases. EMBO J 17:1526–1534

Bickle TA, Kruger DH (1993) Biology of DNA restriction. Microbiol Rev 57:434–450

Bond CS, Kvaratskhelia M, Richard D, White MF, Hunter WN (2001) Structure of Hjc, a Holliday junction resolvase, from *Sulfolobus solfataricus*. Proc Natl Acad Sci USA 98:5509–5514

Bozic D, Grazulis S, Siksnys V, Huber R (1996) Crystal structure of *Citrobacter freundii* restriction endonuclease Cfr10I at 2.15 Å resolution. J Mol Biol 255:176–186

Bujnicki JM (2000) Phylogeny of the restriction endonuclease-like superfamily inferred from comparison of protein structures. J Mol Evol 50:39–44

Bujnicki JM (2001a) A model of structure and action of Sau3AI restriction endonuclease that comprises two MutH-like endonuclease domains within a single polypeptide. Acta Microbiol Pol 50:219–231

Bujnicki JM (2001b) Understanding the evolution of restriction-modification systems: clues from sequence and structure comparisons. Acta Biochim Pol 48:1–33

Bujnicki JM (2003) Crystallographic and bioinformatics studies on restriction endonucleases: inference of evolutionary relationships in the 'midnight zone' of homology. Curr Prot Pept Sci 4 (in press)

Bujnicki JM, Rychlewski L (2001a) Grouping together highly diverged PD-(D/E)XK nucleases and identification of novel superfamily members using structure-guided alignment of sequence profiles. J Mol Microbiol Biotechnol 3:69–72

Bujnicki JM, Rychlewski L (2001b) Identification of a PD-(D/E)XK-like domain with a novel configuration of the endonuclease active site in the methyl-directed restriction enzyme Mrr and its homologs. Gene 267:183–191

Bujnicki JM, Rychlewski L (2001 c) The herpesvirus alkaline exonuclease belongs to the restriction endonuclease PD-(D/E)XK superfamily: insight from molecular modeling and phylogenetic analysis. Virus Genes 22:219–230

Bujnicki JM, Rychlewski L (2001d) Unusual evolutionary history of the tRNA splicing endonuclease EndA: relationship to the LAGLIDADG and PD-(D/E)XK deoxyribonucleases. Protein Sci 10:656–660

Bujnicki JM, Radlinska M, Rychlewski L (2000) Atomic model of the 5-methylcytosine-specific restriction enzyme McrA reveals an atypical zinc-finger and structural similarity to ββαMe endonucleases. Mol Microbiol 37:1280–1281

Bujnicki JM, Radlinska M, Rychlewski L (2001a) Polyphyletic evolution of Type II restriction enzymes revisited: two independent sources of second-hand folds revealed. Trends Biochem Sci 26:9–11

Bujnicki JM, Rotkiewicz P, Kolinski A, Rychlewski L (2001b) Three-dimensional modeling of the I-TevI homing endonuclease catalytic domain, a GIY-YIG member, using NMR restraints and Monte Carlo dynamics. Protein Eng 14:717–21

Carlson K, Kosturko LD (1998) Endonuclease II of coliphage T4: a recombinase disguised as a restriction endonuclease? Mol Microbiol 27:671–676

Carugo O, Pongor S (2002) Protein fold similarity estimated by a probabilistic approach based on Cα-Cα distance comparison. J Mol Biol 315:887–898

Chang HW, Julin DA (2001) Structure and function of the *Escherichia coli* RecE protein, a member of the RecB nuclease domain family. J Biol Chem 276:46004–46010

Cheng X, Balendiran K, Schildkraut I, Anderson JE (1994) Structure of PvuII endonuclease with cognate DNA. EMBO J 13:3927–3935

Chinen A, Uchiyama I, Kobayashi I (2000) Comparison between *Pyrococcus horikoshii* and *Pyrococcus abyssi* genome sequences reveals linkage of restriction-modification genes with large genome polymorphisms. Gene 259:109–121

Daiyasu H, Komori K, Sakae S, Ishino Y, Toh H (2000) Hjc resolvase is a distantly related member of the Type II restriction endonuclease family. Nucleic Acids Res 28:4540–4543

Davies GP, Kemp P, Molineux IJ, Murray NE (1999a) The DNA translocation and ATPase activities of restriction-deficient mutants of EcoKI. J Mol Biol 292:787–796

Davies GP, Martin I, Sturrock SS, Cronshaw A, Murray NE, Dryden DT (1999b) On the structure and operation of Type I DNA restriction enzymes. J Mol Biol 290:565–579

Deibert M, Grazulis S, Janulaitis A, Siksnys V, Huber R (1999) Crystal structure of MunI restriction endonuclease in complex with cognate DNA at 1.7 A resolution. EMBO J 18:5805–5816

Deibert M, Grazulis S, Sasnauskas G, Siksnys V, Huber R (2000) Structure of the tetrameric restriction endonuclease NgoMIV in complex with cleaved DNA. Nat Struct Biol 7:792–799

Finnegan DJ (1997) Transposable elements: how non-LTR retrotransposons do it. Curr Biol 7:R245-R248

Friedhoff P, Lurz R, Luder G, Pingoud A (2001) Sau3AI, a monomeric Type II restriction endonuclease that dimerizes on the DNA and thereby induces dna loops. J Biol Chem 276:23581–23588

Fuxreiter M, Simon I (2002) Protein stability indicates divergent evolution of PD-(D/E)XK Type II restriction endonucleases. Protein Sci 11:1978–1983

Gimble FS (2000) Invasion of a multitude of genetic niches by mobile endonuclease genes. FEMS Microbiol Lett 185:99–107

Grazulis S, Deibert M, Rimseliene R, Skirgaila R, Sasnauskas G, Lagunavicius A, Repin V, Urbanke C, Huber R, Siksnys V (2002) Crystal structure of the Bse634I restriction endonuclease: comparison of two enzymes recognizing the same DNA sequence. Nucleic Acids Res 30:876–885

Grishin NV (1997) Estimation of evolutionary distances from protein spatial structures. J Mol Evol 45:359–369

Grishin NV (2001) Fold change in evolution of protein structures. J Struct Biol 134:167–185

Hadden JM, Convery MA, Declais AC, Lilley DM, Phillips SE (2001) Crystal structure of the Holliday junction resolving enzyme T7 endonuclease I. Nat Struct Biol 8:62–67

Hasson MS, Schlichting I, Moulai J, Taylor K, Barrett W, Kenyon GL, Babbitt PC, Gerlt JA, Petsko GA, Ringe D (1998) Evolution of an enzyme active site: the structure of a new crystal form of muconate lactonizing enzyme compared with mandelate racemase and enolase. Proc Natl Acad Sci USA 95:10396–10401

Heitman J (1993) On the origins, structures and functions of restriction-modification enzymes. Genet Eng N Y 15:57–108

Hickman AB, Li Y, Mathew SV, May EW, Craig NL, Dyda F (2000) Unexpected structural diversity in DNA recombination: the restriction endonuclease connection. Mol Cell 5:1025–1034

Horton NC, Dorner LF, Perona JJ (2002) Sequence selectivity and degeneracy of a restriction endonuclease mediated by DNA intercalation. Nat Struct Biol 9:42–47

Huai Q, Colandene JD, Chen Y, Luo F, Zhao Y, Topal MD, Ke H (2000) Crystal structure of NaeI-an evolutionary bridge between DNA endonuclease and topoisomerase. EMBO J 19:3110–3118

Huai Q, Colandene JD, Topal MD, Ke H (2001) Structure of NaeI-DNA complex reveals dual-mode DNA recognition and complete dimer rearrangement. Nat Struct Biol 8:665–669

Ichiyanagi K, Ishino Y, Ariyoshi M, Komori K, Morikawa K (2000) Crystal structure of an archaeal intein-encoded homing endonuclease PI-PfuI. J Mol Biol 300:889–901

Iyer LM, Koonin EV, Aravind L (2002) Extensive domain shuffling in transcription regulators of DNA viruses and implications for the origin of fungal APSES transcription factors. Genome Biol 3:RESEARCH0012

Janosi L, Yonemitsu H, Hong H, Kaji A (1994) Molecular cloning and expression of a novel hydroxymethylcytosine-specific restriction enzyme (PvuRts1I) modulated by glucosylation of DNA. J Mol Biol 242:45–61

Janscak P, Sandmeier U, Bickle TA (1999) Single amino acid substitutions in the HsdR subunit of the Type IB restriction enzyme EcoAI uncouple the DNA translocation and DNA cleavage activities of the enzyme. Nucleic Acids Res 27:2638–2643

Janscak P, Sandmeier U, Szczelkun MD, Bickle TA (2001) Subunit assembly and mode of DNA cleavage of the Type III restriction endonucleases EcoP1I and EcoP15I. J Mol Biol 306:417–431

Jeltsch A, Kroger M, Pingoud A (1995) Evidence for an evolutionary relationship among type-II restriction endonucleases. Gene 160:7–16

Jeltsch A, Pingoud A (1996) Horizontal gene transfer contributes to the wide distribution and evolution of Type II restriction-modification systems. J Mol Evol 42:91–96

Johnson MS, Sutcliffe MJ, Blundell TL (1990) Molecular anatomy: phyletic relationships derived from three-dimensional structures of proteins. J Mol Evol 30:43–59

Jurica MS, Stoddard BL (1999) Homing endonucleases: structure, function and evolution. Cell Mol Life Sci 55:1304–1326

Jurica MS, Monnat RJJ, Stoddard BL (1998) DNA recognition and cleavage by the LAGLIDADG homing endonuclease I-CreI. Mol Cell 2:469–476

Kim Y, Grable JC, Love R, Green PJ, Rosenberg JM (1990) Refinement of EcoRI endonuclease crystal structure: a revised protein chain tracing. Science 249:1307–1309

Kinch LN, Grishin NV (2002) Evolution of protein structures and functions. Curr Opin Struct Biol 12:400–408

Kobayashi I (2001) Behavior of restriction-modification systems as selfish mobile elements and their impact on genome evolution. Nucleic Acids Res 29:3742–3756

Kong H, Lin LF, Porter N, Stickel S, Byrd D, Posfai J, Roberts RJ (2000) Functional analysis of putative restriction-modification system genes in the *Helicobacter pylori* J99 genome. Nucleic Acids Res 28:3216–3223

Kostrewa D, Winkler FK (1995) Mg^{2+} binding to the active site of EcoRV endonuclease: a crystallographic study of complexes with substrate and product DNA at 2 Å resolution. Biochemistry 34:683–696

Kovall RA, Matthews BW (1998) Structural, functional, and evolutionary relationships between lambda-exonuclease and the Type II restriction endonucleases. Proc Natl Acad Sci USA 95:7893–7897

Kowalski JC, Belfort M, Stapleton MA, Holpert M, Dansereau JT, Pietrokovski S, Baxter SM, Derbyshire V (1999) Configuration of the catalytic GIY-YIG domain of intron endonuclease I-TevI: coincidence of computational and molecular findings. Nucleic Acids Res 27:2115–2125

Kruger DH, Bickle TA (1983) Bacteriophage survival: multiple mechanisms for avoiding the deoxyribonucleic acid restriction systems of their hosts. Microbiol Rev 47:345–360

Kvaratskhelia M, Wardleworth BN, Norman DG, White MF (2000) A conserved nuclease domain in the archaeal holliday junction resolving enzyme Hjc. J Biol Chem 275:25540–25546

Lagunavicius A, Siksnys V (1997) Site-directed mutagenesis of putative active site residues of MunI restriction endonuclease: replacement of catalytically essential carboxylate residues triggers DNA binding specificity. Biochemistry 36:11086–11092

Lagunavicius A, Sasnauskas G, Halford SE, Siksnys V (2003) The metal-independent Type IIS restriction enzyme BfiI is a dimer that binds two DNA sites but has only one catalytic centre. J Mol Biol 326(4):1051–64

Li H, Abelson J (2000) Crystal structure of a dimeric archaeal splicing endonuclease. J Mol Biol 302:639–648

Li H, Trotta CR, Abelson J (1998) Crystal structure and evolution of a transfer RNA splicing enzyme. Science 280:279–284

Lin LF, Posfai J, Roberts RJ, Kong H (2001) Comparative genomics of the restriction-modification systems in *Helicobacter pylori*. Proc Natl Acad Sci USA 98:2740–2745

Lukacs CM, Kucera R, Schildkraut I, Aggarwal AK (2000) Understanding the immutability of restriction enzymes: crystal structure of BglII and its DNA substrate at 1.5 Å resolution. Nat Struct Biol 7:134–140

Lunnen KD, Barsomian JM, Camp RR, Card CO, Chen SZ, Croft R, Looney MC, Meda MM, Moran LS, Nwankwo DO (1988) Cloning type-II restriction and modification genes. Gene 74:25–32

Luria SE, Human ML (1952) A nonhereditary, host-induced variation of bacterial viruses. J Bacteriol 64:557–569

Martin JL (1995) Thioredoxin -a fold for all reasons. Structure 3:245–250

Matveyev AV, Young KT, Meng A, Elhai J (2001) DNA methyltransferases of the cyanobacterium *Anabaena* PCC 7120. Nucleic Acids Res 29:1491–1506

May AC (1999) Toward more meaningful hierarchical classification of protein three-dimensional structures. Proteins 37:20–29

McClarin JA, Frederick CA, Wang BC, Greene P, Boyer HW, Grable J, Rosenberg JM (1986) Structure of the DNA–Eco RI endonuclease recognition complex at 3 Å resolution. Science 234:1526–1541

Morgan RD, Xiao JP, Xu SY (1998) Characterization of an extremely thermostable restriction enzyme, PspGI, from a *Pyrococcus* strain and cloning of the PspGI restriction- modification system in Escherichia coli. Appl Environ Microbiol 64:3669–3673

Mucke M, Grelle G, Behlke J, Kraft R, Kruger DH, Reuter M (2002) EcoRII: a restriction enzyme evolving recombination functions? EMBO J 21:5262–5268

Murzin AG (1998) How far divergent evolution goes in proteins. Curr Opin Struct Biol 8:380–387

Murzin AG, Brenner SE, Hubbard T, Chothia C (1995) SCOP: a structural classification of proteins database for the investigation of sequences and structures. J Mol Biol 247:536–540

Nastri HG, Evans PD, Walker IH, Riggs PD (1997) Catalytic and DNA binding properties of PvuII restriction endonuclease mutants. J Biol Chem 272:25761–25767

Nei M (1996) Phylogenetic analysis in molecular evolutionary genetics. Annu Rev Genet 30:371–403

Newman M, Lunnen K, Wilson G, Greci J, Schildkraut I, Phillips SE (1998) Crystal structure of restriction endonuclease BglI bound to its interrupted DNA recognition sequence. EMBO J 17:5466–5476

Newman M, Strzelecka T, Dorner LF, Schildkraut I, Aggarwal AK (1994) Structure of restriction endonuclease BamHI and its relationship to EcoRI. Nature 368:660–664

Nishino T, Komori K, Tsuchiya D, Ishino Y, Morikawa K (2001) Crystal structure of the Archaeal Holliday junction resolvase Hjc and implications for DNA recognition. Structure 9:197–204

Nobusato A, Uchiyama I, Kobayashi I (2000a) Diversity of restriction-modification gene homologues in *Helicobacter pylori*. Gene 259:89–98

Nobusato A, Uchiyama I, Ohashi S, Kobayashi I (2000b) Insertion with long target duplication: a mechanism for gene mobility suggested from comparison of two related bacterial genomes. Gene 259:99–108

Nolling J, de Vos WM (1992) Characterization of the archaeal, plasmid-encoded Type II restriction- modification system MthTI from *Methanobacterium thermoformicicum* THF: homology to the bacterial NgoPII system from *Neisseria gonorrhoeae*. J Bacteriol 174:5719–5726

Pingoud A, Jeltsch A (1997) Recognition and cleavage of DNA by type-II restriction endonucleases. Eur J Biochem 246:1–22

Pingoud A, Jeltsch A (2001) Structure and function of Type II restriction endonucleases. Nucleic Acids Res 29:3705–3727

Pingoud V, Kubareva E, Stengel G, Friedhoff P, Bujnicki JM, Urbanke C, Sudina A, Pingoud A (2002) Evolutionary relationship between different subgroups of restriction endonucleases. J Biol Chem 277:14306–14314

Reuter M, Schneider-Mergener J, Kupper D, Meisel A, Mackeldanz P, Kruger DH, Schroeder C (1999) Regions of endonuclease EcoRII involved in DNA target recognition identified by membrane-bound peptide repertoires. J Biol Chem 274:5213–5221

Rigden DJ, Setlow P, Setlow B, Bagyan I, Stein RA, Jedrzejas MJ (2002) PrfA protein of *Bacillus* species: prediction and demonstration of endonuclease activity on DNA. Protein Sci 11:2370–2381

Roberts RJ, Macelis D (2001) REBASE-restriction enzymes and methylases. Nucleic Acids Res 29:268–269

Roberts RJ, Belfort M, Bestor T, Bhagwat AS, Bickle TA, Bitinaite J, Blumenthal RM, Degtyarev SK, Dryden DT, Dybvig K, Firman K, Gromova ES, Gumport RI, Halford SE, Hattman S, Heitman J, Hornby DP, Janulaitis A, Jeltsch A, Josephsen J, Kiss A, Iaenhammer T, Kobayashi I, Kong H, Kruger D, Lacks S, Marinus MG, Miyahara M, Morgan RD, Murray NE, Nagaraja V, Piekarowicz A, Pingoud A, Raleigh E, Rao DN, Reich N, Repin V, Selker E, Shaw PC, Stein DC, Stoddard BL, Szybalski W, Trautner TA, Van Etten JL, Vitor JM, Wilson GG, Xu SY (2003a) A nomenclature for restriction enzymes, DNA methyltransferases, homing endonucleases and their genes. Nucleic Acids Res 31(7):1805–1812

Roberts RJ, Vincze T, Posfai J, Macelis D (2003b) REBASE: restriction enzymes and methyltransferases. Nucleic Acids Res 31:418–420

Rocha EP, Danchin A, Viari A (2001) Evolutionary role of restriction/modification systems as revealed by comparative genome analysis. Genome Res 11:946–958

Rost B (1997) Protein structures sustain evolutionary drift. Fold Des 2:S19-S24

Rost B (2002) Enzyme function less conserved than anticipated. J Mol Biol 318:595–608

Sapranauskas R, Sasnauskas G, Lagunavicius A, Vilkaitis G, Lubys A, Siksnys V (2000) Novel subtype of Type IIs restriction enzymes. J Biol Chem 275:30878–30885

Siksnys V, Zareckaja N, Vaisvila R, Timinskas A, Stakenas P, Butkus V, Janulaitis A (1994) CAATTG-specific restriction-modification MunI genes from *Mycoplasma*: sequence similarities between R.MunI and R.EcoRI. Gene 142:1–8

Siksnys V, Timinskas A, Klimasauskas S, Butkus V, Janulaitis A (1995) Sequence similarity among type-II restriction endonucleases, related by their recognized 6-bp target and tetranucleotide-overhang cleavage. Gene 157:311–314

Silva GH, Dalgaard JZ, Belfort M, Van Roey P (1999) Crystal structure of the thermostable archaeal intron-encoded endonuclease I-*Dmo*I. J Mol Biol 286:1123–1136

Skirgaila R, Grazulis S, Bozic D, Huber R, Siksnys V (1998) Structure-based redesign of the catalytic/metal binding site of Cfr10I restriction endonuclease reveals importance of spatial rather than sequence conservation of active centre residues. J Mol Biol 279:473–481

Stephenson FH, Ballard BT, Boyer HW, Rosenberg JM, Greene PJ (1989) Comparison of the nucleotide and amino acid sequences of the RsrI and EcoRI restriction endonucleases. Gene 85:1–13

Tamulaitis G, Solonin AS, Siksnys V (2002) Alternative arrangements of catalytic residues at the active sites of restriction enzymes. FEBS Lett 518:17–22

Thielking V, Selent U, Kohler E, Wolfes H, Pieper U, Geiger R, Urbanke C, Winkler FK, Pingoud A (1991) Site-directed mutagenesis studies with EcoRV restriction endonuclease to identify regions involved in recognition and catalysis. Biochemistry 30:6416–6422

Todd AE, Orengo CA, Thornton JM (2001) Evolution of function in protein superfamilies, from a structural perspective. J Mol Biol 307:1113–1143

Todone F, Weinzierl RO, Brick P, Onesti S (2000) Crystal structure of RPB5, a universal eukaryotic RNA polymerase subunit and transcription factor interaction target. Proc Natl Acad Sci USA 97:6306–6310

Tsutakawa SE, Jingami H, Morikawa K (1999a) Recognition of a TG mismatch: the crystal structure of very short patch repair endonuclease in complex with a DNA duplex. Cell 99:615–623

Tsutakawa SE, Muto T, Kawate T, Jingami H, Kunishima N, Ariyoshi M, Kohda D, Nakagawa M, Morikawa K (1999b) Crystallographic and functional studies of very short patch repair endonuclease. Mol Cell 3:621–628

van der Woerd MJ, Pelletier JJ, Xu S, Friedman AM (2001) Restriction enzyme BsoBI-DNA complex: a tunnel for recognition of degenerate DNA sequences and potential histidine catalysis. Structure (Camb) 9:133–144

Van Roey P, Meehan L, Kowalski JC, Belfort M, Derbyshire V (2002) Catalytic domain structure and hypothesis for function of GIY-YIG intron endonuclease I-TevI. Nat Struct Biol 9:806–811

Venclovas C, Timinskas A, Siksnys V (1994) Five-stranded beta-sheet sandwiched with two alpha-helices: a structural link between restriction endonucleases EcoRI and EcoRV. Proteins 20:279–282

Viadiu H, Aggarwal AK (1998) The role of metals in catalysis by the restriction endonuclease BamHI. Nat Struct Biol 5:910–916

Wah DA, Hirsch JA, Dorner LF, Schildkraut I, Aggarwal AK (1997) Structure of the multimodular endonuclease FokI bound to DNA. Nature 388:97–100

Wang J, Chen R, Julin DA (2000) A single nuclease active site of the *Escherichia coli* RecBCD enzyme catalyzes single-stranded DNA degradation in both directions. J Biol Chem 275:507–513

Wilson GG (1991) Organization of restriction-modification systems. Nucleic Acids Res 19:2539–2566

Wilson GG, Murray NE (1991) Restriction and modification systems. Annu Rev Genet 25:585–627

Winkler FK, Banner DW, Oefner C, Tsernoglou D, Brown RS, Heathman SP, Bryan RK, Martin PD, Petratos K, Wilson KS (1993) The crystal structure of EcoRV endonuclease and of its complexes with cognate and non-cognate DNA fragments. EMBO J 12:1781–1795

Withers BE, Ambroso LA, Dunbar JC (1992) Structure and evolution of the XcyI restriction-modification system. Nucleic Acids Res 20:6267–6273

Xu Y, Lunnen KD, Kong H (2001) Engineering a nicking endonuclease N.AlwI by domain swapping. Proc Natl Acad Sci USA 98:12990–12995

Yang J, Malik HS, Eickbush TH (1999) Identification of the endonuclease domain encoded by R2 and other site-specific, non-long terminal repeat retrotransposable elements. Proc Natl Acad Sci USA 96:7847–7852

Sliding or Hopping? How Restriction Enzymes Find Their Way on DNA

A. Jeltsch, C. Urbanke

1 Introduction

In this contribution, we discuss target site location of restriction endonucleases, which are extremely common among prokaryotes occurring in almost every species. These enzymes specifically recognize and cleave short, often palindromic, sequences on bacteriophage or other kinds of foreign DNA (reviews: Pingoud and Jeltsch 1997, 2001). By cleaving incoming DNA, they protect bacteria against bacteriophage infections acting like an immune system. In addition, restriction endonucleases have an important role in the control of horizontal gene transfer and bacterial evolution (Arber 2000; Jeltsch 2003). The cellular DNA is protected against nucleolytic attack by an accompanying DNA methyltransferase which recognizes the same nucleotide sequence and methylates one adenine or cytosine residue within the site (review: Jeltsch 2002). Since restriction endonucleases and DNA methyltransferases are present at the same time in the cell, there is a kinetic competition between them, which necessitates that the restriction enzyme finds its target site on an invading DNA faster than the methyltransferase. In addition, fast target site location is also important, because the invading bacteriophage DNA has to be degraded before it gains control over the cellular metabolism. Therefore, restriction enzymes were among the earliest examples where facilitated diffusion has been shown to be effective in target site location (Jack et al. 1982; Ehrbrecht et al. 1985; Terry et al. 1985). However, the actual mechanism of this process is still under debate: whereas most authors had proposed a sliding mechanism, this has recently been disputed by Halford and colleagues who favored a hopping process (Stanford et al. 2000; Halford and Szczelkun 2002; Gowers and Halford 2003).

A. Jeltsch
Institut für Biochemie, FB 8, Heinrich-Buff-Ring 58, Justus-Liebig-Universität, 35392 Giessen, Germany
C. Urbanke
Zentrale Einrichtung Biophysikalisch-Biochemische Verfahren, Medizinische Hochschule, 30625 Hannover, Germany

Nucleic Acids and Molecular Biology, Vol. 14
Alfred Pingoud (Ed.)
Restriction Endonucleases
© Springer-Verlag Berlin Heidelberg 2004

2 Mechanisms of Facilitated Target Site Location by Proteins on DNA

The first step in enzymatic catalysis is the binding of the enzyme to its substrate(s), which occurs by diffusional motion. The random motion of a particle itself is described by its mean square displacement (Eq. 1):

$$\left\langle y^2 \right\rangle_i = 2nDt \tag{1}$$

where $\left\langle y^2 \right\rangle$ =average (squared) distance travelled by the particle, D=diffusion coefficient, n=dimensionality of motion (1–3), and t=duration of the random walk.

Describing diffusion as Brownian motion with a frequency of v steps of length l per time unit defines the diffusion coefficient as

$$D = \frac{vl^2}{2n} \tag{2}$$

Then, the number of diffusional steps N required to move a certain distance d is given by:

$$N = tv = \left\langle y^2 \right\rangle / l2 = d / l2 \tag{3}$$

The rate of association can be accelerated significantly by a reduction in dimensionality (Adam and Delbrück 1968; Richter and Eigen 1974), because the size of the search space is proportional to x in one-dimensional search, but to x^2 and x^3 in two- and three-dimensional search. If, for example, the protein starts diffusion at a site and the target site is at a distance d, after travelling d the protein has a 50 % chance to bind to the target site in a 1D movement, because it can have moved to one of only two sides away from the starting point on the 1D line. In contrast, in a 3D movement, the end points of a random walk over a distance d from the target site are located on the surface of a sphere with the radius of d. The actual number of sites depends on the size of the target site and d, but the number grows with d^3. It has been proposed in the 1960s that the reduction in dimensionality can facilitate target site location by DNA-binding proteins (Adam and Delbrück 1968; Richter and Eigen 1974). The concept has been theoretically and experimentally developed in the early 1980s by pioneering work of von Hippel and Berg (Berg et al. 1981; Winter et al. 1981; Winter and von Hippel 1981). In their treatment, these authors distinguished between three different types of facilitated diffusion by which the search process of the enzyme is confined basically to DNA molecules and not the solution surrounding them: sliding, "jumping" or "hopping"

3D/1D diffusion   **3D diffusion („hopping", „jumping")**

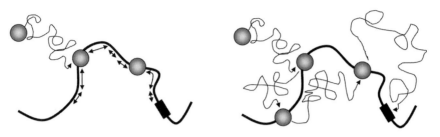

Fig. 1. Schematic figure of target site location by coupled 3D/1D diffusion or by 3D diffusion ("hopping", "jumping")

and intersegment transfer (Fig. 1; reviews: Berg and von Hippel 1985; von Hippel and Berg 1989):

2.1 Sliding

Sliding of a DNA-binding protein on the DNA takes place in a nonspecifically bound state (Fig. 1). The process describes the movement of the protein by one base pair. The sliding rate depends on the size of the barrier ($\Delta G^{\#}_{slide}$) between the protein being bound to the DNA in one register and in the next (Fig. 2). Nonspecific DNA binding is mediated by electrostatics and by hydrogen bonds to the phosphate groups. Since electrostatic forces are not directional, an attractive electrostatic potential will appear smooth with $\Delta G^{\#}_{slide}$ being very small. Hence, if the electrostatics is properly balanced, the protein can be trapped on the DNA, but nevertheless move with very low friction. In addition, hydrogen bonds between the protein and the DNA contribute to DNA affinity. These hydrogen bonds have to be broken for the movement of the protein by one base pair, a process that contributes to $\Delta G^{\#}_{slide}$. However, many of these bonds are mediated by water in nonspecific complexes leading to the presence of a large amount of water in the interface of nonspecific protein–DNA complexes that has to be released upon specific complex formation (Sidorova and Rau 1996; Robinson and Sligar 1998). This water serves as lubricant for the sliding process, since it supports hydrogen bonds with changing geometry and fluctuating interaction partners. In addition, proteins do not behave as rigid bodies during the movement, such that hydrogen bonds may form already at the front of the movement while still not being broken at the rear.

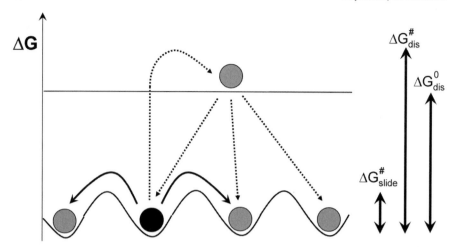

Fig. 2. Compilation of the changes in Gibbs energy associated with a sliding and hopping movement of a protein on DNA. To slide on the DNA by one base pair (*filled arrows*) the protein has to overcome the activation barrier $\Delta G^{\#}_{slide}$. For a hopping movement (*dotted arrows*), the protein must dissociate from the DNA, which requires to pass the activation barrier for dissociation $\Delta G^{\#}_{dis}$. Thermodynamically the energy of the unbound state is increased by ΔG^{0}_{dis}

2.2 Hopping

Facilitated diffusion by hopping occurs, because after any micro-dissociation from the DNA, the protein is still located next to the DNA (Fig. 1). Therefore, it has a high chance to rebind the same DNA molecule again very fast. Only after diffusion to a critical distance, which depends on the concentration of the DNA and the protein, one can consider the protein to be released into the bulk solvent, which macroscopically corresponds to a dissociation process. These short-lived fluctuations of the protein being bound and unbound are called "hopping" or "jumping" movement, depending on the distance between the original position and the new one. Since the size of these steps are randomly distributed, with small steps being more probable than large ones, hopping and jumping refer to the same basic process. It should be recognized that the hopping movement is a direct consequence of the dynamics of three-dimensional diffusion, because if a particle starts diffusion at a certain place, volume elements nearby have a much higher chance to be occupied transiently than ones that are more distant. In the case of the hopping movement, this leads to preferential rebinding to the DNA, because nearby volume elements also have a higher propensity to contain the DNA molecule. Therefore, hopping does not alter the search space and search pathway with respect to 3D diffusion.

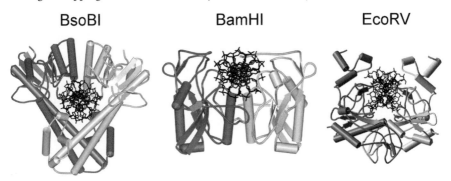

Fig. 3. Comparison of the structures of different restriction endonucleases in complex with specific DNA

The difference between sliding and hopping is illustrated in Figs. 1 and 2: both processes require a certain loosening of the contacts between the DNA and the protein. However, in the case of sliding the protein still stays in contact with the DNA, such that there remains a strong enthalpic interaction that mediates the binding of the protein. In contrast, for the hopping process all molecular interactions must be broken to release the protein from the DNA, leading to a ΔG^0_{dis} that must be larger than $\Delta G^{\#}_{slide}$. In addition to this thermodynamic barrier to dissociation, kinetic barriers also exist that decelerate dissociation of the protein from the DNA. For example, many DNA-binding proteins and restriction enzymes enwrap the DNA, such that their arms have to open up prior to the dissociation process (Fig. 3). Therefore, the activation energy for the dissociation process $\Delta G^{\#}_{dis}$ can be significantly larger than the binding energy itself. Since after the dissociation process, the protein stays in the vicinity of the DNA, reassociation to the same DNA molecule at the same or another site is more likely than binding to another DNA molecule keeping in mind that reassociation to the same site is the most likely of all reassociation events. This situation can be described as an entropic "interaction" between the DNA molecule and the protein.

Since DNA is a flexible molecule with a persistence length of approx. 150 base pairs, by looping DNA segments separated by some hundreds of base pairs can approach each other in 3D space much closer than their distance is on the DNA itself. Therefore, by 3D diffusion the enzyme has a certain propensity to jump over larger distances that is not possible by sliding.

2.3 Intersegment Transfer

Intersegment transfer is a special mode of movement that is restricted to proteins with two or more DNA binding sites. Binding of DNA to both DNA binding sites will produce loops in the DNA. If one binding site releases the DNA, the enzyme still binds to the DNA with the second site. If rebinding of DNA to the free site is faster than the release of the DNA from the bound site, the protein will crawl within the DNA molecule and hardly ever move away from it. Therefore, intersegement transfer is an alternative way to achieve diffusion in 3D but it also reduces the search space in the 3D movement. Since the 3D movement is confined to the DNA, intersegment transfer is a clear reduction of dimensionality and the movement can be considered as taking place in a "space" of some fractal dimension smaller than 3. When combined with linear diffusion to search for the target sites in each DNA binding site, it is a very efficient strategy that combines an exhaustive coverage at the local level with occasional translocatoins over far distances. Maybe this is the reason why so many restriction enzymes are of types E or F and contain two DNA binding sites that are occupied under physiological conditions (Roberts et al. 2003).

3 Critical Factors Determining the Efficiency of Target Site Location by Sliding and Hopping Processes

For the sliding process, we can consider one diffusion step as a movement by one bp corresponding to 3.3 Å. The average number of step required to travel a distance of N base pairs is in the order of N^2. The usual weight of restriction enzyme dimers is between 40 and 70 kDa which would correspond to a typical D^3 (for 3D diffusion) of about $7x10^{-11}$ m^2 s^{-1} ($7x10^{-7}$ cm^2 s^{-1}). The number of bp scanned by restriction enzymes during one binding event can be accurately determined experimentally (see below). These numbers are around 10^6 bp leading to an efficient search length of approx. 1000 bp (Ehbrecht et al. 1985; Jeltsch et al. 1994, 1996; Jeltsch and Pingoud 1998). These 10^6 diffusional steps take place during the dwell time of the protein on the DNA, which is determined by k_{off} from nonspecific DNA. However, in this case the microscopic dissociation constant must be used, which is higher than the experimentally accessible macroscopic dissociation constant, because the macroscopic constant is distorted by the same rapid reassociation events we have already discussed in the context of a hopping movement. To estimate the dissociation constant, we can use binding data to short oligonucleotide substrates, which are of sufficient length to fill the DNA binding site of the enzyme but which do not support hopping. Such nonspecific substrates are bound with K_{ass} of 10^6 M^{-1} by different restriction enzymes (Thielking et al. 1990; Alves et al. 1995). Typical association rates to such substrates are close to 10^8 M^{-1} s^{-1} (Alves et al. 1989) which is slightly lower than the absolute limit of diffusional control (Cantor and Schimmel 1980). With these assumptions, k_{off} should be in the order of 100/s. Then 10^6 diffusion steps are performed in 10 ms and v in Eq. (2) turns out to be 10^8 steps/s.

Again, using a step size of 3.3×10^{-10} m in Eq. (2), these estimates lead to a D^1 of approximately 5×10^{-12} m^2 s^{-1}. These estimations show that in a first approximation, the 1D diffusion constant is about ten times smaller than the 3D value. The reduced diffusion constant for sliding in part is due to the fact that the sliding movement is not straight, but that it follows the helical pitch of the DNA, such that the pathway of the protein is a spiral. If we assume the center of mass of the restriction enyzme to be about 10 Å away from the center of the DNA helix, then the spiral pathway of the protein is approximately two times longer than a straight movement. In addition, one has to consider that every movement by 10.5 bp requires one rotation of the protein. These differences account for most of the variation between D^3 and D^1. This result is in agreement to the general consideration that the friction experienced by the molecule during the 3D and (water mediated) 1D movement may be comparable such that the diffusion constants need not be very different.

According to the above estimation, one 1D-diffusional step requires approximately 10 ns and the total search for a single site in the DNA will take approximately $(N/2)2 \times 10^{-8}$ s which is 2.5 ms for 1000 bp, 0.25 s for 10^4 bp and 25 s for 10^5 bp. This very simplistic model demonstrates that 1D diffusion is very fast over short distances (which is due to the reduction in the search space) but it makes no sense to extend linear diffusion over a certain length to accelerate target site location. The reason for this outcome is the great redundancy in the search process by linear diffusion, where most sites are occupied several times. In contrast, in 3D diffusion the chance of repetitive visits of the same site is much smaller. Experimental results demonstrate that 1D diffusion usually takes about 10^6 steps, covering 1000 base pairs. This search process is completed within few milliseconds.

Our treatment of 3D diffusion for target site location so far contains two principal oversimplifications: (1) it does not consider that on the DNA a large excess of nontarget sites surrounds the target site and (2) it assumes that 3D diffusion is not affected by the presence of nonspecific DNA. In a more realistic scenario, we also have to take into account that most times the enzyme will not bind to its target site, but to any other nonspecific site, because in each encounter nonspecific binding is preferred by a large statistical factor. If we consider, for example, a DNA comprising 10,000 bp, in each binding event the enzyme has only a 1/10,000 chance of binding the target site. In the pure 3D (hopping) model, dissociation from the DNA must occur after each binding to a non-target site. In the above example on average 10,000 dissociation events of the protein from the DNA are required, which if we use our estimate for k_{off} of 100 s^{-1} would require 100 s of search time. This additional time is not required in the 1D search. Therefore, apart from the reduction of the search space, linear diffusion leads to accelerated target site location, because it allows movement of the enzyme to the target site after binding to a non-target site without necessity of dissociation from the DNA. Disregarding the time requirement for dissociation from the DNA after binding to non-target sites is a critical point in the theoretical treatment of Halford and colleagues. In their work, they come to the conclusion that a 3D "hopping" or "jumping" movement should be superior to a sliding model (Halford and Szczelkun 2002). However, they implicitly assume that the dissociation rate of the protein from the DNA is very fast such that the time required for dissociation of the enzyme from the DNA is comparable to the time required for one 1D or 3D diffusion step. Given this assumption, i.e. that the dwell time of unspecifically bound protein on the DNA is of the order of the time required for a single diffusion step and considering association rates being determined by diffusion of the protein to the DNA (the fastest association rate possible in this case) leads to the conclusion that the unspecific affinity of the protein for DNA vanishes. This makes the treatment of Halford and colleagues very questionable.

The conclusion of this consideration is that it is impossible to locate a specific binding site on the DNA by pure 3D diffusion in a reasonable period of time. It should be noticed that the advantage of 3D/1D diffusion over pure 3D diffusion (including "hopping") is not critical dependent on the estimation of the k_{off}-value from nonspecific DNA used here. If k_{off} is smaller, our estimate for D^1 will become smaller and hence 1D diffusion slower. Then, the 1D diffusion process will require more time. However, the time required to release a non-target site during 3D diffusion also would increase, such that the relative advantage of the 3D/1D model remained similar.

4 Sliding or "Hopping" – A Survey of Experimental Data

4.1 Structures of Restriction Endonucleases

Several structures of restriction endonucleases in complex with DNA have been solved (review: Pingoud and Jeltsch 2001). In all the cases, the DNA is deeply buried in a binding cleft of the enzyme that closes after DNA binding such that in most cases the DNA could not be released from the enzyme without a considerable conformational change (Fig. 3). For example, BsoBI totally enwraps the target DNA. These complexes are suggestive for a sliding movement on the DNA. The structures of the few nonspecific complexes solved so far show the DNA bound by electrostatic interactions, also partially enwrapped by the enzyme (Winkler et al. 1993; Viadiu and Aggarwal 2000).

4.2 Accurate Scanning of the DNA for the Presence of Target Sites

If one considers that hopping mainly consists of short jumps a distinction between hopping and sliding is difficult. One experimental setup is to use substrates that contain two recognition sites close to one end. Then, most enzyme molecules will bind on one side of both target sites and the enzymes will diffuse from one direction to the targets site. It has been shown for EcoRI and EcoRV that under these conditions the target site first encountered by the enzyme is cleaved much faster than the second (Jeltsch et al. 1994; Jeltsch and Pingoud 1998). These results demonstrate that both enzymes do not overlook any target site, a result that has been confirmed also for BssHII (Berkhout and van Wamel 1996). In addition, it has been shown for EcoRI that it pauses at sites whose sequence resembles the recognition site, again demonstrating an accurate scanning process (Jeltsch et al. 1994). These findings provide a very strong case for a sliding movement, because in a pure hopping mechanism, target sites will be reached in a random order whilst in a sliding mechanism, in which the protein follows the helical geometry of the DNA, these sites will be recognized sequentially.

4.3 Length Dependence of Linear Diffusion

In the case of a sliding mechanism the dependence of the association rate constant to the specific site on the length of the DNA molecule is given by Eq. (4) (Ehbrecht et al. 1985):

$$k_{on}/k_{+1} = \sum_{n=1}^{L} e^{\frac{-1(n-s)^2}{P_{dif}}} \qquad (4)$$

where: k_{on}=association rate constant to the specific site, k_{+1}=association rate constant to a nonspecific site, L=length of the DNA molecule, s=position of the specific site and P_{dif}=probability of linear diffusion versus dissociation ($=k_{dif}/k_{off}$)

Basically, this relationship means that the rate of association to the specific site is increased by binding events to nonspecific sites. If binding occurs close to the specific site, the enzyme has a very high chance to reach the site: if binding is far away, the probability to reach the specific site in this binding event is low. The equation describes a saturation curve, because increasing the DNA length over the length typically searched by the enzyme does not lead to a further rate enhancement. At short DNA lengths, there is an almost linear increase of the association rate with the length of the DNA. It has been shown in two examples that experimental data can be nicely fitted by Eq. (4), which supports the sliding model (Fig. 4). In addition, a detailed kinetic analysis revealed that the steady-state kinetic parameters of EcoRI are dominated by the sliding process (Wright et al. 1999) and that the efficiency of sliding correlates to the in vivo efficiency of restriction enzymes, demonstrating that sliding also takes place in the cell (Jeltsch et al. 1996).

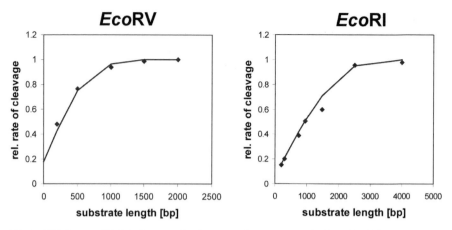

Fig. 4. Efficiency of DNA cleavage by EcoRI and EcoRV on the length of the DNA substrate (data taken from Ehbrecht et al. 1985 and Jeltsch et al. 1996). The *lines* are fits of the data to Eq. (4)

4.4 Processivity of DNA Cleavage

After cleaving one site, the restriction enzyme can stay bound to the same DNA molecule and cleave it at a second site nearby. The processivity of enzymatic turnover depends on the movement of the enzyme along the DNA. Thus, measuring the degree of processivity of the cleavage of two sites, which have different distances from each other, can be used to determine the efficiency and mechanism of the movement of the enzyme on the DNA. In the case of EcoRI, for example, it has been shown that the enzyme leaves its target site by sliding along the DNA and not by direct dissociation (Jack et al. 1982).

A similar approach has been chosen by Halford and colleagues who studied the dependence of the processivity on the distance of the two sites for EcoRV (Stanford et al. 2000). The probability P_{dif} of the enzyme to make one step of linear diffusion instead of dissociating from the DNA is given by k_{dif}/k_{off}. Then, the probability (P_l) to make one step of linear diffusion is:

$$P = k_{diff} / \left(k_{diff} + k_{off}\right) = P_{dif} / \left(P_{dif} + 1\right)$$

Since it takes N^2 steps to come from one site to another that is N bp apart the processivity factor (f_p) is defined by Eq (5) (Stanford et al. 2000):

$$f_p = f_{max} \times \left[Pdif / \left(Pdif + 1\right)\right]^{N^2} \tag{5}$$

where: f_{max}=maximal possible processivity.

Since after cleavage of one site the enzyme can bind to only one of the two cleavage products the maximal possible processivity is 0.5. This value has been used by Stanford et al. to fit theoretical curves to their experimentally determined data (curve 1 and 2 in Fig. 5; Fig. 6 in Stanford et al. 2000). They provide two curves: one fitting the first data point and the other the last data point and conclude that no curve can fit all data such that processes distinct from sliding must be operative. They continue arguing that jumping might be one model, where they would expect a 1/n dependence of f_p from the distance of the two sites (n). They show that two of their four data points indeed can be fitted by a straight line in a f_p vs. 1/distance plot (Fig. 7 in Stanford et al. 2000) and argue that a jumping mechanism of dissociation/reassociation must be operative.

However, the assumption of Stanford et al. that f_{max} should be 0.5 relies on the assumption that EcoRV will leave the cleavage site by sliding in one of the two possible directions. This assumption is not justified, since these authors later argue their data would disprove a sliding movement. In fact, this is a critical point, because EcoRV cleaves the DNA to produce blunt ends. Thereby, two new phosphomonoester end groups which both carry approximately 1/2 extra negative charge are produced in close proximity to each other which further increase the natural electrostatic repulsion of the DNA. In the case of EcoRI the DNA is cleaved to produce sticky ends with a 4 bp overhang. Therefore, in this case, the extra negative charge is separated and the cleavage products are still connected by four base pairs. Since the DNA is still bound by the enzyme, the movement of the DNA in the DNA binding tunnel of the enzyme is restricted, and the DNA has a good chance to retain a double-stranded conformation such that the enzyme can leave the target site by sliding as experimentally observed (Jack et al. 1982). In case of EcoRV, the DNA ends are not staggered, thus both DNA ends are likely to diffuse rapidly out of the DNA binding tunnel of EcoRV. After this first dissociation step, the enzyme can rebind to one of the cleavage fragments. This automatically makes the first step in relocation of the enzyme on the DNA a hopping event and it is evident that a hopping mechanism followed by sliding can describe this special kind

Fig. 5. Dependence of the processivity of DNA cleavage by EcoRV on the distance of two EcoRV sites (data were taken from Stanford et al. 2000, for the definition of f_p cf. Eq. 5). *Curves 1* and *2* are fits of the data to Eq. (5), provided by Stanford et al. to demonstrate that Eq. (5) does not fit their data. *Curve 3* is a fit of the data (excluding the first data point) to Eq. (5), in which the maximally possible processivity was also varied

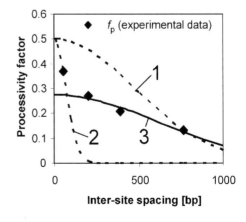

of processivity. In this mechanism, nothing can guarantee that the enzyme always will bind to one of the fragments but one rather has to consider the possibility that the enzyme dissociates from the DNA completely after cleavage. Therefore, the maximal possible processivity can be much lower than the value of 0.5, which has been used by Stanford et al. for their analysis. In fact, the maximal processivity depends on many factors including enzyme and DNA concentration, viscosity of the medium and temperature.

As shown by curve 3 in Fig. 5, the maximal possible processivity critically influences the fit of the experimentally determined processivity at different distances of cleavage sites. Here the data of Stanford et al. are fitted to the equation derived by Stanford et al. with the important exception that f_{max} was also varied. The data were fitted best by an f_{max} of 0.27 and a P_{dif} of 7.4x10^5. It is notable that the P_{dif} value derived here is very close to the value obtained in a different study for EcoRV (10^6) under conditions of low Mg^{2+} (Jeltsch and Pingoud 1998). This model with the exception of the first data point nicely fits the data. The inability to fit the processivity of the first data point can be easily explained: In all cases except the first, the sites are separated by distances large enough to ensure that processive cleavage can only occur after rebinding of the enzyme to one cleavage fragment and sliding on this fragment, i.e. by a coupled 3D/1D process exactly as initial target site location takes place. In the case of the sites separated by only 56 bp, the enzyme also has a reasonable chance to find the second site by pure 3D diffusion. Thus, 3D association adds to the 3D/1D association, which leads to a higher processivity in this case. Taken together, the results reported by Stanford et al. can well be explained by a model in which after cleavage a single hopping, i.e. dissociation-3D diffusion reassociation event, is followed by 1D scan of the DNA. The data may point to a difference in the kinetic mechanism between restriction enzymes which produce blunt ends, where the DNA can rapidly diffuse out of the DNA binding tunnel of the enzyme, and the enzyme has to find the next substrate by 3D diffusion, and those that produce staggered ends which can directly leave their target sites after DNA cleavage by sliding.

4.5 DNA Cleavage by a Covalently Closed EcoRV Variant

By site directed mutagenesis and chemical cross-linking an EcoRV variant was prepared in which the arms grasping around the DNA are covalently connected at their tips (Fig. 6; Schulze et al. 1998). This prevents direct binding of the DNA by opening of the arms but only allows threading of the DNA into the enzyme from one end. In addition, release of the DNA can only occur via the ends, which absolutely prevents any kind of "hopping" or "jumping" movement. Nevertheless, if the cross-linking is carried out with a circular DNA bound to EcoRV in the absence of Mg^{2+} where EcoRV binds nonspecifically to DNA (Taylor et al. 1991; Thielking et al. 1992), after addition of Mg^{2+} the vari-

Fig. 6. Schematic representation of the covalently closed EcoRV-DNA complex prepared by covalent linking the arms of EcoRV, which enwrap the DNA (Schulze et al. 1998). Note that, in this complex, DNA binding and release can only occur at the ends of linear DNA

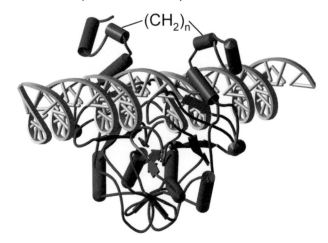

ant is even more active than uncross-linked enzyme preincubated with the same amount of plasmid. This result shows that the cross-linked variant can efficiently find its cleavage sites after addition of Mg^{2+} demonstrating that it efficiently slides along the DNA and that hopping is not necessary for target site location (Schulze et al. 1998).

4.6 Cleavage of Topological Connected Plasmid Molecules

Recently, topologically connected circular plasmids have been prepared and used as substrates for EcoRV. It was shown that the rate of cleavage of a mini-circle could be stimulated by concatenating with larger nontarget circles. This result demonstrates that after binding to the long circular plasmid, the EcoRV enzyme can jump to the small minicircle (Gowers and Halford 2003). It confirms that 3D diffusion contributes to the location of targets sites by EcoRV over longer distances.

5 Conclusions

Restriction enzymes must locate their target site as fast as possible. They make use of a combined 3D/1D strategy, in which the DNA is bound non-specifically in the first step and then scanned for the presence of the cleavage site by sliding on the DNA. Sliding mainly accelerates target site location, because it strongly increases the chance to locate a target site during each binding event. Typical ranges of linear diffusion by restriction enzymes are around 1000 bp, which is close to the statistical occurrence of the target sites of 4 to 6 bp cutter. Thus, usually every binding event will lead to target site location and restriction enzymes are evolutionarily optimized with respect to this property. 3D/1D diffusion should be a property of most, if not all, proteins binding DNA at specific and rare sites because of basic laws of physical chemistry: for their biological function, a certain affinity of the protein to the target site is necessary. Since target sites and non-target sites are very similar in terms of gross molecular properties (shape, charge), it is not possible to create a protein with a sufficient affinity to the specific site but no nonspecific DNA-binding properties. Therefore, the rate of dissociation from nonspecific DNA cannot be lower than given by a certain threshold, which is around 100 per second for restriction enzymes. Since DNA contains many non-target sites, the limited size of k_{off} from nonspecific DNA would strongly decelerate target site location by trapping the protein at these nonspecific sites. Sliding is the only mechanism by which a protein can effectively scan longer stretches of DNA still being nonspecifically bound to the nucleic acid. However, sliding alone only allows scanning a limited range of a DNA sequence. To locate single target sites on large DNA molecules a 3D diffusion within the space occupied by the DNA will also be required.

References

Adam G, Delbrück M (1968) In: Rich A, Davidson N (eds) Structural chemistry and molecular biology. Freeman, San Francisco, pp 198–215

Alves J, Urbanke C, Fliess A, Maass G, Pingoud A (1989) Fluorescence stopped-flow kinetics of the cleavage of synthetic oligodeoxynucleotides by the EcoRI restriction endonuclease. Biochemistry 28:7879–7888

Alves J, Selent U, Wolfes H (1995) Accuracy of the EcoRV restriction endonuclease: binding and cleavage studies with oligodeoxynucleotide substrates containing degenerate recognition sequences. Biochemistry 34:11191–11197

Arber W (2000) Genetic variation: molecular mechanisms and impact on microbial evolution. FEMS Microbiol Rev 24:1–7

Berg OG, von Hippel PH (1985) Diffusion-controlled macromolecular interactions. Annu Rev Biophys Biophys Chem 14:131–160

Berg OG, Winter RB, von Hippel PH (1981) Diffusion-driven mechanisms of protein translocation on nucleic acids, vol 1. Models Theory. Biochem 20:6929–6948

Berkhout B, van Wamel J (1996) Accurate scanning of the BssHII endonuclease in search for its DNA cleavage site. J Biol Chem 271:1837–1840

Cantor CR, Schimmel PR (1980) Biophysical chemistry. Freeman, San Francisco

Ehbrecht H-J, Pingoud A, Urbanke C, Maass G, Gualerzi C (1985) Linear diffusion of restriction endonucleases on DNA. J Biol Chem 260:6160–6166

Gowers DM, Halford SE (2003) Protein motion from non-specific to specific DNA by three-dimensional routes aided by supercoiling. EMBO J 22:1410–1418

Halford SE, Szczelkun MD (2002) How to get from A to B: strategies for analysing protein motion on DNA. Eur Biophys J 31:257–267

Jack WE, Terry BJ, Modrich P (1982) Involvement of outside DNA sequences in the major kinetic path by which EcoRI endonuclease locates and leaves its recognition sequence. Proc Natl Acad Sci USA 79:4010–4014

Jeltsch A (2002) Beyond Watson and Crick: DNA methylation and molecular enzymology of DNA methyltransferases. Chem Bio Chem 3:274–293

Jeltsch A (2003) Maintaining of species identity and controlling speciation of bacteria: a new function for restriction/modification systems? Gene (in press)

Jeltsch A, Pingoud A (1998) Kinetic characterization of linear diffusion of the restriction endonuclease EcoRV on DNA. Biochemistry 37:2160–2169

Jeltsch A, Alves J, Wolfes H, Maass G, Pingoud A (1994) Pausing of the restriction endonuclease EcoRI during linear diffusion on DNA. Biochemistry 33:10215–10219

Jeltsch A, Wenz C, Stahl F, Pingoud A (1996) Linear diffusion of the restriction endonuclease EcoRV on DNA is essential for the in vivo function of the enzyme. EMBO J 15:5104–5111

Pingoud A, Jeltsch A (1997) Recognition and cleavage of DNA by type-II restriction endonucleases. Eur J Biochem 246:1–22

Pingoud A, Jeltsch A (2001) Structure and function of Type II restriction endonucleases. Nucleic Acids Res 29:3705–3727

Richter PH, Eigen M (1974) Diffusion controlled reaction rates in spheroidal geometry. Application to repressor–operator association and membrane bound enzymes. Biophys Chem 2:255–263

Roberts RJ, Belfort M, Bestor T, Bhagwat AS, Bickle TA, Bitinaite J, Blumenthal RM, Degtyarev S, Dryden DT, Dybvig K, Firman K, Gromova ES, Gumport RI, Halford SE, Hattman S, Heitman J, Hornby DP, Janulaitis A, Jeltsch A, Josephsen J, Kiss A, Klaenhammer TR, Kobayashi I, Kong H, Kruger DH, Lacks S, Marinus MG, Miyahara M, Morgan RD, Murray NE, Nagaraja V, Piekarowicz A, Pingoud A, Raleigh E, Rao DN, Reich N, Repin VE, Selker EU, Shaw PC, Stein DC, Stoddard BL, Szybalski W, Trautner TA, Van Etten JL, Vitor JM, Wilson GG, Xu SY (2003) A nomenclature for restriction enzymes, DNA methyltransferases, homing endonucleases and their genes. Nucleic Acids Res 31:1805–1812

Robinson CR, Sligar SG (1998) Changes in solvation during DNA binding and cleavage are critical to altered specificity of the EcoRI endonuclease. Proc Natl Acad Sci USA 95:2186–2191

Schulze C, Jeltsch A, Franke I, Urbanke C, Pingoud A (1998) Crosslinking the EcoRV restriction endonuclease across the DNA-binding site reveals transient intermediates and conformational changes of the enzyme during DNA binding and catalytic turnover. EMBO J 17:6757–6766

Sidorova NY, Rau DC (1996) Differences in water release for the binding of EcoRI to specific and nonspecific DNA sequences. Proc Natl Acad Sci USA 93:12272–12277

Stanford NP, Szczelkun MD, Marko JF, Halford SE (2000) One- and three-dimensional pathways for proteins to reach specific DNA sites. EMBO J 19:6546–6557

Taylor JD, Badcoe IG, Clarke AR, Halford SE (1991) EcoRV restriction endonuclease binds all DNA sequences with equal affinity. Biochemistry 30:8743–8753

Terry BJ, Jack WE, Modrich P (1985) Facilitated diffusion during catalysis by EcoRI endonuclease: nonspecific interactions in EcoRI catalysis. J Biol Chem 260:13130–13137

Thielking V, Alves J, Fliess A, Maass G, Pingoud A (1990) Accuracy of the EcoRI restriction endonuclease: binding and cleavage studies with oligodeoxynucleotide substrates containing degenerate recognition sequences. Biochemistry 29:4682–4691

Thielking V, Selent U, Kohler E, Landgraf Z, Wolfes H, Alves J, Pingoud A (1992) Mg^{2+} confers DNA binding specificity to the EcoRV restriction endonuclease. Biochemistry 31:3727–3732

Viadiu H, Aggarwal AK (2000) Structure of BamHI bound to nonspecific DNA: a model for DNA sliding. Mol Cell 5:889–895 von Hippel PH, Berg OG (1989) Facilitated target location in biological systems. J Biol Chem 264:675–8

Winkler FK, Banner DW, Oefner C, Tsernoglou D, Brown RS, Heathman SP, Bryan RK, Martin PD, Petratos K, Wilson KS (1993) The crystal structure of EcoRV endonuclease and of its complexes with cognate and non-cognate DNA fragments. EMBO J 12:1781–1795

Winter RB, von Hippel PH (1981) Diffusion-driven mechanisms of protein translocation on nucleic acids, vol 2. The *Escherichia coli* repressor–operator interaction: equilibrium measurements. Biochemistry 20:6948–6960

Winter RB, Berg OG, von Hippel PH (1981) Diffusion-driven mechanisms of protein translocation on nucleic acids, vol 3. The *Escherichia coli* lac repressor–operator interaction: kinetic measurements and conclusions. Biochemistry 20:6961–6977

Wright DJ, Jack WE, Modrich P (1999) The kinetic mechanism of EcoRI endonuclease. J Biol Chem 274:31896–31902

The Type I and III Restriction Endonucleases: Structural Elements in Molecular Motors that Process DNA

S.E. McClelland, M.D. Szczelkun

The Type I and III restriction endonucleases are large, multimeric protein complexes with four enzyme activities; DNA methyltransferase, DNA endonuclease, ATPase and DNA translocase. It has been demonstrated that ATP-dependent protein motion along DNA is necessary for endonuclease activity. Studies have shown that Type I enzymes remain bound to their recognition sites whilst simultaneously translocating adjacent non-specific dsDNA past a stationary complex. This occurs bi-directionally so that two DNA loops are extruded. An equivalent unidirectional mechanism has been suggested for the Type III enzymes. DNA cleavage generally results when the enzymes stall against another restriction enzyme complex. Both the HsdR subunits of the Type I enzymes and the Res subunits of the Type III enzymes carry amino acid motifs characteristic of superfamily 2 helicases. In this review, the structural and mechanistic implications of this relationship are discussed and models suggested for how the ATP-dependent restriction enzymes might couple chemical energy to mechanical motion on DNA.

1 Energy-Dependent DNA Processing

The cellular control of complex genomes requires a fair degree of molecular gymnastics from large consortiums of proteins. In this regard, enzymatic dsDNA cleavage may seem a rather straightforward affair, involving the recognition of a specific sequence and the alignment of a pair of active sites proximal to a pair of scissile phosphodiester bonds. Glancing through commercial catalogues at lists of restriction enzymes may certainly give this impression. However, it is becoming increasingly clear that even here what may seem like simple DNA recognition and cleavage may in fact involve more

S.E. McClelland, M.D. Szczelkun
DNA-Protein Interactions Group, Department of Biochemistry, School of Medical Sciences, University of Bristol, Bristol, BS8 1TD, UK

Nucleic Acids and Molecular Biology, Vol. 14
Alfred Pingoud (Ed.)
Restriction Endonucleases
© Springer-Verlag Berlin Heidelberg 2004

complex molecular communication events such as DNA looping (e.g., see A.J. Welsh et al., this Vol.). The NTP-dependent restriction endonucleases, which comprise the modification-dependent restriction (MMR), Type I and Type III systems, are less well known than the Type II enzymes principally because they have not been exploited commercially. However, these enzymes are widely distributed in nature with almost half of the current sequenced microorganism genomes containing one or more putative NTP-dependent system (Roberts et al. 2003). Their dependence on an external energy source is key to their mechanism – communication between sequence-specific DNA recognition and non-specific DNA cleavage requires one-dimensional DNA translocation (reviewed comprehensively in Murray 2000; Rao et al. 2000; Szczelkun 2000; Dryden et al. 2001; Bourniquel and Bickle 2002). Nearly every genetic processes including replication, recombination, transcription and repair relies on enzymes which consume NTPs. In doing so, these proteins convert chemical energy into mechanical energy and hence they have come to be considered as "molecular motor proteins" (West 1996). The work done could be the unwinding of a DNA duplex (Maluf et al. 2003), backtracking of a disrupted replication fork (Mahdi et al. 2003), remodelling of chromatin (Whitehouse et al. 2003), or ousting of a stalled RNA polymerase (Park et al. 2002). In a great many cases these proteins share common amino acid sequences and folds, such as helicase or AAA+ motifs, that are also found in the NTP-dependent restriction enzymes. Hence, a relatively straightforward process – DNA cleavage – is, in many cases, being undertaken by complex molecular machines.

Much of our understanding of restriction-modification enzymes arose from pioneering genetic work carried out on NTP-dependent systems (e.g., Luria and Human 1952; Bertani and Weigle 1953; Arber and Dussoix 1962). This culminated in the purification of the first restriction endonuclease, the Type I enzyme EcoKI (Meselson and Yuan 1968). With the subsequent purification of Type III enzymes (Haberman 1974) there has followed over 35 years of elegant work on these systems. A number of recent reviews have covered the history, biology and mechanism of these enzymes in detail (Murray 2000; Rao et al. 2000; Szczelkun 2000; Dryden et al. 2001; Bourniquel and Bickle 2002). Nonetheless, a great many questions remain unanswered, principally in the details of how the chemical energy from ATP is converted into mechanical work during DNA processing. Given the depth of coverage in the earlier reviews, the aim of this article is to focus on the structural features of the "motor" domains from the Type I and III restriction endonucleases to help elucidate mechanistic features of these enzymes. We will also discuss how this information can be informed by, and inform, our understanding of helicases, AAA+ ATPases and related motor enzymes.

2 Motor Enzyme Architecture of the ATP-Dependent Restriction Endonucleases

For all NTP-dependent restriction endonucleases, DNA cleavage is carried out by a large macromolecular complex built from multiple copies of specific subunits encoded by genetically-linked genes. Nonetheless, once assembled these complexes can undertake their reactions without the necessity for other protein factors (it should be noted, however, that other proteins may play a role in vivo – for instance, topoisomerases may be important in relieving topological strain induced by DNA translocation). The ATP binding and hydrolysis functions are carried out in each case by one particular subunit: the HsdR subunit in Type I systems and the Res subunit in Type III systems. The catalytic centre that utilises ATP is constructed from a series of discontinuous amino acid motifs that are common to DNA helicases and to related AAA+ ATPases (Gorbalenya and Koonin 1991; Murray et al. 1993; Titheradge et al. 1996; Caruthers and McKay 2002). Because of their self-contained activity and relative simplicity, the Type I and III restriction endonucleases are attractive test systems with which to explore helicase mechanism.

2.1 Motor Enzyme Motifs in the Type I and III Restriction Endonucleases

A wide variety of proteins from bacteria to man can be classified on the basis of characteristic amino acid motifs into one of four helicase superfamilies – SF1, SF2, SF3 or SF4 (Hall and Matson 1999). Whilst some members are undoubtedly *bona fide* helicases (i.e. many SF1 enzymes unwind dsDNA into the corresponding ssDNA), there is mounting evidence that other members, particularly those in SF2, do not *directly* unwind DNA (Singleton and Wigley 2002). Instead, these enzymes couple chemical energy from NTP hydrolysis to other mechanical events such as processive dsDNA translocation, remodelling of protein-DNA complexes such as chromatin or stalled polymerases, etc. The HsdR and Res subunits fall into this SF2 class (Gorbalenya and Koonin 1991; Murray et al. 1993; Titheradge et al. 1996). With ongoing genome sequencing programmes, putative Open Reading Frames (ORFs) have been identified in a broad range of microorganisms which may encode HsdR or Res subunits as part of restriction endonuclease operons (see rebase.neb.com; Roberts et al. 2003). In combination with characterised ORFs, we undertook an alignment of 143 HsdR sequences and of 39 Res sequences. Using ClustalW (Higgins et al. 1994), we re-checked the consensus sequences for the previously identified helicase motifs and identified a number of new motifs which may be involved in ATP binding and hydrolysis and/or DNA translocation (Table 1).

Table 1. Helicase motifs in HsdR and Res proteins. Amino acid sequence alignments were carried out using ClustalW (Higgins et al. 1994) at the European Bioinformatics Institute (http://www.ebi.ac.uk/clustalw/index.html). HsdR and Res Protein sequences were obtained from REBASE (Roberts et al. 2003). 143 Type I HsdR sequences were aligned, of which 121 were translations of uncharacterised ORFs (this includes some duplication between strains). For the Type IIIA enzymes (see below), 21 Res Sequences were aligned, of which 16 were translations of uncharacterised ORFs. For the Type IIIB enzymes (see below), 18 Res sequences were aligned, of which 15 were translations of uncharacterised ORFs. In all cases, sequences were discarded which had <20 % identity, or which were truncated, across the helicase region. The name and consensus sequences of the SF2 helicase motifs (Hall and Matson 1999) are shown in the left-hand column. Regions which are likely to be characteristic of the restriction endonuclease family alone are indicated in italics. The order of the motifs are as shown; note, however, that the relative spacing will vary. For the consensus sequences, black shaded abbreviations indicates >80 % identity, uppercase abbreviations indicate >60 % identity, a "+" indicates a hydrophobic residue with >80 % identity, an "o" indicates a hydrophilic residue with >80 % identity, and an "x" represents a variable residue. Underlined amino acids are those residues mutated and shown to affect ATPase, endonuclease and translocation activities (see main text). [**Type IIIA Res:** EcoPlI (Humbelin et al. 1988), PhaBI (Ryan and Lo 1999), StyLTI (Dartois et al. 1993), BceSI (Hegna et al. 2001), HpyAXI (Lin et al. 2001), *Bacteroides fragilis* ORFC143 and ORFC196, *Burkholderia pseudomallei* K ORFC742, *Chlorobium tepidum* T ORF908, *Clostridium thermocellum* ORFS4, *E. coli* CFT073 ORF5372, *Fusobacterium nucleatum* Knorr ORF416, *Helicobacter pylori* 26695 ORF1522 and J99 ORF1411, *Magnetococcus* species MC-1 ORFC186, *Neisseria gonorrhoeae* FA1090 AX, *Neisseria meningitidis* Z2491 ORF1467 and B strain MC58 ORF1261, *Pasteurella multocida* Pm70 ORF698, *Salmonella typhimurium* CT18 ORF388 and LT2 ORF357. **Type IIIB Res:** LlaFI (Su et al. 1999), PstII (G.G.Wilson and M.D.S, unpublished observations), Hpy1061 (De Vries et al. 2002), *Chlorobium tepidum* TLS ORF1729, *Ferroplasma acidarmanus* ORFC158, *Geobacter metallireducens* ORFC20, *Helicobacter pylori* 26695 ORF1370 and J99 ORF1296, *Mycoplasma pulmonis* UAB CTIP ORF3960, *Nitrosomonas europeaea* ORFC219, *Neisseria gonorrhoeae* FA1090 ORFC707, *Neisseria meningitidis* Z2491 ORF1590, *Neisseria meningitidis* B strain MC58 ORF1375, *Rhodospirillum rubrum* MORFS3P, *Thermosynechococcus elongatus* BP-1 ORF1481, *Thermoplasma volcanium* ORF1464 and ORF1476, *Xanthomonas campestris* C ORF1068]

Q-Tip, α_Q Helix		Recognition and stabilisation of the base in ATP (*e.g.*, Theis et al. 1999)
Type IIIA	LxoQxxA+oA+xx+FoG	
Type IIIB	xQooA+oo+	
Type I	RxxQxxA	

Motif	Type	Sequence	Description
Motif I (Walker A) GxGK(T/S)	Type IIIA	+D+oMETGTGKTYxXxoTxFELH	Highly conserved sequence throughout helicases and related NTPases. Interaction with phosphates of MgATP/MgADP and coordination of the Mg²⁺ ion.
	Type IIIB	+xFoMAVGxGKTx+MAx+I	
	Type I	+++xHooGSGKSTxT	
Motif Ia +++xPoo	Type IIIA	GxxKF++VVPSxAIKEG(x)₉₋₁₁ EHF	Possibly involved in contacting DNA? For the type IIIA group, Motifs I and Ia are contiguous. This sequence is weakly conserved within the IIIB group.
	Type IIIB	oF+++ooxx	
	Type I	+++xDRxoLooQ	
Motif II (Walker B) +++DExH	Type IIIA	E+ox+xxxRP+++LI+DEPHO (Q)	Highly conserved sequence throughout helicases. Coordination of Mg²⁺ ion and ATP catalysis. Interactions with Motif III?
	Type IIIB	o+V++xDEAIH	
	Type I	xx++++DEAHR (C)	
Motif III +x+(T/S)(A/G)(T/G)	Type IIIA	ExxxLRYGATF (S H)	Role of Motif III in SF2 enzymes yet to be resolved fully.
	Type IIIB	LoFSATxx (T)	
	Type I	+GFTCTP	
region IIIa	Type IIIA	oLoA+DAxNoxLVRx+x+	Region IIIa is immediately adjacent to Motif III. Possibly involved in contacting DNA?
	Type IIIB	Lxo+xoxGxVK	

Motif / Region	Type	Sequence	Description
Region Y	Type IIIB	$\overset{\text{E}}{\text{E}}$xxxxxxYxxxxxxxD	Alternative to Motif IV in type I enzymes (Titheradge et al., 1996) and some of the type IIIB enzymes. Possibly involved in contacting DNA?
	Type I	EGox+oxVo+ooA+oD (81/121); $\overset{\text{E}}{\text{E}}(x)_{7-25}$ VoxooAIoD (40/121)	
Motif IV ++(F/Y)xxoxo	Type IIIA	$\overset{\text{T}}{\text{I}}$KxLSLFFIDo+xxYR	Possibly involved in contacting DNA? No significant alignment was observed in the type I or type IIIB enzymes (see Region Y).
Region IV_a	Type IIIA	VExGYFSoDoo	Between Motifs IV and V but not observed in the B group; may represent a non-essential sequence?
Region V_a	Type IIIA	ILxoKEKLxS+	Immediately adjacent to motif V. Possibly involved in contacting DNA?
	Type IIIB	xxxooooLxox	
Motif V +x(T/S)xxxxxG+o+xo+	Type IIIA	PxRFIFSooALxREGWDNPNVFxxICoL	Possibly involved in contacting DNA and ATP? These sequences are strongly conserved amongst the type I and III enzymes but do not clearly fit into a pattern with other SF2 enzymes.
	Type IIIB	P+R+I+oVooIoEGWDVxN+xoIVxL	
	Type I	xxx++IVxox+LTGFD+Pxxoxxxx+DK	
Motif VI (arginine finger) QxxGRxxR	Type IIIA	KxQEVGRGLRLxxVNooGoR	Highly conserved sequence throughout helicases and related NTPases; implicated in stabilisation of the transition state during ATP hydrolysis.
	Type IIIB	xoQ++GRG+R	
	Type I	+xQx+xRxNR	
Region VI_a	Type IIIA	F++ooLQoEI	An extended region of homology beyond Motif VI which may represent an additional DNA contact loop.

2.1.1 Gross Organisation of the Type I HsdR Subunits

The HsdR subunits are typically 900–1050 aa in length and have theoretical molecular weights of 100–120 kDa. These values compare well with the characterised values for enzymes such as EcoKI, EcoR124I, EcoAI, etc. The subunit domain organisation was similar throughout (Fig. 1A). There are two regions of significant sequence homology; the N-terminal Region X characteristic of endonuclease activity (Davies et al. 1999a; Janscak et al. 1999a) and the centrally located collection of helicase motifs (Gorbalenya and Koonin 1991; Murray et al. 1993; Titheradge et al. 1996). Some variation in distance between helicase motifs Ia and III and between IV(Y) and V was observed (Fig. 1A), but this can be easily accommodated in the helicase fold (see Sect. 2.2; Fig. 2). A new, putative helicase motif, the Q-tip helix, was also identified N-terminal to Motif I (see Sect. 2.1.4). There was no significant sequence homology elsewhere. Moreover, the biggest variations in protein length were seen outside of

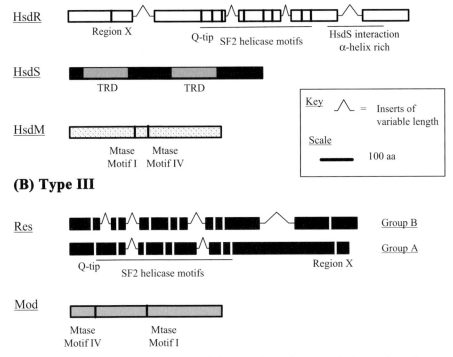

Fig. 1. Subunit domain architecture of the ATP-dependent restriction endonucleases. Proteins are represented as rectangles approximately to scale with inserts of variable length indicated (see key, *inset*). *Shaded rectangles* or *vertical lines* indicate regions with a defined mechanistic function (see main text) which may or may not show sequence conservation. A *horizontal line* indicates a collection of motifs which form a specific catalytic site or fold. **A** The HsdR, HsdS and HsdM proteins of the Type I enzymes. **B** The Res and Mod proteins of the Type IIIA and IIIB enzymes. *TRD* Target recognition domain

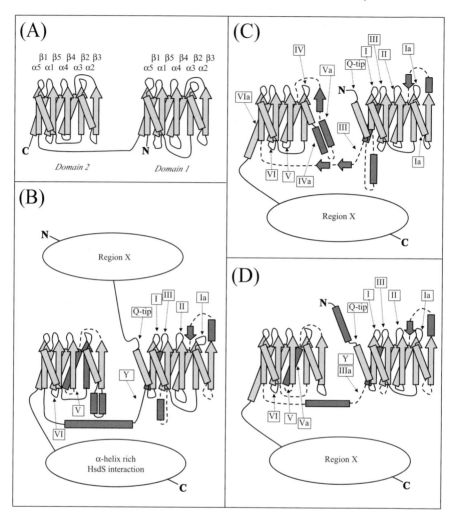

Fig. 2. Topological folds in the HsdR and Res subunits. Schematic diagrams showing α-helices and β-strands, with conserved structural features in *light grey*, variable structural features in *dark grey*, and variable inserts as *dashed lines*. Domains outside of the helicase motor are shown as ellipses. The locations of conserved motifs are indicated. Topologies of the HsdR and Res subunits were determined with PSIPRED (McGuffin et al. 2000) using at least four representative sequences in each case. **A** Representative arrangement of the fold and structural domains in a helicase (Caruthers and McKay 2002). The structural elements are named above and below the diagram. **B** Representative arrangement of the fold and structural domains in HsdR. **C** Representative arrangement of the fold and structural domains in Type IIIA Res. **D** Representative arrangement of the fold and structural domains in Type IIIB Res

the conserved regions (Fig. 1). Proteolytic digestion of the HsdR subunit of EcoKI identified that interactions with the HsdS subunit occur within the C-terminal domain (Davies et al. 1999a), but no sequence homology was observed in this region. However, secondary structure conservation was observed (Fig. 2; Sect. 2.2). Alignments of subsets of HsdR sequences from phylogentically related organisms did not reveal any additional motifs (data not shown). Some ORFs lack the N- or C-terminal regions which suggest that they would not be able to cleave DNA or assemble into active complexes. However, the majority of the sequences appears intact and could potentially participate in active Type I systems.

2.1.2 Gross Organisation of the Type III Res Subunits

The Res subunits are typically 950–1000 aa in length and have theoretical molecular weights of ~110 kDa. These values compare well with the characterised values for enzymes such as EcoPI1 and EcoP15I. As with the Type I subunits, there were two regions of significant sequence homology (Fig. 1B). In this case, however, the relative order was switched; the helicase motifs (Gorbalenya and Koonin 1991) were located in the N-terminal domain whilst Region X (Janscak et al. 2001) was located near the C-terminus. Initial attempts to align all 39 sequences were unsuccessful but convincing alignments were obtained by splitting the sequences into two subsets which we have termed IIIA and IIIB (Table 1, Fig. 1). The IIIA sequences were very closely related; in addition to the seven classical helicase motifs, four other regions of homology were identified which may play roles in the motor mechanism (Table 1; Sect. 2.1.5). In comparison, the IIIB sequences were less well conserved and with greater variability in the spacing of the motifs. Intriguingly, the IIIB sequences have a greater sequence identity across the helicase region with the HsdR sequences than with the IIIA Res sequences – in particular, Region Y (Tiheradge et al. 1996; Davies et al. 1998, 1999b) appears conserved in some Type IIIB enzymes (Table 1). A structural characterisation also suggests that Type I and IIIB may be related (Sect. 2.2). However, representatives of both A and B groups have DNA cleavage activity typical of Type III enzymes rather than Type I enzymes (L.J. Peakman, G.G.Wilson and M.D.S., unpubl. data). The Q-tip helix motif was also identified in both groups, but was located further from Motif I than in the Type I enzymes (Sect. 2.1.4).

2.1.3 Core Helicase Motifs in ATP Binding and Catalysis

The crystal structures of a number of SF1 and SF2 DNA and RNA helicases have revealed that the superfamily motifs cluster into an NTP binding and catalysis pocket between two structurally-related domains (Singleton and Wigley 2002; Caruthers and McKay 2002). A similar structural fold is also seen in other superfamilies and in related AAA+ ATPases such as RuvB (Putnam et al. 2001; Caruthers and McKay 2002). The central core of the NTP pocket comprises Motifs I and II [the so-called Walker A and B boxes first identified in ATP synthase (Walker et al. 1982)], and Motif VI. These motifs coordinate the magnesium ion and phosphate groups of MgATP/MgADP. Accordingly, these motifs are the most strongly conserved in the HsdR and Res subunits (Table 1) and it is highly likely that these core regions function in a very similar manner to many other ATPases. In all helicase structures to-date (Singleton and Wigley 2002; Caruthers and McKay 2002): The amino group of the conserved lysine of Motif I [GxG$\underline{K}$(T/S) for SF2] interacts with the mononucleotide phosphates whilst the hydroxyl group of the conserved serine/threonine [GxGK($\underline{T/S}$) for SF2] coordinates the Mg^{2+} ion; the conserved aspartate of Motif II [$\underline{D}$ExH for SF2] also coordinates the Mg^{2+} ion whilst the conserved glutamate [D$\underline{E}$xH] is a catalytic residue; and, the first conserved arginine in Motif VI [Qxx$G\underline{R}$xxR for SF2] is implicated in binding the γ phosphate group (note however that in some helicase structures the side chain is not ideally placed to bind the phosphate; Theis et al. 1999; Caruthers et al. 2000; Putnam et al. 2001). Mutations at some of these key residues affect the coupling of ATP hydrolysis to DNA translocase and cleavage by the Type I enzymes (Table 1, Webb et al. 1996; Davies et al. 1998, 1999b) and the ATPase and nuclease activities of the Type III enzymes (Table 1, Saha and Rao 1997; Saha et al. 1998). A number of other contacts beyond Motifs II and VI may play additional structural roles. Residues immediately C-terminal to the DExx sequence of Motif II may be involved in forming salt bridges with other structural domains; e.g., with motif VI (Yao et al. 1997; Theis et al. 1999) and/or with Motif III (Theis et al. 1999; Putnham et al. 2001).

2.1.4 The "Q-Tip Helix" – A New Helicase Motif?

In aligning the HsdR subunits around Motif I, it was noted that a highly-conserved glutamine and small hydrophobic residue were present in 99 % of the sequences, located ~30 aa N-terminal to the first conserved G residue of Motif I (Table 1). A similar motif was also found in the Type IIIA and IIIB enzymes, albeit with more variable spacing to Motif I. In structures of the PcrA (Subramanya et al. 1996; Velankar et al. 1999), rep (Korolev et al. 1997), UvrB (Theis et al. 1999), and RecG (Singleton et al. 2001) helicases and the AAA+ helicase RuvB (Putnam et al. 2001), a Qxx+ motif is also present at the N-terminal end of an α-helix which packs against the recA fold in domain 1 (the helix, called here $α_Q$, leads directly into $α_1$ of Motif I; see Fig. 2). A similarly-located conserved Q-residue has been implicated in ATPase activity of DEAD-box RNA helicases (Tanner et al. 2003). In the UvrB and RecG structures the glutamine side chain points into the ATP binding pocket, packs against the adenine and forms hydrogen bond with the cyclic N7 and exocyclic N6 positions of the base (Theis et al. 1999; Singleton et al. 2001). In other ATPase structures the residue is located close to the ATP binding pocket but is not ideally oriented to contact the base. However, mutants at the glutamine residue in RuvB and the BLM helicase have been identified as deleterious to function (Iwasaki et al. 2000; Bahr et al. 1998). Applying secondary structure prediction algorithms to the Type I and III sequences in this region indicate that the conserved glutamine also resides at the end of a helix (Fig. 2). We have noted that this residue and helix are common to a broad range of SF1 and SF2 helicases and some AAA+ ATPases (Szczelkun, manuscript in preparation) and thus represents a new helicase motif – we suggest that the motif is named the 'Q-tip helix' as it shows both sequence and structural conservation. Further testing will be required to probe its importance to the Type I and III restriction endonucleases.

2.1.5 The DNA Binding Motifs – Family Specific Deviations

Many of the remaining helicase motifs are implicated in DNA/RNA interactions and in domain–domain interactions that both stabilise the motor structure and communicate conformational changes during ATP binding/hydrolysis (Caruthers and McKay 2002; Singleton and Wigley 2002). Analysis of the Res and HsdR subunits reveals that whilst some of these motifs are conserved, others are either weakly conserved or absent altogether (Table 1). In some cases, alternative conserved sequences can be identified that may undertake similar roles to the 'missing' motif.

Motif Ia: In most helicases, the amino acid backbone of motif Ia primarily makes contact to the sugar-phosphate backbone of the polynucleotide substrate (Caruthers and McKay 2002). The close packing of motifs Ia, II and III suggest that changes in the ATP/ADP binding can be communicated to the DNA via Ia and III. In UvrB, a salt bridge is also made to motif V (Theis et al. 1999). In the Type I and III enzymes, this region is poorly conserved. Since the DNA contacts may be made via amino acid backbone contacts, it follows that Motif Ia could accommodate variations in amino acid sequence. Strong sequence conservation *is* observed in the Type IIIA enzymes, but as Motifs I and Ia are contiguous this may simply reflect closely related sequences rather than a conserved motif *per se*. Conversely, this region is not well conserved in the Type IIIB group. Because Motif Ia is a less well-defined consensus sequence overall (+++xPoo, Hall and Matson 1999), it has not been explored as fully in structure-functions studies as some other motifs. However, mutations in this motif in the Type I enzyme EcoKI revealed its importance in coupling ATPase activity to DNA cleavage and translocation (Table 1; Davies et al. 1998, 1999b).

Motifs III and III_a. In all SF2 helicase structures to-date (Caruthers and McKay 2002), the serine/threonine residues in either position of the conserved (T/S)(A/G)(T/G) sequence of Motif III (Hall and Matson 1999) make hydrogen bond contacts to motif II. In the SF1 enzymes, the unrelated Motif III sequence also makes contacts to the phosphates of ATP/ADP and to the polynucleotide substrate (Korolev et al. 1997; Velankar et al. 1999) – the same contacts have yet to be implicated for the SF2 enzymes. Unlike the contacts made by motif Ia, Motif III of the SF1 enzymes makes hydrogen bond and stacking interactions with the DNA bases. Although the Type I and III enzymes carry helicase motifs, DNA unwinding activity has never been observed (Szczelkun 2000) and a growing number of SF2 enzymes have been designated dsDNA translocases as opposed to unwindases. Consequently, the DNA contacts made by HsdR and Res may differ from those seen in SF1 helicases. Nonetheless, the importance of Motif III has been proven in the Type I enzymes (Table 1; Davies et al. 1998, 1999b). Motif IIIa was identified in the Type IIIA and B groups (Table 1), immediately adjacent to Motif III (except in *Salmonella typhimurium* sequences – not shown). Given the possible role of Motif III, this conserved region could also be involved in contacting the DNA.

Motif IV, IV$_a$ and Region Y. In SF1 enzymes, Motif IV is implicated in mononucleotide binding (Hall and Matson 1999) whereas in SF2 enzymes a role in polynucleotide binding is suggested (Yao et al. 1997). However, this motif is not always identified unambiguously in SF2 enzymes. Moreover, the consensus is sufficiently accommodating that many alignments have inappropriately assigned functions to random sections of sequence. In fact, Motif IV from the SF2 helicase NS3 was shown to have a high sequence similarity to a region of the SF1 helicase Rep revealed to be in contact with DNA (Kim et al. 1998). Therefore, there may be considerable family-specific variation in how the polynucleotide substrate is contacted. No significant similarity to Motif IV was observed in alignments of the Type I or Type IIIB enzymes. For the Type I enzymes, an alternative region of similarity, Region Y (Titheradge et al. 1996), is the more likely candidate (Davies et al. 1998, 1999b). A subset of the Type I enzymes have a highly variable non-specific insert in this region (Table 1) – intriguingly, 12 of this subset are archaeal sequences. We also found evidence that at least some of the Type IIIB enzymes contain a Region Y-like sequence (Table 1). This putative motif is in a similar position to region III$_a$ so may not be relevant. However, Type IIIB sequences showed stronger sequence and structural conservation across the helicase domain with Type I sequences than with IIIA sequences. An additional region of homology in the IIIA sequences was also identified (Motif IV$_a$; Table 1). This sequence was located between motifs IV and V in a region of variable structure (Fig. 2).

Motif V and Region V$_a$. The sequence similarity in Motif V was strong across the Type I and III sequences (Table 1). Generally, this region is more divergent in sequence and length amongst the helicase families. Motif V has been implicated in forming a network of interactions which ligate MgATP/MgADP and which stabilise domain contacts (e.g., contacting Motif II) and in making contacts to the polynucleotide (Caruthers and McKay 2002). Mutations in this motif in EcoKI affect ATPase, nuclease and translocase activities (Table 1, Davies et al. 1998, 1999b). An additional region of strong homology N-terminal to Motif V (called Region V_a) was observed in the Type IIIA group (Table 1). A weakly-conserved lysine-rich region was also observed in some of the IIIB group. These sequences may simply be extensions of motif V in the Type III enzymes.

Family-Specific DNA Binding Loops. Outside of the helicase motifs identified by amino acid sequence homology are a number of other structural motifs that are involved in DNA binding: e.g., the QxxR motif of DEA(D/H) box RNA helicases (Caruthers et al. 2000), the TxGx ssDNA binding motif (Korolev et al. 1997; Velankar et al. 1999; Yao et al. 1999) and, the TRG motif of RecG (Mahdi et al. 2003) and Mfd (N. Savery, pers. comm.). These regions do not show the same conservation of sequence between families as seen in the classical motifs but instead relate to structural elements found in specific locations within the conserved topological fold (Sect. 2.2). It has been suggested that the dsDNA translocation mechanism of RecG and Mfd may be a prototype for other SF2 enzymes (Singleton et al. 2001; Mahdi et al. 2003). However, the TRG motif which is found C-terminal to Motif VI following a helix-turn-helix structure is not found in either the HsdR or Res subunits (R. Lloyd, personal communication). An additional region of homology, Motif VIa, was also found in the Type IIIA enzymes (Table 1). This sequence is located in α_5 of domain 2 (Fig. 2), but does not have the same relative location as the TRG in RecG. Aligning the sequence motifs onto a structural fold gives some indications of alternative regions that may be important (see Sect. 2.2, below), but it remains to be seen if these structurally-conserved features can be confirmed as critical parts of the restriction enzyme motors.

2.2 A Motor Enzyme Fold in the Type I and III Restriction Endonucleases

The growing number of motor enzyme structures available in the PDB have revealed that for helicases and many related ATPases, the core sequence motifs occupy common positions within a shared protein fold (Singleton and Wigley 2002; Caruthers and McKay 2000). In every case, an α-β domain is found that is similar in topology and structure to the RecA recombination protein (Story and Steitz 1992). In many SF1 and SF2 enzymes, a second domain is also present which is essentially a repeat of the first, but which lacks the classical ATP binding motifs. It is the combination of the two domains which arrays the helicase motifs into an ATP/ADP binding pocket (Fig. 2A). The arrangement of each domain is such that the amino acid sequence can make excursions into separate topological folds or domains without disrupting the RecA fold (Singleton and Wigley 2002; Caruthers and McKay 2002). The result is a core motor enzyme architecture onto which any other functionality can be 'bolted' – in the case of the Type I and III enzymes, domains involved in DNA cleavage and protein-protein contacts. This modular design makes the helicases highly adaptable.

Secondary structure predictions using PSIPRED (McGuffin et al. 2000) were undertaken on a selection of HsdR and Res sequences (Fig. 2). In combination with previous analyses (Davies et al. 1999a), it is clear that the core β-sheet components of the motor fold are largely present in the HsdR and Res subunits and that they are arranged into a two-domain structure. Based on the topology diagram for a representative helicase proposed by Caruthers and McKay (2002) (Fig. 2A), putative model architectures were constructed for the HsdR and Res subunits (Fig. 2B–D). The positions of the motor motifs and α-helices were matched according to the positions of the predicted β-sheets and by reference to the structure of the SF2 helicase RecG (Singleton et al. 2001).

For the Type I enzymes (Fig. 2B), the N-terminal third of the protein contains region X (Davies et al. 1999a; Janscak et al. 1999a; Fig. 1) but no conservation of structure was observed in this domain. The C-terminal third is proposed to interact with the HsdS subunit (Davies et al. 1999a) and in this case an α-helix rich structure was conserved which may be important in the protein-protein contacts. Within the central domains 1 and 2 of the representative helicase fold, a number of variable regions exist, some of which have distinct structural elements. α_Q (Sect. 2.1.4) was positioned according to the packing of similar structures in RecG (Singleton et al. 2001) and UvrB (Theis et al. 1999). The region between $\alpha 2$ and $\alpha 3$ is particularly changeable; the alignment as shown (Fig. 2B) puts the variable loop between $\beta 3$ and $\alpha 3$, but it is equally likely that this could be rearranged with the variable loop positioned between $\alpha 2$ and $\beta 3$. For some enzymes, such as EcoAI, this region is truncated which may downplay its importance (EcoAI is smaller overall and may be a more compact structure). Nonetheless, it is worth noting that the region between $\beta 3$ and $\alpha 3$ coincides with the TxGx motif implicated in RNA interactions by eIF4A (Caruthers et al. 2000). An additional helical domain may be present between $\alpha 4$ and $\beta 5$ (Fig. 2B). Region Y is located after Motif III but appears to have mixed structure from the predictions. However, following this motif is a conserved, α-helix rich region which we suggest forms the link between domains 1 and 2. The positions of the conserved motifs in domain 1 relative to the α-β fold are exactly as predicted for a SF2 helicase (Caruthers and McKay 2002).

Whilst the repeated α-β structure of HsdR domain 1 appears remarkably well conserved, there is more variability in domain 2 (Fig. 2B). Core structural components such as $\beta 2$ and $\alpha 3$ are absent in some cases and additional structures, mainly helical in nature, may exist between $\alpha 2$ and $\beta 3$ (Fig. 2B). It should be noted that domain 2 is less well conserved in other helicases; for example, α3 is absent in RecG such that β3 and β4 are linked by an extended loop carrying Motif V (Singleton et al. 2001). Nonetheless, Motifs V and VI are positioned in HsdR as expected (Caruthers and McKay 2002). In many helicases, the β2 and α3 structural elements position Motifs IV and/or QxxR proximal to the DNA. The variability seen here suggests that alternative parts of the structure may be used instead, possibly region Y located closer to domain 1.

For the Type III enzymes (Fig. 2C, D), the N-terminal halves of the Res subunits comprise domains 1 and 2 of the helicase fold whilst the C-terminal halves are less well conserved α-β structures containing region X. The arrangement of the Q-tip motif in the III enzymes is somewhat different to the Type I – a variable loop structure with conserved α-helical elements is inserted between $α_Q$ and β1 (Fig. 2B, C) such that the Q-tip might be part of a helix-turn-helix. We assumed that $α_Q$ still packs against the RecA fold as observed in other helicase structures. Beyond that difference, both Type III structures follow a very similar topology in domain 1 to that observed for the Type I enzymes, with a similar extended variable region present between β3 and α3. Thus, this region may play a conserved role in the restriction endonucleases. In contrast, the topology of domain 2 was distinct in each case. For the Type IIIA enzymes (Fig. 2C), an extended loop with some β-sheet structure connects domains 1 and 2. Following Motif IV, a variable α-helical region was observed which, in some cases, lacked β3. The additional motifs IVa and Va could be found in this region. Motif VIa was located in $α_5$ of domain 2. For the Type IIIB enzymes (Fig. 2D), the overall topology of domain 2 appears more similar to that of the Type I enzymes with the absence of α3 most notable. Domains 1 and 2 are linked by an α-helix rich stretch containing the putative Region Y/Motif IIIa in a similar manner to HsdR. As the region encompassing Motif IV is less well defined, it may be that region Y plays a similar role to the Type I enzymes.

Given the results of this exercise and results from similar studies on EcoKI (Davies et al. 1999a), it is highly likely that the Type I and III motor subunits share the same topological fold and arrangement of amino acid motifs as seen in many other helicases and related motor enzymes. We await a crystal structure of either HsdR or Res to confirm these predictions.

2.3 Macromolecular Assembly of the Type I and III Restriction Endonucleases

Neither the HsdR nor the Res subunits have appreciable DNA-binding, ATPase or motor activity in isolation. Instead, the subunits must assemble as part of a larger macromolecular complex (Murray 2000; Rao et al. 2000; Szczelkun 2000; Dryden et al. 2001; Bourniquel and Bickle 2002). Similarly, many helicase show relatively poor enzyme activity when separated from other protein cofactors (Delagoutte and von Hippel 2003). The Type I and III endonucleases must remain bound, or at least close to, their recognition sequences whilst also translocating expanding loops of non-specific DNA. For neither system has the architecture of the complex that undertakes this convoluted DNA manoeuvring been resolved.

For the Type I enzymes, a single HsdS subunit (Fig. 1A) is required for DNA recognition. This subunit contains two variable Target Recognition Domains (TRDs) within a conserved protein structure with two-fold rotational symmetry. Each TRD has a motif comprising two loops and a β-strand (Sturrock and Dryden 1997; O'Neill et al. 1998) which is likely to recognise, via the major groove, one half of the bipartite Type I recognition sequence. This subunit forms a scaffold for the addition of two HsdM subunits (Dryden et al. 1997) which contain motifs characteristic of adenine methyltransferases (Fig. 1A). There is some evidence that this interaction occurs through the C-terminal region of HsdM (Cooper and Dryden 1994).

Two HsdR subunits are then required for DNA cleavage and translocation activities although complexes with a single HsdR can translocate DNA in a non-processive manner (Dryden et al. 1997; Janscak et al. 1998; Firman and Szczelkun 2000). There has been a great deal of debate in the helicase field as to whether SF1 and SF2 enzymes operate as monomers or as dimers (see Maluf et al. 2003, for recent arguments in favour of dimer activity; see Nanduri et al. 2002, for recent arguments in favour of monomer activity). One way to assemble a Type I endonuclease is with each HsdR placed at opposite ends of the complex, attached to non-specific DNA (Fig. 3A). This allows for the bi-directional DNA translocation from both sides of the site. There is at least some evidence for this arrangement from DNA footprinting studies (Powell et al. 1998). In this case the monomers are likely to act independently of each other. However, translocation studies using complexes with a single HsdR suggest that when a subunit releases DNA during motion, it can bind and translocate the DNA on the *opposite* side of the complex (Firman and Szczelkun 2000). Therefore, an alternative proposition is that the HsdR subunits are cheek-by-jowl (Fig. 3A). In this case the non-specific DNA may need to wrap around the complex to contact the motors. DNA distortion during binding may assist in this process (Mernagh and Kneale 1996). This alternative arrangement could give several advantages: Firstly, the expanding loop would be pre-formed. This may assist in the initiation of motion when under-twisting needs to be accommodated in the loop (Szczelkun et al. 1996); Secondly, the HsdR subunits could directly communicate, which may explain why two subunits produce processive motion but one alone produces non-processive motion; and thirdly, cleavage of circular DNA with a single recognition site would be facilitated as the HsdR subunits could assist in producing a double strand break. In conflict with this model is recent evidence from translocation studies on circular DNA (S.E.McM. and M.D.S., unpubl. data) which suggests that DNA twisting accumulates at much less than one turn per 10 bp,

(A) Type I

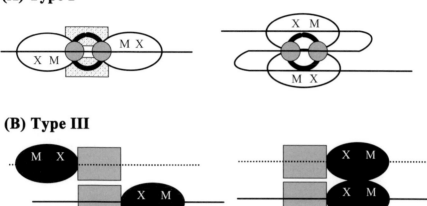

(B) Type III

Fig. 3. Subunit assembly of the Type I and III restriction endonucleases. **A** Assembly of the Type I enzymes. The HsdS subunit is illustrated as *two grey circles* (TRDs) connected by a *black circular line*. The HsdM subunits are illustrated as *rectangles* filled with dots. The HsdR subunits are illustrated as *ellipses* with the motor (*M*) and region X (*X*) domains indicated. DNA is indicated by a *black line*. Two arrangements are shown: *left* the HsdR subunits are independent and bound at opposite ends of the complex; *right* although still bound at opposite ends of the DNA, the HsdR subunits are in contact with each other and can cooperate. This requires a wrapping of the DNA around the complex. **B** Assembly of the Type III enzymes. The Mod subunits are illustrated as *two grey rectangles*. The Res subunits are illustrated as *black ellipses* with the motor (*M*) and region X (*X*) domains indicated. DNA is indicated by a *black line*. Putative DNA contacts are indicated by a *dotted line*. Two arrangements are shown: *left* the Res subunits are independent and bound at opposite ends of the complex. This might allow bi-directional translocation although this is less likely with Type III enzymes; *right* the HsdR subunits are cooperative. Either both Res subunits translocate one DNA, or the second Mod and Res may be involved in non-specific contacts to stabilise the complex

so loop topology may not be as limiting as previously thought. Furthermore, there is some evidence that a single HsdR can cut both strands so that cooperation is not required (Janscak et al. 1999b). To-date, there is no unambiguous experimental data that maps where the HsdR subunits start translocating DNA, although we do know that it cannot occur over a long distance via DNA looping (Szczelkun et al. 1996). Nor is there any clear evidence as to whether processive translocation occurs as part of an HsdR dimer or is undertaken by a single HsdR. What is clear is that processive DNA translocation by this class of helicase motor occurs as part of a larger complex and therefore the role of the complex *as a whole* must not be ignored.

DNA sequence recognition by the Type III enzymes is undertaken by a single subunit, Mod (Figs. 1B and 3B), which contains motifs characteristic of adenine methyltransferases (Humbelin et al. 1988). Surprisingly, this occurs as a Mod$_2$ complex (Ahmad et al. 1995). Unlike the Type I complexes where two target adenines are methylated, only one base is targeted for methylation in Type III sites. Thus this arrangement appears to result in a spare Mod subunit. Similarly, when Res binds to form the endonuclease, two subunits attach (Janscak et al. 2001). Two alternative arrangements are suggested in Fig. 3B. Given the stoichiometry of the complex, each Mod-Res unit could interact with a separate cognate DNA site. However, on a two-site substrate two Res$_2$Mod$_2$ complexes are required for maximum DNA cleavage (L.J. Peakman, unpubl. data). It is possible that non-specific DNA binding by the auxiliary Mod stabilises the complex. A similar recognition subunit redundancy has been observed with the Type IIS enzyme BspMI, which is a tetramer that binds two asymmetric sites (Gormley et al. 2002). The translocation model for Type III enzymes is based largely on that proposed for the Type I enzymes (Rao et al. 2000; Dryden et al. 2001; Bourniquel and Bickle 2002) and it has yet to be proven unambiguously whether translocation is uni- or bi-directional, or as efficient as Type I motion. It is possible that each Res subunit in a Res$_2$Mod$_2$ complex could translocate DNA. Alternatively, the Type III enzymes could be examples of dimeric motors with the two Res subunits cooperating to translocate DNA unidirectionally.

It has been long appreciated that the relative arrangement of the DNA motor and DNA cleavage domains in HsdR and Res determines whether the enzymes cut non-specific DNA at distant or proximal sites. If both motors share the motor polarity common to SF2 enzymes and move along dsDNA using one strand in a 3'-5' direction (e.g., Whitehouse et al. 2003), then for the Type I enzymes Region X would be ahead of the motor, near to distant DNA (Figs. 3A and 4), whilst for the Type III enzymes Region X would be behind the motor, near to the proximal DNA (Fig. 3B).

3 Future Directions

3.1 Coupling Chemical Energy to Mechanical Motion

The Type I and III restriction endonucleases contain motifs and structural elements typical of SF2 DNA helicases and, like many of the family (Delagoutte and von Hippel 2003), they are active as part of a larger macromolecular machine. Therefore, how does DNA translocation occur? What has become increasingly clear from helicase studies is that conserved motifs and structure (Table 1, Fig. 2) do not define helicase activity per se (Hall and Matson 1999; Singleton and Wigley 2002). Instead they define a mechanocoupling device that, upon binding and hydrolysing NTPs, can undergo domain motions and rearrangements which can then drive specific processes. For the helicases, the minimal motor fold (Fig. 2A) produces translocation of a polynucleotide: In some cases, extra domains are involved in unwinding dsDNA such that the polynucleotide transported is ssDNA (Korolev et al. 1997; Velankar et al. 1999); In other cases it is more likely that the polynucleotide transported is intact dsDNA and that the extra domains play alternative roles (Singleton et al. 2001; Mahdi et al. 2003). DNA translocation has been proven unambiguously for the Type I enzymes and there is convincing evidence for a similar overall mechanism for the Type III enzymes (Murray 2000; Rao et al. 2000; Szczelkun 2000; Dryden et al. 2001; Bourniquel and Bickle 2002). However, in neither case has any evidence for strand separation been obtained. Therefore, it is more likely that, in common with many other SF2 enzymes, the Type I and III endonucleases are dsDNA translocase in which the motor domains are coupled to an endonuclease/protein-protein interaction domain(s).

A model to demonstrate how DNA motion may occur is illustrated in Fig. 4. This scheme is based around recent evidence from other SF2 helicases, in particular the RecG helicase (Singleton et al. 2001). Despite the mechanistic complexities of the Type I and III enzymes, it is likely that this simple model will be the underlying basis for DNA translocation in both systems. The model shows an HsdR subunit attached via HsdS to a Type I complex (a similar mechanism is likely to occur for a Res subunit in a Type III complex.) The

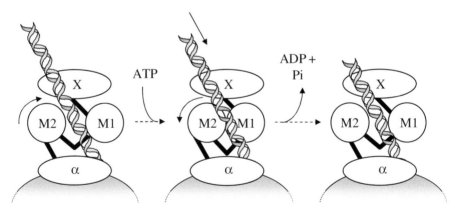

Fig. 4. Model for DNA translocation by a restriction endonuclease. A model is illustrated for Type I enzymes, but a similar model can be envisaged for Type III enzymes. Only one side of the bi-directional motor is shown for clarity. The HsdR subunit is illustrated as three domains with the connectivity as shown in Fig. 2: an N-terminal region X (*X*), a helicase domain (*M1* and *M2*), and a C-terminal protein-protein domain (α). HsdS is shown in *grey* (the rest of the complex is omitted). Upon binding ATP, the motor domains 1 and 2 (Fig. 2), come together. This causes conformation rearrangements which transports the DNA past HsdR into the expanding loop. The hydrolysis of ATP and release of products causes the motor domains to relax and reset to a new position further along the DNA. An alternative view might be that the same domain motions occur but that the DNA is passed back-and-forth between two closely arranged HsdR subunits (see Fig. 3). The direction of the DNA has not been confirmed for SF2 enzymes, and may be more horizontally-placed across M1 and M2

HsdR subunit contacts non-specific DNA adjacent to the recognition site. Upon binding of ATP and Mg^{2+} ions into the cleft between motor domains 1 and 2, the cleft closes and a conformational change occurs which transports the DNA from ahead of the complex into the expanding loop. Binding of S-adenosyl methionine, most likely by the methyltransferase domains, also produces conformational changes which stabilise the translocation state. To allow the system to repeat the mechanical transport of DNA, the motor domain must reset. This is achieved by hydrolysing the ATP, releasing ADP and inorganic phosphate and re-opening the cleft. To prevent the DNA simply sliding back, it must be held in contact at all times. There are a number of putative loops and motifs in the HsdR and Res structures which could undertake the role of a 'molecular ratchet'; for instance Motif Y in the Type I and IIIB enzymes (Table 1). However we should not discount the possibility that the HsdS, HsdM, or Mod subunits may also be involved in these processes. The model assumes a monomeric motor, but an alternative possibility is that the second HsdR or Res in the complex is also involved. In this case the DNA could be passed back-and-forth between the two motor subunits (the rolling model of helicase motion proposed by Wong and Lohman 1992). This is a less likely scenario given the current data on the Type I enzymes (Bourniquel and Bickle 2002), but nonetheless it needs to be considered.

The DNA in Fig. 4 is shown as an intact duplex, but contacts need not be made to both strands. Indeed, other SF2 helicases which translocate intact DNA do so using one strand with a defined polarity (e.g., Whitehouse et al. 2003). Furthermore, there is growing evidence that the SF2 enzymes utilise sugar-phosphate backbone contacts rather than the bases and there is evidence from DNA crosslinking studies using psoralen that this is likely to be the case for the Type I enzymes (Murray 2000). The step size, or amount of DNA transported per physical step, is not defined by our model. From studies of SF1 helicases, this could be anything from 1–2 bp (Dillingham et al. 2000) to 3–5 bp (Ali and Lohman 1997) or even 23 bp (Bianco and Kowalczykowski 2000). Consequently, the coupling efficiency – the number of nucleotides translocated per NTP consumed – is also not defined. This has been estimated for SF1 helicases as between 3–0.3 nucleotides/NTP (Roman and Kowalczykowski 1989; Dillingham et al. 2000; Kim et al. 2002). One complication of ascertaining the latter has been that the Type I enzymes have at least two ATP hydrolysis states; one during translocation and another that occurs after collision and DNA cleavage. The size of the step and the coupling efficiency will, of course, have important implications for the mechanism in terms of how a domain attaches to a *forward* section of DNA and draws it past the enzyme. Nevertheless, the general features of the model in Fig. 4 are likely to remain.

In our Type I model, the domain carrying Region X is sited in front of the motor such that, following multiple steps, DNA cleavage will occur at a distant site (Fig. 4). DNA cleavage during translocation may be prevented by: (1) Region X being unable to catalyse a phosphoryl transfer reaction on a 'moving' substrate (the DNA translocation rate being several orders of magnitude faster than the cleavage rate); (2) Region X being inactive until a second HsdR from a separate complex comes into contact after collision between two translocating enzymes; or, (3) Region X being disconnected from the DNA (unlike the model in Fig. 4) until collision between two translocating enzymes brings it back into contact. In contrast, the Res subunits lack an N-terminal endonuclease domain. Instead Region X is located in the C-terminal domain which also makes protein-protein contacts with the rest of the complex and is thus behind the motor. In this situation, DNA cleavage would occur close to the recognition site. Similar mechanisms may apply to prevent promiscuous DNA cleavage by the Res subunits during translocation.

3.2 Tools for Nanotechnology Rather than Biotechnology?

As illuminated elsewhere in this volume, the molecular biology revolution was built on the precision and reliability of the Type II restriction enzymes. In turn their usefulness as laboratory tools has driven the search for new activities. At first sight the NTP-dependent enzymes may seem like poor relations – their complex activities and relatively imprecise DNA cleavage loci have found only limited use in the laboratory (Moncke-Buchner et al. 2002). On the other hand, the complexity of these reactions may find a wider role in the burgeoning field of nanotechnology. The goal of nanotechnology is to design microscopic apparatus that can carry out specific tasks. A framework for such technology is provided by biological materials such as proteins and DNA and the ATP-dependent restriction enzymes may be ideal candidates for exploitation. Given the modular nature of the motor domains illustrated in Figs. 2 and 4, one can readily see how the motors could be adapted; for instance, by replacing Region X, HsdR subunits could carry an alternative enzyme activity or an attached cargo. The Type I enzymes can translocate tens of thousands of bp without dissociation at speeds in excess of 450 bp/s (Szczelkun et al. 1997; Firman and Szczelkun 2000). This corresponds to a motor which can travel >3.4 μm – a vast distance in nanomachine terms – in just over 20 s. They can also translocate DNA of any topology without loss of speed or processivity (unpubl. data) and can displace molecular roadblocks such as DNA-binding proteins (Dreier et al. 1996). As we learn more about the biology of the ATP-dependent restriction endonucleases, in particular their motor mechanisms, we will also be well-placed to harness their potential as miniaturised machines.

Acknowledgements. We thank Elaine Davies, David Dryden, Steve Halford, Pavel Jan-scak, Luke Peakman and Nigel Savery for discussions and technical assistance. We apologise to those colleagues whose work we have not had the space to cite fully. Work in our laboratory is supported by the Wellcome Trust and by the BBSRC. MDS is a Wellcome Trust Senior Research Fellow in Basic Biomedical Science.

References

Ahmad I, Krishnamurthy V, Rao DN (1995) DNA recognition by the EcoP15I and EcoPI modification methyltransferases. Gene 157:143–147

Ali JA, Lohman TM (1997) Kinetic measurement of the step size of DNA unwinding by *Escherichia coli* UvrD helicase. Science 275:377–380

Arber W, Dussoix D (1962) Host specificity of DNA produced by *Escherichia coli*. I. Host controlled modification of bacteriophage lambda. J Mol Biol 5:18–36

Bahr A, De Graeve F, Kedinger C, Chatton B (1998) Point mutations causing Bloom's syndrome abolish ATPase and DNA helicase activities of the BLM protein. Oncogene 17:2565–2571

Bertani G, Weigle, JJ (1953) Host-controlled variation in bacterial viruses. J Bacteriol 65:113–121

Bianco PR, Kowalczykowski SC (2000) Translocation step size and mechanism of the RecBC DNA helicase. Nature 405:368–372

Bourniquel AA, Bickle TA (2002) Complex restriction enzymes: NTP-driven molecular motors. Biochimie 84:1047–1059

Caruthers JM, McKay DB (2002) Helicase structure and mechanism. Curr Opin Struct Biol 12:123–33

Caruthers JM, Johnson ER, McKay DB (2000) Crystal structure of yeast initiation factor 4A, a DEAD-box RNA helicase. Proc Natl Acad Sci USA 97:13080–13085

Cooper LP, Dryden DT (1994) The domains of a Type I DNA methyltransferase. Interactions and role in recognition of DNA methylation. J Mol Biol 236:1011–1021

Dartois V, De Backer O, Colson C (1993) Sequence of the *Salmonella typhimurium* StyLT1 restriction-modification genes: homologies with EcoP1 and EcoP15 type-III R-M systems and presence of helicase domains. Gene 127:105–110

Davies GP, Powell LM, Webb JL, Cooper LP, Murray NE (1998) EcoKI with an amino acid substitution in any one of seven DEAD-box motifs has impaired ATPase and endonuclease activities. Nucleic Acids Res 26:4828–4836

Davies GP, Martin I, Sturrock SS, Cronshaw A, Murray NE, Dryden DT (1999a) On the structure and operation of Type I DNA restriction enzymes. J Mol Biol 290:565–579

Davies GP, Kemp P, Molineux IJ, Murray NE (1999b) The DNA translocation and ATPase activities of restriction-deficient mutants of EcoKI. J Mol Biol 292:787–796

Delagoutte E, von Hippel PH. (2003) Helicase mechanisms and the coupling of helicases within macromolecular machines. Part II: Integration of helicases into cellular processes. Q Rev Biophys 36:1–69

de Vries N, Duinsbergen D, Kuipers EJ, Pot RG, Wiesenekker P, Penn CW, van Vliet AH, Vandenbroucke-Grauls CM, Kusters JG (2002) Transcriptional phase variation of a Type III restriction-modification system in *Helicobacter pylori*. J Bacteriol 184:6615–6623

Dila D, Sutherland E, Moran L, Slatko B, Raleigh EA (1990) Genetic and sequence organisation of the mcrBC locus of *Escherichia coli* K-12. J Bacteriol 172:4888–4900

Dillingham MS, Wigley DB, Webb MR (2000) Demonstration of unidirectional single-stranded DNA translocation by PcrA helicase: measurement of step size and translocation speed. Biochemistry 39:205–212

Dreier J, MacWilliams MP, Bickle TA (1996) DNA cleavage by the Type IC restriction-modification enzyme EcoR124II. J Mol Biol 264:722–733

Dryden DT, Cooper LP, Thorpe PH, Byron O (1997) The in vitro assembly of the EcoKI Type I DNA restriction/modification enzyme and its in vivo implications. Biochemistry 5:1065–1076

Dryden DT, Murray NE, Rao DN (2001) Nucleoside triphosphate-dependent restriction enzymes. Nucleic Acids Res 29:3728–3741

Firman K, Szczelkun MD (2000) Measuring motion on DNA by the Type I restriction endonuclease EcoR124I using triplex displacement. EMBO J 19:2094–2102

Gorbalenya AE, Koonin EV (1991) Endonuclease (R) subunits of Type I and Type III restriction enzymes contain a helicase-like domain, FEBS Lett 291:277–281

Gormley NA, Hillberg AL, Halford SE (2002) The Type IIs restriction endonuclease BspMI is a tetramer that acts concertedly at two copies of an asymmetric DNA sequence. J Biol Chem 277:4034–4041

Haberman A (1974) The bacteriophage P1 restriction endonuclease. J Mol Biol 89:545–563

Hall MC, Matson SW (1999) Helicase motifs: the engine that powers DNA unwinding. Mol Microbiol 34: 867–877

Hegna IK, Bratland H, Kolsto AB (2001) BceS1, a new addition to the Type III restriction and modification family. FEMS Microbiol Lett 202:189–193

Higgins D, Thompson J, Gibson T, Thompson JD, Higgins DG, Gibson TJ (1994) CLUSTAL W: improving the sensitivity of progressive multiple sequence alignment through sequence weighting, position-specific gap penalties and weight matrix choice. Nucleic Acids Res 22:4673–4680

Humbelin M, Suri B, Rao DN, Hornby DP, Eberle H, Pripfl T, Kenel S, Bickle TA (1988) Type III DNA restriction and modification systems EcoP1 and EcoP15. Nucleotide sequence of the EcoP1 operon, the EcoP15 mod gene and some EcoP1 mod mutants. J Mol Biol 200:23–29

Iwasaki H, Han YW, Okamoto T, Ohnishi T, Yoshikawa M, Yamada K, Toh H, Daiyasu H, Ogura T, Shinagawa H (2000) Mutational analysis of the functional motifs of RuvB, an AAA+ class helicase and motor protein for holliday junction branch migration. Mol Microbiol 36:528–538

Janscak P, Bickle TA (2000) DNA supercoiling during ATP-dependent DNA translocation by the Type I restriction enzyme EcoAI. J Mol Biol 295:1089–1099

Janscak P, Dryden DT, Firman K (1998) Analysis of the subunit assembly of the typeIC restriction-modification enzyme EcoR124I. NucleicAcids Res 26:4439–4445

Janscak P, Sandmeier U, Bickle TA (1999a) Single amino acid substitutions in the HsdR subunit of the Type IB restriction enzyme EcoAI uncouple the DNA translocation and DNA cleavage activities of the enzyme. Nucleic Acids Res 27:2638–2643

Janscak P, MacWilliams MP, Sandmeier U, Nagaraja V, Bickle TA (1999b) DNA translocation blockage, a general mechanism of cleavage site selection by Type I restriction enzymes. EMBO J 18:2638–2647

Janscak P, Sandmeier U, Szczelkun MD, Bickle TA (2001) Subunit assembly and mode of DNA cleavage of the Type III restriction endonucleases EcoP1I and EcoP15I. J Mol Biol 306:417–431

Kim JL, Morgenstern KA, Griffith JP, Dwyer MD, Thomson JA, Murcko MA, Lin C, Caron PR (1998) Hepatitis C virus NS3 RNA helicase domain with a bound oligonucleotide: the crystal structure provides insights into the mode of unwinding. Structure 6:89–100

Kim DE, Narayan M, Patel SS (2002) T7 DNA helicase: a molecular motor that processively and unidirectionally translocates along single-stranded DNA. J Mol Biol 321:807–819

Korolev S, Hsieh J, Gauss GH, Lohman TM, Waksman G (1997) Major domain swivelling revealed by the crystal structures of complexes of *E. coli* Rep helicase bound to single-stranded DNA and ADP. Cell 90:635–647

Lin LF, Posfai J, Roberts RJ, Kong H (2001) Comparative genomics of the restriction-modification systems in *Helicobacter pylori*. Proc Natl Acad Sci USA 98:2740–2745

Luria SE, Human ML (1952) A nonhereditary, host-induced variation in bacterial viruses. J Bacteriol 64:557–569

Mahdi AA, Briggs GS, Sharples GJ, Wen Q, Lloyd RG (2003) A model for dsDNA translocation revealed by a structural motif common to RecG and Mfd proteins. EMBO J 22:724–734

Maluf NK, Fischer CJ, Lohman TM (2003) A dimer of *Escherichia coli* UvrD is the active form of the helicase in vitro. J Mol Biol 325:913–935

McGuffin LJ, Bryson K, Jones DT (2000) The PSIPRED protein structure prediction server. Bioinformatics 16:404–405

Mernagh DR, Kneale GG (1996) High resolution footprinting of a Type I methyltransferase reveals a large structural distortion within the DNA recognition site. Nucleic Acids Res 24:4853–4858

Meselson M, Yuan R (1968) DNA restriction enzyme from *E. coli*. Nature 217:1110–1114

Moncke-Buchner E, Reich S, Mucke M, Reuter M, Messer W, Wanker EE, Kruger DH (2002) Counting CAG repeats in the Huntington's disease gene by restriction endonuclease EcoP15I cleavage. Nucleic Acids Res 30:e83

Murray NE (2000) Type I restriction systems: sophisticated molecular machines (a legacy of Bertani and Weigle). Microbiol Mol Biol Rev 64:412–434

Murray NE, Daniel AS, Cowan, GM, Sharp PM (1993) Conservation of motifs within the unusually variable polypeptide sequences of Type I restriction enzymes. Mol Microbiol 9:133–143

Nanduri B, Byrd AK, Eoff RL, Tackett AJ, Raney KD (2002) Pre-steady-state DNA unwinding by bacteriophage T4 Dda helicase reveals a monomeric molecular motor. Proc Natl Acad Sci USA 99:14722–14727

O'Neill M, Dryden DT, Murray NE (1998) Localization of a protein-DNA interface by random mutagenesis. EMBO J 17:7118–7127

Park JS, Marr MT, Roberts JW (2002) *E. coli* Transcription repair coupling factor (Mfd protein) rescues arrested complexes by promoting forward translocation. Cell 109:757–767

Powell LM, Dryden DT, Murray NE (1998) Sequence-specific DNA binding by EcoKI, a Type IA DNA restriction enzyme. J Mol Biol 283:963–976

Putnam CD, Clancy SB, Tsuruta H, Gonzalez S, Wetmur JG, Tainer JA (2001) Structure and mechanism of the RuvB Holliday junction branch migration motor. J Mol Biol 311:297–310

Rao DN, Saha S, Krishnamurthy V (2000) ATP-dependent restriction enzymes. Prog Nucleic Acid Res Mol Biol 64:1–63

Roberts RJ, Vincze T, Posfai, J, Macelis D (2003) REBASE – restriction enzymes and methylases. Nucleic Acids Res 31:418–420

Roman LJ, Kowalczykowski SC (1989) Characterization of the adenosinetriphosphatase activity of the *Escherichia coli* RecBCD enzyme: relationship of ATP hydrolysis to the unwinding of duplex DNA. Biochemistry 28:2873–2881

Ryan KA, Lo RY (1999) Characterization of a CACAG pentanucleotide repeat in *Pasteurella haemolytica* and its possible role in modulation of a novel Type III restriction-modification system. Nucleic Acids Res 27:1505–11

Saha S, Rao DN (1997) Mutations in the Res subunit of the EcoPI restriction enzyme that affect ATP-dependent reactions. J Mol Biol 269:342–354

Saha S, Ahmad I, Reddy YV, Krishnamurthy V, Rao DN (1998) Functional analysis of conserved motifs in Type III restriction-modification enzymes. Biol Chem 379:511–517

Singleton MR, Wigley DB (2002) Modularity and specialization in superfamily 1 and 2 helicases. J Bacteriol 184:1819–1826

Singleton MR, Scaife S, Wigley DB (2001) Structural analysis of DNA replication fork reversal by RecG. Cell 107:79–89

Story RM, Steitz TA (1992) Structure of the recA protein-ADP complex. Nature 355:374–376

Sturrock SS, Dryden DT (1997) A prediction of the amino acids and structures involved in DNA recognition by Type I DNA restriction and modification enzymes. Nucleic Acids Res 25:3408–3414

Su P, Im H, Hsieh H, Kang'A S, Dunn NW (1999) LlaFI, a Type III restriction and modification system in *Lactococcus lactis*. Appl Environ Microbiol 65:686–693

Subramanya HS, Bird LE, Brannigan JA, Wigley DB (1996) Crystal structure of a DExx box DNA helicase. Nature 384:379–383

Szczelkun MD (2000) How do proteins move along DNA? Lessons from type-I and type-III restriction endonucleases. Essay Biochem 35:131–143

Szczelkun MD, Dillingham MS, Janscak P, Firman K, Halford SE (1996) Repercussions of DNA tracking by the Type IC restriction endonuclease EcoR124I on linear, circular and catenated substrates. EMBO J 15:6335–6347

Szczelkun MD, Janscak P, Firman K, Halford SE. (1997) Selection of non-specific DNA cleavage sites by the Type IC restriction endonuclease EcoR124I. J Mol Biol 271:112–123

Tanner NK, Cordin O, Banroques J, Doere M, Linder P (2003) The Q motif: a newly identified motif in DEAD box helicases may regulate ATP binding and hydrolysis. Mol Cell 11:127–138

Theis K, Chen PJ, Skorvaga M, Van Houten B, Kisker C (1999) Crystal structure of UvrB, a DNA helicase adapted for nucleotide excision repair. EMBO J 18:6899–6907

Titheradge AJB, Ternent D, Murrary NE (1996) A third family of allelic *hsd* genes in *Salmonella enterica*: sequence comparisons with related proteins identify regions implicated in restriction of DNA. Mol Microbiol 22:437–447

Velankar SS, Soultanas P, Dillingham MS, Subramanya HS, Wigley DB (1999) Crystal structures of complexes of PcrA DNA helicase with a DNA substrate indicate an inchworm mechanism. Cell 97:75–84

Walker JE, Saraste M, Runswick MJ, Gay NJ (1982) Distantly related sequences in the alpha- and beta-subunits of ATP synthase, myosin, kinases and other ATP-requiring enzymes and a common nucleotide binding fold. EMBO J 1:945–951

Webb JL, King G, Ternent D, Titheradge AJ, Murray NE (1996) Restriction by EcoKI is enhanced by co-operative interactions between target sequences and is dependent on DEAD box motifs. EMBO J 15:2003–2009

West SC (1996) DNA helicases: new breeds of translocating motors and molecular pumps. Cell 86:177–180

Whitehouse I, Stockdale C, Flaus A, Szczelkun MD, Owen-Hughes T (2003) Evidence for DNA translocation by the ISWI chromatin-remodeling enzyme. Mol Cell Biol 23:1935–1945

Wong I, Lohman TM (1992) Allosteric effects of nucleotide cofactors on *Escherichia coli* Rep helicase-DNA binding. Science 256:350–355

Yao N, Hesson T, Cable M, Hong Z, Kwong AD, Le HV, Weber PC (1997) Structure of the hepatitis C virus RNA helicase domain. Nat Struct Biol 4:463–467

The Integration of Recognition and Cleavage: X-Ray Structures of Pre-Transition State Complex, Post-Reactive Complex, and the DNA-Free Endonuclease

A. Grigorescu, M. Horvath, P.A. Wilkosz, K. Chandrasekhar, J.M. Rosenberg

1 Introduction

DNA has been selected as the biological information storage molecule for many reasons; one of which is its stability. The rate of spontaneous hydrolysis of DNA (at 24 °C and pH 7.4) was estimated at 5.7×10^{-14} s^{-1} (Bunton et al. 1960; Kumamoto et al. 1956; Serpersu et al. 1987); more recently, Radzicka and Wolfenden have estimated this value as 1.7×10^{-13} s^{-1}. (Radzicka and Wolfenden 1995). This corresponds to an estimated half-life of 130,000 years for a DNA phosphodiester bond in solution, placing DNA hydrolysis among the slowest of biochemical reactions in the absence of enzymes.

Consequently, the rate enhancements produced by nucleases are among the largest known; even the relatively slower restriction endonucleases achieve remarkable accelerations. EcoRI endonuclease can hydrolyze DNA at a rate of 0.34 s^{-1} (based on single turnover kinetics) (Lesser et al. 1990a) corresponding to an acceleration of 2×10^{12} over the uncatalyzed rate. This corresponds to a reduction of the activation energy by approximately 17 kcal/mol. How is this achieved?

The rate enhancement is especially noteworthy in view of the high specificity achieved by this restriction enzyme. It recognizes the inverted repeat, GAATTC, and the most common "error" occurs at sequences that match this site at five of the six base positions; these sites are known as EcoRI* sites for historical reasons. Under physiological conditions, the rate of single-strand cleavage (nicking) at EcoRI* sites is 10^7 slower than the cleavage rate at the "canonical" GAATTC; the rate of double-strand cleavage at EcoRI* sites is

A. Grigorescu, M. Horvath, P.A. Wilkosz, K. Chandrasekhar, J.M. Rosenberg
Department of Biological Sciences, University of Pittsburgh; Pittsburgh, Pennsylvania 15260, USA

Nucleic Acids and Molecular Biology, Vol. 14
Alfred Pingoud (Ed.)
Restriction Endonucleases
© Springer-Verlag Berlin Heidelberg 2004

undetectable under these conditions. The only other examples of macromolecules with this level of specificity are other restriction enzymes and the "proofreading" systems, such as DNA polymerase. How is the catalytic rate controlled so precisely to maintain the extraordinary specificity required of a restriction enzyme?

Here, we report of our structural analysis of three structures containing EcoRI endonuclease: (1) A high resolution (1.85 Å) pre-transition state EcoRI-DNA recognition complex (refined from the older 2.7 Å version; Choi 1994). (2) A post-reactive EcoRI-DNA complex that includes Mn^{2+} at 2.7 Å resolution. (3) The structure of the protein free of DNA (apo EcoRI endonuclease) that suggests an important order-disorder transition. The discussion here will focus on the results of these analyses, crystallographic methods and additional details will be reported elsewhere (summary refinement statistics are presented in Table 1).

Table 1. Refinement statistics
Summary statistics

Statistic	Post-reactive complex	Pre-transition state complex
Asymmetric unit	1 protein subunit	1 DNA strand
Resolution (Å)	8–2.5	8–1.87
Cutoff for structure factors	3 σ	2σ
Number of observed reflections (8–2.5 Å; 8–1.85)	7596	28,956
Number of possible reflections (8–2.5 Å; 8–1.85)	13,662	31125
R-factor/R_{free}	0.177	0.21/0.25
RMS δ for bond distances (Å)	0.008	0.003
RMS δ for bond angles (°)	1.653	1.026
Total number of bound solvent (water) molecules	42	156
Average B-factors ($Å^2$):		
All atoms (entire structure)	24.29	38
Protein:		
All protein atoms	25.3	39
Main chain atoms	24.0	37
Side chain atoms	25.9	42
DNA	16.9	34
Bound solvent (water)	29.4	43

2 The Pre-Transition State Complex

The "Results and Discussion" are organized such that the cleavage reaction follows an analysis of the detailed structural features of the pre-transition state complex. This will facilitate the ensuing discussion of how these features are integrated to achieve catalysis. We also discuss broader issues such as relationship to other restriction endonucleases and evolutionary significance.

2.1 General Features

One of the salient features of the new crystal structure is the enhancement in resolution, from 2.7 to 1.85 Å. Three factors contributed to the enhancement: (1) the crystals were grown in microgravity. (2) A long-standing technical problem was resolved; the crystals had proven difficult to cryoprotect requiring room temperature data collection using several different crystals. Successful crytoprotection enabled a complete data set to be collected from a single crystal. (3) This facilitated effective use of synchrotron data (which were collected on the F1 beamline of CHESS at 100 K). Figure 1 shows a representative section of the electron density.

The complex appears to be a single, somewhat irregular globular entity approximately 50 Å across. It is composed of a DNA duplex and two identical protein subunits related by a central axis of two-fold rotational symmetry. The

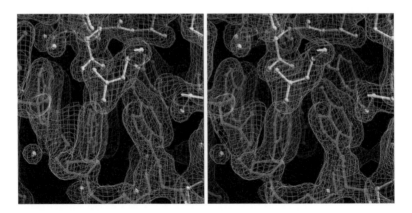

Fig. 1. Pre-transition state structure electron density map: This 2Fo-Fc electron density map shown in stereo is contoured at 2σ near the AATT tetramer ($A_{-2}A_{-1}T_1T_2$) where the DNA is *green* and the protein is *yellow*. The water is represented by *yellow spheres* engulfed in a mesh of density. These DNA moieties are surrounded by may of the key residues involved in recognition and cleavage. The protein shown here is part of the chain from Asn[141] to Arg[145]. The quality of density seen here is representative of that for the entire structure

DNA is distorted significantly near the symmetry axis into an altered structure referred to as the EcoRI kink. The protein has α/β architecture where each subunit consists of a central β sheet surrounded by α helices (see Kim et al. 1994) for a more complete description of the general aspects of the structure).

The DNA–protein interface is formed around the major groove and the neighboring sugar–phosphate backbones where the DNA has many interactions with amino acid side chains and main chain amide groups. A significant amount of bound solvent is located at the DNA–protein and at the inter-subunit protein–protein interfaces where it forms bridging hydrogen bonding interactions.

2.2 DNA Numbering Scheme

In the following discussion, we refer to DNA bases, which are numbered outwards from the center of the EcoRI site. This numbering scheme was chosen to facilitate comparison with cocrystals containing oligonucleotides of differing length and sequence. Nucleotides with the same number would then always be positioned the same with respect to the EcoRI site. Following this convention, the oligonucleotide used is labeled $T_{-7} C_{-6} G_{-5} C_{-4} G_{-3} A_{-2} A_{-1} T_1 T_2 C_3 G_4 C_5 G_6$.

2.3 Secondary Structure

The higher resolution resulted in a slight modification in the secondary structure assignments reported previously (Kim et al. 1994). The principal change is that the extended chain motif now contains an isolated strand of β sheet (β_r) described below. This is actually a very subtle change since the extended chain motif was always in the β region of the Ramachandran plot; it became more regular during the final stages of refinement.

The secondary structure is indicated schematically in Fig. 2; the core of each subunit is a largely parallel five-stranded β sheet. However, the second strand (β_2) is antiparallel to the other four strands of the sheet. Consequently, β_1 through β_3 forms a β meander motif that contains many of the amino acid residues active in the cleavage reaction, including Asp[91], Glu[111] and Lys[113]. Similarly, β_3 through β_5 is folded like one-half of the well known nucleotide binding fold. The crossover helices (α_4 and α_5) of this "half Rossmann motif" form the core of both the DNA–protein and protein–protein (inter-subunit) interfaces. Specifically, we refer to α_4 and α_5 as the inner and outer recognition helices, respectively, because of their recognition role.

The inner recognition helix interacts with the DNA backbone as well as the bases. Backbone interactions include hydrogen bonds between the phosphate

Fig. 2. Secondary structure: the major elements of secondary structure are shown; α helices are shown as *rods* or *coils*, while β sheets are represented as *arrows*. **A** One subunit of the protein and the DNA. The locations of the secondary structure elements are indicated. **B** A stereo view of one subunit of the protein and of the DNA

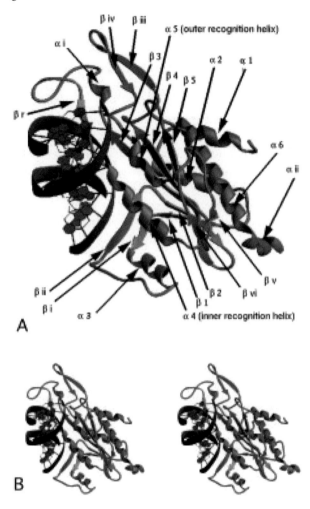

at G_{-3} and the side chains of both Lys[148] and Asn[149]. Arg[145] has a "dual" role, hydrogen bonding to both the N_7 of A_{-1} and the scissile phosphate at A_{-2}. Similarly, Arg[203] on the outer recognition helix participates indirectly in the recognition of G_{-3} via a bound water molecule; it also hydrogen bonds directly to the phosphate of the nucleotide immediately adjacent to the recognition site, C_{-4}.

There are several notable protrusions from the domain core, including the β bridge, the inner and outer arms and a β hairpin (Lys[221] through Leu[232]) that projects out the back of the complex. The β bridge is part of the subunit interface and the arms hold the DNA against the recognition-cleavage site. Part of the inner arm participates in the sequence-specific contacts to the bases; this is the extended chain motif, Met[137] to Ala[142]. Together, the recognition helices and the extended chain motif position the amino acid side chains that form the interaction site complementary to the DNA bases; they are Met[137], Asn[141], Ala[143], Glu[144], Arg[145], Arg[200], and Arg[203]. There are also protein–pyrimidine hydrogen bonds to the backbone amides at Ala[138] and Ala[142]

2.4 The EcoRI Kink

The DNA within the complex possesses a torsional kink (or dislocation) in the middle of the recognition sequence that facilitates cleavage and has been described in detail for the 2.7-Å pre-transition state structure (Kim et al. 1994). The values for the structure after refinement to 1.85 Å are extremely similar to this report. Most fall within or very close to the error limits stated therein. The principal features of the *EcoRI* kink include: (1) large departures of roll (ρ) from the neighborhood of 0°; these distortions are located at the central ApA, ApT, and TpT steps where ϱ takes the values 32°, –50°, and 21°, respectively. (2) Increased rise (D_z) of 4.3, 4.4, and 4.1 Å each at the same three base-pair steps. (3) The DNA is underwound by approximately 30°; the unwinding is distributed over the entire EcoRI recognition site. (4) The DNA backbone appears to have a distinct kink or jog at the ApA step because the vector connecting the C1′ atoms of the adjacent adenines is nearly parallel to the average helix axis. (5) The unwinding and increased rise widen both grooves – the major groove by approximately 3 Å and the minor groove by about 5 Å.

2.5 Recognition Overview

The sequence-specific interactions can be divided into the usual three categories: (1) direct interactions are the sequence-specific contacts to the bases in the EcoRI recognition site. (2) Indirect interactions are intra DNA interactions that can be identified as contributing to the sequence-specific deformability of DNA containing the EcoRI site. (3) "Buttressing" interactions are hydrogen bonds between amino acid side chains that contact the DNA (or between side chains and the main chain amide group). In this case they are remarkably extensive, forming the extended recognition network. All of the moieties that directly participate in the recognition also hydrogen bond to each other either directly or through this complex network of hydrogen bonded groups that includes virtually all of the important moieties involved at the DNA–protein interface.

Specificity arises because any base-pair substitution would disrupt some of the direct interactions (and alter the sequence-dependent deformability of the DNA); such disruptions would propagate through the extended network. While this does provide an answer to the specificity question, two caveats are indicated: First, biochemical data clearly shows that the DNA–protein inter-face has adaptive characteristics (Jen-Jacobson et al. 1991; Lesser et al. 1990a, 1993); i.e. the structure shifts in response to even small perturbations in ways that are difficult to predict. The structural basis of this adaptability could be similar to the plasticity noted in the ED144 structure (see below). A further extension of these ideas by Jen-Jacobson and coworkers advances the idea that molecular distortion, i.e., strain plays a significant role in determining the thermodynamics of the interaction and specificity (Jen-Jacobson et al. 2000b).

However, a full assessment of these issues is beyond the scope of this dis-cussion. "Plasticity" takes in a large territory and a quantitative discussion would require quantitative measures of "distortability." Indeed, establishing quantitative connections between molecular distortion and strain, on the one hand, and functional properties on the other is the domain of statistical mechanics. Qualitative efforts to achieve this have been underway form some time, e.g., Kumar et al. (1994) demonstrated that the EcoRI kink is signifi-cantly strained and Duan et al. (1996) demonstrated dynamic contributions to the binding entropy. However, these efforts, which are ongoing in a number of laboratories including our own, are still more qualitative than quantitative. Therefore a rigorous physical test of the degree of applicability of the highly intriguing strain concept is very challenging, given the current state of knowl-edge. Here therefore, we focus on those features we can observe directly in the structures and emphasize that this is not to imply either that they are the whole story or that indirect effects are not important because they clearly are.

The second caveat is the need for comparison(s) with complexes between the protein and DNA containing pseudo recognition sites termed EcoRI* sites for historical reasons. They match the "canonical" or true EcoRI site (GAATTC) at five out of the six positions. Thus, AAATTC is an EcoRI* site. There are nine unique EcoRI* sites due to the inverted repeat in the canonical site. We are currently determining the structure of the EcoRI* complex with the site GACTTC; the most striking differences on preliminary analysis are the DNA structure and its placement relative to the protein. We will report further on this structure when complete.

2.6 Sequence-Specific Hydrogen Bonds

The direct interactions are diagrammed in Fig. 3 for the pre-transition state structure, which shows the interactions for one-half of the EcoRI site i.e. the asymmetric unit. They include seven protein–base hydrogen bonds, two water-mediated hydrogen bonds and an elaborate set of van der Waals con-tacts. (The distances between donor and acceptor atoms are somewhat vari-able hence the inferred strengths of the hydrogen bonds are expected to vary). The seven hydrogen bonds to the purines all involve amino acid side chains

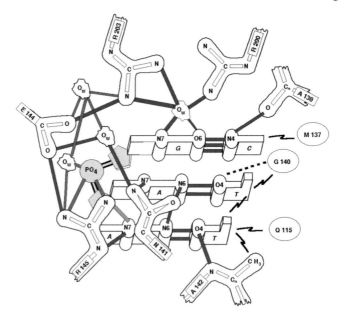

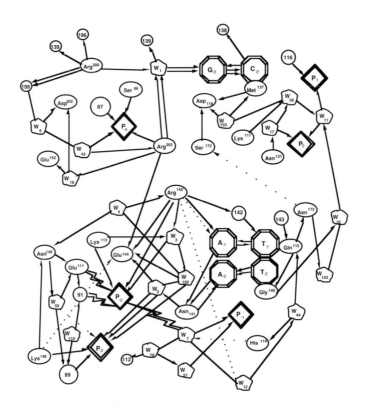

(either directly or via the water molecule) while the two hydrogen bonds to the pyrimidines are both from the protein backbone. For the complete EcoRI site there is a total of 18 protein–base hydrogen bonds; all of them are specific because they would be disrupted by base changes.

It is interesting to note that EcoRI endonuclease discriminates purines by pairs of hydrogen bonds emanating (directly or indirectly) from amino acid side chains while it uses single hydrogen bonds from the polypeptide backbone to recognize pyrimidines (it also uses van der Waals contacts). One reason for this was provided by Seeman, Rosenberg, and Rich who noted that a single hydrogen bond to O4 of thymine or N4 of cytosine would be completely sequence specific only if the protein moiety is unable to shift its position by approximately 1 Å (Seeman et al. 1976). This rigidity is more likely for the polypeptide backbone than it is for amino acid side chains. Indeed, these considerations lead to an energetic hierarchy of levels of sequence-discrimination that could be anticipated from protein–base hydrogen bonds: Backbone–base >side chain–base >protein–water–base. Here, the most strongly discriminating interactions are those on the left.

The involvement of backbone amide groups in the recognition mechanism has implications for genetic/molecular biological investigations of this system

---

Fig. 3. Schematic of EcoRI endonuclease recognition interactions. *Above* The bases are seen *edge on*. Hydrogen-bonding moieties rounded and van der Waals moieties (e.g., thymine methyl groups) are indicated as *squared surfaces*. The scissile linkage is shown as *pentagons* (ribose moieties) and a *circle* (phosphates). The functional elements of the amino acid side chains that interact with them are also shown. Hydrogen bonds are indicated by *bars*. The hydrogen bonds from Arg[145] to the scissile phosphate are also shown; they emphasize the dual role of this residue and the integration of recognition and cleavage (see text). Some of the critical, bridging water molecules are also shown. They have roles both in recognition and in forming the extended network (see text and below). The *wavy lines* indicate "hydrophobic" e.g., the methylene portion of the Gln 115 side chain is in van der Waals contact with the methyl group of the inner thymine. The *thick dashed line* indicates the CH–O hydrogen bond between Gly[140] Cα and the O4 of T_2. A preliminary version of this figure has been published (Kim et al. 1994). Note the introduction of bound solvent and the interactions with the scissile linkage. Note especially the dual role of Arg[145]. This shows the integration of recognition and cleavage, one of the major conclusions to emerge from this analysis. *Below* In this schematic representation, DNA bases are indicated by *octagons*; phosphate moieties are shown as *diamonds*; amino acid side chains are represented by *ovals* with the residue named, e.g., Glu[144] refers to hydrogen bonds to the side chain carboxyl oxygens. However, when only the polypeptide backbone participates in an interaction, it is shown as a *circle* labeled by the residue number, e.g. the carbonyl of residue 138 (alanine) receives a hydrogen bond from the N4 of cytosine. (When both main and side chains participate in hydrogen bonding, the side chain is named.) Hydrogen bonds are indicated by *arrows* pointing from donor to acceptor and van der Waals contacts are indicated where the symbols for two residues are in contact. The numbering scheme for the bases is outwards from the center, i.e., $G_{-3}A_{-2}A_{-1}T_1T_2C_3$

because it is extremely difficult to genetically modify a protein such that one or two amide groups are moved (without introducing a drastic rearrangement of the entire structure). These interactions would therefore tend to be genetically "invisible", i.e., they could not be detected in a genetic experiment.

Sequence-specific hydrogen bonds between DNA bases and protein amide groups are not unique to EcoRI endonuclease. For example, similar hydrogen bonds were seen with EcoRV endonuclease, where they also are major determinants of its specificity (Winkler et al. 1993a), they are not limited to restriction enzymes.

The pre-transition state structure shows evidence of a 3.1 Å CH–O "hydrogen bond" where the hydrogen on the Gly[140] Cα atom acts as the hydrogen bond donor and the O4 of T_2 acts as the acceptor (Fig. 4). This is the only major groove DNA base hydrogen bond acceptor in the recognition sequence that is not satisfied by conventional hydrogen bonding. Structural biologists have directed much attention towards the role of CH–O hydrogen bonds as possibly a new, unrealized structural feature stabilizing nucleic acid and protein structures. (Bella and Berman 1996; Derewenda et al. 1995; Mandel-Gutfreund et al. 1998) The putative CH–O bond in the EcoRI-DNA complex is of an appropriate length for such an interaction. CH–O bond lengths surveyed by Mandel-Gutfreund et. al. (1998) were between 2.8 and 3.3 Å long. There is functional support for this interaction from the base-analog studies of Jen-Jacobson and colleagues (Jen-Jacobson 1995; Jen-Jacobson et al. 1991; Lesser et al. 1990b). They reported that the binding is weakened if T_2 is replaced with a base-analog lacking the exocyclic O_4 moiety suggesting that it interacts favorably with the protein. This CH–O interaction has been a stable feature of

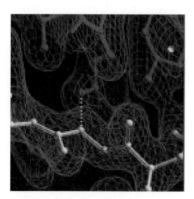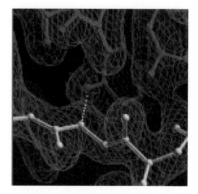

Fig. 4. CH–O bonding: This is a 2Fo-Fc electron density map contoured at 2σ showing a 3.1-Å CH–O contact (*dashed line*) between the protein (*yellow*) Gly[140] Cα and the DNA (*green*) T_2 O4. The hydrogen on the glycine Cα acts as a hydrogen bond donor while the thymine is the acceptor. This outer thymine carbonyl is the only DNA base hydrogen bond acceptor in the recognition sequence that is not satisfied through conventional hydrogen bonding

refinement in both the older room temperature data as well as the high resolution, synchrotron data reported here. Electron density consistent with this feature is also clear in simulated annealing omit maps. We therefore conclude that it is probably a real feature of the structure that contributes to specificity.

2.7 Sequence-Specific Interactions via Bound Water

Arg^{200} and Arg^{203} both bind an isolated water molecule (W_1) that is approximately 3 Å away from N_7 and O_6 of guanine. The interaction with Arg^{200} is complicated by the fact that both the guanidinium group and W_1 also form apparent hydrogen bonds to the carbonyl oxygen of Ala^{139}. We interpret this as two bifurcated hydrogen bonds, one in which Arg^{200} donates one proton jointly to the carbonyl oxygen and W_1 and the other in which W_1 donates to the same carbonyl and to O_6 of G_{-3}. This, coupled with the hydrogen bonds from Arg^{203} directs the protons of W_1 towards the base, requiring the latter to have hydrogen bond acceptors at N_7 and O_6 positions. This is sequence-specific since guanine is the only base with two such hydrogen bond acceptors in the major groove. The recognition interface is riddled with such interactions. In the pre-transition state structure their are 20 water molecules at the interface making bridging contacts (40 with symmetry) which facilitate interactions between distally located protein or DNA residues (see Fig. 3). Fifteen of these water molecules are seen in at least one other independently solved cognate EcoRI-DNA crystallographic structure. The five remaining molecules are from the 1.85-Å pre-transition state complex and have higher than average B-factors. As a result, they are only visible at high resolution. The high correspondence of waters in this region suggests that they are conserved features of the recognition interface and are important to fastening the protein tightly and specifically to the DNA substrate.

2.8 Sequence-Specific van der Waals Interactions

A broad area of hydrophobic interaction is formed where $-CH_2-$ and $-CH_3$ groups of the protein contact the methyl groups of both thymines as well as the hydrogen atoms at the 5- and 6-positions of cytosine. These interactions are also sequence-specific because only the cognate bases would fit properly against this hydrophobic patch on the protein; noncognate bases would either not fit or they would require shifts of the DNA backbone in order to fit. These shifts would be expected to disrupt phosphate-protein interactions thereby incurring an energetic penalty.

These contacts are indicated schematically in Fig. 3. The side chain of Met^{137} packs against the C_5 and C_6 positions of cytosine C_3. This is cytosine-specific because substitution of thymine would lead to steric clash with the thymine methyl group while substitution of either purine would lead to blockage of the hydrogen-bonding site at N_7. Similarly, the methyl groups of thymine residues T_1 and T_2 are in van der Waals contact with the side chain of Met[137], the α-CH_2 of Gly[140], and the side chains of Gln[115] and Ala[142]. These interactions are also sequence-specific because thymine is the only natural base possessing an exocyclic methyl group at this (or any other) position.

The interaction with Ala[142] side chain is particularly important because, as noted by Kumar et al. the β-CH_3 is the "alanine wedge" that is partially inserted between the central base pairs at the EcoRI kink. It stabilizes the unusual roll values at the center of the EcoRI recognition sequence (Kumar et al. 1994). Ala[142] is very close to the molecular two-fold symmetry axis and the side chains from both symmetry-related residues participate in the formation of the wedge.

2.9 Redundancy of Direct Sequence-Specific Interactions

Each base pair is over determined because the recognition contacts are redundant. Any single hydrogen bond or van der Waals interaction could be removed and the correct base pair would still be specified by the remaining interactions, although the expected level of discrimination would be reduced. This is dramatically different from the pattern that emerges from repressor–operator cocrystals (Steitz 1990) where the hydrogen bonding and van der Waals contacts are close to or less than the minimal number predicted by Seeman et al. 1976).

The redundant recognition contacts seen here provide a structural explanation for a series of genetic analyses based on both random and site-directed mutagenesis (Alves et al. 1989; Hager et al. 1990; Heitman and Model 1990a, b; Needels et al. 1989; Oelgeschlager et al. 1990; Steitz 1990; Thielking et al. 1990, 1991; Wolfes et al. 1986; Yanofsky et al. 1987). To summarize, any of the critical side chains, such as Glu[144], Arg[145] and Arg[200], can be conservatively replaced (Glu→Asp or Arg→Lys) with the result that while the protein's affinity for DNA is reduced by several kcal/mol, the specificity is unchanged. Most other substitutions completely inactivate the enzyme, with the curious exception of Cys. Similar results have been obtained at Asn[141] (Fritz et al. 1998) and Arg[203] (Greene et al., unpubl.). As confirmed by our structural analysis of the ED144 and RK145 mutants (manuscript in preparation), the mutant complexes retain sufficient interactions with the bases to determine specificity for the EcoRI site.

No instances have been reported in which mutations alter the specificity of the protein from that of the EcoRI site to a different sequence. There is a series of "promiscuous" mutations that increase the level of cleavage at EcoRI* sites, however proteins carrying these mutations still prefer the canonical GAATTC over EcoRI* sites (Heitman and Model 1990a). In effect, the promiscuous mutations lower the level of sequence discrimination without altering the target sequence.

2.10 Bound Solvent

The pre-transition state structure includes 152 bound solvent molecules that are represented as water although we cannot exclude the possibility that they may be hydroxide, hydronium, or ammonium ions. Most have more than one hydrogen bond to the protein and/or the DNA; many are "sandwiched" between the protein and the DNA while others bridge different groups on the protein, filling either "voids" or "cracks". Here, we use the term "void" as a small hole big enough for one isolated water molecule whose hydrogen bonding potential is completely satisfied by contacts to the protein and/or DNA and "cracks" as regions filled by a chain of three or four solvent molecules that hydrogen bond to each other as well as bridging different macromolecular components.

Eighty-two of the 1.85-Å pre-transition state solvent molecules have multiple apparent hydrogen bonds to protein and/or DNA, i.e., they participate in bridging interactions. (The hydrogen bonds are inferred from oxygen–oxygen or oxygen–nitrogen distances.) Nine of the bound solvent molecules are within hydrogen bonding distance of five donors/acceptors on the protein and/or DNA. The most likely explanation is bifurcated or three-center hydrogen bonds. Twenty bound solvent molecules appear to be hydrogen bonded only to other bound solvent, i.e., they are in the second solvation shell. The bridging water molecules have two roles in the recognition by EcoRI endonuclease: (1) they provide specificity by bridging one of the bases (guanine) and amino acid side chains (Arg^{200} and Arg^{203}). (2) They are significant components of the extended recognition network described above. These roles are somewhat different from that of the bound solvent in the trp-repressor-operator complex where the bound solvent has a central role in the protein–base contacts (Otwinowski et al. 1988). In the EcoRI case, the recognition role(s) of the bound solvent are more like that of a "supporting actor" than that of "leading actor".

2.11 "Buried" DNA Phosphate Groups

A section of the DNA backbone becomes buried in a cleft of the protein when the protein and the DNA bind. The cleft is formed between the β bridge and the recognition helices. The phosphate groups of C_{-4} and G_{-3} are completely occluded from the bulk solvent (G_{-3} is the first base of the recognition sequence). The scissile phosphate at A_{-2} is partially buried, with an open channel where Mg^{2+} can diffuse to the cleavage site.

Not surprisingly, the surface of the protein lining this cleft is highly charged and polar, a classical example of complementarity. Critical interactions include hydrogen bonds from Arg^{203}, Asn^{149}, Lys^{148}, Arg^{145}, and Lys^{113} as well as those from the amide nitrogen atoms in the β bridge (see Fig. 5 for a schematic representation of these interactions). These contacts do not produce a surface on the protein that, in and of itself, precisely matches that of the DNA; the remaining "cracks and crevices" are filled by some of the tightly bound solvent. They contribute additional hydrogen bonds to the protein–DNA interface and from a structural viewpoint they are integral members of that interface.

The phosphate contacts at C_{-4} and G_{-3} are critical to the formation of the complex because they are central "hubs" in the recognition network described below. It is therefore not surprising that they are also sites of strong ethylation interference in solution, leading Jen-Jacobson to identify them as critical

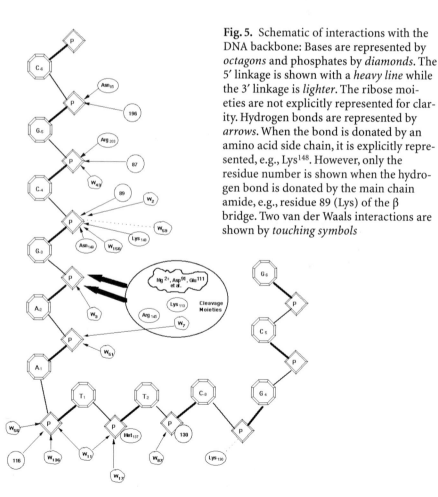

Fig. 5. Schematic of interactions with the DNA backbone: Bases are represented by *octagons* and phosphates by *diamonds*. The 5′ linkage is shown with a *heavy line* while the 3′ linkage is *lighter*. The ribose moieties are not explicitly represented for clarity. Hydrogen bonds are represented by *arrows*. When the bond is donated by an amino acid side chain, it is explicitly represented, e.g., Lys^{148}. However, only the residue number is shown when the hydrogen bond is donated by the main chain amide, e.g., residue 89 (Lys) of the β bridge. Two van der Waals interactions are shown by *touching symbols*

"clamps" (Lesser et al. 1990a). The protein also makes numerous contacts at other phosphate groups although the interactions at most of the other phosphates are not as extensive as those described above.

2.12 Solvent Mediated Contacts to the DNA Bases Flanking the Recognition Site

The pre-transition state structure disclosed a protein–water–DNA interface 5′ to the recognition site involving three protein backbone functional groups, one protein side chain, five water molecules and three DNA bases (C_{-4}, G_{-5}, and C_{-6}) that form a hydrogen bonding network. At this location the DNA major groove formed by residues flanking the recognition site is passing through a deep polar cleft in the protein that appears to be filled with well ordered solvent. The solvent facilitates bridging interactions between the protein and DNA. Figure 6 shows the most important of these interactions. C_{-4} donates an amino group hydrogen bond to W_{41}, which makes two strong bonds to Ala^{139} N and Ile^{197} O. G_{-5} is bridged twice to both of these backbone

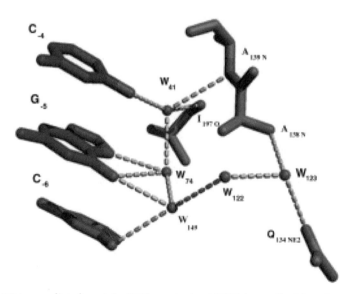

Fig. 6. Water-mediated protein–DNA contacts at DNA bases flanking the recognition site. Water mediated protein–DNA contacts can be seen in the 1.85-Å structure that may contribute to the tendency of the EcoRI to cleave its DNA recognition sites at different rates depending on their flanking sequence context. Protein residues are in *green*, DNA is in *blue*, and water is in *red*. Functional groups participating in the network are highlighted in *red* and labeled for the protein. The most important protein–DNA–water hydrogen bonds are represented by *dashed lines*

atoms as well by W_{41} and W_{74}. Both G_{-5} and C_{-6} are also connected to Ala^{138} N and Gln^{134} $N_{\varepsilon 2}$ through W_{149}, W_{122}, and W_{123}.

EcoRI endonuclease is known to cleave its recognition sequence with different efficiencies dependent on the sequence of bases flanking the hexanucleotide recognition site (Thomas and Davis 1975). The intimacy between ordered solvent, DNA bases, and the protein backbone seen in our structure naturally raises the question of the significance of these waters to the flanking sequence effect.

These questions are underscored by another structure, which we will report separately (Grigorescu et al., in prep.), in which G_{-5} has been replaced by A (and C_5 by T). There are three principle differences between that structure and the high-resolution pre-transition state structure reported here: (1) the base-pair substitution itself. (2) The solvent structure in the vicinity of the base-pair substitution. (3) The apparent dynamics of the protein and DNA in the neighborhood of the base-pair substitution as indicated by spatially localized variations in the average B-factors. Otherwise, the structures are highly similar.

There are thus local structural changes associations with these flanking sequence changes; the question remains, are they causative? This is more difficult to assess for reasons that will be discussed in the full report (Grigorescu et al., in prep.). For example, Jen-Jacobson and coworkers have suggested that differential strain plays a significant role in the flanking sequence effect (Jen-Jacobson et al. 2000b).

2.13 DNA Minor Groove

The DNA minor groove contains a pseudo-spine of hydration involving 18 water molecules along the entire length of the helix. Berman and coworkers have analyzed patterns of DNA hydration across numerous B-DNA crystallographic structures from the Nucleic Acid Database in order to predict water positions (Schneider et al. 1993). They noted that in B-DNA the most frequent water bridge between bases in the minor groove are N_2/O_2-water-O_4' linkages. (Schneider et al. 1993) The famous Dickerson-Drew dodecamer crystallographic structure, CGCGAATTCGCG, which houses the EcoRI recognition site has a spine of hydration in the minor grove that consists mainly of O_2/N_3–water–O_2 bridges. (Drew and Dickerson 1981) The entirety of the DNA helix in our structure interacts with minor groove waters in a classical manner as described by the Berman and Drew analyses, but does so less thoroughly. For example, the G_{-3} N_2--W_{27}- A_{-2} O_4' bridging distances are 3.35 and 2.98 Å, respectively, but the next water down the chain (W_{30}) that makes a strong contact to A_{-2} N_3 is 4.04 Å away from A_{-2} O_4'. We call this a pseudo-spine because it portrays typical hydration spine characteristics but is incomplete due to the 5 Å widening of the minor groove by the EcoRI kink.

2.14 Contrasting the 1.85 and 2.7 Å Pre-Transition State Complexes

The two structures were remarkably similar given the large difference in resolution between them. Simulated annealing omit maps were calculated in
critical regions as were a complete set of omit maps that systematically
spanned the structure (see methods). All showed that no major adjustments
were required of either the protein or the DNA chains. The high-resolution
structure has given a more detailed look at the same general view the room
temperature structure provides, especially in the way of bulk solvent. Superposition of the protein chains shows that there are no differences in the chain
tracing but only slight repositioning of the amino acid side chains in areas on
the surface of the protein. Table 2 shows that the rms difference between the
two coordinate sets is 1.13 Å, but the estimated standard deviation (ESD) for
that value is 1.55 Å which means the RMS difference is within error. The ESD
calculation comes from Stroud and coworkers who suggested the following
empirical equation (Perry et al. 1990):

$$\sigma = 0.75Rr(0.0015B^2 - 0.203B + 0.359)$$

where R is the conventional crystallographic R-factor (R_{work}) and r is the resolution of the data (Å).

 There are differences however in the positioning of some of the sugar and
phosphate moieties of the DNA which can be attributed to the use of the most
recent nucleic acid force field parameters in the refinement (Parkinson et al.
1996). None of these differences alter the interpretation of the recognition
interface. They maintain the most critical protein–DNA interactions first
characterized in the 2.7-Å room temperature pre-transition state structure
(Choi 1994; Kim et al. 1994). Any changes seen are due to slight repositioning
of individual DNA, protein, or water atoms. There is an overall translational
shift of 0.166 Å and an overall net rotation of 0.137° between the two structures. Table 2 shows that the RMS deviations between the two structures at
both the backbone and side chain level is within error, which suggests that

Table 2. RMS differences between the 1.85- and 2.7-Å pre-transition state complexes

Structural region	RMS deviation (Å)	Estimated standard deviation (ESD) (Å)	RMS/ESD
All atoms	1.13	1.55	0.73
Protein and DNA backbone	1.14	1.41	0.81
Side chains and DNA bases	1.10	1.75	0.63

overall, the differences are not significant. Only six residues had RMS/ESD >2 (Lys[44], Leu[198], Asn[208], Asn[220], Thr[260], and Leu[263]) due to side chain repositioning, but these do not have a role at the recognition interface or in the catalytic mechanism. Overall the DNA residues show RMS/ESD >3 in the sugar-phosphate backbone due to the difference in DNA force field parameters used.

3 The Post-Reactive Complex and the Cleavage Mechanism

3.1 EcoRI Endonuclease Crystal Packing

It is remarkable that we are able to observe the in situ cleavage without a major disruption of the lattice. Crystalline enzymes are sometimes inactive; more commonly, the crystals shatter when the catalytic machinery is activated (lattice forces are generally weaker than those responsible for the conformational changes involved). EcoRI endonuclease is an exception because fortuitously none of the catalytically critical regions are involved in lattice contacts. It is in fact the unique packing arrangement which makes the *in crystallo* reaction possible.

We have obtained over a dozen different crystal forms in four different space groups by varying the oligonucleotide employed in the crystallization. All of these turn out to be variations on the same packing scheme (Grable et al. 1984; Samudzi 1990; Wilkosz et al. 1995). EcoRI endonuclease has a strong tendency to associate in sheets; the principal difference between the crystal forms is how the invariant two-dimensional sheets are "stacked" upon each other to form the three-dimensional lattice. Within a sheet, the lattice contacts are on the "flanks" of the complex, well away from the DNA–protein interface. The complex has twofold rotational symmetry axes perpendicular to the threefold axes, i.e., in the plane of the sheets.

The contacts between different sheets mainly involve the ends of the DNA (which forms continuous rods through the crystals) and the "top" surface of the protein. These surfaces are also well away from the cleavage and recognition sites. These crystals are approximately 60 % solvent that is organized into large channels running parallel to the threefold axes. The central point here is that they completely bathe the critical regions of the complex in bulk solvent, facilitating the movements required to hydrolyze the DNA. The product DNA cannot diffuse away because it is tightly bound to the enzyme and because its ends are held by DNA–DNA lattice contacts. These fortuitous circumstances make it possible to go from the pre-transition state complex to that of the post-reactive within the same lattice, facilitating the comparative study reported here.

3.2 Catalytic Site

The binding of Mn^{2+} and the bond scission event were readily discernible in the initial difference electron density map and remained prominent in all subsequent electron density maps. The manganese ion is octahedrally coordinated by the post-reactive phosphate of A_{-2}, the side chains of Asp[91] and Glu[111], the main chain carbonyl of Ala[112] and two water molecules as shown in Fig. 7. The residues critical for bond scission, Asp[91], Glu[111], and Lys[113], had also been identified genetically (King et al. 1989) and by comparison with EcoRV endonuclease (Winkler et al. 1993a). Table 3 shows the coordination distances and angles of the cation. They are consistent with the properties of manganese, as determined in high resolution (small molecule) crystal structures (Cheng and Wang 1991; Chmelikova et al. 1986; Cudennec et al. 1989; Garrett et al. 1983; Lightfoot and Cheetham 1987; Lis 1982, 1983, 1992; Ross et al. 1991), where Mn–O distances range from 1.9 to 2.33 Å, with a mean of 2.2 Å.

Table 3. Manganese coordination geometry

Mn^{2+} – ligand distances: Ligand	Distance to Mn^{2+} (Å)
O2P of A_{-2}	2.14
Oδ2 of Asp[91]	2.21
Oϵ1 of Glu[111]	2.23
Carbonyl O of Ala[112]	2.33
O of W_A	2.25
O of W_B	2.13

Ligand – Mn^{2+} – ligand angles Ligands	Angle (°)
A_{-2}, O2P – Mn^{2+} – Oδ2, Asp[91]	170.4
A_{-2}, O2P – Mn^{2+} – Oϵ1, Glu[111]	109.6
A_{-2}, O2P – Mn^{2+} – O, Ala[112]	90.8
A_{-2}, O2P – Mn^{2+} – O, W_A	97.4
A_{-2}, O2P – Mn^{2+} – O, W_B	67.2
Asp[91], Oδ2 – Mn^{2+} – Oϵ1, Glu[111]	97.9
Asp[91], Oδ2 – Mn^{2+} – O, Ala[112]	79.4
Asp[91], Oδ2 – Mn^{2+} – O, W_A	92.2
Asp[91], Oδ2 – Mn^{2+} – O, W_B	85.3
Oϵ1, Glu[111] – Mn^{2+} – O, Ala[112]	99.4
Oϵ1, Glu[111] – Mn^{2+} – O, W_A	80.2
Oϵ1, Glu[111] – Mn^{2+} – O, W_B	176.8
O, Ala[112] – Mn^{2+} – O, W_A	171.4
O, Ala[112] – Mn^{2+} – O, W_B	80.3
O, W_A – Mn^{2+} – O, W_B	100.5

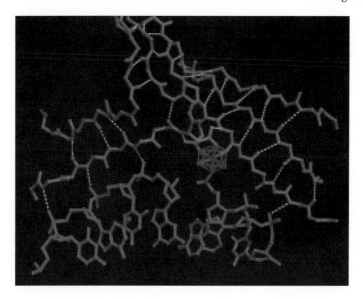

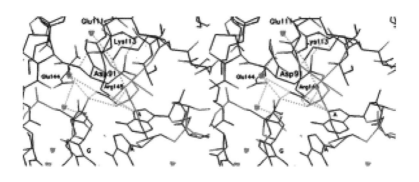

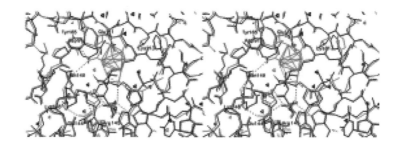

Similarly, these high-resolution studies show that the O–Mn–O angles are within 15° of either 90° or 180°.

Several critical hydrogen bonds buttress the cleavage site. The Glu[111] side chain receives a hydrogen bond from the main chain amide of Asp[91] (see Fig. 7). This hydrogen bond is to the oxygen atom not coordinated to Mn^{2+}. Similarly, the side chain of Asp[91] is buttressed by the main chain amide of Ala[112], whose carbonyl oxygen atom forms one of the other ligands for manganese. This interlocking network precisely positions three of the critical ligands at the corners of one of the triangular faces of the octahedron, determining the location of the cation with respect to the protein.

Further buttressing of the cleavage site is accomplished by hydrogen bonds between the post reactive phosphate and a water molecule coordinated to the Mn^{2+} cation (see Fig. 7). The phosphate also receives hydrogen bonds from Lys[113] and Arg[145]; similar hydrogen bonds are formed in the pre-transition state complex. These two basic amino acid residues have central roles in the catalytic mechanism proposed below.

The pre-transition state and post-reactive complexes are generally very similar, apart from the necessary changes at the cleavage site. The significant differences in the active site region can be seen in stereo in Fig. 7. Besides the binding of the cation itself, they include the movement of the post-reactive phosphate, Asp[91] and Glu[111]. These shifts are larger than the expected error of the two structures suggesting that there may be subtle differences between the two protein–DNA complexes that are beyond the current resolution of the post-reactive structure.

The hydrogen bonding to Glu[111] is consistent with a genetic result obtained by Modrich and coworkers who found that the EQ111 mutant (Glu[111]→Gln) bound DNA more tightly than wild-type in the absence of divalent cation, but did not cleave it when the latter was present (King et al. 1989). Clearly, the side

---

Fig. 7. The cleavage site. *Above* The chimeric β sheets and the manganese ligands. The protein is shown as *yellow-orange*, the DNA is *cyan* and the manganese octahedron is *red* (the cation is shown connected to its ligands with *"solid" bonds* for clarity). The backbone–backbone NH–O hydrogen bonds are *yellow*. The primary β sheet of this subunit angles from the top-center to the lower right of the figure. The first strand of the sheet is closest to the viewer and subsequent strands are displaced backwards and to the right. The two-stranded β sheet on the left is the β bridge. *Center* The stereo pair shows the post-reactive structure in the vicinity of the cleavage site. "Atomic" coloring is used where carbon is *black*, nitrogen *blue*, oxygen *red*, phosphorus *green*, and the manganese cation is *magenta*. The divalent cation is joined by *solid lines* to its six oxygen ligands (octahedral coordination). Selected hydrogen bonds are shown and the principal residues are labeled. *Below* This stereo pair is a superposition of the pre-transition state and post-reactive complexes shown in *blue* and *red*, respectively. The differences in the active site region are primarily the shift of the post-reactive phosphate and the Asp[91] and Glu[111] side chains

chain nitrogen of glutamine could not coordinate a divalent cation even if it were oriented as in the wild-type. The environment within the protein eliminates the alternative (where the side chain carbonyl coordinated the cation) because hydrogen bonds are received by both carboxyl oxygen atoms of the wild-type side chain, including the buttressing hydrogen bond described above.

Interestingly, a minor modification in the position of the side chain of residue 111 would enable a glutamine to donate a hydrogen bond to the phosphate oxygen of A_{-2}. In this pre-transition state model the 111 side chain is in a conformation similar to that of the post-reactive complex, while the rest of the structure is similar to the X-ray structure pre-transition state complex (Rosenberg, unpubl.). The extra DNA–protein hydrogen bond may contribute to the tighter DNA binding of the mutant.

Arg^{145} appears to have a dual role, polarizing the scissile phosphate (along with Lys^{113}) as well as forming critical hydrogen bonds with the adenine A_{-1}, bound solvent molecules W_3 and W_4 and the amide of Ala^{142}. This double duty explains the unusual sensitivity of residue 145 to amino acid substitution; the RK145 ($Arg^{145} \rightarrow Lys$) has significantly lower activity than similar conservative mutations of the recognition residues including Glu^{144} and Arg^{200} (Heitman and Model 1990b).

All these residues are located near the scissile phosphate, P_{-2}, as the preceding discussion implies. Asp^{91} is at the carboxyl terminus of the bridge loop, between β_{ii} and β_2 while Glu^{111} and Lys^{113} are in β_3. This places these critical residues at a switch point of the primary β sheet. This is consistent with the observation that active sites are frequently found at switch points in many α/β structures (Richardson 1981).

3.3 The Proposed Cleavage Mechanism

Eckstein and coworkers showed that the reaction inverted the configuration of the reactive phosphate; they suggested that the simplest mechanism would involve nucleophilic in line attack of the phosphorus by an activated water molecule (Connolly et al. 1984). Pingoud and coworkers suggested a specific S_N2 mechanism where one of the DNA phosphates (A_{-1}) served as the Lewis base that abstracted a proton to activate the water (Jeltsch et al. 1992; Pingoud and Jeltsch 1997). Common threads in many phosphoryl-transfer mechanisms include metal ion activation of a water molecule or hydroxide ion, which then attacks the phosphorus as the reaction proceeds through a pentacoordinate phosphorus transition state. These ideas can be readily incorporated into the structural results reported here.

As we have shown, the post-reactive and pre-transition state complexes are generally very similar, apart from the differences we have noted at the cleavage site. Since we are able to conduct the reaction *in crystallo,* it is reasonable to assume that the reaction proceeds via a simple, direct route from substrate to product. This is a parsimony argument, which can be combined with the results described above to give the proposed reaction mechanism shown in Fig. 8.

The proposed mechanism begins with species I (Fig. 8), which is simply the pre-transition state structure. The first step of the reaction is the binding of the divalent cation, forming species II. The model for this, shown in the figure, is essentially a hybrid of the two X-ray structures; the DNA is from the pre-transition state structure while the protein and cation used the coordinates of the post-reactive structure. Therefore the metal ion is coordinated by Asp[91], Glu[111] and the carbonyl of Ala[112]; additionally it is coordinated by an oxygen atom of the scissile phosphate and two water molecules. The phosphate-metal coordination of species II required a very modest adjustment of the position of the scissile phosphate that is within the estimated RMS error of the pre-transition state structure (0.3 Å, based on a Luzatti plot).

This model for species II naturally places one of the waters in the manganese coordination sphere in excellent position for in line nucleophilic attack on the phosphorus, directly opposite the 3′ leaving group, as can be seen in Fig. 8. It is merely using Occam's Razor to suggest that this is the attacking nucleophile.

One of the mysteries of the mechanism had been the identity of the Lewis base that abstracts a proton from the attacking nucleophile; we suggest that it is another water molecule. In the model for species II, water molecule W_7 of the pre-transition state complex is approximately 2.9 Å from the attacking water molecule bound to the manganese ion. The geometry is good for hydrogen bonding and the proposed hydrogen bond is indicated in Fig. 8. Again, it is Occam's Razor to suggest that W_7 is the Lewis base. Once the proton has been transferred, "W_7" acquires a positive charge and the manganese-bound hydroxide is activated for attack on the phosphorus. This arrangement is electrostatically stabilized by an adjacent phosphate group (A_{-1}) and the manganese ion. The sequence of charges running from the adjacent phosphate to the Mn^{2+} is –, +, – 2+. Thus, the pk_a of W_7 would be raised by its proximity to the negative charge of the phosphate. An additional electrostatic element derives from Lys[113] and Arg[145], which further polarize the scissile phosphate.

The penta-coordinate intermediate, species III, is the next step. The model for this (Fig. 8) was constructed with 1.5- and 1.9-Å bond lengths for the axial and apical oxygen atoms, respectively. The penta-coordinate intermediate fits naturally into the active site and reasonable geometry could be maintained via a small adjustment of the covalently linked DNA. Interestingly, it appeared that this transition state formed better hydrogen bonds with Lys[113] and Arg[145] than the other species, consistent with the concept of transition state stabilization.

Indeed, the positions of all five transition state oxygen atoms are stabilized in this model. The attacking hydroxide is stabilized by the manganese ion, as

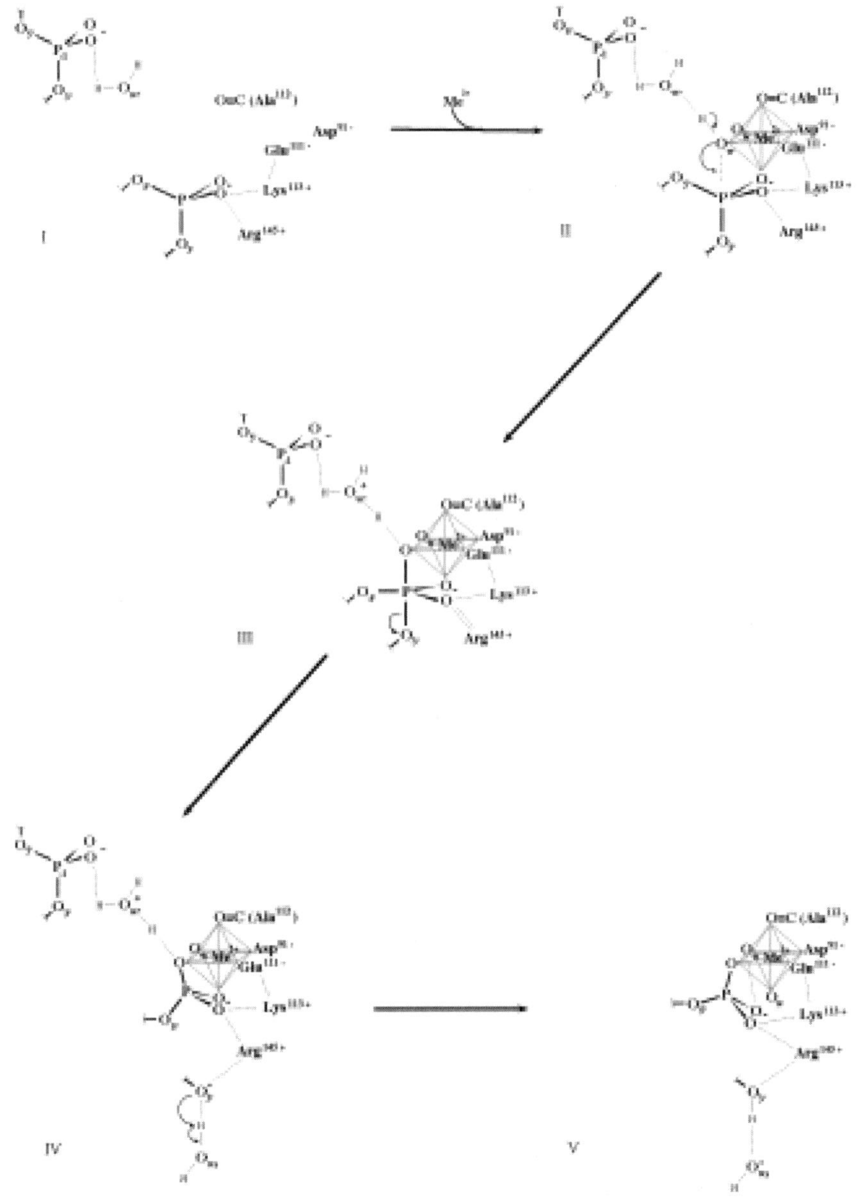

Fig. 8. The proposed reaction mechanism. **A** The comparison between the pre-transition state- and post-reactive-complexes combined with results from several laboratories (see text) suggested the S_N2 mechanism shown here. *Species I* and *V* are the X-ray structures of the pre-transition state and post-reactive complexes; detailed models have been constructed for intermediates *II, III* and *IV*. The pentacoordinate transition state is species *III*. **B** X-ray structures and models corresponding to the reaction mechanism. Each of the stereo pairs corresponds to one of the steps of the mechanism shown in **A**. The first and last stages are the pre-transition state and post-reactive structures as indicated

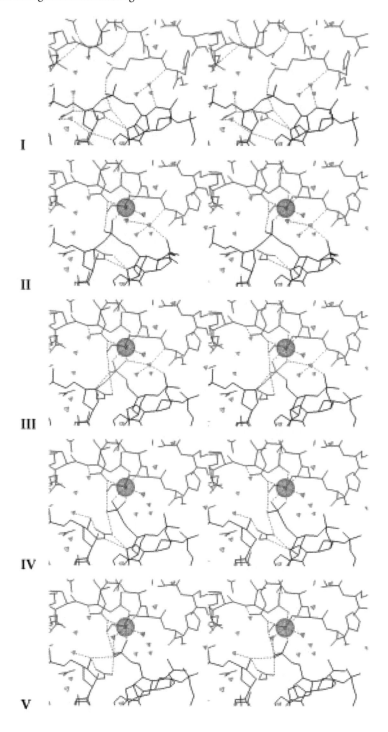

I

II

III

IV

V

well as W_7. One of the non-esterified oxygen atoms is stabilized by Lys^{113} and Arg^{145}, as described, while the other is stabilized by the divalent cation. The positions of the $O_{3'}$ and $O_{5'}$ atoms are stabilized by the DNA itself.

An additional proton must be transferred to the O_3' of the leaving group to produce the 3' hydroxyl. Here too we find an appropriately positioned water molecule, W_3, in both the pre-transition state and post-reactive complexes. Arg^{145} is the only other moiety near $O_{3'}$ that could serve as a proton donor, which seems less likely than W_3, in view of its pk_a. W_3 and Arg^{145} are within hydrogen bonding distance of each other in the X-ray structure so that even if the arginine is the initial proton donor, it would quickly abstract a proton from W_3. Both pathways may be utilized, but for either, the $O_3'^-$ oxyanion is electrostatically drawn towards W_3 and Arg^{145}. The movement is accompanied by a repuckering of the deoxyribose moiety of G_{-3}. This forms species IV of Fig. 8 and allows the completion of the reaction, as shown in the figure.

The conversion from species IV to species V is a relaxation where one of the non-esterified phosphate oxygen atoms is replaced by a water in the manganese coordination sphere. It should be noted that species IV can be readily built into the post-reactive structure, but it does not fit the electron density maps as well as the final complex, species V.

This mechanism has some points of similarity with one proposed earlier by Pingoud and coworkers (Jeltsch et al. 1992; Pingoud and Jeltsch 1997), although it differs in detail. They invoked Lys^{113} as a polarizing residue but did not invoke Arg^{145} and they suggested that the adjacent phosphate was the Lewis base. In fairness, it should be noted that their mechanism was proposed before the structural data were available.

3.4 The Integration of EcoRI-Catalyzed DNA Cleavage with Substrate Recognition

Secondary or buttressing interactions link all the key moieties at the DNA–protein interface into a complex network that is shown in Fig. 3. This network (built from the pre-transition state structure) is critical to the recognition because it stabilizes the spatial relationships of these groups, which include the bases within the EcoRI site, amino acid side chains, the polypeptide backbone, bound solvent and the DNA phosphates. The network also includes all the groups critical to the cleavage reaction, thereby integrating recognition and cleavage.

Glu^{144} is one member of this network; it receives hydrogen bonds from $Arg^{145'}$, Lys^{148}, Arg^{203}, and water molecules W_2 and W_{150}; the latter are sandwiched between Glu^{144} on the one hand and Asn^{141} and $Arg^{145'}$, respectively, on the other. (The side chain for Arg^{145} reaches across the inter-subunit interface so that the interactions described here are with amino acid side chain from the symmetry-related subunit). These interactions maintain the correct orientations of some of the side chains that directly interact with the bases. The experiments of Heitman and Model (1990b) and of Hager et al. (1990) show that these stabilizing interactions are so vital that by genetic means they are virtually indistinguishable from the direct contacts.

The hydrogen bonding network includes all of the recognition residues, six phosphates and the catalytic residues, Asp[91], Glu[111], Arg[145] and Lys[113]. It is organized around six "hub" amino acid residues, Gln[115], Glu[144], Arg[145], Asn[149], Arg[200], and Arg[203] and the two phosphate groups at P_{-3} and P_{-4}. The hub moieties bind multiple members of the network and most of the other members of the net are either bound to the hubs themselves or to groups bound to the hubs. Note the inclusion in the network of all critical amino acid residues in the previously described cleavage site/cleavage mechanism. This strongly suggests that one cannot form without the other and especially that a perturbation in the recognition site (i.e., an incorrect base pair) would propagate via the network to the cleavage site and thus prevent cleavage at incorrect sites.

Another obvious implication of the extended recognition network is that it suggests that the specific DNA–protein interface forms cooperatively. Indeed, it is hard to see how formation of the interface could not be strongly driven to completion once a nucleus of DNA–protein interaction had been formed.

On a broader level, the network can be subdivided into an inner and an outer region that interact with the two A·T and with the G·C base pairs, respectively. The inner region forms the lower half of Fig. 3, while the outer region forms the upper half. The inner network is more intricately interconnected; in producing Fig. 3B we were unable to find an arrangement that did not have some of the lines representing hydrogen bonds crossing each other. In contrast, the outer network could be easily drawn in two dimensions without intersections. The connections between the two regions are relatively limited.

There is an interesting correlation between the level of interconnection of the network and base analog studies performed by Jen-Jacobson and others. Each analog (e.g., 7-deaza-deoxyguanosine) perturbed one of the specificity interactions described above. The analogs reduced the binding constant, K_A, by a factor of from 6.7 to 17 (increased the binding free energy ΔG_s by 1 to 2 kcal/mol) (Jen-Jacobson et al. 1991; Lesser et al. 1990a). Many of the substitutions were "additive" in the sense that the ΔG_s^0 of the binding to oligonucleotides containing multiple analogs was the sum of the ΔG_s^0 values for oligonucleotides containing each substitution separately. Some of the substitutions were "isosteric" in that they showed the same ethylation interference pattern as the natural DNA, while others were not (Lesser et al. 1990a). Similar studies have been conducted by Gumport, McLaughlin and colleagues (Aiken et al. 1991; Brennan et al. 1986; McLaughlin et al. 1987).

Substitutions within the inner AATT tetrad were generally additive and isosteric whereas modifications to the outer G·C pair were non-additive and non-isosteric. For example, substitutions of 7-deaza-deoxyadenosine for adenine were additive energetically with other inner substitutions; their ethylation interference patterns were like those of the native, canonical complex. In contrast, the 7-deaza-deoxyguanosine substitution was not additive and its ethylation interference pattern was more like EcoRI* substitutions (Lesser et al. 1990a). The structural and functional data taken together suggest that the inner region is more robust and possibly somewhat plastic (adaptable) while the outer region is less robust and rather brittle.

3.5 Chimeric β Sheets

The DNA cleavage site is flanked by two examples of a structural motif we refer to as a "chimeric β sheet." Examination of the post-reactive structure emphasizes the role of chimeric β sheets in facilitating the proper positioning of the cleavage site residues Asp[91], Glu[111] and Lys[113] (see Fig. 8). A chimeric β sheet satisfies the following four conditions: (1) the protein and DNA backbones are either approximately parallel or antiparallel. (2) The protein portion consists of at least two strands of β sheet. (3) There are NH⋯O hydrogen bonds between peptide amide and the phosphate groups of the DNA. (4) All the backbone segments of both protein and DNA can be placed on a surface that maintains the usual β twist.

There are two chimeric β sheets in all solved EcoRI endonuclease-DNA complexes: The first chimeric β sheet is formed by the "β bridge" and the DNA backbone in the region G_{-5} through G_{-3}. The β bridge has been described (Kim et al. 1990; Rosenberg 1991); it is a two-stranded β ribbon that spans a cleft in the complex. The other chimeric β sheet includes the third strand of the primary β sheet, which extends forward and twists to make the NH⋯O hydrogen bond between the amide of Gly[116] and the phosphate of T_1. The Gly[116]–T_1 interaction has been characterized chemically by the Jen-Jacobson group who incorporated chiral phosphorothioates at T_1; the S_p-containing phosphorothioate bound more tightly (by 1.8 kcal/mole) than the R_p-isomer, consistent with the mode of hydrogen bonding described here (Lesser et al. 1992).

Chimeric β sheets are possible because the distance between adjacent phosphates in DNA (7 Å) is approximately twice the spatial separation of amide groups in a strand of β sheet. This correspondence was first noted by Carter and Kraut (1974) and by Church et al . (1977) who proposed models where β hairpins would bind in the minor groove of double helical RNA or the major groove of DNA, respectively. Tanaka et al. subsequently proposed that the E. coli DNA-binding protein, hU, used the mode of interaction proposed by Church et al. to bind DNA nonspecifically (Tanaka et al. 1984; White et al. 1989). The chimeric β sheets of EcoRI endonuclease are somewhat different in that only one DNA strand participates and the polypeptide chains are not embedded in one of the grooves; however the EcoRI structure does underscore the periodicity relationships first noted by Carter and Kraut and Church et al.

This periodicity relationship reinforces the notion of an inherent complementarity between protein and DNA that has obvious evolutionary implications. It is now very well established that α helices fit "comfortably" within the major groove of DNA, i.e. that the "outer" radius of a generic α helix matches (is complementary to) the approximate curvature of the major groove. β sheets can now be seen to be complementary to DNA as well.

Given the complementarity between β sheets and DNA, it is somewhat surprising that chimeric β sheets are not more commonly observed in DNA–protein complexes. Indeed, they do appear to be present in other restriction enzymes (Hager et al. 1990; Heitman and Model 1990a; Heitman and Model 1990b; Needels et al. 1989; Steitz 1990; Wolfes et al. 1986; Yanofsky et al. 1987). An arrangement similar to the second chimeric sheet is seen in the EcoRV endonuclease (Anderson 1993; Winkler 1992; Winkler et al. 1993b). Indeed, Winkler et al. note that the two restriction enzymes have similar active sites and Siksnys and coworkers noted the similarity of the general arrangement of β sheets and DNA in the two structures (Venclovas et al. 1994). Similarly, BamHI and PvuII have a similar arrangement around the active site (Athanasiadis et al. 1994; Cheng et al. 1994; Newman et al. 1994). Newman et al. (1994) show figures for EcoRI, BamHI and EcoRV that correspond to the centermost region of Fig. 7A; a simple extension of their Fig. 3 shows that the chimeric β sheet towards the right of our Fig. 7A (and the left of their Fig. 3) is present in the three restriction enzymes. Similarly, Cheng et al. (1994) present figures showing it is possible to align the primary β sheets of the four enzymes, including PvuII. The common features of the four sheets include a five-stranded mixed β sheet with the cleavage site adjacent to the third (middle) strand of the sheet. Although the DNA is not shown in the figures of Cheng et al. the other results show it would align with the third β strand as for EcoRI, i.e., a chimeric β sheet appears to be present in all four cases.

As described above, the conserved carboxylates (Asp[91] and Glu[111] in EcoRI) receive hydrogen bonds from the polypeptide backbone when they coordinate manganese; a backbone carbonyl from the β sheet also coordinates the divalent cation and a critical water molecule is sandwiched between the protein and DNA backbones. These features create a requirement for a β strand at the cleavage site providing a structural explanation for the conservation of this feature. The chimeric β sheets then form a natural way of aligning the DNA backbone with the cleavage site. This alignment of secondary structure with the DNA and the cleavage site suggests it could be very ancient, i.e., going back to the time when proteins began to take over catalytic functions from the RNA world.

Are chimeric β sheets unique to restriction enzymes? Venclovas and Siksnys noted the similarities between restriction endonucleases and the superfamily of polynucleotidyl transferases (PNT) which include HIV-1 integrase,

RNase H, and RuvC (Venclovas and Siksnys 1995). Specifically, they noticed that the geometry of the β sheets in the vicinity of the active site is similar for both classes of enzymes. PNTs not only require a Mg^{2+} cofactor for catalysis but they have active site residues corresponding to EcoRI's Asp^{91} and Glu^{111} located on a central strand of β sheet. Given, however, the different ordering of the β sheets (e.g., parallel vs. antiparallel) between these two classes of enzymes, they do not appear to represent divergent evolution from a common ancestor. But the structural analogy is striking and the catalytic involvement of the extended strand of β sheet suggests a functional reason for the similarities of the structures. Thus, it would appear that the structural analogy represents convergent evolution driven by the functional requirements of the mechanism.

4 Apo-EcoRI Endonuclease (The Protein in the Absence of DNA)

Efforts to crystallize EcoRI endonuclease without DNA and determine its structure were initiated almost 25 years ago. However the crystals diffracted weakly, were highly radiation sensitive, and, until recently, were resistant to cryoprotection. Despite these difficulties, a 3.3-Å resolution dataset was collected from 13 different crystals. It enabled a preliminary structure based on molecular replacement methods (Chandrasekhar, Wilkosz and Rosenberg, unpublished). Subsequently, ongoing efforts at cryoprotection were successful enabling the collection of a synchrotron (CHESS) dataset from a single crystal to 2.7-Å resolution. The synchrotron data were markedly better (in terms of completeness and signal-to-noise ratio) than the room temperature data and the resulting structure represents a significant improvement. In this section, we present an initial description of that structure and a preliminary analysis of its implications; a full account will be reported elsewhere (Grigorescu, et al., manuscript in preparation).

4.1 Three Polypeptide Chains in the Asymmetric Unit

EcoRI endonuclease (in the absence of DNA) crystallizes in space group C_2 with unit cell dimensions a=206.4 Å, b=125.4 Å, c=48.4 Å, α=90°, β=98°, γ =90°. The asymmetric unit contains three protein chains designated A, B, and C. Chains A and B form a dimer about a noncrystallographic twofold rotation axis, chain C dimerizes with a symmetry-related C subunit about a crystallographic twofold axis. The three independent subunits are also related by a noncrystallographic threefold axis of rotational symmetry which is inclined by approximately 20° with respect to the crystallographic c-axis.

4.2 Disordered "Arm" Regions

The three chains exhibit very similar structural features with the notable exceptions of regions 116–140 and 170–196. These regions of the protein form each "arm;" there are two such arms per dimer and in the cognate DNA-complex they surround the DNA (Rosenberg 1991).

Residues 116–140 and 170–196 of subunits A and C could not be modeled into the electron density, which is completely broken up and un-interpretable there. In contrast, the density for the rest of the structure was very clear and easily interpretable. We conclude that segments 116–140 and 170–196 do not have well ordered conformations in chains A and C. These regions are shown in Fig. 9. However, subunit B has good density for most of this region, including residues 116–129, 133–140, and 170–196. In summary, the "arms" of EcoRI endonuclease are disordered in chains A and C, but reasonably well ordered in chain B (though not as well ordered as in the DNA-containing complexes).

When ordered arms were modeled into subunits A and C (based on that of subunit B), constrained to be identical to subunit B and refined, the real space R factors are significantly higher in the modeled regions of chains A and C than those of the corresponding segment of subunit B; indeed, they are unacceptably large. The crystallographic R_{free} increased whenever we attempted to model ordered arms in subunits A and C.

4.3 Relation to Crystallographic Packing

The differential ordering of arms in the three protein chains appears to originate from differences in their packing environments. A segment of the β hairpin portion of the arm in chain B makes a series of van der Waals contacts and hydrogen bonds with an identical region of a symmetry-related equivalent the same chain, including four main chain–main chain hydrogen bonds made between residues Asp[185], Gly[186], Arg[187]. The buried surface area associated with these contacts is 315 Å^2.

Modeling efforts show no *a priori* reason for the disordered arms in subunits A and C. They would not form crystallographic close contacts or unfavorable interactions that would specifically prevent their formation. Indeed, each of the modeled arms is fully exposed in large solvent channels, i.e., completely bathed in solvent. In other words, the complete absence of tertiary and/or symmetry-related contacts is strikingly associated with the disordered segments of chains A and C.

These observations indicate that the intermolecular interactions made by subunit B in the crystal lattice are most likely the factor responsible for stabilizing the conformation of its arms. Similar effects of packing interactions on stabilizing relatively large protein regions have been previously reported in other crystallographic studies (Freitag et al. 1997; Perona and Martin 1997a;

A

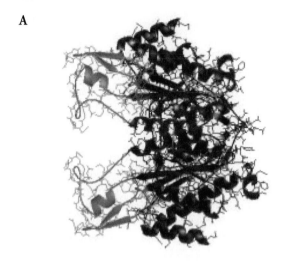

B

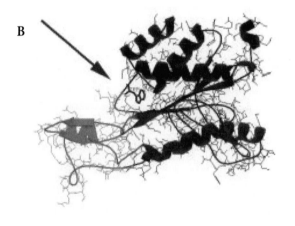

C

Fig. 9. Regions of EcoRI endonuclease disordered in the absence of DNA. **A** A protein dimer (using coordinates from the pre-transition state complex) color-coded to show (in *red*) the disordered regions, i.e., the arms, in subunits A and C of the apo-protien (see text). Regions shown in *blue* remain highly similar with and without DNA bound. **B** The *arrow* indicates the location of Trp[246]. It is shown on one subunit, which has been rotated slightly from that in **A** to show its location clearly. Trp[246] shows altered fluorescence properties when DNA binds (see text). **C** The location of two classes of mutants isolated by Heitman and collaborators are shown on this stereo diagram that is color-coded consistent with **A** and **B**. All the atoms of the wild-type alleles of the mutants are shown as van der Waals spheres. Temperature-sensitive mutations are shown in *light blue*; they disrupt folding and/or dimerization of the subunits (see text). Promiscuous mutants, which reduce the specificity of the enzyme, are shown in *orange*

Raghunathan et al. 1994; Vigil et al. 2001). Particularly relevant from the perspective of the present study is the case of Type II restriction endonuclease EcoRV. An analysis of the conformation of the unliganded protein in different crystal forms revealed that a 28 amino acid segment localized at the C terminal of the protein is stabilized only when packing interactions are made in the crystal (Perona and Martin 1997b). In this study Perona and Martin have suggested that folding of the C terminal region of EcoRV endonuclease is controlled by local ordering of an adjacent segment in response to DNA binding.

As discussed above, all known EcoRI endonuclease crystal packing arrangements are based on highly similar sheets with the principal packing differences being how the sheets are stacked upon each other (and, of course, the effects of DNA, if present). This includes the apoprotein described here, where the sheets are stacked through the packing interactions made by the outer arms of the B subunits, i.e., they form inter-sheet contacts. In the cocrystals containing the specific EcoRI-DNA complex the corresponding inter-sheet contacts are made by the ends of the DNA molecule. In this case the arms do not make inter-sheet contacts because the sheets have been displaced parallel to their planes such that the arms are too far apart (this displacement is required so that the ends of the DNA molecules make more stable inter-sheet lattice contacts). In the specific complex the arms are wrapped around the DNA and the regions connecting the arms to the main domain of the protein make a series of highly specific interactions with the DNA bases of the cognate site. In other words, in the cognate complexes, protein–DNA contacts stabilize the arms.

4.4 Order-Disorder Transition Associated with DNA Binding

The comparative analysis of the free and DNA-bound structures of the protein as revealed by X-ray crystallography indicates that binding of EcoRI to its specific DNA site is accompanied by folding of the arms. This model is supported by previous thermodynamic and biophysical studies. In fact, the magnitude and location of this folding transition were predicted by Spolar and Record (1994) based on thermodynamic premises.

Other solution studies are consistent with an order-disorder transition. Early in their analysis of the EcoRI endonuclease system, it was noted by Jen-Jacobson and coworkers (1986) that the DNA-free protein is very susceptible to proteolysis while the specific DNA–protein complex is much more resistant. More recently, spectroscopic studies performed by Watrob and Barkley have suggested a conformational change associated with DNA-binding (Watrob et al. 2001). They monitored changes in the fluorescence of Trp[246] upon binding of the enzyme to the specific DNA site; they indicated a major structural change in the protein taking place in the environment of this residue. Trp[246] is located approximately 12 Å away from the DNA binding interface but only 3–4 Å from residues 170–173 of the arm which – according to our data- become ordered upon DNA binding. Indeed, residue Trp[246] is for-tuitously placed in an ideal position to report on the order-disorder transi-tion, as can be seen in Fig. 9B.

A re-examination of molecular architecture of the EcoRI protein in the cognate complex reveals that each subunit consist of two major structural regions: A relatively large, well structured globular domain (referred to as the primary domain or the main domain of EcoRI) and a smaller, region with low structural stability in the apo enzyme corresponding to the arms. The arms of each monomer form long protrusions from the main domain, which are ori-entated away from the protein–protein interface. In the specific complexes, or more generally when ordered enough to be visible in the crystal structure, the arms contain elements of secondary structure, including a 3:5-IG hairpin (formed by residues 179–191) and 1.5 turns of an α helix (formed by residues 118–123). Within each monomer, the arms make tertiary interactions with each other but not with the rest of the protein. These observations suggest that the arms of EcoRI exhibit local rather than global cooperativity and might represent a separate folding unit (i.e., a minidomain).

The regions connecting the arms with the main domain appear to have increased flexibility in the absence of DNA – as revealed by their conforma-tional heterogeneity and elevated B factors in the three subunits of the apo-enzyme. Examination of the hydrogen bonding patterns at the protein–DNA interface shows that the polypeptide segments 138–145 and 200–203 are mak-ing all the hydrogen bonds with the functional groups of the GAATTC bases (see Fig. 3). Both these regions form the boundary between the arms and the main domain. They appear to be relatively flexible in the unbound protein but become very well ordered in the specific complex; they are also central to the extended recognition network (see Fig, 3). One hypothesis is that ordering of these segments in response to DNA binding "triggers" the ordering of the arms. A very similar hypothesis has been proposed for EcoRV (Perona and Martin 1997b).

The idea that the uneven distribution of structural stability in the EcoRI apo-enzyme plays a key role in molecular recognition is supported by a re-examination of two classes of mutants previously isolated by Muir et. al. (1997) and Heitman and Model (1990a). In a genetic screen for temperature sensitive mutants Heitman and Model (1990a) have isolated 11 single amino acid, tem-perature-sensitive substitutions; in vivo and in vitro characterization of the phenotype and properties of these mutants indicated that these substitutions impair correct folding and/or dimerization of the protein at non-permissive temperatures. As shown in Fig. 9C all 11 sites identified in this study cluster in the primary domain of EcoRI endonuclease. Most of these substitutions are predicted to disrupt local structure in the main domain or alter the dimeriza-tion of the protein (Heitman and Model 1990a). This is consistent with the idea that many of the residues located in the main domain have the primary role of providing and maintaining the correct architectural foundation, the scaffold for the three dimensional structure of the dimeric protein.

In contrast, the amino acid substitutions that reduce the sequence specificity of the enzyme (promiscuous mutants) (Heitman and Model 1990b) are all localized in the regions with low structural stability in the apo-protein: Glu^{192} and Tyr^{193} are in the outer segment of the arm, Ala^{138} and His^{114} are at the boundary between the arm and the primary domain (see Fig. 9C). A mutation (as well as solution conditions) that alter the native distribution of structural stability in the protein could reduce the overall sequence discrimination of the enzyme – if the sequence discrimination mechanism depends on the correct "triggering" of the cooperative pathway for stabilizing these interactions. Our hypothesis is that the promiscuous mutants perturb the conformational equilibrium between various states. Such perturbations could manifest in the unliganded protein, any of the DNA-bound forms and/or any combination of the forgoing. The functional effect of the perturbations would be to reduce the specificity of the enzyme. Because the effects can be manifest at multiple levels, untangling them will require a careful study on a mutant-by-mutant basis.

Finally, we note that, in aggregate, approximately 100 amino acid residues undergo an order-disorder transition in the protein dimer upon site-specific binding; this conformational change is likely to make a significant energetic contribution to complex formation. The issue here is the number of conformers that populate the disordered state. This is difficult to assess because it relates to that part of the structure, which we do not see in the apo-protein. However, the appearance of the electron density there is such that it appears the number of conformers is not small; a reasonable model for the disordered arm would therefore be a molten globule.

As mentioned above, based on thermodynamic data and the structural information on the EcoRI-DNA complex (Kim et al. 1990), Spolar and Record (1994) predicted a major local folding reaction coupled with specific binding for EcoRI endonuclease; these authors estimated the number of residues involved in this order–disorder transition to be about 78. Subsequently, Jen-Jacobson and coworkers questioned the basis of Spolar and Record's calculations on a number of grounds, including the experimental observation of significant heat capacity differences associated with changes in the flanking sequence environment (Jen-Jacobson 1997; Jen-Jacobson et al. 1996, 2000a, b); indeed, these workers assign a central role to strain in the determination of the thermodynamics (Jen-Jacobson et al. 2000a). Strain clearly contributes, however, in the light of this controversy, and of the many other effects that also contribute to the thermodynamics, it is difficult to make a quantitative statement at this time concerning the magnitude of the thermodynamic effects associated with the folding of the arms.

We do want to point out, however, that the differences in the thermodynamic parameters observed between the complexes with various flanking sequences (Jen-Jacobson et al. 2000a, b) might contain a "hidden" contribution from the coupled folding transition. The GAATTC sites embedded in various flanking contexts represent for EcoRI endonuclease a series of homologous ligands for which the enzyme exhibits a 300-fold range in binding affinity preferences. Thus, according to the theoretical model of Eftnik and coworkers (Eftink et al. 1983) the measured heat capacity, enthalpy and entropy changes upon formation of each of these complexes most likely contain *different* contributions from the coupled transition.

5 Summary and Conclusions

Enzymatic cleavage of cognate DNA within the crystalline environment was achieved by soaking Mn^{2+} into pre-transition state–EcoRI endonuclease cocrystals (grown in the absence of divalent cation). The product was readily visualized in electron density maps; the (single) Mn^{2+} ion was coordinated by the post-reactive phosphate, the side chains of Asp^{91}, Glu^{111}, the main chain carbonyl of Ala^{112} and two water molecules. Hydrogen bonds donated by peptide amides stabilized the acidic side chains. Arg^{145} and Lys^{113} hydrogen bond with the scissile phosphate, suggesting that they have a role in polarizing it. The pre-transition state and post-reactive cocrystals show no significant differences outside the immediate region of the cleavage site. The *in crystallo* reaction is possible because the critical regions of the protein and DNA are not involved in lattice contacts.

These results also suggest that minimal atomic movements are required to affect the transition from the pre-transition state- to post-reactive-complex; utilizing this, we propose a detailed mechanism for the reaction. The proposed mechanism preserves previous suggestions of an in line (S_N2) attack by a metal-activated water molecule (hydroxide ion) with the observed Mn^{2+} being that metal. A novel feature of the mechanism is the role of an additional bound water molecule (W_7) as the Lewis base which abstracts a proton from the attacking water. The second water is appropriately positioned by hydrogen bonds to the protein and the DNA. The latter is a hydrogen bond to a phosphate which would raise the pK_a of the second water. The proposed mechanism therefore includes an element of pre-transition state assisted catalysis.

Salient features of the DNA–EcoRI endonuclease pre-transition state complex include the redundancy of the interactions between the protein and the bases. These overdetermine the recognition, providing a structural interpretation for the observation that single amino acid substitutions do not alter the DNA sequence targeted by this protein.

All the critical groups at the DNA–protein interface are linked by an extended recognition network of hydrogen bonds. This network can be subdivided into an inner region including the central tetranucleotide of the EcoRI site while the outer network incorporates the outer G•C pairs. The inner region is highly cross-linked while the outer region is less so. This appears to correlate with Jen-Jacobson's findings that base analog substitutions within the inner region are characterized by additive and isosteric properties while the outer region is non-additive and adaptive in its response to these changes (Jen-Jacobson et al. 1991; Lesser et al. 1990a).

Recognition and cleavage are integrated at two levels: (1) the amino acid side chains with an identified role in the cleavage reaction also participate in the recognition network; perturbations of that network would therefore perturb the catalytic site. (2) Some residues, such as Arg[145], have a dual role, simultaneously donating a sequence-specific hydrogen bond to one of the bases while polarizing the scissile phosphate. This integration of function provides an obvious mechanism for enhancing the sequence-specificity of this restriction enzyme.

The arms of EcoRI endonuclease are disordered in the absence of DNA (or another stabilizing interaction, such as the lattice contact made by subunit B. This crystallographic evidence in accordance with several previous solution experiments indicate that a major localized folding transition accompanies the formation of the complex between the EcoRI enzyme and its specific DNA site. The arms of the protein do not posses a well ordered conformation in the unliganded form but become ordered in the cognate complex. The correlation with previous thermodynamic results indicates that the folding transition does not take place when the protein binds non-specifically to DNA. Specific interactions of the protein with the GAATTC site appear therefore to promote the folding/stabilization of the arms – but the mechanisms and pathways that propagate the binding interactions to distant regions of the protein are yet to be identified.

Acknowledgements. This work was supported by grants from NIH (GM62221), the CHESS synchrotron facility at Cornell, and the Pittsburgh Supercomputing Center. Supercomputer time was provided by the Pittsburgh Supercomputer Center through grant (DMB890026P).

References

Aiken CR, McLaughlin LW, Gumport RI (1991) The highly homologous isoschizomers RsrI enodonuclease and EcoRI endonuclease do not recognize their target sequence identity. J Biol Chem 266:19070–19078

Alves J, Ruter T, Geiger R, Fliess A, Maass G, Pingoud A (1989) Changing the hydrogen-bonding potential in the DNA binding site of EcoRI by site-directed mutagenesis drastically reduces the enzymatic activity, not, however, the preference of this restriction endonuclease for cleavage within the site -GAATTC-. Biochemistry 28:2678–2684

Anderson JE (1993) Restriction endonucleases and modification methylases. Curr Opin Struct Biol 3 24–30

Athanasiadis A, Vlassi M, Kotsifaki D, Tucker PA, Wilson KS, Kokkinidis M (1994) Crystal structure of PvuII endonuclease reveals extensive structural homologies to EcoRV. Nature Struct Biol 1:469–475

Bella J, Berman HM (1996) Crystallographic evidence for Ca-H--O=C hydrogen bonds in a collagen triple helix. J Mol Biol 264:734–742

Brennan CA, Van Cleve MD, Gumport RI (1986) The effects of base analogue substitutions on the cleavage by the EcoRI restriction endonuclease of octadeoxyribonucleotides containing modified EcoRI recognition sequences. J Biol Chem 261:7270–7278

Bunton CA, Mhala MM, Oldham KG, Vernon CA (1960) The reactions of organic phosphates. 3. The hydrolysis of dimethyl phosphate. J Chem Soc 81:3293–3301

Carter CW Jr, Kraut J (1974) A proposed model for interaction of polypeptides with RNA. Proc Natl Acad Sci USA 71:283–287

Cheng C-Y, Wang S-L (1991) Structure acetate dihydrate. Acta Cryst C47:1734–1736

Cheng X, Balendiran K, Schildkraut I, Anderson JE (1994) Structure of PvuII endonuclease with cognate DNA. EMBO J 13:3927–3935

Chmelikova R, Loub J, Petrjcek V (1986) Structure of manganese(II) sodium dihydrogenphosphite monohydrate. Acta Cryst C42:1281–1283

Choi J (1994) Crystal structure analysis of site-directed mutants of EcoRI enodnuclease complexed to DNA, PhD Thesis, University of Pittsburgh

Church GM, Sussman JL, Kim S-H (1977) Secondary structural complementarity between DNA and proteins. Proc Natl Acad Sci USA 74:1458–1462

Connolly BA, Eckstein F, Pingoud A (1984) The stereochemical course of the restriction endonuclease EcoRI-catalyzed reaction. J Biol Chem 259:10760–10763

Cudennec Y, Riou A, Gerault Y (1989) Manganese(II) hydrogenphosphate trihydrate. Acta Cryst C45:1411–1412

Derewenda ZS, Lee L, Derewenda U (1995) The occurrence of C-H–O hydrogen bonds in proteins. J Mol Biol 252:248–262

Drew HR, Dickerson RE (1981) Structure of a B-DNA dodecamer. III. Geometry of hydration. J Mol Biol 151:535–556

Duan Y, Wilkosz P, Rosenberg JM (1996) Dynamic contributions to the DNA binding entropy of the EcoRI and EcoRV restriction endonucleases. J Mol Biol 264:546–555

Eftink MR, Anusiem AC, Biltonen RL (1983) Enthalpy-entropy compensation and heat capacity changes for protein–ligand interactions: general thermodynamic models and data for the binding of nucleotides to ribonuclease A. Biochemistry 22:3884–3896

Freitag S, Le Trong I, Klumb L, Stayton PS, Stenkamp RE (1997) Structural studies of the streptavidin binding loop. Prot Sci 6:1157–1166

Fritz A, Kuster W, Alves J (1998) Asn(141) is essential for DNA recognition by EcoRI restriction endonuclease. FEBS Lett 438:66–70

Garrett TPJ, Guss JM, Greeman HC (1983) trans-Diaquatetrakis(imidazole)manganese(II) dichloride. Acta Cryst C39:1031–1034

Grable J, Frederick CA, Samudzi C, Jen-Jacobson L, Lesser D, Greene P, Boyer HW, Itakura K, Rosenberg JM (1984) Two-fold symmetry of crystalline DNA–EcoRI endonuclease recognition complexes. J Biomol Struct Dyn 1:1149–1160

Hager P, Reich N, Day J, Coch TG, Boyer HW, Rosenberg JM, Greene P (1990) Probing the role of glutamic acid 144 in the EcoRI endonuclease using aspartic acid and glutamine replacements. J Biol Chem 265:21520–21526

Heitman J, Model P (1990a) Mutants of the EcoRI endonuclease with promiscuous substrate specificity implicate residues involved in substrate recognition. EMBO J 9(10): 3369–3378

Heitman J, Model P (1990b) Substrate recognition by the EcoRI endonuclease. Proteins 7:185–197

Jeltsch A, Alves J, Maass G, Pingoud A.(1992) On the catalytic mechanism of EcoRI and EcoRV. A detailed proposal based on biochemical results, structural data and molecular modelling. FEBS Lett 304:4–8

Jen-Jacobson L (1995) Structural-perturbation approaches to thermodynamics of site-specific protein–DNA interactions. In: Johnson ML, Ackers GK (eds) Methods in enzymology, vol 259. Academic Press, San Diego, pp 305–344

Jen-Jacobson L (1997) Protein–DNA recognition complexes: conservation of structure and binding energy in the transition state. Biopolymers 44:153–180

Jen-Jacobson L, Engler LE, Ames JT, Kurpiewski, MR, Grigorescu A (2000a) Thermodynamic paramters of specific and nonspecific protein–DNA binding. Supermol Chem 12(2):143 + Special Issue

Jen-Jacobson L, Engler LE, Jacobson LA (2000b) Structural and thermodynamic strategies for site-specific DNA binding proteins.[erratum appears in Structure Fold Des 2000, Dec 15, 8(12):251 following]. Structure 8:1015–1023

Jen-Jacobson L, Engler LE, Lesser DR, Kurpiewski M R, Yee C, McVerry B (1996) Structural adaptations in the interaction of EcoRI endonuclease with methylated GAATTC sites. EMBO J 15:2870–2882

Jen-Jacobson L, Lesser D, Kurpiewski M (1986) The enfolding arms of EcoRI endonuclease: role in DNA binding and cleavage. Cell 45:619–629

Jen-Jacobson L, Lesser DR, Kurpiewski MR (1991) DNA sequence discrimination by EcoRI endonuclese. In: Eckstein F, Lilley DMJ (eds) Nucleic acids and molecular biology. Springer, Berlin Heidelberg New York, pp 142–170

Kim Y, Choi J, Grable JC, Greene P, Hager P, Rosenberg JM (1994) Studies on the canonical DNA–EcoRI endonuclease complex and the EcoRI kink. In: Sarma RH, Sarma MH (eds) Structural biology: the state of the art. Adenine Press, Schenectady, pp 225–246

Kim YC, Grable JC, Love R, Greene PJ, Rosenberg JM (1990) Refinement of EcoRI endonuclease crystal structure: a revised protein chain tracing. Science 249:1307–1309

King K, Benkovic SJ, Modrich P (1989) Glu-111 is required for activation of the DNA cleavage center of EcoRI endonuclease. J Biol Chem 264:11807–11815

Kumamoto J, Cox JR, Westheimer FH (1956) J Am Chem Soc 77:4858–4860

Kumar S, Duan Y, Kollman PA, Rosenberg JM (1994) Molecular-dynamics simulations suggest that the EcoRI kink is an example of molecular strain. J Biomol Str Dyn 12:487–525

Lesser DR, Grajkowski A, Kurpiewski MR, Koziolkiewicz M, Stec W, Jen-Jacobson L (1992) Stereoselective interaction with chiral phosphorothioates at the central DNA kink of the EcoRI endonuclease-GAATTC complex. J Biol Chem 267:24810–24818

Lesser DR, Kurpiewski MR, Jen-Jacobson L (1990a) The energetic basis of sequence specificity in the interaction of EcoRI endonuclease with DNA. Science 250:776–786

Lesser DR, Kurpiewski MR, Jen-Jacobson L (1990b) The energetic basis of specificity in the EcoRI endonuclease-DNA interaction. Science 250:776–786

Lesse, DR, Kurpiewski MR, Waters T, Connolly BA, Jen-Jacobson L (1993) Facilitated distortion of the DNA site enhances EcoRI endonuclease-DNA recognition [see comments]. Proc Natl Acad Sci USA 90:7548–7552

Lightfoot P, Cheetham AK (1987) Structure of manganese(II) trisodium tripolyphosphate dodecahydrate. Acta Cryst C43:4–7

Lis T (1982) Structure of manganese(II) L-lactate Dihydrate. Acta Cryst B38:937–939

Lis T (1983). Structure of manganses(II) maleate trihydrate, and reeinvestigation of the structure of manganese(II) hydrogen tetrahydrate. Acta Cryst C39:39–41

Lis T (1992) Structure of zinc(II), magnesium(II) and manganese(II) bis(phosphoenolpyruvate) dihydrate. Acta Cryst C48:424–427

Mandel-Gutfreund Y, Margalit H, Jernigan RL, Zhurkin VB (1998) A role for CH–O interactions in protein–DNA recognition. J Mol Biol 277:1129–1140

McLaughlin LW, Benseler F, Graeser E, Piel N, Scholtissek S (1987) Effects of functional group changes in the EcoRI recognition site on the cleavage reaction catalyzed by the endonuclease. Biochem 26:7238–7245

Muir RS, Flores H, Zinder ND, Model P, Soberon X, Heitman J (1997) Temperature-sensitive mutants of the EcoRI endonuclease. J Mol Biol 274:722–737

Needels MC, Fried SR, Love R, Rosenberg JM, Boyer H W, Greene PJ (1989) Determinants of EcoRI endonuclease sequence discrimination. Proc Natl Acad Sci USA 86:3579–3583

Newman M, Strzelecka T, Dorner LF, Schildkraut I, Aggarwal AK (1994) Structure of restriction endonuclease bamhi phased at 1.95 Å resolution by MAD analysis. Structure 2:439–452

Oelgeschlager T, Geiger R, Ruter T, Alves J, Fliess A, Pingoud A (1990) Probing the function of individual amino acid residues in the DNA binding site of the EcoRI restriction endonuclease by analysing the toxicity of genetically engineered mutants. Gene 89:19–27

Otwinowski Z, Schevitz RW, Zhang RG, Lawson CL, Joachimiak A, Marmorstein RQ, Luisi BF, Sigler PB (1988) Crystal structure of trp repressor/operator complex at atomic resolution. Nature 335:321–329

Parkinson G, Vojtechnovsky J, Clowney L, Brunger A T, Berman HM (1996) New parameters for the refinement of nucleic acid containing structures. Acta Cryst D52:57–64

Perona JJ, Martin AM (1997a) Conformational transitions and structural deformability of EcoRV endonuclease revealed by crystallographic analysis. J Mol Biol 273:207–225

Perona JJ, Martin AM (1997b). Conformational transitions and structural deformability of EcoRV endonuclease revealed by crystallographic analysis. J Mol Biol 273:207–225

Perry KM, Fauman EB, Finer-Moore JS, Montfort WR, Maley GF, Maley F, Stroud RM (1990) Plastic adaptation toward mutations in proteins: structural comparison of thymidylate synthases. Proteins: Struct Funct Genet 8:315–333

Pingoud A, Jeltsch A (1997) Recognition and cleavage of DNA by Type-II restriction endonucleases. Eur J Biochem 246:1–22

Radzicka A, Wolfenden R (1995) A proficient enzyme. Science 267:90–93

Raghunathan S, Chandross RJ, Kretsinger RH, Allison TJ, Penington CJ, Rule GS (1994) Crystal structure of human class mu glutathione transferase GSTM2-2. Effects of lattice packing on conformational heterogeneity. J Mol Biol 238:815–832

Richardson JS (1981) The anatomy and taxonomy of protein structure. Adv Protein Chem 34:167–339

Rosenberg JM (1991) Structure and function of restriction endonucleases. Curr Opin Struct Biol 1:104–113

Ross NL, Reynard B, Guyot F (1991) Structure of high-pressure Mn GeO$_3$ ilmenite. Acta Cryst C47:1794–1796

Samudzi CT (1990) Use of the molecular replacement method in structural studies of EcoRI endonuclease, PhD, University of Pittsburgh

Schneider B, Cohen DM, Schleifer L, Srinivasan AR, Olson WK, Berman HM (1993) A systematic method for studying the spatial distribution of water molecules around nucleic acid bases. Biophys J 65:2291–2303

Seeman NC, Rosenberg JM, Rich A (1976) Sequence-specific recognition of double helical nucleic acids by proteins. Proc Natl Acad Sci USA 73:804–808

Serpersu EH, Shortle D, Mildvan AS (1987) Kinetic and magnetic resonance studies of active-site mutants of staphylococcoal nuclease: factors contributing to catalysis. Biochemistry 26:1289–1300

Spolar RS, Record MT Jr (1994) Coupling of local folding to site-specific binding of proteins to DNA. Science 263:777–784

Steitz TA (1990) Structural studies of protein–nucleic acid interaction: the sources of sequence-specific binding. Q Rev Biophys 23:205–280

Tanaka I, Appelt K, Dij KL, White SW, Wilson KS (1984) 3 Å resolution structure of a protein with histone-like properties in prokaryotes. Nature 310:376–381

Thielking V, Alves J, Fleiss A, Maass G, Pingoud A (1990) Accuracy of the EcoRI endonuclease: binding and cleavage studies with oligodeoxynucleotide substrates containing degenerate recognition sequences. Biochemistry 29:4682–4691

Thielking V, Selent U, Kohler E, Wolfes H, Pieper U, Geiger R, Urbanke C, Winkler FK, Pingoud A (1991) Site-directed mutagenesis studies with EcoRV restriction endonuclease to identify regions involved in recognition and catalysis. Biochemistry 30:6416–6422

Thomas M, Davis RW (1975) Studies on the cleavage of bacteriophage lambda DNA with EcoRI restriction endonuclease. J Mol Biol 91:315–328

Venclovas C, Siksnys V (1995) Different enzymes with similar structures involved in Mg^{2+}-mediated polynucleotidyl transfer [letter]. Nat Struct Biol 2:838–841

Venclovas C, Timinskas A, Siksnys V (1994) Five-stranded beta sheet sandwiched with two alpha-helices: a structural link between restriction endonucleases EcoRI and EcoRV. Proteins 20:279–282

Vigil D, Gallagher SC, Trewhella J, Garcia AE (2001) Functional dynamics of the hydrophobic cleft in the N-domain of calmodulin. Biophys J 80:2082–2092

Watrob H, Liu W, Chen Y, Bartlett SG, Jen-Jacobson L, Barkley MD (2001) Solution conformation of EcoRI restriction endonuclease changes upon binding of cognate DNA and Mg^{2+} cofactor. Biochemistry 40:683–692

White SW, Appelt K, Wilson KS, Tanaka I (1989) A protein structural motif that bends DNA. Proteins 5:281–288

Wilkosz PA, Chandrasekhar K, Rosenberg JM (1995) Preliminary characterization of EcoRI*-DNA Co-crystals: factorial design strategies for oligonucleotide sequences used in protein–DNA cocrystals. Acta Crsyt D 51:938–945

Winkler FK (1992) Structure and function of restriction endonucleases. Curr Opin Struct Biol 2:93–99

Winkler FK, Banner DW, Oefner C, Tsernoglou D, Brown RS, Heathman SP, Bryan RK, Martin PD, Petratos K, Wilson KS (1993a) The crystal structure of EcoRV endonuclease and of its complexes with cognate and noncognate DNA. EMBO J 12:1781–1795

Winkler FK, Banner DW, Oefner C, Tsernoglou D, Brown RS, Heathman SP, Bryan RK, Martin PD, Petratos K, Wilson KS (1993b) The crystal structure of EcoRV endonuclease and of its complexes with cognate And non-cognate DNA fragments. EMBO J 12:1781–1795

Wolfes H, Alves J, Fliess A, Geiger R, Pingoud A (1986) Site directed mutagenesis experiments suggest that Glu 111, Glu 144 and Arg 145 are essential for endonucleolytic activity of EcoRI. Nucleic Acids Res 14:9063–9080

Yanofsky SD, Love R, McClarin JA, Rosenberg JM, Boyer HW, Greene PJ (1987) Clustering of null mutations in the EcoRI endonuclease. Proteins 2:273–282

Structure and Function of EcoRV Endonuclease

F.K. Winkler, A.E. Prota

1 Introduction

The homodimeric, orthodox Type II restriction endonuclease EcoRV cleaves double-stranded DNA with high specificity at hexameric GAT↓ATC sites in a blunt-ended fashion. Like most of the restriction enzymes used as tools in recombinant DNA technology it can be considered as a simple hydrolytic enzyme of relatively small size with precisely defined substrate specificity and the rather common requirement for Mg^{2+} ions. Yet, these apparently simple enzymes become fascinating objects for structural and functional studies, and their molecular simplicity becomes an advantage once we start asking detailed mechanistic questions. For example, it is quite obvious that these enzymes have been engineered by evolution to deal with the problem of efficiently locating a specific site on DNA, which is immersed in a huge molar excess of other, partly very similar sites.

Initially, the exceptionally high and diverse sequence specificity of these enzymes was the major driving force for their study, and they were considered as model systems for gaining general insights into DNA recognition by proteins. The large number of protein–DNA complexes analyzed to date has shown that there is no simple code to classify or predict such interactions (Luscombe et al. 2001). Nevertheless, structural studies with a number of restriction endonucleases, in particular with EcoRI and EcoRV, have become a rich source for analyzing the diversity and detailed stereochemistry of such interactions. To date, the crystal structures of 14 Type II restriction endonucleases have been determined in one or more states, unliganded or bound to double-stranded DNA fragments. Their characteristic functional and structural features have been discussed in several reviews (Pingoud and Jeltsch 1997, 2001; Perona 2002). In the course of these studies, many new and more

F.K. Winkler, A.E. Prota
Biomolecular Research, Paul Scherrer Institut, 5232 Villigen, Switzerland

Nucleic Acids and Molecular Biology, Vol. 14
Alfred Pingoud (Ed.)
Restriction Endonucleases
© Springer-Verlag Berlin Heidelberg 2004

precisely defined questions have become apparent. They relate to the full functional cycle of these enzymes which involves binding to nonspecific DNA, searching for target sites, formation of the productive complex with Mg^{2+} ions at the target site, double-strand cleavage and product dissociation. The chemical rate of phosphodiester hydrolysis has turned out to be very sensitive to DNA substrate–enzyme interactions remote from the active site. The understanding of the structural basis of this remarkable coupling between recognition and catalysis and of the details of the divalent ion catalyzed hydrolytic mechanism have become of particular interest. Considerable adaptability in tertiary and quaternary structure is a hallmark of EcoRV endonuclease and appears to be needed to proceed through the stages of this cycle.

In the first part of this chapter, we will give an overview of the crystallographic work on EcoRV endonuclease, and we will describe the salient structural features that have emerged. These rather comprehensive and detailed studies have been aided greatly by the readiness of this enzyme to crystallize in several different crystal forms. This is perhaps best illustrated by the fact that 18 structures from 7 different crystal forms have been analyzed and deposited (Table 1). EcoRV is a telling example on how structural and functional studies, carried out largely in parallel, can inspire each other. In the second part, the emphasis will be on the biochemical characteristics of this enzyme and on the mechanistic insights that have emerged from combining the structural and biochemical data. It illustrates the pitfalls of our usually simple mechanistic interpretations that tend to assign functional properties to single functional groups or amino acid residues and neglect the complexity of networks of interactions. While not all secrets have been revealed and some issues remain controversial, the mechanistic insights that have been gained certainly bear relevance beyond the specific features of EcoRV endonuclease.

2 Structural Characteristics of EcoRV Endonuclease

The first three crystal structures of the EcoRV endonuclease, jointly published in 1993 (Winkler et al. 1993), were that of the free enzyme containing 244 residues per subunit, that of a complex with a cognate decamer DNA fragment and that with a nonspecific DNA fragment.[1] The three structures represent significantly different structural states, quite obvious at the quaternary struc-

---

[1] Following (Horton and Perona 1998) the term 'noncognate' denotes DNA sites that differ from the specific site GATATC by one or two base pairs (Gewirth and Sigler 1995) and the term 'nonspecific' denotes sites that have three or fewer base pairs in common with this site. The terms specific and cognate are used interchangeably to denote the EcoRV 5'-GATATC recognition sequence.

Table 1. Crystal structures of EcoRV endonuclease

Structure-type	Oligonucleotide	Bound ions	Mutation	Space group	Resolution (Berman et al. 2000)	PDB code	Reference
–	–	–	–	–	–	–	–
Unliganded	–	–	wt	$P2_12_12_1$ (form A)	2.50	1RVE	Winkler et al. (1993)
Unliganded	–	–	wt	$P2_12_12_1$ (form B)	2.40	1AZ3	Perona and Martin (1997)
Unliganded	–	–	T93A	$P2_12_12_1$ (form B)	2.40	1AZ4	Perona and Martin (1997)
–	–	–	–	–	–	–	–
Nonspecific	CGAGCTCG	–	wt	$P2_1$	3.0	2RVE	Winkler et al. (1993)
–	–	–	–	–	–	–	–
Specific (I)[a]	GGGATATCCC	–	wt	$C222_1$	3.0	4RVE	Winkler et al. (1993)
Specific (III)	AAAGATATCTT	Mg^{2+}	wt	P1	2.0	1RVA	Winkler et al. (1993)
Specific (III)	AAAGATATCTT	Mg^{2+}	wt	P1	2.1	1RVB	Winkler et al. (1993)
Specific (III)	AAAGAT/pATCTT	Ca^{2+}	wt	P1	2.1	1RVC	Winkler et al. (1993)
Specific (III)	AAAGATATCTT		wt	P1	2.0	1AZ0	Perona and Martin (1997)
Specific (III)	CGGGATATCCC	–	wt	P1	2.0	1BGB	Horton and Perona (1998a)
Specific (III)	AAAGATATCTT	Ca^{2+}	T93A	P1	2.0	1BSS	Horton et al. (1998)
Specific (III)	AAGA(5mC[b])(I[c])TCTT TTCT(I)(5mC)AGAAA	Ca^{2+}	wt	P1	2.0	1BSU	Martin et al. (1999)
Specific (III)	AAAGAC(I)TCTT	Ca^{2+}	wt	P1	2.15	1BUA	Martin et al. (1999)
Specific (III)	AAAGAT/ATCTT	–	wt	P1	2.10	1RV5	Horton and Perona (1998b)
Specific (III)	AAAGATATCTT	Ca^{2+}	wt	P1	1.90	1B94	Thomas et al. (1999)
Specific (III)	AAAGATATCTT	–	wt	P1	2.05	1B95	Thomas et al. (1999)
Specific (III)	AAAGATATCTT	–	Q69E	P1	2.30	1B96	Thomas et al. (1999)
Specific (III)	AAAGATATCTT	–	Q69L	P1	1.90	1B97	Thomas et al. (1999)
Specific (III)	CAAGA(Tsp[d])ATCTT	Mg^{2+}	wt	P1	2.00	1EO3	Horton et al. (2000)
Specific (III)	CAAGA(Tsp)ATCTT	Mn^{2+}	wt	P1	1.90	1EO4	Horton et al. (2000)
Specific (III)	CAAGA(Tsp) A TCTT TTCT A (Tsp)AGAAA	Ca^{2+}	wt	P1	1.60	1EON	Horton et al. (2000)
Specific (II)	GAAGATATCTTC	–	wt	$C222_1$	2.16	1EOO	Horton and Perona (2000)
Specific (IV)	AAGATATCTTA	–	wt	$P4_12_12$	2.60	1EOP	(Horton and Perona 2000)

[a] Numbers in brackets indicate the crystal forms of specific complexes according to (Horton and Perona 2000)
[b] 5mC: 5-methyl-cytosine
[c] I: Inosine
[d] Tsp: 3'-thio-thymidine

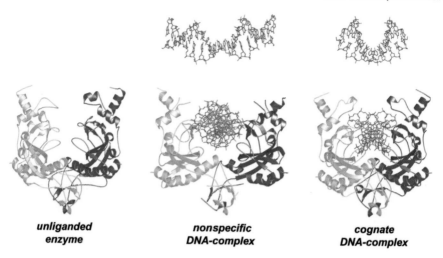

Fig. 1. Structural states of EcoRV endonuclease as observed in the crystal structures 1RVE (unliganded enzyme), 2RVE (complex with nonspecific DNA) and 1RVA (complex with cognate DNA). The two subunits are shown in *yellow* and *blue*, respectively, and the DNA in *red*. On top of the complexes the DNA is shown at right angle from the view below to illustrate the different degree of bending

ture level (Fig. 1). Many additional crystallographic studies have since been reported and are summarized in Table 1. They have addressed a variety of specific structural questions, such as the interactions with different DNA substrates, the structural changes in mutant enzymes and the binding sites of different divalent ions. Moreover, they have revealed further structural states whose characteristics have been systematically analyzed and compared (Perona and Martin 1997; Horton and Perona 2000).

The asymmetrically shaped EcoRV monomer (Fig. 2A) has a mixed α/β structure and shows no subdivision into distinct globular folding domains. However, comparison of the monomer structure observed in the different structural states (Winkler et al. 1993; Perona and Martin 1997) suggests that it is useful to discriminate two domains which we will term dimerization domain (DIM-domain) and DNA binding domain (DB-domain), respectively (Fig. 2A). These two subdomains are linked by four chain segments (L1–L4) that show a variable degree of disorder. In view of the multitude of observed relative orientations between the DIM and DB domain of one monomer the variable order in these linking segments is not surprising. As shown in Fig. 2B, the DNA lies in a deep cleft between the two subunits that form a U-shaped homodimer.

Two separate parts of the protein–protein dimer interface can be discriminated. The first one, formed between the two very small DIM-domains at the bottom of the U-shaped dimer, consists of a short and rigid four-stranded antiparallel β-sheet interaction and is supplemented by a number of hydro-

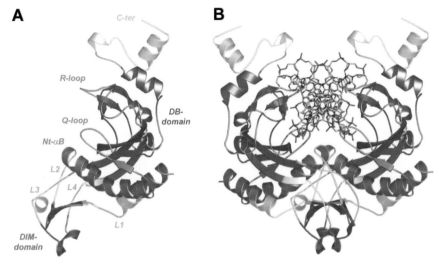

Fig. 2. Ribbon representation of the EcoRV endonuclease monomer and dimer structure. A. Cartoon of the EcoRV monomer (244 residues, 2–245) showing the very small but rigid dimerization DIM-domain (*red*) and the rigid DNA binding DB-domain (*blue*). Structural elements undergoing some order–disorder transitions upon DNA binding are shown in *gray*, *green* and *yellow*. The N-terminal portion of α-helix B (Nt-αB) is shown in *orange*. The R-loop (residues 182–187), the Q-loop (resides 68–71) and a segment of the C-terminus (residues 221–228) are better or fully ordered in complexes with DNA. The four linker segments L1–L4 connecting the DIM- and DB-domain can assume different conformations and are generally better ordered in complexes with DNA. B. Cartoon of the EcoRV dimer (same color code as in A) complexed with a cognate DNA fragment shown in ball-and-stick representation (1RVA, Table 1). The minor groove of the DNA is seen to face the two Nt-αB's at the molecular dyad

phobic side chain interactions. This dimeric substructure serves as a useful reference for superposing and analyzing the different quaternary structure states (Perona and Martin 1997). The second component of the dimer interface is variable in its detailed interactions. It involves hydrophobic side chains from the N-terminal half of the αB helix (Nt-αB) (Fig. 2B) of each subunit and some residues of the DIM domain of each subunit. Analysis of αB in the different structures shows that Nt-αB bends away from the C-terminal portion to different degrees (Perona and Martin 1997).

To first approximation, the EcoRV dimer can be described as consisting of three rigid parts, the dimeric substructure at the bottom of the U-shaped dimer and the two DB-domains (Perona and Martin 1997). The relative orientation and position of the latter two with respect to the former correlates with the functional state (unliganded, bound to specific or nonspecific DNA). The deviations from twofold symmetry are small for all complexes with specific DNA fragments but are substantial in the nonspecific complex and in the unliganded state.

The DB domain contains all the catalytic residues and all structural elements interacting with DNA. It has three segments that are partly or completely disordered at least in some of the crystal structures. Two of these segments, the short Q and R surface loops (Fig. 2), are disordered in the unliganded enzyme but form well-defined interactions with cognate DNA in the minor and major groove, respectively (Winkler et al. 1993). The third segment comprises the 25 C-terminal residues 221–245. In solution, this segment is possibly completely disordered in the unliganded state and becomes ordered only upon binding to DNA. Its partial ordering seen in the crystals of the unliganded enzyme may thus be induced by crystal contacts (Perona and Martin 1997). Furthermore, the degree of order observed for this segment in the complexes with DNA depends on the length of the bound DNA fragment (Winkler et al. 1993; Kostrewa and Winkler 1995).

The combination of flexible and rigid structural elements in EcoRV appears well suited to cope with the functional requirement to effectively screen nonspecific DNA for the target DNA sequence and cleave it with high specificity.

2.1 DNA–Protein Interactions in Specific and Nonspecific Complexes

The steps involved in DNA recognition and cleavage by restriction endonucleases require that these enzymes interact in various ways with DNA (Pingoud and Jeltsch 1997, 2001). At least two generic binding modes, specific and nonspecific, are usually discriminated. The latter is thought to be functionally important for initial binding to large DNA substrates and for efficient location of the target sequence through facilitated diffusion on DNA (see Sect. 3.2.2). Knowledge of at least these two binding modes is needed to understand the structural basis of the high sequence specificity of these enzymes. With substrate DNA, complexes are only stable in the absence of the essential cofactor Mg^{2+}. Such complexes with short cognate DNA fragments have been crystallized for 12 different restriction endonucleases (Pingoud and Jeltsch 2001; Perona 2002). The observed interactions with the diverse target sequences are extensive and tight in all these cases indicating that the absence of the metal cofactor does not interfere with the formation of the specific complex. It has been much more difficult to obtain crystals of nonspecific complexes, probably because many different nonspecific sites can be occupied at roughly equal stoichiometries even in short duplexes. Thus far, the only two examples are EcoRV (Winkler et al. 1993) and *Bam*HI (Viadiu and Aggarwal 2000). Both complexes serve as good models for an enzyme conformation competent for facilitated diffusion but incompetent for cleavage.

2.1.1 The Specific DNA Binding Mode

The specific interaction of a double-stranded DNA sequence with a protein is usually divided into direct readout interactions with the bases in the major and minor groove and indirect readout interactions with the DNA backbone. In most cases, the direct amino acid–base interactions in the major groove are of primary importance for discrimination. The general aspects of such interactions have been analyzed using more than 100 protein–DNA complex structures (Luscombe et al. 2001).

The direct protein–DNA hydrogen bonding interactions and selected, base-specific hydrophobic contacts that are observed in the specific complexes of EcoRV (Table 1) are summarized in Fig. 3. One striking feature of the specific binding mode, first revealed in the complex with the decamer oligonucleotide GGGATATCCC (Table 1, 4RVE), is that the protein dimer tightly embraces the recognition part of the DNA fragment that is bound in a strongly kinked conformation (Fig. 1). The highly localized central bend of about 50° is accomplished by a positive roll into the major groove. A comparable bend angle of about 50° has also been derived from solution studies with longer DNA fragments (Stover et al. 1993). The bound DNA further deviates from canonical B-DNA in being unwound by about 45° between the central four base pairs and in having two sugar rings 5′ to the cleavage site in C3′-endo-like rather than C2′-endo-like conformation (Winkler et al. 1993). The distorted DNA conformation makes the major groove narrower and deeper while the minor groove becomes shallower and wider.

EcoRV binds the DNA from the minor groove side and contacts the bases in the major groove through the R-loops that reach into the groove from both sides (Fig. 2B). The relative orientation of DNA and protein along the twofold symmetry axis appears to be one distinctive feature between the two main branches proposed on the basis of structural comparisons for the restriction endonuclease superfamily (Bujnicki 2000; Pingoud and Jeltsch 2001). An EcoRI-like and an EcoRV-like family had been proposed before, partly based on the functional distinction of producing 5′-overhangs and blunt ends respectively upon cleavage (Anderson 1993; Pingoud and Jeltsch 1997). The two R-loops, comprising residues 182–187 of each subunit, are responsible for all base-specific interactions (Fig. 3) and are only ordered in the specific complexes. Each R-loop forms six specific hydrogen bonds to the outer two base pairs of one half-site (GAT) of the hexameric recognition sequence. Three hydrogen bonds involve main-chain donor and acceptor groups, the other three small polar side chains (N185 and T186). The formation of a highly cooperative hydrogen bond network is a characteristic feature of protein–DNA interactions in restriction endonucleases. However, their detailed nature and architecture is found to be highly variable.

The central two base pairs are not read out by hydrogen bonds to the protein, yet they are also shielded from solvent access by the tips of the two

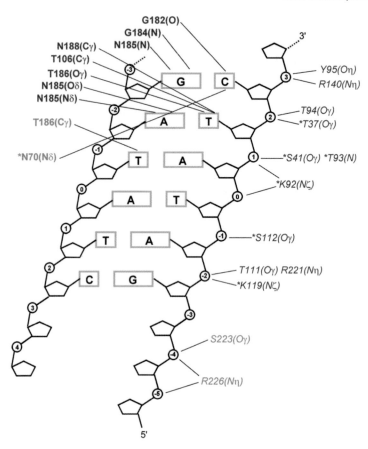

Fig. 3. Schematic representation of the direct protein–DNA interactions observed in the specific complexes (hydrogen bonds, salt bridges, selected hydrophobic contacts). The interactions with the DNA bases are shown in *bold* for one recognition half-site (GAT), and those with the DNA backbone in *italics* for one of the two strands. Residues of the two subunits A and B are in blue and red, respectively. Hydrophobic interactions are only shown for the two thymine 5-methyl groups. One R-loop (residues 182–187) reads the outer two base pairs of one half-site and contacts the 5-methyl group of the inner thymine of the other half-site. N70 forms the only direct hydrogen bond to a base in the minor groove. The interactions of residues marked with an asterisk (*) are not conserved in all structures. The scissile phosphodiester group at the TA step is marked as *0* and *numbers* are increasing in the 5′ to 3′ direction

recognition loops. The domed arrangement of these two base pairs, resulting from the large positive roll at this step, is such that the N6 nitrogen of each adenine can form two hydrogen bonds, one to its Watson-Crick base-paired thymine, the other to the O4 oxygen of the intrastrand thymine on its 5′ side (Winkler et al. 1993). Although this hydrogen-bonding arrangement does not appear optimal, it certainly helps to reduce the free energy cost of desolvation.

Deformations of bound DNA from the canonical B-form are observed in many of the specific complexes of restriction endonucleases, yet EcoRV represents the most extreme case known. Likewise, the absence of direct major groove readout for two of the six recognition base pairs is unique.

The importance of individual major groove interactions has been probed extensively using both mutant enzymes (Thielking et al. 1991; Vermote et al. 1992) and oligonucleotide substrates containing star sites (differing in only one base pair from the cognate sequence) (Alves et al. 1995) or modified bases (Fliess et al. 1988; Newman et al. 1990a, b; Waters and Connolly 1994; Martin et al. 1999b; Parry et al. 2003). As will be discussed later, no or strongly reduced catalytic activity is observed for nearly all modifications that directly perturb the recognition interface.

Like the R-loop, the Q-loop (residues 68–71) is disordered in the unliganded enzyme and becomes better ordered upon DNA binding. It inserts into the minor groove at about the second base pair of one recognition half-site and makes van der Waals contacts with the backbone of both DNA strands. Some conformational flexibility of this loop is observed even in the specific complexes. In particular, the conformation of the turn appears to easily switch between Type I and Type II in response to small perturbations. The side chain of Q69 extends across the minor groove and interacts with T37 of the opposing subunit in different ways. The side chain of N70 is observed in two conformations that correlate with the Q-loop turn conformation. In the Type I turn conformation, N70 forms a hydrogen bond to the recognition site cytosine in the minor groove (Fig. 3). In the central part of the widened minor groove there are no direct hydrogen bonds between the bases and the protein. A number of water molecules and the side chains of the two symmetry related K38 residues occupy this loosely packed part of the protein–DNA interface. The side chain of T37 forms a hydrogen bond to P_{+2} from the minor groove side (Fig. 3) in all high resolution structures. As discussed later, this hydrogen bond is not observed in the decamer complex where the methyl group of T37 packs tightly against the deoxyribose moiety on the 5′ side of P_{+2}.

A third chain segment, which becomes more ordered in the specific complexes with undecamer and dodecamer DNA fragments, comprises residues 221–229. Interactions with the DNA backbone are made by residues R221, S223, and R226. R226 forms salt bridges with the P_{-4} and P_{-5} phosphodiester groups (Fig. 3). The absence of P_{-5} in the shorter decamer duplex explains why such an ordering is not observed in this complex.

Further important contacts to the sugar phosphate backbone are made by a number of residues belonging to the rigid part of the DB-domain. Most importantly, the chain segment 90–95 of β-strand βE runs almost parallel to the DNA backbone and contacts four phosphodiester groups (P_0 to P_{+3}, Fig. 3). On the other side of the scissile phosphate, T111 and S112 make direct contacts with P_{-2} and P_{-1} respectively. The hydrogen bonding and salt bridge contacts are only one aspect of the DNA-backbone–protein interface. On top of the direct interactions there is a network of hydrogen bonds mediated by highly ordered water molecules. In the case of $P_{-2}/P_{-1}/P_0$ this network interconnects the three phosphates and it joins them to residues of the recognition loop, to catalytic residues, and bound metal ions (Kostrewa and Winkler 1995). The protein–DNA-backbone contacts have been probed by site-directed mutagenesis (Wenz et al. 1996), by using dodecamer oligonucleotides substituted by either an R_P- or S_P-phosphorothioate group from P_{-4} to P_{+4} (Thorogood et al. 1996a, b) and by ethylation interference footprinting (Engler et al. 1997). Overall, the results confirm that important interactions occur with the six phosphodiester groups from P_{-2} P_{+3}. Of the mutant enzymes probing these contacts (Wenz et al. 1996), the T37A mutation shows the most dramatic effect with a rate reduction of nearly 10^3. K119, R140, and R226, which all make salt bridges to DNA phosphates, reduce binding but do not affect the steady-state turnover rate (k_{cat}).

In summary, the majority of direct protein–DNA interactions in the specific complexes are provided by four chain segments of the protein, three of which undergo order–disorder transitions upon binding. The detailed analysis of this interface with a contact area of 2200–2400 $Å^2$ (Winkler et al. 1993; Pingoud and Jeltsch 1997) reveals that about 55 % are polar contacts and that each protein monomer interacts more extensively with the strand it cleaves (70 %).

2.1.2 The Nonspecific DNA Binding Mode

The structure of EcoRV complexed with two duplexes of the self-complementary octamer CGAGCTCG, stacked end-to-end at the twofold axis, can be considered to first approximation as a complex with a 16 base pair duplex with missing central phosphodiester linkages and has been assumed to be representative for nonspecific binding. The pseudohexadecamer has no base pair in common with the recognition sequence (5′-TCG/CGA-3′ as compared to GATATC). The helical axes of the two octamer duplexes are dislocated at the center by about 4 Å and there are significant deviations in this complex from twofold symmetry (see Fig. 4 in Winkler et al. 1993). This and the noncontinuous nature of the phosphodiester backbone between abutting octamers has raised some doubts on the validity of the above assumption (Erie et al. 1994). However, several other observations support it: (1) the undistorted B-like DNA structure makes sense, (2) many of the protein–sugar phosphate backbone contacts are similar to those in the specific complex and would permit a smooth transition to the specific binding mode (3) a model with continuous B-DNA bound in a symmetric fashion can be generated without stereochemical problems (Winkler et al. 1993) and (4) missing scissile phosphates do not disrupt the structure of the specific complex either (Horton and Perona 1998). Perhaps most importantly, the structure yields a very plausible explanation why no cleavage can occur in this binding mode. The well-defined contacts between the protein and the sugar–phosphate backbone on the 5′ side of the missing scissile phosphate suggest that these contacts would be similarly formed with a continuous strand. Modeling the 3′ extension of a strand fixed in this way shows that the scissile phosphodiester group is located at least 3 Å away from that observed in the cognate complex (Winkler et al. 1993; Kostrewa and Winkler 1995). The nonspecific complex reported for BamHI (Viadiu and Aggarwal 2000) shows the closest phosphate more than 6 Å displaced from the scissile position in the specific complex. Here, the structural change needed to go from nonspecific to specific binding appears more dramatic than for EcoRV and involves primarily an adaptation of the protein conformation.

The protein–DNA interface in the nonspecific complex is about 800 Å² smaller than in the specific complex and the interface is much less tightly packed as indicated by a high gap volume index (Jones et al. 2001). One might therefore expect nonspecific binding to be considerably weaker than specific binding. Despite a number of experimental studies on this question this has remained a controversial issue as will be discussed later.

2.2 The Structure of the Active Site

2.2.1 The Structure in the Absence of Divalent Cations

Sequence comparison and preliminary structural results had indicated that EcoRV shares an active site motif with EcoRI, which is characterized by PD...(D/E)XK (Thielking et al. 1991; Winkler 1992). The two acidic side chains of D74 and D90, belonging to this motif, are close to the scissile phosphate, while that of another acidic residue E45, not part of this conserved motif, is somewhat more remote (Fig. 4A). The clustering of the conserved acidic residues near the scissile phosphate suggested them to be involved in binding the essential Mg^{2+} ion. The amino function of the conserved basic residue of this motif (K92) is positioned between P_0 and P_{+1} and is assumed to be protonated.

The conformation of the two acidic side chains (Fig. 4A) is very similar in all structures in the absence of divalent cations because each forms a conserved hydrogen bond to a main chain N–H. In contrast, the side chains of K92 and E45 show conformational flexibility. In the different specific complexes, the position of the DNA backbone relative to that of the rigid part of the protein active site varies up to 1 Å. This explains why the hydrogen bonding network that involves the scissile phosphate, the nearby acidic and basic groups and bridging water molecules also shows variation. For example, K92 is found within hydrogen bonding distance of either P_0 or P_{+1}, or in between in some cases. The adaptability of this network suggests that rearrangements upon metal binding or during catalysis can proceed rather smoothly. It

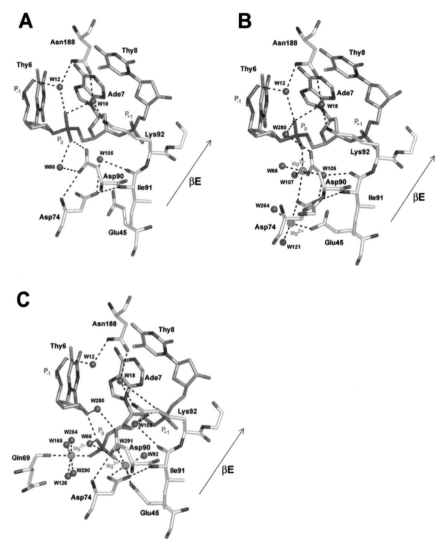

Fig. 4. The active site of EcoRV endonuclease in different complexes with DNA. The three nucleotides TAT (from GA**TAT**C) with the scissile phosphodiester (P_0) at the TA step are shown together with selected residues (catalytic motif and E45) and bound water molecules. β-strand βE runs parallel to the DNA backbone up to P_{+3}. The color code for half bonds is as follows: oxygen (*red*), nitrogen (*blue*), phosphorus (*magenta*) and carbon (*white* and *orange* for protein and DNA, respectively). Water molecules are shown as *red* and bound Mg^{2+} ions as *green spheres*. Hydrogen bonds are indicated with *dashed lines*. **A** Complex with the undecamer DNA fragment in the absence of divalent cations (1RVA). **B** The same complex in the presence of Mg^{2+} (1RVB). **C** Product complex of the undecamer substrate DNA in the presence of Mg^{2+} (1RVC)

appears possible that D90 is protonated in these structures as all were crystallized at slightly acidic pH (6.5–7.0). In most structures, three water molecules are observed within hydrogen bond distance of the nonbridging oxygens of the scissile phosphate. Two of them connect to the side chain of N188 at the C-terminal end of the recognition loop which represents the most direct hydrogen-bond-mediated contact between the R-loop and the scissile phosphate.

2.2.2 The Location of Bound Divalent Metal Ions

Three distinct divalent metal ion sites involving the active site residues have been observed in the crystallographic studies. Soaking of the triclinic crystals with the catalytically active cations Mg^{2+}, Mn^{2+} or Co^{2+} does lead to metal binding in the active site without cleavage. Two binding sites were identified (Kostrewa and Winkler 1995) and, on the basis of the ligating acidic residues, they were characterized as the 90/74 and the 74/45 site (Fig. 4B). Mg^{2+} and Mn^{2+} ions, at 30 mM concentrations, only bind in one of the two active sites of the noncrystallographic dimer and the DNA backbone at this site shifts by about 1 Å in the 5′ to 3′ direction. With Ca^{2+}, both active sites bind one metal at the 90/74 site and no DNA shift is observed. However, in complexes of the mutant enzyme T93A with the same DNA fragment, two different Ca^{2+} binding sites have been observed (Horton et al. 1998). One is very similar to the 74/45 site and a new one, labeled site I, is ligated by D74, by the main-chain carbonyl oxygen of I91 and by three water molecules. It is located more towards the 3′-side of the scissile group and has been implicated in catalysis (Horton et al. 1998; see Sect. 3.3.5). Two similar sites are occupied in complexes with the refractory phosphorothiolate 5′-AAAGATsATCTT (Horton et al. 2000). Soaking studies with crystals of the complex with AAAGAUATCC, which has deoxyuridine in place of the first thymine, showed predominantly site I occupied in one subunit but sites II (the 74/45 site) and III (the 90/74 site) in the other (Winkler and Kostrewa, unpubl.). All these observations indicate that there are at least three distinct divalent cation binding sites in the active site region, and that small perturbations introduced by a mutation, by a modified substrate or by crystal packing constraints can significantly shift their occupancies. In no case, however, more than two sites have been seen occupied at the same time.

2.2.3 The Structure of the Enzyme Product Complex

With large DNA substrate molecules, product dissociation from EcoRV endonuclease has been found to be some 200-fold slower than cleavage, while comparable rates for the two steps are observed with small oligonucleotide substrates (Baldwin et al. 1995; Erskine et al. 1997). It was not a priori obvious whether the DNA binding mode in a product complex would be specific-like or nonspecific-like. From a concentrated reaction mixture that contained enzyme, undecamer fragment 5′-AAAGATATCTT-3′ and $MgCl_2$, crystals iso-morphous to the substrate complex (1RVA) were obtained (Kostrewa and Winkler 1995). The product DNA is bound in a very similar fashion as the substrate, and the structural adaptations are confined to the vicinity of the cleavage site although the 5′-phosphate is nearly 5 Å away from its previous position. Two metal ions are bound and directly coordinated by oxygens of the doubly charged 5′-phosphate group (Fig. 4C). The site involving the main-chain carbonyl oxygen of Q69 requires a Type I turn of the Q-loop. The loca-tion of both of these metal ions is different from that observed with uncleaved DNA. The other interesting structural change is a reorientation of the two central adenine bases provoked by the dislocation of their 5′ sugar–phosphate moieties upon cleavage. The two bases become involved in a cross-strand stacking at the twofold symmetry axis and the corresponding two A–T base pairs become strongly buckled.

A very similar, product-like complex that lacks the 5′-phosphate of the cleaved pentamer fragment, has been crystallized in the same crystal form (Table 1, 1RV5; Horton and Perona 1998). The DNA structure at the active site is intermediate between substrate and product complex but is much closer to that with substrate. The fact that the same specific binding mode is observed in this noncontinuous, 5′-phosphate and metal-free complex suggests that metal binding cannot be critical for stabilizing the product complex in this binding mode.

3 Biochemical Characteristics of *Eco*RV Endonuclease and Structure–Function Relationships

EcoRV endonuclease is probably the most thoroughly studied of the Type II restriction endonucleases. Many of the thermodynamic and kinetic studies employing modified substrates and mutant enzymes were carried out after the first crystal structures had become available, and there has been a very fruitful and inspiring interaction between the structural and functional investigations.

In the presence of Mg^{2+}, Type II restriction endonucleases cleave DNA at their respective recognition sequence with extremely high specificities (Roberts and Halford 1993). DNA sequences differing from the recognition site by just one base pair are typically cleaved over 1 million times more slowly (Taylor and Halford 1989). This is much more than one would expect from the loss of two to three hydrogen bonding interactions with a single base pair in a substrate enzyme complex. Restriction endonucleases exercise sub-strate specificity both in the DNA binding step (K_D) and in the subsequent chemical rate of hydrolysis (k_c). This has first been systematically investigated for EcoRI endonuclease with cognate and noncognate substrates, and both steps were found to contribute similarly to total discrimination quantified as the k_c/K_D value relative to that for a cognate substrate (Lesser et al. 1990). Studies with a number of other restriction endonucleases have shown that total discrimination is similarly large, but the relative contribution of binding and catalysis can vary substantially (Pingoud and Jeltsch 1997).

To understand the structural and energetic basis of this high discrimination one would ideally want to compare the structures of the corresponding enzyme–substrate–cofactor complexes at the transition state. As this is experimentally impossible, the question is to what extent the observed ground state complexes can serve as substitutes or as starting points for extrapolation. At least for good substrates, it appears reasonable to assume that most of the interactions observed in the large interfaces of the ground state complexes are maintained at the transition state. The observed and thermodynamically characterized specific complexes are thus likely to be on the path to the transition state (Jen-Jacobson 1997). However, we must keep in mind that the ground state structures can never reveal all the interactions of the transition state and that small changes in the details of critical attractive or repulsive interactions may have profound effects on the energetics.

3.1 Thermodynamics and Energetics of DNA Binding

In general, we must be happy if we can account in a qualitative or semiquantitative way for larger differences in the binding constants of different substrates, except perhaps in cases where the changes in interactions are highly localized. Apart from the intrinsic difficulties in estimating free energy differences between different structural states, the observed structures (mostly from crystal structure analysis) are often not strictly comparable to the systems or conditions for which the thermodynamic data have been measured.

3.1.1 Specific and Nonspecific Binding to DNA

The determination of the specific and nonspecific binding constants for EcoRV has led to conflicting results in the absence of Mg^{2+}. Mainly three experimental techniques have been used for determining thermodynamic binding constants of restriction endonuclease–DNA complexes: (1) gel mobility shift, (2) filter binding, and (3) steady-state fluorescence anisotropy (Connolly et al. 2001). The first systematic studies using the gel shift method led to the conclusion that this enzyme binds to all sequences with essentially equal affinity of about 10^6 M^{-1} (Taylor et al. 1991). This surprising finding suggested that the energy difference between the two structurally characterized, distinct binding modes is very small (Winkler et al. 1993). It was also in strong contrast to the situation with EcoRI endonuclease, which had been found to discriminate 10^3 to 10^4-fold in binding against sequences with one incorrect base pair (Lesser et al. 1990; Thielking et al. 1990). EcoRV was suggested to exercise sequence specificity solely in the catalytic step in the presence of Mg^{2+} and thus in a fundamentally different way than EcoRI (Taylor et al. 1991; Thielking et al. 1992; Vipond and Halford 1993; Vipond and Halford 1995; Szczelkun and Connolly 1995). This conclusion was later challenged based on filter binding and gel retardation experiments carried out with shorter DNA fragments (22 base pairs) under a range of pH and salt conditions (Engler et al. 1997). Specific binding was found to be strongly pH-dependent but even at pH 7.5 the selectivity value (ratio of specific to nonspecific binding constant) was postulated to be as high as 120. As neither gel retardation nor filter binding techniques are done under true equilibrium conditions the issue was reexamined by binding studies at pH 7.5 in solution using fluorescently labeled oligonucleotides for FRET (fluorescence resonant energy transfer) and FD (fluorescence depolarization) measurements (Erskine and Halford 1998; Reid et al. 2001). The selectivity value found in these experiments was below 10 with little variation between different oligonucleotides, apparently confirming the previous reports on no or only weak discrimination of EcoRV in DNA binding in the absence of cations. Similarly low specificity values were also observed in a later study using the gel retardation method modified in a way to minimize protein–DNA dissociation during the experiment (Martin et al. 1999a). The majority of studies and particularly those less prone to experimental artifacts show only weak selectivity at pH 7.5. Its increase at lower pH, which is not controversial, appears structurally plausible in view of the energetically unfavorable clustering of negative charges at the active site that would be reduced by protonation of one of the acidic residues. EcoRV and some other restriction endonucleases which show a similar behavior (Pingoud and Jeltsch 2001) may thus be considered as examples with a larger unfavorable accumulation of negative charge in the specific binding mode rather than representing a fundamentally distinct mechanistic class. In accordance with such a view, EcoRI mutants like K130A, K130E or R131E make the specific binding of this enzyme Ca^{2+} dependent (Windolph and Alves 1997).

From a structural point of view, it is against our intuition that the specific and nonspecific binding mode of EcoRV should hardly differ in binding energy (in the absence of divalent cations). A large interface with many tight interactions showing almost ideal stereochemistry stands against a much smaller and in detail much looser interface. A number of factors may help to explain this apparent discrepancy although their absolute or relative contribution is difficult to estimate. One important such factor is entropic in nature and is usually termed configurational entropy. In our case, it concerns the ordering of several structural elements associated with the specific binding mode, in particular the R-loop, residues from the C-terminus and to a lesser extent the Q-loop. In addition, the looser interface of the nonspecific complex is much less restrictive for the local and global mobility of both DNA and protein, which further increases the relative configurational entropy cost of the specific complex. In contrast, the favorable entropic contribution to binding through the release of water molecules from nonpolar surfaces should be considerably larger for the specific complex (Winkler et al. 1993). There are two enthalpic contributions that may destabilize the specific complex and can thus be considered as strain. One is the abundance of negatively charged groups near the scissile phosphate that, as discussed before, makes specific binding strongly pH and metal cation dependent. The other is the deformation of the DNA with an almost complete unstacking of the central two base pairs which may cost 5 kcal/mol or more (Delcourt and Blake 1991; Hunter 1993; SantaLucia et al. 1996; Hunter and Lu 1997). A detailed analysis of enthalpic and entropic factors contributing to site-specific DNA binding based on data for 10 different proteins, but not including EcoRV, has been presented by Jen-Jacobson et al. (2000). It shows that isothermal enthalpy–entropy compensation is important in these systems.

A direct assessment of the effect of Mg^{2+} on binding is not possible with normal DNA substrates because of rapid cleavage. A number of studies have addressed this question indirectly through investigating catalytically inactive mutants (Thielking et al. 1992), using Ca^{2+}, which does not support catalysis (Vipond and Halford 1995; Vipond et al. 1995), or using poor or noncleavable substrate analogues (Szczelkun and Connolly 1995; Engler et al. 1997; Martin et al. 1999a, b). At pH 7.5 large increases in binding affinity up to 10^4-fold are observed with Ca^{2+} for DNA fragments containing a cognate site. Smaller increases are seen when the cognate site is substituted by cleavable sites containing modified bases (for example GAUATC) or when it contains base analogues conferring resistance to hydrolysis (Martin et al. 1999a). Only marginal increases are observed with nonspecific DNA fragments. These results are consistent with the view that optimal binding of Ca^{2+} takes place at cognate sites and that this binding is perturbed directly (impaired metal coordination) or indirectly (impaired protein–DNA interface) with modified substrates. At noncognate sites, differing in just one base pair from the cognate site, the specific binding mode is already so strongly disfavored that Ca^{2+} binding cannot overcome this energy penalty. For some analogues that contain very slowly cleaved sites and for hydrolysis resistant analogues gel shift binding assays in the presence of Mg^{2+} were possible (Martin et al. 1999a). For the former, the binding constants measured with Mg^{2+} were nearly as strong as with Ca^{2+} suggesting that Ca^{2+} can be a useful substitute. However, Mg^{2+} hardly improved the binding affinity for the noncleavable, sulfur-modified substrates. There are clearly metal specific effects whose magnitude depends on the particular nature of a perturbation and how it affects the formation of the metal binding sites.

The structural interpretation of the binding experiments carried out with Mg^{2+} and Ca^{2+} ions is complicated. Both the occupancy and the detailed coordination of the binding sites may differ depending on the metal ion used and on whether a mutant enzyme or modified substrate is used. In wild-type enzyme–cognate DNA complexes, site III has been seen occupied with each of the metals Mg^{2+}, Mn^{2+}, Co^{2+}, and Ca^{2+} (Kostrewa and Winkler 1995), and crystallization in the presence of Ca^{2+} has yielded the same binding site as soaking experiments (Perona and Martin 1997). Both site III and site I require the kinked DNA conformation for their formation and do not exist in the non-specific binding mode. Most likely, site III is primarily responsible for the increase in specific binding seen in the presence of divalent cations. With the soft metal ions Mn^{2+} and Co^{2+} the situation may be even more complex due to the presence of additional binding sites distal from the active site (Winkler, unpubl.; Horton et al. 2000).

3.1.2 The Energetics of Direct and Indirect Readout Interactions

The detailed analysis of the energetic basis of specificity in restriction endonucleases has been pioneered for EcoRI endonuclease by the Jen-Jacobson group (Lesser et al. 1990, 1993). EcoRI shows strong specific binding in the absence of divalent metal ions. This has permitted to accurately assess the energy penalty of perturbing selected interactions on binding and catalysis by using suitable base analogs. Most single or double hydrogen bond deletions did hardly reduce the cleavage rate and thus have about equal effects on the ground state and transition state complex. In contrast, large reductions in both binding and cleavage rate occurred with noncognate DNA that differed in one base pair. An altered conformation of the ground state complex was indicated from phosphate ethylation interference experiments in these cases (Lesser et al. 1990), and this state was interpreted to be off the direct path to the transition state (Jen-Jacobson 1997).

For EcoRV, early studies of the steady-state kinetics using self-complementary dodecamer oligonucleotides that contain modified bases within the recognition sequence (Newman et al. 1990b; Waters and Connolly 1994) revealed large reductions in cleavage rates but not much change in apparent affinity (K_m). All nine major groove hydrogen bonding functions were probed with suitable base analogs. Rate reductions up to one hundred fold were also observed for some analogs carrying modifications in the minor groove. Possible explanations include the perturbation of water mediated minor groove contacts but these are all rather speculative (Szczelkun and Connolly 1995). More recently, the major groove contacts of the outer two base pairs have been reinvestigated in a much more thorough way (Parry et al. 2003). Dissociation constants (K_D) in the presence of Ca^{2+} and single turnover rate constants (k_c) using Mg^{2+} or Mn^{2+} were determined for heteroduplexes and homoduplexes formed from oligonucleotides of length 19 and 22 carrying a base modifications in one or both DNA strands. The results are completely consistent with the previous findings but reveal new and more quantitative aspects. With the A:T base pair, binding of DNA with a single modification was reduced by 0.9–1.7 kcal/mol as would be expected for the simple loss of a single contact. About twice the effect is measured with DNA modified in both strands and a similar behavior is observed for the reductions in the catalytic rates. With the G:C base pair, the binding penalty is about 2 kcal/mol for the single site d⁴ᴴC modification while it amounts to 3.5–4 kcal/mol for the single dG modifications. In contrast to the A:T base pair, binding is similarly reduced with DNA carrying a G:C base pair modification in both strands. The d⁴ᴴC modification is unique in having no effect on the catalytic rate. Thus, a single perturbation in one of the two G:C base pairs appears to cause symmetric structural adaptations, but only for the dG modifications this structural change propagates to the active site.

A unique aspect of specific recognition by EcoRV is the lack of direct hydrogen bond-mediated readout of the two central base pairs. Discrimination at these base pairs is as high as for the outer base pairs, as confirmed with an oligonucleotide substrate containing all possible single base pair substitutions (Alves et al. 1995). Cleavage rates were also determined for all possible mismatch substrates (Alves et al. 1995). Very heterogeneous double strand cleavage rate reductions ranging from 40 to several million-fold were observed. The rates for first and second strand cleavage of a given such substrate varied up to 1000-fold. This demonstrates that the relaxation of strain is dependent on the detailed structural context and may introduce significant asymmetry.

The structural and energetic origin of the readout at the central TA step has been investigated in great detail via functional analysis of a series of symmetric base analogs (Martin et al. 1999a, b). Four modified versions of the undecamer AAAGATATCTT with the central TA step replaced by UA, MI, CI and CG were investigated (Fig. 5). The TA–UA and MI-CI pairs probe the loss of

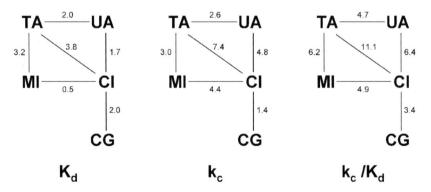

Fig. 5. Free energy component analysis for DNA substrates with a central TA, UA, MI, CI and CG step (Martin et al. 1999). (U=deoxyuracil, M=5-methylcytosine, I=inosine a guanine-like base lacking the minor groove 2-amino group) . Thermodynamic dissociation constants (K_d) were determined in the presence of Ca^{2+} by gel shift analysis, and single-turnover reactions were used to determine the chemical rate constants (k_c) for phosphodiester hydrolysis. The ratios of these values have been converted to free energy differences ΔΔG given in kcal/mol

the only direct readout interaction, the hydrophobic contact between the methyl groups of the thymine base and the T186 side chain, in two different contexts. The relative changes in k_c/K_d transform into very similar $\Delta\Delta G$ values of 4.7 and 4.9 kcal/mol but they are quite differently partitioned into K_d and k_c. The crystal structure of the MI complex (1BSU, Table 1) is essentially identical to that of the cognate complex. A number of distinct structural differences are, however, observed for the CI complex (1BUA) although its binding is only marginally weaker. The small but numerous structural adaptations of the CI analog suggest that considerable enthalpy entropy compensation may be involved. However, the structural relaxation in the ground state is not on the path to the transition state and the corresponding gain in binding energy must be paid back during catalysis.

Most of the total discrimination of 6.2 kcal/mol of the MI analog must be due to indirect readout as its interactions with the protein are essentially identical to those with cognate DNA. It is well known from theoretical and thermodynamic studies that TA steps are among the least stable (Delcourt and Blake 1991; Hunter 1993; Hunter and Lu 1997), and the differential energetic cost for base pair unstacking has been suspected to be an important factor in the indirect readout of EcoRV (Winkler et al. 1993). As the central base pair becomes essentially unstacked but not solvated in the complex with EcoRV the energetic cost must be mostly enthalpic. The difference in binding energy between TA and MI (3.2 kcal/mol) should include all of this differential unstacking cost as a 50° bend amounts to the full loss of stacking energy. Stacking enthalpies differ up to 5 kcal/mol for the natural base pairs and are highest for CG and GC steps (SantaLucia et al. 1996). Differential unstacking cost could thus fully account for the reduced binding of the MI analog. However, there also is the possibility of a differential energy cost for major groove desolvation at the central two base pairs. We have already discussed the intrastrand hydrogen bonding/electrostatic contacts between the exocyclic 4-oxo (dT) and 6-amino (dA) groups at the TA bend which may partly compensate for the loss of major groove desolvation. In a CG step and likewise in the MI and CI analogs the positions of the exocyclic atoms are preserved within experimental error, but the amino and oxo groups are interchanged. The differential desolvation cost may thus be small in this case. However, for steps like AT or CT with unstacking enthalpies as small as that of TA (SantaLucia et al. 1996) other contributions to this indirect readout must exist to explain the similarly large total discrimination. A differential cost of major groove desolvation may therefore contribute significantly to indirect readout for noncognate base steps where a favorable intrastrand donor/acceptor arrangement is not possible.

The additional discrimination seen in the catalytic step cannot be assigned to differential unstacking energies. As indicated by the analysis of the different specific complexes (Horton and Perona 2000) the central DNA bend and the resulting compression in the major groove may need to be further increased towards the transition state. The anticipated structural changes in the recognition interface are small and nonbonded interactions probably change by at most a few tenths of one Å. Discrimination may therefore be the result of few critical interactions that become more strongly repulsive at noncognate sequences.

3.2 Chemistry and Kinetics of DNA Cleavage

3.2.1 General Aspects of Phosphodiester Hydrolysis

Like all restriction endonucleases, EcoRV cleaves DNA yielding 3'-OH and 5'-phosphate ends and the reaction can be considered as a phosphoryl transfer to water. Hydrolysis occurs with inversion of configuration at the phosphorus atom (Grasby and Connolly 1992), suggesting an attack of a water molecule in line with the cleaved P-3'O bond. Two principal mechanisms, associative or dissociative, represent two extremes of possible intermediate mechanisms (Pingoud and Jeltsch 2001). Despite the lack of experimental evidence, a more associative-type mechanism is usually assumed to be operative in DNA cleavage by restriction endonucleases. This mechanism requires (1) a general base to generate the active hydroxide nucleophile for attack at the scissile phosphodiester group, (2) a Lewis acid stabilizing the extra negative charge at the pentacoordinated phosphorane transition state and (3) a general acid that stabilizes or protonates the alcoholate leaving group. No detailed mechanism of DNA cleavage has thus far been elucidated for any restriction endonucleases but it appears that there is considerable variation in detail despite striking similarities in the active site residues and architecture (Pingoud and Jeltsch 2001). In the case of EcoRV, different hypotheses that will be discussed in section 3.3 have been developed and experimentally probed using mutant enzymes and modified substrates.

3.2.2 Steady State and Rapid Reaction Kinetics

Steady-state kinetics for EcoRV cleavage at a single site on a plasmid yields a k_{cat} of $0.015\ s^{-1}$ (Halford and Goodall 1988) and a K_m of 0.5 nM (Taylor and Halford 1989). With small oligonucleotides k_{cat} is found to be about 50-fold faster ($0.7\ s^{-1}$) and K_m values are typically 0.1–1 μM (Waters and Connolly 1994; Baldwin et al. 1995). As steady-state kinetics provides little information about intermediates in reaction pathways, rapid reaction analysis has been carried on a small 12 bp duplex (Baldwin et al. 1995) and on plasmid DNA that contains a single site (Erskine et al. 1997) using stopped-flow fluorescence and quench-flow methods under single turnover reaction conditions. With both substrates, similar rate constants of 2–3 s^{-1} for the DNA cleavage step are measured and they appear to be determined by the chemical reactions (k_c). For the 12-bp duplex this rate is within an order of magnitude of k_{cat} which is thus partly determined by the cleavage step and partly by product dissociation (Baldwin et al. 1995). For the plasmid substrate, the open-circle nicked form is transiently formed and the rate constants for cleavage of the two strands are very similar. The rate-limiting step for the complete catalytic cycle occurs after cleavage and must be slow product release as proposed earlier (Halford and Goodall 1988). If product release rather than product formation is rate limiting, the K_D of the enzyme–substrate complex is expected to be larger than K_m (Gutfreund 1995). In agreement with this, determination of association and dissociation rate constants yields a K_D of 10 nM for the plasmid substrate (Erskine et al. 1997), 20-fold higher than the K_m. Intramolecular transfer from nonspecific sites to the recognition site, known to occur from competition experiments (Taylor et al. 1991; Jeltsch et al. 1996), was too fast to be measurable in these experiments. The mechanism of this transfer has been assumed to be linear diffusion by sliding on DNA, based on the observed cleavage rate enhancement with increasing length of the DNA as compared to a small reference substrate (Taylor et al. 1991; Jeltsch et al. 1996; Jeltsch and Pingoud 1998). This view has been challenged based on the measurement of processive cleavage at two sites separated by varied distances (Stanford et al. 2000) and using higher DNA/enzyme ratios to avoid having more than one enzyme bound to one substrate molecule. A model that combines multiple dissociation/reassociations through three-dimensional space with one-dimensional diffusion over relatively short distances (~25 base pairs) has been proposed instead. This model explains the slower release from product with large DNA substrate equally well because dissociation/reassociation events can continue for a relatively long time before the protein eventually escapes the product DNA molecule that contains no more cleavage sites. Clearly, the two models are mechanistically quite different as linear diffusion proceeds without breaking all interactions.

The about 200-fold difference between the rates of product release and product formation with plasmid DNA means that steady state kinetics is inappropriate to assess mechanistic effects of mutants. For example, a 2000-fold reduction of the cleavage rate can result in just a ten-fold reduction of k_{cat} (Erskine et al. 1997). Such a mutant, Q69L, has indeed been found (Hancox and Halford 1997).

Different divalent metal ions show differential effects on the kinetics and specificity of the EcoRV catalyzed reaction. The relative activity (k_{cat}) using a plasmid DNA substrate has been found to vary in the following order $Mg^{2+} > Co^{2+} > Mn^{2+} >> Cd^{2+} > Zn^{2+} > Ni^{2+} >> Ca^{2+}$ (Vipond et al. 1995). With Ca^{2+} there is virtually no reaction and the reaction with Mg^{2+} is inhibited in its presence. Using a 12-bp oligonucleotide substrate, steady-state experiments showed that Co^{2+} has a very similar turnover rate (k_{cat}) as Mg^{2+} while that of Mn^{2+} is about tenfold reduced (Baldwin et al. 1999). The reaction with these three metal ions was also analyzed under single turnover conditions after rapid mixing. While Mg^{2+} and Co^{2+} yielded very similar cleavage rate constants around 1 s^{-1}, that for Mn^{2+} was found to be about five- to sevenfold faster (Baldwin et al. 1999; Sam and Perona 1999b). However, product release is nearly 30 times slower with Mn^{2+}. A very similar pattern is also observed with plasmid substrates. In the presence of Mn^{2+}, but not Co^{2+}, EcoRV looses much of its ability to discriminate between cognate and noncognate sequences (Vermote and Halford 1992; Baldwin et al. 1999). The combination of increased phosphodiester hydrolysis rate and increased lifetime of the enzyme–substrate complex may at least be partly responsible for this reduced specificity. In the case of EcoRV the rationalization of these metal specific effects is complicated by the fact that at least two metal ions appear to be involved in catalysis (see below). Furthermore, the additional binding sites of soft divalent ions (Mn^{2+}, Co^{2+}) outside the active site may contribute to metal specific effects (Sam et al. 2001).

3.3 Mechanistic Hypotheses

The mechanistic studies of EcoRV have been largely driven by structure-based hypotheses. The identification of the essential catalytic residues, the understanding of their specific role and the determination of the number, location and coordination of the essential metal binding site(s) have been important issues. Ultimately one would like to have a chemically and stereo-chemically convincing model of the transition state and of the path leading from the productive ground state through the transition state to product. Structural information is critical to develop such detailed models but in most cases only low-energy, ground state structures are experimentally accessible. Unavoidably, this raises the question of how close these are to the unknown transition state. The structure of a high-energy intermediate close to the transition state has recently been determined for the phospho-group transfer reaction of phospoglucomutase (Knowles 2003; Lahiri et al. 2003). It shows a stretched pentacoordinate trigonal bypiramidal oxyphosphorane whose axial bond lengths of 2.0 Å suggest a partly associative in-line displacement. A similar geometry may be expected for the pentavalent phosphorane transition state of restriction endonucleases.

For EcoRV, studies of steady-state, single turnover and rapid kinetics with mutant enzymes and modified substrates, studies of the effects of different metal ions and studies of the pH dependence of catalysis have provided us, together with the abundance of structural studies, with a wealth of data. Several mechanisms, differing in the number and location of the metal binding sites, in the nature of the general base and in the conformation of the DNA near the scissile group have been considered and experimentally explored (Kovall and Matthews 1999; Pingoud and Jeltsch 2001). Although some aspects remain controversial, the more recent two or three metal ion mechanism proposed by the Perona group (Horton et al. 1998; Sam and Perona 1999a) appears to us as the most plausible.

3.3.1 The Role of the Conserved, Catalytic-Motif Residues

The structurally conserved nature and essential role of D74, D90 and K92 has been recognized early and has been verified by mutagenesis (Thielking et al. 1991; Winkler 1992). They are part of the catalytic sequence motif PD...(D/E)XK that occurs in many if not most Type II restriction endonucleases (Pingoud and Jeltsch 2001). A clear indication of the critical role of the two conserved acidic residues is that their mutants are equally inactive with Mg^{2+} and Mn^{2+}.

Based on its location on the 3′ side of the scissile phosphodiester group, the amine function of K92 appeared as a primary candidate for the general base (Winkler et al. 1993). It has been argued that a general base is not an absolute requirement for water activation, and that a hydroxide ion from the bulk solution could fulfill this role as in the case of the 3′-5′ exonuclease activity of DNA polymerase I (Fothergill et al. 1995). For EcoRV, this can almost certainly be ruled out on the basis of the pH dependence of the maximal rate of cleavage (Stanford et al. 1999; Sam and Perona 1999a). However, the pK_a of K92 would have to be lowered by perhaps 2–3 units to make it an efficient base at intracellular pH. This seems unreasonable at first sight in view of its proximity to three negatively charged groups (D90 and the two phosphodiester groups at P_0 and P_{+1}). However, the binding of divalent cations during catalysis may more than compensate for this. The lysine residue of the catalytic motif is not strictly conserved, and BamHI for example has an acidic residue at this position (E113). In the absence of decisive data, other candidates for the general base have been considered. One such hypothesis is based on modeling studies using the structures of the substrate complexes of EcoRV and EcoRI (Jeltsch et al. 1992) and has been substantiated by experimental evidence with modified oligonucleotide substrates (Jeltsch et al. 1993). It suggests a substrate-assisted mechanism in which the P_{+1} phosphate (on the 3′ side of the scissile group) acts as the general base. A major problem with this mechanism is the low intrinsic pK_a of this group (<2). It seems more likely that this group is critically involved in generating the optimal geometry of the hydrogen bonding network that positions the functional groups and water molecules for metal binding and water activation.

3.3.2 The Role of Metal Ions and the Need of Conformational Changes in Different Mechanisms

It is now generally accepted that at least two metal ion sites are essential for catalysis by EcoRV. As described, the crystallographic metal binding studies have revealed three distinct sites, but at most two of them have been seen occupied together in the same structure. Evidence for more than one catalytically important binding site has also been derived from studies using mixtures of divalent cations (Vipond et al. 1995) and from stopped-flow fluorescence studies revealing two Mg^{2+} concentration dependent rates (Baldwin et al. 1995). As binding site I was unknown at that time, these were assigned to sites III and II respectively. It was, however not obvious why binding to these two sites in the triclinic crystal form did not cause DNA cleavage even under drastic conditions. Therefore, a larger conformational change that would be inhibited by crystal packing forces seemed indicated (Baldwin et al. 1995; Kostrewa and Winkler 1995).

The uncertainty about the need of a larger conformational change inspired a molecular dynamics (MD) simulation starting from the observed structure with two metals bound at sites III and II as observed in the triclinic crystal form (Baldwin et al. 1999). A spontaneous transformation of the DNA conformation into a new state in which both nonbridging phosphodiester oxygens had become ligated to a metal ion was observed. The displacement of the scissile group was produced by a change in the sugar pucker of the thymidine deoxyribose at the central TpA step from C3′-endo to C2′-endo. The side chain of E45 changed to an outer shell ligand of the site II metal and thus fulfilled two important roles in the postulated mechanism: first, it participates in the initial binding of the site II metal and secondly, it activates the attacking water after the conformational change. Such an important role appears, however, difficult to reconcile with the fact that E45 is not a conserved residue.

The above mechanistic hypothesis has been further refined by looking at the pH dependence of the cleavage reaction under single-turnover conditions and by assigning an important role to another acidic residue, D36 (Stanford et al. 1999). In the different structures, the carboxylate of D36 of one subunit is located 5–7 Å away from that of E45 in the other subunit. Mutations of D36 cause large cleavage rate reductions of up to 10^4-fold but show normal specific binding in the presence of Ca^{2+} (Stanford et al. 1999) suggestive of a role in catalysis. The pH dependence of the wild-type enzyme was studied with Mn^{2+} to guarantee full metal occupancy at all pH values up to pH 8.5. An asymptotic increase of the cleavage rate reaching a plateau above pH 7.5 was best fitted with a model requiring the deprotonation of two ionizing groups with similar pK_a values of about 7. The close interaction between the carboxylates of D36 and E45 was postulated to raise the pK_a of E45 and make it more effective in deprotonating the attacking water. This modification of the 'conformational change' mechanism did explain the observed pH profile and the large effects seen with D36 mutations but has a number of unsatisfactory aspects. Neither D36 nor E45 are conserved, and the two acidic groups are not in direct contact as to justify a large pK_a shift. Furthermore, the essential roles of the conserved residues D90 and K92 remain insufficiently defined.

Yet another modification of the 'conformational-change' mechanism was provoked by the observation that mutants of the Q-loop residue Q69 displayed strongly reduced cleavage rates (Hancox and Halford 1997). Crystal structure analysis of DNA complexes of the Q69E and Q69L mutants did not reveal large structural changes and failed to provide an obvious explanation for their strongly impaired catalytic activity (Thomas et al. 1999). An MD simulation provided a structural link between the Q-loop conformation and the stabilization of the postulated pentavalent transition state and thus a possible explanation for the impaired catalytic rate of the Q69 mutants (Thomas et al. 1999). It must be emphasized though that these MD simulations are open to several kinds of criticism and, in our opinion, they can at best provide ideas on possible structural states. We think that the analysis of active and inactive conformations of EcoRV as discussed below provides an alternative and more plausible explanation for the rate reductions of the Q69 mutations.

3.3.3 Catalytically Active and Inactive Conformational States

The postulate of a larger conformational change during catalysis originated from the observation that no DNA cleavage was observed in the triclinic crystals despite binding of the metal cofactor (Kostrewa and Winkler 1995). Four different crystal forms that contain specific complexes with substrate DNA (Table 1) have been reported (Winkler et al. 1993; Kostrewa and Winkler 1995; Horton and Perona 2000). Only form I crystals, containing a decamer DNA duplex, show *in crystallo* cleavage activity upon soaking with Mg^{2+} and will be referred to as active. The detailed comparison of the protein and substrate DNA conformations in all four crystal lattices has revealed a very significant variation of the DNA-bending angle and a correlation of this variation with protein structural changes (Horton and Perona 2000). Very importantly, however, the active (I) and inactive (II–IV) crystal forms cluster into two distinct states with respect to the B helix interface across the molecular twofold, as illustrated in Fig. 6. At the quaternary structure level the two states differ to first approximation by (1) a rotation of the two DB-D's (together with the bound DNA) by about 10° about the dimer axis and (2) by a relative translocation of the two antiparallel B helices of about one turn. The translocation is such that symmetry equivalent residues on this helix are 4–5 Å further apart in the active state. As helix B extends out to the surface of the molecule it is not surprising that such an expansion can become forbidden in a crystal lattice. Apart from the change in the direct interactions between the symmetry related B helices one also observes a number of distinct structural differences between the two states. They mainly concern the N-terminal turn of helix B with residues D36, T37 and K38 that are located near the molecular twofold on the minor groove side of the DNA. The coupling of this change in protein structure to one in DNA conformation may be mediated by the contact of T37 with the ribose sugar of the thymine nucleotide of each ATC half-site (Horton and Perona 2000). In the two states, T37 also interacts in different ways with

A   **B**

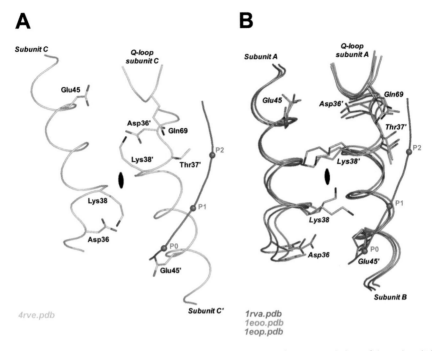

Fig. 6. B-helix interface viewed along the dyad axis in the active (**A**) and inactive (**B**) conformational states (superimposed on the basis of their R-loops). The side chains of selected residues are shown in *sticks* (color code according to indicated PDB code). The Q-loop with Q69 is shown for only one subunit of each dimer and the course of the DNA backbone of part of one strand (phosphate positions as indicated) is shown for 4RVE and 1RVA in *magenta*. The position of the crystallographic (4RVE, C and C′ subunit) or noncrystallographic dyad (1RVA, A and B subunit) is marked. The three inactive state structures (**B**) are superimposed based on their DIM-domains. Note the relative translocation of the two helices by about one helical turn. Furthermore, the Q-loop with Q69 as well as the DNA strand are differently positioned relative to the adjacent B helix in the two states

the side chain of Q69 from the Q-loop of the symmetry related subunit and this contact appears important for intersubunit communication.

A superposition of the active and inactive state structures based on the atomic positions of the central TpA step of one DNA strand reveals that the catalytic residues D90, D74 and K92 are very differently positioned with respect to the scissile phosphate. The positional differences are in the range 1.5–2.5 Å and must severely affect metal binding. In particular, the ligating groups of site I are not properly positioned in the inactive state. A significant difference is also observed for the position and orientation of P_{+1}, whose chemical modification has large effects on catalysis and whose *pro-R*$_p$ oxygen has been implicated in substrate-assisted catalysis (Jeltsch et al. 1993). The restriction endonuclease BglI, belonging to the EcoRV-related enzymes

(Bujnicki 2000), has been cocrystallized with a cognate DNA fragment in the presence of $CaCl_2$ and shows two Ca^{2+} binding sites similar to sites I and III, respectively (Newman et al. 1998). Only for the active EcoRV complex does the superposition of the local DNA backbone of the BglI and EcoRV complexes reveal a very close match for all conserved active site residues.

Surprisingly large effects on the chemical cleavage rate but only minor effects on specific binding have been reported for mutants of residues D36, T37 and K38 (Wenz et al. 1996) (Stahl et al. 1998a, b; Stanford et al. 1999; Horton et al. 2002) and of the Q-loop residue Q69 (Hancox and Halford 1997). The observed effects with D36 and Q69 have been assigned a catalytic role in the 'conformational-change' mechanisms postulated by the Halford group (Thomas et al. 1999; Stanford et al. 1999). However, the direct minor groove interaction between the Q69 side chain and T37 suggests that this contact may affect the catalytic rate indirectly through influencing the stability of the active state.

In the active state conformation, the D36 aspartate caps helix B through a hydrogen bond to the main chain N-H group of K38. In addition, the K38 side chain forms a salt bridge to D36. In the inactive state conformation, these interactions are lost, and the residues become less well ordered. The interactions of T37 with its neighboring groups differ considerably and its hydrogen bond to P_{+2} is only observed in the inactive state conformation. The severe effect on catalysis observed for mutants of these three residues indicates that their interactions are critical for defining the active conformation. From studies with heterodimers in which only one subunit carries a T37A, K38A or D36A mutation (Stahl et al. 1998a; Stahl et al. 1998b) it was concluded that these mutations only affect the activity of their own subunit. These results were in contradiction to the role of D36 in the Halford mechanism where it exerts its effect on the active site of the other subunit (Stanford et al. 1999). However, if the interaction between T37 and the thymidine sugar is critical for transmitting the conformational switch to the active site, this would represent an intrasubunit process consistent with the heterodimer results.

Starting from the idea that all charged residues in the vicinity of the scissile phosphate could exert some effect on catalytic activity through influencing the active site electrostatics, Perona and coworkers examined the role of eight charged residues within 12 Å of the scissile phosphate in single-turnover experiments (Horton et al. 2002). Most interestingly they observed that the double mutant K38M/D36N, with both charges removed was 100 times less impaired in catalytic activity than the single mutant D36N. The authors suggest that the larger charge imbalance introduced by the single mutations may be responsible for their increased rate reductions. The observed effects are strong evidence against a direct role of D36 in catalysis as proposed by the Halford group (Stanford et al. 1999).

An important question concerns the relative stability of the two states. With the exception of the decamer substrate, which makes fewer contacts with the protein and cannot induce the ordering of segment 221–229, the other complexes with undecamer and dodecamer DNA duplexes are observed in the inactive state. At least in the absence of catalytically competent divalent metal ions this state must be significantly populated in solution and cannot be considered as a crystallization artifact. Crystals grown in the presence of Ca^{2+} ions also trap the inactive state and it must be assumed that this inhibitory metal ion preferentially stabilizes this state also in solution. Factors that destabilize the active state relative to the inactive one may be the increased DNA bending (Horton and Perona 2000) and the less extended and less tightly packed B helix interface. On the other hand, increased DNA bending appears to correlate with an increase of the surface area buried between the Q-loop/Nt-αB segments and DNA, which may partly compensate for the destabilizing components (Horton and Perona 2000). An important factor for tipping the balance in favor of the active conformation may be the binding of the catalytically competent metals. In view of these newer data, the rapid reaction analysis of the catalytic cycle of EcoRV that suggested the binding of two essential metal ions may now be reinterpreted. Addition of $MgCl_2$ to the preformed complex with the dodecamer cognate DNA fragment leads to tryptophan fluorescence enhancement on a 2-s time scale followed by an even slower fluorescence decay that was shown to reflect product formation (Baldwin et al. 1995). The initial cofactor dependent conformational change has a time constant of about 7 s^{-1} and may well represent the transition from the inactive to the active state. This pre-catalytic ternary complex then binds the second metal ion needed for catalysis. The metal binding studies in the inactive state would suggest that site III, possibly together with the structural site II, is first occupied and that site I is the one associated with catalysis. However, when EcoRV binds to DNA in the presence of Mg^{2+} the initial enhancement is not seen suggesting that the inactive state is not an obligatory intermediate for the wild type enzyme (Baldwin et al. 1995). Interestingly, transient kinetics on Q69L indicated that its reaction involved a slow conformational change with a rate constant of 0.025 s^{-1} (Hancox and Halford 1997).

The tight protein–DNA interface of the active conformation means that any perturbation introduced by a mutation or substrate modification cannot easily be relaxed without distorting the optimal stereochemistry for productive Mg^{2+} binding. The Mn^{2+} catalyzed reaction may be more tolerant to such perturbations because Mn^{2+} has a higher affinity for the binding site (Vipond et al. 1995) and is more tolerant to deviations from an optimal inner ligand arrangement (Winkler et al. 1993; Perona 2002). For the Mg^{2+}-catalyzed reaction of the Q69L mutant and of other such perturbations, the less tightly packed inactive state may be further stabilized with respect to the active one and may thus become an obligatory intermediate in the reaction.

3.3.4 The Three Metal Ion Mechanism

The mechanism elaborated by Perona and coworkers (Horton et al. 1998; Sam and Perona 1999a) defines the active state as the pretransition state and therefore no longer requires a larger conformational change. A first important postulate is that a metal bound at site I (Fig. 7) plays an essential catalytic role.

As previously discussed, this site is only seen in crystalline complexes of a mutant enzyme (T93A) or with poor or refractory DNA substrates. This illustrates that relatively small structural perturbations in the protein–DNA interface can have profound effects on metal binding. One of the water ligands of site I bridges to the P_{+1} pro-S_P oxygen while the other is suitably positioned for in-line attack and is within hydrogen bonding distance to K92. A very similarly located metal binding site (occupied by Ca^{2+} or Mn^{2+} in crystal soaking studies) has been observed in all six restriction endonucleases for which metal complexes have been determined (Pingoud and Jeltsch 2001). Two variants of a transition-state model invoking three distinct metal sites but differing in the role of K92 have been proposed (Fig. 7; Horton et al. 1998; Sam and Perona 1999a). A purely structural role is assigned to site II, consistent with a nonessential role of E45. Quite different results from those obtained by the Halford group (Stanford et al. 1999) have been reported with respect to the pH dependence of the reaction (Sam and Perona 1999a). Using Mg^{2+} and a wider pH range, a bell-shaped dependence of the cleavage rate with a relatively sharp optimum at pH 8.5 was observed. This and the fact that a linear slope of log (k_c) against pH was observed has led to a different interpretation with respect to the identity of the titrated groups. The curve is interpreted as resulting from the titration of an acidic and a basic group with the pK_a values differing by at most one unit from the observed pH optimum. The acidic function is postulated to be a site III bound water molecule that protonates the leaving alcoholate. A water molecule in the inner sphere of the site I metal is postulated to become activated for in-line attack. Mg^{2+}, Mn^{2+} and Co^{2+} bound water molecules have pK_a values of 11.4, 10.5, and 9.7, respectively, for the first deprotonation of their hexaquo-complexes (Dean 1985). It must be emphasized, however, that the pK_a of a specific water molecule bound to a metal coordinated by negatively charged ligands in a protein environment may be quite different. K92 is a natural candidate for the general base in this mechanism, as it makes a hydrogen bond to the attacking water molecule. As discussed previously, its pK_a needs to be lowered, which appears structurally possible in view of the close site I metal ion. The site III and possibly site I metal provide the stabilization of the doubly charged phosphorane transition state, which is generally required for phosphoryl transfer reactions.

Altogether, the three-metal ion mechanism is consistent with the majority of the experimental data, it is stereochemically convincing and, equally important, it shares essential features with the other restriction endonucleases (Kovall and Matthews 1999).

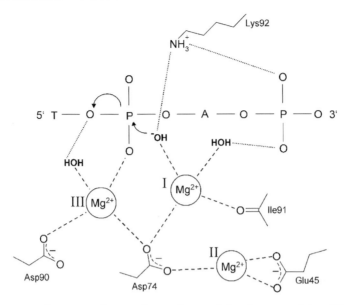

Fig. 7. Proposed state on the reaction pathway of the three metal ion DNA cleavage mechanism after water activation (Sam and Perona 1999). An inner shell water of the site I metal ion has become activated for nucleophilic attack, most likely by K92 acting as the general base. The metal in site III stabilizes the increasing negative charge at the scissile group on the way to the transition state. A water ligand of the site III metal serves as the general acid to protonate the 3'-alcoholate upon dissociation. The role of the metal in site II is proposed to be structural. *Dashed lines* connect metal ions with inner sphere ligands (some additional water ligands are omitted). *Dotted lines* indicate hydrogen bonds. Schematic diagram adapted from (Kovall and Matthews 1999)

4 Conclusions

After its initial characterization (Schildkraut et al. 1984; Kuz'min et al. 1984), EcoRV endonuclease has been extensively investigated over the last twenty years through a combination of detailed mechanistic and structural studies by a number of research groups. We have gained insight into the time course of the different processes along the kinetic pathway from the initial association with DNA to target site location, DNA cleavage and product release. The specificity of the enzyme with respect to many naturally occurring or artificially introduced changes at the hexameric target sequence has been quantified and correlated with the extensive structural data. With the exception of EcoRI, other restriction endonucleases have not been investigated in such detail. Apart from some features of the hydrolytic mechanism that appear shared by all investigated Type II restriction endonucleases, the EcoRV appears unique in most of its detailed mechanistic aspects.

Perhaps the most intriguing such aspect is how the catalysis is made highly sensitive to perturbations in the spatially remote recognition interface with cognate DNA. In the case of EcoRV, and in contrast to EcoRI endonuclease (Jen-Jacobson 1997), hardly any modifications within the tightly packed part of the interface are known to reduce binding but not the rate of cleavage. This indicates that strain relaxation in this part of the interface almost inevitably requires structural adaptations that propagate to and severely distort the active site. At the transition state of the reaction this distortion must be recorrected and the relaxed strain energy must be paid back. The structural basis of this general coupling mechanism may be different for each restriction endonuclease, and only for EcoRI and EcoRV have some important elements been identified. At most DNA sites, EcoRV will be bound transiently in the nonspecific binding mode. To probe the DNA sequence at cognate and at noncognate sites the complex must pass through intermediate stages with only partial formation of the recognition interactions. The inactive conformation is deceptively similar to the active one and permits the formation of all base specific interactions. Its less tightly packed interface with DNA in the major groove is compatible with a range of bent DNA conformations. It is tempting to speculate that the inactive conformation is an obligatory intermediate for cleavage of large DNA substrates. Probing recognition in an inactive conformation could be a safeguard against undesired cleavage.

Can we get even deeper insight or resolve some of the open questions in the near future? It is certainly possible that similarly detailed studies with other restriction endonucleases will help to clarify some of the open questions. Sound molecular dynamics simulations for such a complex system on the necessary time scales of milliseconds will remain a dream for some time. On the experimental side, single molecule spectroscopy with suitably labeled probes appears to us as a possible new avenue that could yield some exciting new information on the relevance of the different structural states along the reaction path.

References

Alves J, Selent U, Wolfes H (1995) Accuracy of the EcoRV restriction endonuclease: binding and cleavage studies with oligodeoxynucleotide substrates containing degenerate recognition sequences. Biochemistry 34:11191–11197

Anderson JE (1993) Restriction endonucleases and modification methylases. Curr Opin Struct Biol 3:24–30

Baldwin GS, Vipond IB, Halford SE (1995) Rapid reaction analysis of the catalytic cycle of the EcoRV restriction endonuclease. Biochemistry 34:705–714

Baldwin GS, Sessions RB, Erskine SG, Halford SE (1999) DNA cleavage by the EcoRV restriction endonuclease: roles of divalent metal ions in specificity and catalysis. J Mol Biol 288:87–103

Berman HM, Westbrook J, Feng Z, Gilliland G, Bhat TN, Weissig H, Shindyalov IN, Bourne PE (2000) The protein data bank. Nucleic Acids Res 28:235–242

Bujnicki JM (2000) Phylogeny of the restriction endonuclease-like superfamily inferred from comparison of protein structures. J Mol Evol 50:39–44

Connolly BA, Liu HH, Parry D, Engler LE, Kurpiewski MR, Jen-Jacobson L (2001) Assay of restriction endonucleases using oligonucleotides. Methods Mol Biol 148:465–490

Dean JA (1985) Lange's handbook of chemistry. McGraw Hill, New York

Delcourt SG, Blake RD (1991) Stacking energies in DNA. J Biol Chem 266:15160–15169

Engler LE, Welch KK, Jen-Jacobson L (1997) Specific binding by EcoRV endonuclease to its DNA recognition site GATATC. J Mol Biol 269:82–101

Erie DA, Yang G, Schultz HC, Bustamante C (1994) DNA bending by Cro protein in specific and nonspecific complexes: implications for protein site recognition and specificity. Science 266:1562–1566

Erskine SG, Halford SE (1998) Reactions of the EcoRV restriction endonuclease with fluorescent oligodeoxynucleotides: identical equilibrium constants for binding to specific and non-specific DNA. J Mol Biol 275:759–772

Erskine SG, Baldwin GS, Halford SE (1997) Rapid-reaction analysis of plasmid DNA cleavage by the EcoRV restriction endonuclease. Biochemistry 36:7567–7576

Fliess A, Wolfes H, Seela F, Pingoud A (1988) Analysis of the recognition mechanism involved in the EcoRV catalyzed cleavage of DNA using modified oligodeoxynucleotides. Nucleic Acids Res 16:11781–11793

Fothergill M, Goodman MF, Petruska J, Warshel A (1995) Structure-energy analysis of the role of metal-ions in phosphodiester bond hydrolysis by DNA polymerase I. J Am Chem Soc 117:11619–11627

Gewirth DT, Sigler PB (1995) The basis for half-site specificity explored through a non-cognate steroid receptor-DNA complex. Nat Struct Biol 2:386–394

Grasby JA, Connolly BA (1992) Stereochemical outcome of the hydrolysis reaction catalyzed by the EcoRV restriction endonuclease. Biochemistry 31:7855–7861

Gutfreund H (1995) Kinetics for the life sciences. Cambridge University Press, Cambridge

Halford SE, Goodall AJ (1988) Modes of DNA cleavage by the EcoRV restriction endonuclease. Biochemistry 27:1771–1777

Hancox EL, Halford SE (1997) Kinetic analysis of a mutational hot spot in the EcoRV restriction endonuclease. Biochemistry 36:7577–7585

Horton NC, Perona JJ (1998) Role of protein-induced bending in the specificity of DNA recognition: crystal structure of EcoRV endonuclease complexed with d(AAAGAT) + d(ATCTT). J Mol Biol 277:779–787

Horton NC, Perona JJ (2000) Crystallographic snapshots along a protein-induced DNA-bending pathway. Proc Natl Acad Sci USA 97:5729–5734

Horton NC, Newberry KJ, Perona JJ (1998) Metal ion-mediated substrate-assisted catalysis in Type II restriction endonucleases. Proc Natl Acad Sci USA 95:13489–13494

Horton NC, Connolly BA, Perona JJ (2000) Inhibition of EcoRV endonuclease by deoxyribo-3′-S-phosphorothiolates: a high-resolution X-ray crystallographic study. J Am Chem Soc 122:3314–3324

Horton NC, Otey C, Lusetti S, Sam MD, Kohn J, Martin AM, Ananthnarayan V, Perona JJ (2002) Electrostatic contributions to site specific DNA cleavage by EcoRV endonuclease. Biochemistry 41:10754–10763

Hunter CA (1993) Sequence-dependent DNA structure. The role of base stacking interactions. J Mol Biol 230:1025–1054

Hunter CA, Lu XJ (1997) DNA base-stacking interactions: a comparison of theoretical calculations with oligonucleotide X-ray crystal structures. J Mol Biol 265:603–619

Jeltsch A, Alves J, Maass G, Pingoud A (1992) On the catalytic mechanism of EcoRI and EcoRV. A detailed proposal based on biochemical results, structural data and molecular modelling. FEBS Lett 304:4–8

Jeltsch A, Alves J, Wolfes H, Maass G, Pingoud A (1993) Substrate-assisted catalysis in the cleavage of DNA by the EcoRI and EcoRV restriction enzymes. Proc Natl Acad Sci USA 90:8499–8503

Jeltsch A, Pingoud A (1998) Kinetic characterization of linear diffusion of the restriction endonuclease EcoRV on DNA. Biochemistry 37:2160–2169

Jeltsch A, Wenz C, Stahl F, Pingoud A (1996) Linear diffusion of the restriction endonuclease EcoRV on DNA is essential for the in vivo function of the enzyme. EMBO J 15:5104–5111

Jen-Jacobson L (1997) Protein-DNA recognition complexes: conservation of structure and binding energy in the transition state. Biopolymers 44:153–180

Jen-Jacobson L, Engler LE, Lesser DR, Kurpiewski MR, Yee C, McVerry B (1996) Structural adaptations in the interaction of EcoRI endonuclease with methylated GAATTC sites. EMBO J 15:2870–2882

Jen-Jacobson L, Engler LE, Jacobson LA (2000) Structural and thermodynamic strategies for site-specific DNA binding proteins. Structure Fold Des 8:1015–1023

Jones S, Daley DT, Luscombe NM, Berman HM, Thornton JM (2001) Protein-RNA interactions: a structural analysis. Nucleic Acids Res 29:943–954

Knowles J (2003) Chemistry. Seeing is believing. Science 299:2002–2003

Kostrewa D, Winkler FK (1995) Mg^{2+} binding to the active site of EcoRV endonuclease: a crystallographic study of complexes with substrate and product DNA at 2 Å resolution. Biochemistry 34:683–696

Kovall RA, Matthews BW (1999) Type II restriction endonucleases: structural, functional and evolutionary relationships. Curr Opin Chem Biol 3:578–583

Kuz'min NP, Loseva SP, Beliaeva RK, Kravets AN, Solonin AS (1984) [EcoRV restrictase: physical and catalytic properties of homogenous enzyme]. Mol Biol (Mosk) 18:197–204

Lahiri SD, Zhang G, Dunaway-Mariano D, Allen KN (2003) The pentacovalent phosphorus intermediate of a phosphoryl transfer reaction. Science 299:2067–2071

Lesser DR, Kurpiewski MR, Jen-Jacobson L (1990) The energetic basis of specificity in the EcoRI endonuclease–DNA interaction. Science 250:776–786

Lesser DR, Kurpiewski MR, Waters T, Connolly BA, Jen-Jacobson L (1993) Facilitated distortion of the DNA site enhances EcoRI endonuclease-DNA recognition. Proc Natl Acad Sci USA 90:7548–7552

Luscombe NM, Laskowski RA, Thornton JM (2001) Amino acid-base interactions: a three-dimensional analysis of protein-DNA interactions at an atomic level. Nucleic Acids Res 29:2860–2874

Martin AM, Horton NC, Lusetti S, Reich NO, Perona JJ (1999a) Divalent metal dependence of site-specific DNA binding by EcoRV endonuclease. Biochemistry 38:8430–8439

Martin AM, Sam MD, Reich NO, Perona JJ (1999b) Structural and energetic origins of indirect readout in site-specific DNA cleavage by a restriction endonuclease. Nat Struct Biol 6:269–277

Newman M, Lunnen K, Wilson G, Greci J, Schildkraut I, Phillips SE (1998) Crystal structure of restriction endonuclease BglI bound to its interrupted DNA recognition sequence. EMBO J 17:5466–5476

Newman PC, Nwosu VU, Williams DM, Cosstick R, Seela F, Connolly BA (1990a) Incorporation of a complete set of deoxyadenosine and thymidine analogues suitable for the study of protein nucleic acid interactions into oligodeoxynucleotides. Application to the EcoRV restriction endonuclease and modification methylase. Biochemistry 29:9891–9901

Newman PC, Williams DM, Cosstick R, Seela F, Connolly BA (1990b) Interaction of the EcoRV restriction endonuclease with the deoxyadenosine and thymidine bases in its recognition hexamer d(GATATC). Biochemistry 29:9902–9910

Parry D, Moon SA, Liu HH, Heslop P Connolly BA (2003) DNA recognition by the EcoRV restriction endonuclease probed using base analogues. J Mol Biol 331:1005–1016

Perona JJ (2002) Type II restriction endonucleases. Methods 28:353–364

Perona JJ, Martin AM (1997) Conformational transitions and structural deformability of EcoRV endonuclease revealed by crystallographic analysis. J Mol Biol 273:207–225

Pingoud A, Jeltsch A (1997) Recognition and cleavage of DNA by type-II restriction endonucleases. Eur J Biochem 246:1–22

Pingoud A, Jeltsch A (2001) Structure and function of Type II restriction endonucleases. Nucleic Acids Res 29:3705–3727

Reid SL, Parry D, Liu HH, Connolly BA (2001) Binding and recognition of GATATC target sequences by the EcoRV restriction endonuclease: a study using fluorescent oligonucleotides and fluorescence polarization. Biochemistry 40:2484–2494

Roberts RJ, Halford SE (1993) Type II restriction endonucleases. In: Linn SM, Lloyd R S, Roberts R J (eds) Nucleases. Cold Spring Harbor Laboratory Press, Cold Spring Harbor, New York, pp 35–88

Sam MD, Perona JJ (1999a) Catalytic roles of divalent metal ions in phosphoryl transfer by EcoRV endonuclease. Biochemistry 38:6576–6586

Sam MD, Perona JJ (1999b) Mn^{2+}-dependent catalysis by restriction enzymes: pre-steady state analysis of EcoRV endonuclease reveals burst kinetics and the origins of reduced activity. J Am Chem Soc 121:1444–1447

Sam MD, Horton NC, Nissan TA, Perona JJ (2001) Catalytic efficiency and sequence selectivity of a restriction endonuclease modulated by a distal manganese ion binding site. J Mol Biol 306:851–861

SantaLucia J Jr, Allawi HT, Seneviratne PA (1996) Improved nearest-neighbor parameters for predicting DNA duplex stability. Biochemistry 35:3555–3562

Schildkraut I, Banner CD, Rhodes CS, Parekh S (1984) The cleavage site for the restriction endonuclease EcoRV is 5'-GAT/ATC-3'. Gene 27:327–329

Stahl F, Wende W, Jeltsch A, Pingoud A (1998a) The mechanism of DNA cleavage by the Type II restriction enzyme EcoRV: Asp36 is not directly involved in DNA cleavage but serves to couple indirect readout to catalysis. Biol Chem 379:467–473

Stahl F, Wende W, Wenz C, Jeltsch A, Pingoud A (1998b) Intra- vs intersubunit communication in the homodimeric restriction enzyme EcoRV: Thr 37 and Lys 38 involved in indirect readout are only important for the catalytic activity of their own subunit. Biochemistry 37:5682–5688

Stanford NP, Halford SE, Baldwin GS (1999) DNA cleavage by the EcoRV restriction endonuclease: pH dependence and proton transfers in catalysis. J Mol Biol 288:105–116

Stanford NP, Szczelkun MD, Marko JF, Halford SE (2000) One- and three-dimensional pathways for proteins to reach specific DNA sites. EMBO J 19:6546–6557

Stover T, Kohler E, Fagin U, Wende W, Wolfes H, Pingoud A (1993) Determination of the DNA bend angle induced by the restriction endonuclease EcoRV in the presence of Mg^{2+}. J Biol Chem 268:8645–8650

Szczelkun MD, Connolly BA (1995) Sequence-specific binding of DNA by the EcoRV restriction and modification enzymes with nucleic acid and cofactor analogues. Biochemistry 34:10724–10733

Taylor JD, Halford SE (1989) Discrimination between DNA sequences by the EcoRV restriction endonuclease. Biochemistry 28:6198–6207

Taylor JD, Badcoe IG, Clarke AR, Halford SE (1991) EcoRV restriction endonuclease binds all DNA sequences with equal affinity. Biochemistry 30:8743–8753

Thielking V, Alves J, Fliess A, Maass G, Pingoud A (1990) Accuracy of the EcoRI restriction endonuclease: binding and cleavage studies with oligodeoxynucleotide substrates containing degenerate recognition sequences. Biochemistry 29:4682–4691

Thielking V, Selent U, Kohler E, Wolfes H, Pieper U, Geiger R, Urbanke C, Winkler FK, Pingoud A (1991) Site-directed mutagenesis studies with EcoRV restriction endonuclease to identify regions involved in recognition and catalysis. Biochemistry 30:6416–6422

Thielking V, Selent U, Kohler E, Landgraf A, Wolfes H, Alves J, Pingoud A (1992) Mg^{2+} confers DNA binding specificity to the EcoRV restriction endonuclease. Biochemistry 31:3727–3732

Thomas MP, Brady RL, Halford SE, Sessions RB, Baldwin GS (1999) Structural analysis of a mutational hot-spot in the EcoRV restriction endonuclease: a catalytic role for a main-chain carbonyl group. Nucleic Acids Res 27:3438–3445

Thorogood H, Grasby JA, Connolly BA (1996a) Influence of the phosphate backbone on the recognition and hydrolysis of DNA by the EcoRV restriction endonuclease. A study using oligodeoxynucleotide phosphorothioates. J Biol Chem 271:8855–8862

Thorogood H, Waters TR, Parker AW, Wharton CW, Connolly BA (1996b) Resonance Raman spectroscopy of 4-thiothymidine and oligodeoxynucleotides containing this base both free in solution and bound to the restriction endonuclease EcoRV. Biochemistry 35:8723–8733

Vermote CL, Halford SE (1992) EcoRV restriction endonuclease: communication between catalytic metal ions and DNA recognition. Biochemistry 31:6082–6089

Vermote CL, Vipond IB, Halford SE (1992) EcoRV restriction endonuclease: communication between DNA recognition and catalysis. Biochemistry 31:6089–6097

Viadiu H, Aggarwal AK (2000) Structure of BamHI bound to nonspecific DNA: a model for DNA sliding. Mol Cell 5:889–895

Vipond IB, Halford SE (1993) Structure-function correlation for the EcoRV restriction enzyme: from non-specific binding to specific DNA cleavage. Mol Microbiol 9:225–231

Vipond IB, Halford SE (1995) Specific DNA recognition by EcoRV restriction endonuclease induced by calcium ions. Biochemistry 34:1113–1119

Vipond IB, Baldwin GS, Halford SE (1995) Divalent metal ions at the active sites of the EcoRV and EcoRI restriction endonucleases. Biochemistry 34:697–704

Waters TR, Connolly BA (1994) Interaction of the restriction endonuclease EcoRV with the deoxyguanosine and deoxycytidine bases in its recognition sequence. Biochemistry 33:1812–1819

Wenz C, Jeltsch A, Pingoud A (1996) Probing the indirect readout of the restriction enzyme EcoRV. Mutational analysis of contacts to the DNA backbone. J Biol Chem 271:5565–5573

Windolph S, Alves J (1997) Influence of divalent cations on inner-arm mutants of restriction endonuclease EcoRI. Eur J Biochem 244:134–139

Winkler FK (1992) Structure and function of restriction endonucleases. Curr Opin Struct Biol 2:93–99

Winkler FK, Banner DW, Oefner C, Tsernoglou D, Brown RS, Heathman SP, Bryan RK, Martin PD, Petratos K, Wilson KS (1993) The crystal structure of EcoRV endonuclease and of its complexes with cognate and non-cognate DNA fragments. EMBO J 12:1781–1795

Two of A Kind: BamHI and BglII

É. Scheuring Vanamee, H. Viadiu, C.M. Lukacs, A.K. Aggarwal

1 Introduction

Among the more than 3500 Type II restriction endonucleases (REases) identified to date (Roberts et al. 2003), fourteen have been structurally characterized so far, including EcoRI (McClarin et al. 1986; Kim et al. 1990), EcoRV (Winkler et al. 1993; Kostrewa and Winkler 1995; Perona and Martin 1997; Horton and Perona 1998; Thomas et al. 1999), BamHI (Newman et al. 1994a, 1995; Viadiu and Aggarwal 1998, 2000) PvuII (Athanasiadis et al. 1994; Cheng et al. 1994; Horton and Cheng 2000) FokI (Wah et al. 1997, 1998), Cfr10I (Bozic et al. 1996), BglI (Newman et al. 1998), BglII (Lukacs et al. 2000, 2001), BsoBI (van der Woerd et al. 2001), NaeI (Huai et al. 2000, 2001), NgoMIV (Deibert et al. 2000), Bse634I (an isoschizomer of Cfr10I), (Grazulis et al. 2002), MunI (Deibert et al. 1999), and HincII (Horton et al. 2002). All except Cfr10I and Bse634I have been found to be bound to their cognate DNA sites. Despite the lack of sequence homology, all REases consist of a central β-sheet that is flanked by α-helices on both sides. Interestingly, a similar α/β core is also present in other DNA-acting enzymes such as λ-exonuclease (Kovall and Matthews 1997, 1998), MutH (Ban and Yang 1998), Vsr endonuclease (Tsutakawa et al. 1999a, b), and TnsA (Hickman et al. 2000) from the Tn7 transposase. In the common core only three β-strands are absolutely conserved, two of these strands contain the amino acid residues directly involved in catalysis. The similarity at the tertiary structure level is strongest between endonucleases that share a similar cleavage pattern, such as between BamHI and EcoRI which cleave DNA to leave four-base 5′ overhangs, or between EcoRV and PvuII which cleave DNA to produce blunt ends (Aggarwal 1995). An exception is BglI that has a similar fold as EcoRV and PvuII but cleaves

É. Scheuring Vanamee, H. Viadiu, C.M. Lukacs, A.K. Aggarwal
Structural Biology Program, Department of Physiology and Biophysics, Mount Sinai School of Medicine 1425 Madison Ave, New York, New York 10029, USA

Nucleic Acids and Molecular Biology, Vol. 14
Alfred Pingoud (Ed.)
Restriction Endonucleases
© Springer-Verlag Berlin Heidelberg 2004

DNA to leave 3′ overhangs (Newman et al. 1998). Overall, the similarity reflects constraints in positioning of the active sites: 17–19 Å apart to produce four-base 5′ overhangs and ~2 Å apart to produce blunt ends (Newman et al. 1995). The active sites occur at one end of the central β-sheet and contain at least three superimposable residues that are critical for catalysis. Two of these residues are acidic, while the third residue is usually a lysine, except in BamHI which has a glutamate (Newman et al. 1994b) and in BglII which has a glutamine (Lukacs et al. 2000).

Here, we focus on the structure–function of two Type IIP REases, BamHI and BglII. BamHI and BglII embody the flexibility and constraints in REase design in the recognition of similar DNA sites.

2 BamHI Endonuclease

BamHI is a 213-amino acid REase isolated from *Bacillus amyloliquefaciens* H (Brooks et al. 1991). It recognizes the palindromic sequence 5′-GGATCC-3′ and cleaves after the first nucleotide leaving four-base 5′ overhangs. BamHI is one of the best structurally studied REase, with structures of the free enzyme and complexes with cognate and noncognate sites available (Newman et al. 1994a, 1995; Viadiu et al. 2000). In addition, the cocrystal structures with divalent metals provide snapshots of the pre- and post-reactive states of BamHI (Viadiu and Aggarwal 1998). Together, these structures grant important insight into DNA recognition, selectivity, and the mechanism of cleavage by this endonuclease.

2.1 The Structure of Free BamHI

The subunit structure of BamHI (Fig. 1) consists of a large six-stranded mixed β-sheet, which is sandwiched on both sides by α-helices (Newman et al. 1994a) Strands $β^3$, $β^4$, and $β^5$ are antiparallel and form a β-meander; strands $β^5$, $β^6$, and $β^7$ are parallel and resemble a Rossman fold, with $α^4$ and $α^6$ acting as the crossover helices. The dimer interface is formed primarily by helices $α^4$ and $α^6$, which pair with the corresponding helices from the symmetry related subunit to form a parallel four-helix bundle. The catalytic residues of Asp94, Glu111, and Glu113 are clustered at one end of the β meander. BamHI is the only enzyme that contains a glutamate at the position of the third catalytic residue. This difference is critical: there is a severe loss of cleavage activity when Glu113 in BamHI is substituted by a lysine residue (Dorner and Schildkraut 1994), or conversely, when Lys92 in EcoRV (Selent et al. 1992) or Lys113 in EcoRI (Grabowski et al. 1995) is substituted by a glutamate residue. At the bottom of the dimer is a cleft that can accommodate B-form DNA.

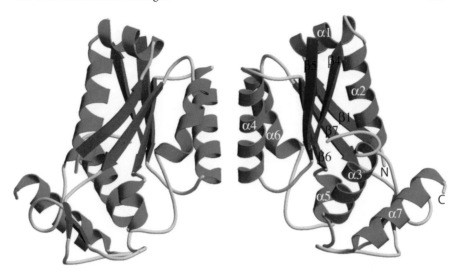

Fig. 1. Overall structure of the Type IIP endonuclease BamHI. Secondary structural elements, along with the N-terminus and C-terminus are labeled on one monomer. Produced with the programs MOLSCRIPT (Kraulis 1991) and RASTER3D (Merritt and Bacon 1997)

2.2 BamHI Nonspecific Complex

Restriction enzymes have evolved stringent specificity for their recognition sequence, such that even a single base-pair change can reduce their cleavage activity by over a millionfold (Thielking et al. 1990). For instance, for every BamHI cognate DNA site in the *B. amyloliquefaciens* H genome, there are ~18 sites that differ by only a single base pair. Only the cognate sites are protected from cleavage by methylation produced by the partner methyltransferase enzyme (Roberts and Halford 1993). The structure of BamHI bound to the 5′-GAATCC-3′ site, differing by only a single base pair from the cognate sequence (5′-GGATCC-3′), has uncovered the basis of this remarkable selectivity (Viadiu and Aggarwal 2000). The structure reveals the enzyme in a conformation that is incompetent for cleavage but competent for sliding (Fig. 2). The noncognate DNA is accommodated loosely, protruding out of the cleft at the bottom of the BamHI dimer. The cleft is only slightly more closed compared to the free enzyme. The protein does not make any base-specific contact with the DNA, and only a few water-mediated contacts to the phosphate backbone. Compared to the free enzyme, residues 79–91 become ordered and cross over to make contacts with residues from the other monomer. The C-terminal residues (194 to 213) are α-helical as in the free enzyme, and their likely role is to stabilize the complex.

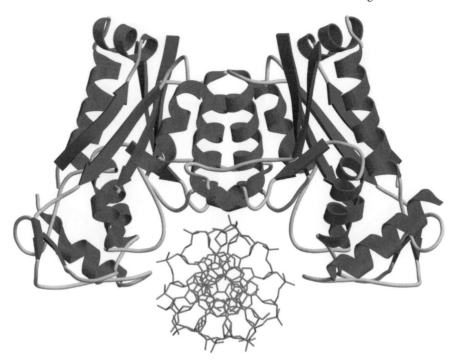

Fig. 2. Overall structure of BamHI in complex with noncognate DNA looking down the DNA axis. Produced with the programs MOLSCRIPT (Kraulis 1991) and RASTER3D (Merritt and Bacon 1997)

BamHI derives its specificity part from binding and part from catalysis. Single base-pair substitutions within the recognition sequence affect binding to a similar extent as a random sequence (Engler 1998), as in the case of EcoRI (Lesser et al. 1990; Thielking et al. 1990). For instance, the single G:C to A:T base-pair change (GAATCC) carries a large energy penalty (11.7 kcal/mol) in forming the activated transition state, with 3.5 kcal/mol due to differences in binding energies in forming the enzyme–DNA complex (E–DNA) and 8.1 kcal/mol due to differences in energies in activating the E-DNA complex to the transition state (Engler 1998; Jen-Jacobson et al. 2000). The basis of this extreme selectivity, important in avoiding potentially lethal cleavages at noncognate sites can be explained by the active site residues (Asp-94, Glu-111, and Glu-113) that are nearly 7 Å away from the scissile phosphodiester, making it impossible for the enzyme to cleave the DNA. Interestingly, the active site residues are also displaced in the structure of EcoRV bound across two short DNA oligomers, which mimics a nonspecific complex (Winkler et al. 1993). However, in EcoRV, the displacement of active site residues is due to a change in DNA conformation, whereas in BamHI it is due mostly to an adaptation in the protein conformation.

Like many DNA binding proteins, BamHI can find its target DNA site faster than the three-dimensional diffusion limit and shows an increase in the cleavage reaction as the length of nonspecific DNA around the cognate site is increased. The nonspecific structure shows how this may occur. The noncognate DNA is bound loosely within a widened DNA binding cleft, making it possible for BamHI to slide along the DNA. The overall lack of DNA backbone contacts may facilitate diffusion by helping to reduce the lifetime of the nonspecific complex and by lowering the activation energy for the breaking and reforming of DNA contacts as the enzyme moves forward (or backward). The nonspecific complex provides a remarkable snapshot of an enzyme poised for linear diffusion, paving the way for single molecule studies that could address whether the enzyme moves by a corkscrew motion (following the DNA major groove) or along one face of the DNA. Alternatively, initial nonspecific binding might be achieved by the protein binding "parallel" to the DNA helical axis (Sun et al. 2003). The protein in this "theoretical" model has a much smaller interaction surface but is favored electrostatically. The enzyme could switch back and forth between these two nonspecific states as it searches the DNA for the target sequence. All in all, the structure of the nonspecific complex provides the most detailed picture yet of how an enzyme selects its cognate site from the multitude of nonspecific sites in a cell.

2.3 BamHI Specific Complex

The cognate DNA is bound "up" in the cleft in concert with residues 79–92 moving up, a movement of ~9 Å as compared to the nonspecific complex (Figs. 3 and 4). The residues are much less mobile in the specific than in the nonspecific complex (C_α average B factor of 23 vs. 41 Å^2), reflecting substantially more electrostatic interactions with the DNA backbone. The cleft is much narrower than in the nonspecific complex (20 vs. 25 Å). Interestingly, the specific complex is asymmetrical. The C terminus (194 to 213), which is α-helical in both the free enzyme and in the nonspecific complex, unfolds to form "arms" with the arm from one subunit (termed R) fitting into the minor groove and the arm from the other (L) subunit following the DNA sugar–phosphate backbone. Because these helices do not unfold in the nonspecific structure, there are no minor groove or DNA backbone contacts from these C-terminal residues. However, since the N termini of these C-terminal helices are directed toward the DNA backbone, their helix dipole moments will contribute to the stabilization of the nonspecific complex. Thus, the BamHI C-terminal residues may fulfill a dual role: first, as helices, they may aid in the initial binding and the diffusion of the enzyme on nonspecific DNA; second, by unfolding they may increase the lifetime of the specific complex for the subsequent cleavage reaction. Overall, the BamHI subunits clamp onto the DNA by a ~10° rotation around the DNA axis moving in a tongs-like motion.

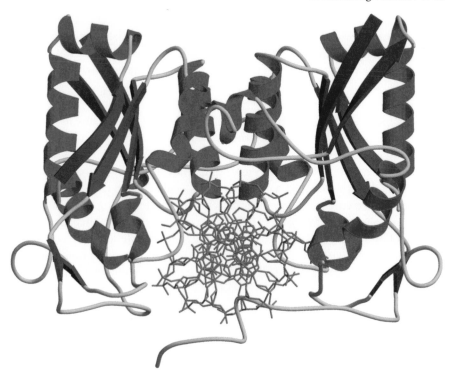

Fig. 3. Overall structure of BamHI in complex with cognate DNA looking down the DNA axis. Produced with the programs MOLSCRIPT (Kraulis 1991) and RASTER3D (Merritt and Bacon 1997)

There are extensive protein–DNA interactions, with side-chain and main-chain atoms, tightly bound water molecules all contributing toward recognition of the BamHI recognition sequence (Newman et al. 1995). Interactions with cognate DNA occur both in the major and minor grooves. In the major groove, every hydrogen bond donor and acceptor group takes part in direct or water-mediated hydrogen bonds with the protein (Fig. 5a). In particular, major groove contacts are made primarily by regions at the NH_2-terminal ends of helices α^4 and α^6. The outer G•C base pair (<u>G</u>GA) is contacted by Arg 155 and Asp154 from the 152–157 loop that precedes helix α^6. The middle G•C base pair (G<u>G</u>A) is contacted by Asp154, Arg 122, and Asn 116. The inner A•T base pair (A<u>T</u>C) is contacted primarily through water mediated hydrogen bonds as well as by Asn 116. The minor groove contacts are formed by the refolded COOH-terminal arm of the R subunit. There is insufficient room to accommodate arms from both subunits if they were to fold symmetrically in the groove. The R arm makes specific interactions in the minor groove with both DNA half-sites via residues Asp196, Gly197, and Met 198. In addition, there are extensive interactions between the enzyme and the DNA

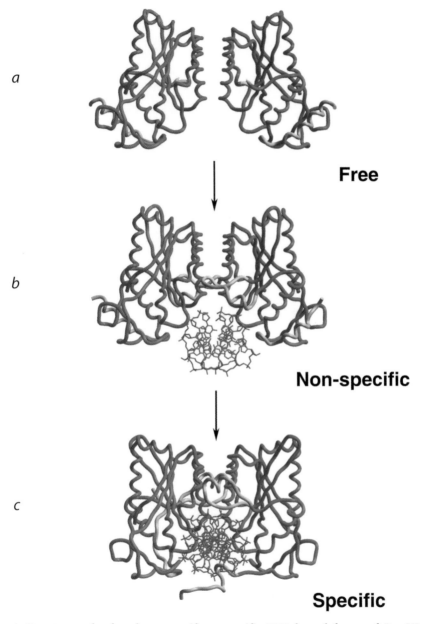

Fig. 4. Structures of **a** free, **b** nonspecific, **c** specific DNA-bound forms of BamHI. Regions that undergo local conformational changes are shown in *yellow* color. The enzyme becomes progressively more closed around the DNA as it goes from the non-specific to the specific DNA binding mode. Residues 79–91 are unstructured in the free enzyme but become ordered in both the nonspecific and the specific complexes, albeit in different conformation. The C-terminal residues unwind in the specific complex to form partially disordered arms, whereas in the free enzyme and the nonspecific complex they remain α-helical. Produced with the programs MOLSCRIPT (Kraulis 1991) and RASTER3D. (Merritt and Bacon 1997)

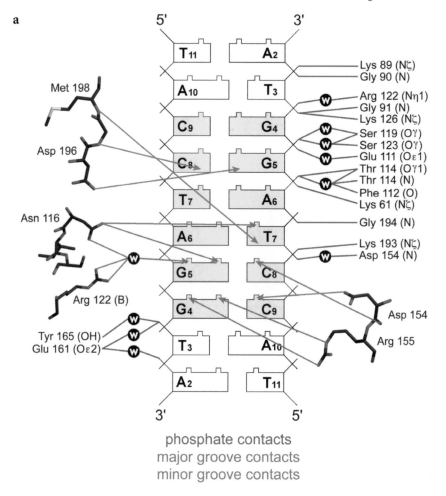

Fig. 5. Schematic representation of the phosphate and base contacts from **a** one monomer of BamHI and **b** one monomer of BglII

sugar–phosphate backbone, from residues at the NH_2-terminal ends of helices α^3, α^4, and α^6, the ordered region before strand β^4 (89 to 91), and the segment preceding the refolded COOH-terminus (193 and 194). The complementarity at the protein–DNA interface ensures that only the BamHI recognition sequence can make all the necessary interactions.

In comparing the cognate and noncognate complexes, it seems remarkable that a single base-pair change could trigger the conformational changes described above for BamHI. However, the introduction of adenine at the second position (GAATCC) in the nonspecific complex will have the effect of disrupting the hydrogen bonds made by Asn-116 with both the middle guanine and the inner thymine of the recognition sequence, while the thymine methyl

b

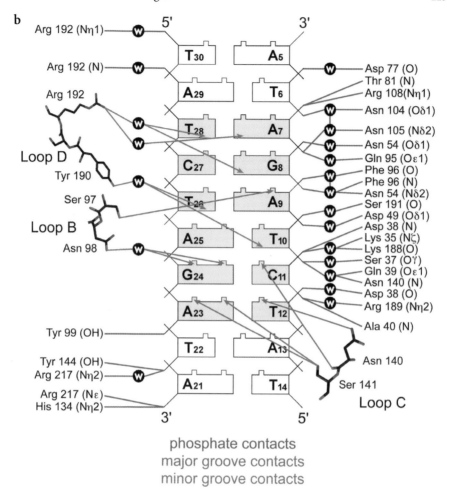

phosphate contacts
major groove contacts
minor groove contacts

group will cause severe steric clashes with both the main chain and side chain atoms of Glu-154, disrupting the hydrogen bonds Asp-154 makes with the cytosines of both the middle and outer base pairs. Thus, it could be argued that the potential loss of four hydrogen bonds and steric repulsion is sufficient to switch the enzyme to a nonspecific mode of DNA binding, in accordance with similarly observed binding constants on single and multiply mutated DNA sites (Engler 1998). The symmetrical, outward movement of both BamHI monomers in response to a single base-pair change is consistent with the loss of cleavage activity observed at both half-sites (Lesser et al. 1990; Engler 1998). Taken together, the remarkable complementarity one sees at the BamHI–DNA interface in the specific complex may be a mechanism to ensure that the introduction of even a single wrong base pair leads to a disruption of the interface, thereby forcing the enzyme to dock in a more open mode, which at the same time ends up displacing the active site residues away from the scissile phosphodiesters, Thus, the enzyme can still bind to the noncognate site (down by 10^2 to 10^3) but will rarely cleave it (down by 10^7 to 10^{10}).

2.4 The Pre-Reactive Complex of BamHI

The pre-reactive complex of BamHI was solved in the presence of Ca^{2+}. Calcium is an inhibitor of REase cleavage but readily binds in place of the natural cofactor Mg^{2+} and has been widely used to study the ground state of restriction enzymes. As expected, the phosphodiester bond stays intact in the presence of Ca^{2+}. Two metals ($Ca^{2+}A$ and $Ca^{2+}B$) are found bound only in the R active site and separated by 4.3 Å (Fig. 6a). $Ca^{2+}A$ is coordinated in an octahe-

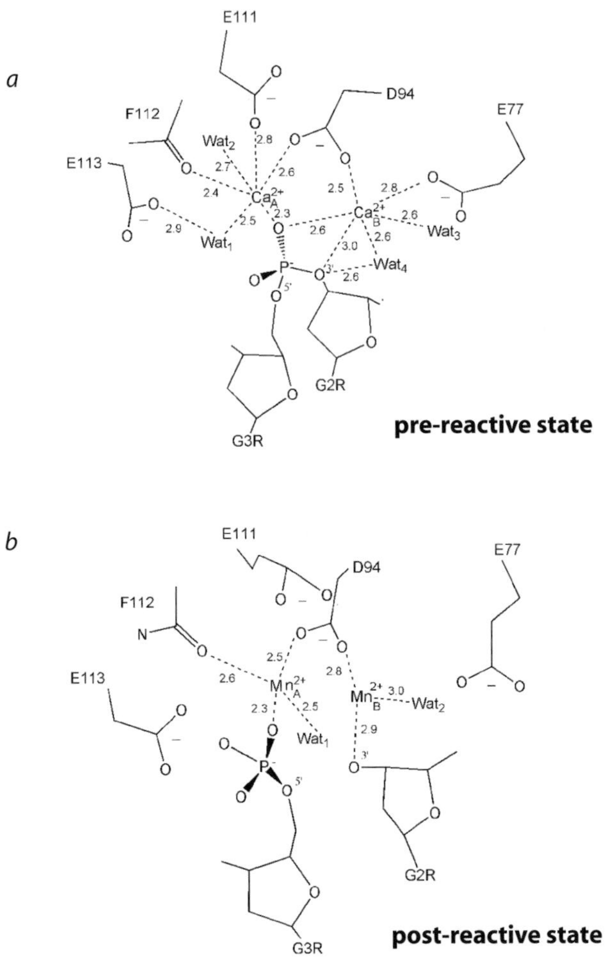

Fig. 6. a The pre-reactive state of BamHI captured in the presence of Ca^{2+}. Both metal ions are coordinated to six ligands in octahedral arrangements. Wat_1–Wat_4 refers to water molecules within the active site. The distance between Ca^{2+}_A and Ca^{2+}_B is 4.3 Å. b The post-reactive state of BamHI captured in the presence of Mn^{2+}. The two Mn^{2+} ions are disposed similarly to the Ca^{2+} ions though they are slightly closer (3.9 Å) and have looser coordination spheres as well. The overall r.m.s.d. between the Cα positions of the pre- and post-reactive complexes is only 0.33 Å. The main difference is in the location of the scissile phosphate group. After cleavage it is displaced by 2.5 Å bringing it within the vicinity of Glu 113

dral arrangement with six ligands: the OD2 of Asp 94 (2.6 Å), the OE2 of Glu 111 (2.8 Å), the carbonyl oxygen of Phe 112 (2.4 Å), the O2 atom of the cleavable phosphate group (2.3 Å), and two water molecules (2.5 Å and 2.7 Å). $Ca^{2+}B$ is also coordinated to six ligands: the OD1 of Asp 94 (2.5 Å), the OE2 of Glu 77 (2.8 Å), the O2 and O3' atoms of the cleavable phosphate group (2.6 and 3.0 Å), and two water molecules (2.6 Å for both). Overall, the Ca^{2+} complex differs by only 0.54 Å r.m.s.d. in Cα positions from the complex determined in the absence of metals. The main differences are in the R active site that binds calcium. The side chain of Asp 94 is coordinated in a bidentate manner to both Ca^{2+} ions, its conformation buttressed by hydrogen bonds to the hydroxyl group of Tyr 65 and the OE1 of Glu 111. The movement of Glu 77 towards the active site is important in creating the binding site for $Ca^{2+}B$. (In the L active site that lacks calcium, Asp 94 and Glu 77 point away from the DNA in a similar manner as in the complex without metal). Amongst the several water molecules at the R active site, two are noteworthy: Wat1, coordinated to $Ca^{2+}A$ and 3.2 Å from the scissile phosphodiester group, is positioned ideally as the attacking water molecule, maintaining in-line orientation with the cleavable P-O3' bond. The molecule is hydrogen bonded to Glu 113 (2.9 Å). The side chain of Glu 113 is fixed in position by a network of hydrogen bonds with the carbonyl oxygen of Ser 119, the Ov atoms of Ser 119 and Ser 123, and the amine of residue 115. On the other side of the scissile phosphodiester group, Wat4, coordinated to $Ca^{2+}B$, forms a hydrogen bond (2.6 Å) with the O3', possibly acting as a general acid in donating a proton to the leaving group. Taken together, the Ca^{2+}-bound structure provides a remarkably detailed view of the BamHI–DNA complex in its ground state, prior to DNA cleavage.

2.5 The Structure of the BamHI Post-Reactive Complex

To BamHI post-reactive complex was solved in the presence of $MnCl_2$. Remarkably, the DNA is cleaved in the cocrystal – showing that BamHI maintains an active conformation in the crystalline state (Fig. 6b). The DNA cleavage occurs only in the R active site that contains two Mn^{2+} ions. The overall r.m.s.d. between the Cα positions of the pre-reactive Ca^{2+} complex and the post-reactive Mn^{2+} complex is only 0.33 Å. The main difference is in the location of the scissile phosphate group after cleavage. Following cleavage, the phosphate group is displaced by 2.5 Å from the position it occupied when the phosphodiester bond was intact, bringing it within the vicinity of Glu 113 (3.0 Å). The two Mn^{2+} ions are disposed similarly to the Ca^{2+} ions though they are slightly closer (3.9 Å as opposed to 4.3 Å). The coordination spheres of the two Mn^{2+} ions are, however, somewhat looser than Ca^{2+}. $Mn^{2+}A$ is coordinated by the OD2 of Asp 94 (2.5 Å), the carbonyl oxygen of Phe 112 (2.6 Å), the O2 atom of the cleaved phosphate group (2.3 Å), and a water molecule (2.5 Å). Glu 111 is no longer coordinated to the metal at site A as in the Ca^{2+} complex, but has moved nearer to the metal at site B. $Mn^{2+}B$ is coordinated by OD1 of Asp 94 (2.8 Å), the O3′ of the Gua 2R (2.9 Å) and a water molecule (3.0 Å), but has lost its contacts with Glu 77. Overall, the metals give the impression of being less tightly bound after the cleavage reaction, which is also reflected by their higher temperature factors (45 and 76 $Å^2$ for Mn^{2+} vs. 31 and 24 $Å^2$ for Ca^{2+} at sites A and B, respectively).

BamHI recognizes DNA in an asymmetric manner, in which the C-terminal arm of the R subunit goes into the DNA minor groove while the corresponding region of the L subunit follows the DNA sugar–phosphate backbone. In the pre- and post-reactive states of BamHI, the metals are only found in the R active site. An intriguing question is whether this asymmetry in metal binding is related to the mechanism of the enzyme. BamHI cleaves DNA in a sequential manner in which the enzyme cleaves first one DNA strand and then the other. It appears that only the subunit with the C-terminal arm in the minor groove is able to bind metals for catalysis. When BamHI binds DNA, it is possible to imagine a competition as to which subunit first manages to get its C-terminal arm into the minor groove. At a later time, as the two subunits exchange their C-terminal arms, the second subunit may then acquire the ability to bind metals for the cleavage of the second DNA strand.

Based on the pre- and post-reactive complexes of BamHI a two metal mechanism has been proposed according to which, metal A activates the attacking water molecule, while metal B stabilizes the build-up of negative charge on the leaving O_3' atom. At the same time, both metals (acting as Lewis acids) help to stabilize the pentacovalent transition state. The positions of the Ca^{2+} and Mn^{2+} ions conform to the classical description of a two-metal mechanism, as discussed by Beese and Steitz (1991) for *E. coli* DNA polymerase I. The two metals lie on a line that is parallel to the apical direction of the trigonal bipyramid geometry that would be expected at the scissile phosphodiester, during the transition state. The distance between two metals (~4 Å) correlates with the anticipated distance of 3.8 Å between the apical oxygens in the transition state. Thus, the ions are in positions to stabilize the 'entering' and 'leaving' oxygens at the apical positions, and to reduce the energy in forming the 90° O–P–O bond angles between the apical and equatorial oxygens. Interestingly, the metal binding sites in BamHI are superimposable to the metal sites found in the NgoMIV-product complex, where metal A is coordinated by Asp140 and Glu201 and metal B is coordinated by Asp140 and Glu70 and the two metal sites are 3.8 Å away from each other (Deibert et al. 2000). This is despite the fact that NgoMIV contains a Lys as the third catalytic residue instead of Glu113 in BamHI, and its second catalytic residue, Glu201 shows only spatial but not sequential alignment. Recent calculations confirm the two metal mechanism for BamHI and favor the transfer of a hydroxide from bulk solution rather than its generation in situ at the active site by a general base (Fuxreiter and Osman 2001;

Mordasini et al. 2003). A two-metal mechanism without the requirement for a general base could indicate a possible common mechanism of cleavage for BamHI and NgoMIV.

3 BglII Endonuclease

BglII isolated from *Bacillus subtilis* (Anton et al. 1997) contains 223 amino acids. It recognizes the sequence 5'-AGATCT-3' and like BamHI cleaves after the first nucleotide to leave four-base 5' overhangs. Although, the BglII and BamHI recognition sites differ at only the outer base pair position, the protein sequences are unrelated.

3.1 The Structure of Free BglII

Despite the lack of sequence homology, the BglII α/β core is similar to that of BamHI, containing a mixed six-stranded β-sheet ($\beta1$, $\beta3$, $\beta4$, $\beta5$, $\beta6$, and $\beta7$) surrounded by five α-helices ($\alpha1$, $\alpha2$, $\alpha3$, $\alpha4$, and $\alpha5$), two of which ($\alpha4$ and $\alpha5$) mediate dimerization (Fig. 7a). However, BglII has an extra β-sandwich ($\beta2$, $\beta8$, $\beta9$, $\beta10$, and $\beta11$) domain outside of the α/β core, which extends the size of the enzyme relative to BamHI. The catalytic residues, Asp84, Glu93, and Gln95 correspond to Asp94, Glu111, and Glu113 in BamHI revealing for the first time a restriction enzyme with a glutamine at the position of the third catalytic residue. Remarkably, BamHI becomes inactive when Glu113 is mutated to a glutamine as in BglII (unpubl. results). The catalytic residues of BglII are sequestered in a way not seen in any other REase structure. Asp84, Glu93, and Gln95 all point into the protein and form hydrogen bonds with residues from helix $\alpha4$.

3.2 The Structure of BglII Bound to its Cognate DNA Site

The DNA is completely encircled by the enzyme; with loops from the β-sandwich subdomain of each monomer (Fig. 7b). Thus, approximately 1300 Å^2 more surface area is buried upon DNA binding than in the BamHI complex (5515 vs. 4200 Å^2). Also, in contrast to the relatively straight DNA in the BamHI complex, the BglII DNA is distorted by bending (~22°) and by local unwinding and overwinding. In particular, the central step of the BglII recognition sequence (AGCT) unwinds by approximately 15°, which is analogous to the unwinding seen at the central steps in DNA complexes of EcoRI, EcoRV, and MunI. (Kim et al. 1990; Winkler et al. 1993; Deibert et al. 1999) The protein–DNA interactions are extensive and tightly organized. The α/β core supplies two loops (loops B and C) that reach into the major groove in a similar

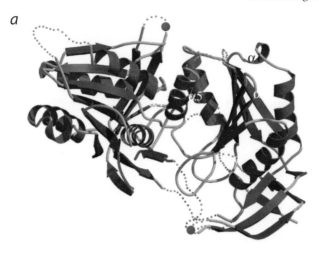

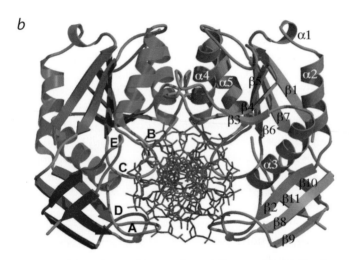

Fig. 7. Comparison of the **a** free and **b** DNA bound dimers of BglII. The free enzyme is shown with its right subunit in the same orientation as the right subunit of the complex. Loops A and D and a part of loop E are disordered in the free enzyme, and are drawn with *dotted lines*, corresponding to the conformation seen in the enzyme–DNA complex. The complex is viewed looking down the DNA axis. Secondary structural elements, along with the N-terminus and C-terminus are labeled on one monomer. *Blue spheres* mark the respective positions of Lys188 in the free and DNA bound dimers. Each monomer swings by as much as ~50°, like the blades of a pair of scissors, to open and close the binding cleft. The sheer magnitude of this motion is reflected by the dramatic increase in distances across the binding cleft. For instance, the distance between symmetrically related Lys188 residues at the rim of the cleft increases from ~17 Å in the complex to ~61 Å in the free enzyme. Produced with the programs MOLSCRIPT (Kraulis 1991) and RASTER3D (Merritt and Bacon 1997)

way to loops in BamHI. The β-sandwich subdomain provides another two loops (loops A and D), which project into the major and minor grooves, respectively. Loops A, D, and E are disordered in the free enzyme but become ordered upon DNA binding.

As the BglII recognition sequence differs from that of BamHI by only the outer base pairs (<u>A</u>GATC<u>T</u> vs. <u>G</u>GATC<u>C</u>), it was suspected that the recognition of the inner and middle base pairs would be very similar. This turns out not to be the case, even though the residues near these base pairs are similar in both complexes. For instance, Asn116 is directly hydrogen bonded to the inner and middle base pairs in BamHI (G<u>GA</u>), but the equivalent Asn98 residue in BglII adopts a different configuration to form dimer contacts and only contributes to the recognition of the middle guanine (A<u>G</u>A) via a water molecule (Fig. 4b). Thus, a mere χ-rotation completely changes the role of these residues, which, at first glance, seem to be structurally equivalent. Also, in BamHI, there are no direct contacts to the adenine of the inner base pair (GG<u>A</u>), whereas in BglII, a direct hydrogen bond is formed between the adenine (AG<u>A</u>) and Ser97 from the twofold related monomer. Overall, there is surprising diversity in how the common base pairs are recognized by these structurally related enzymes.

Recognition of the variable, outer base pair is, as expected, different between BglII and BamHI. Residues Asn140 and Ser141 in BglII substitute for residues Asp154 and Arg155 in BamHI, which effectively reverses the hydrogen-bonding pattern to recognize an A•T base pair (<u>A</u>GA) instead of a G•C base pair (<u>G</u>GA). Because the loops carrying these residues in the two enzymes are highly superimposable, it would seem, at first sight, that the specificity of BamHI could be changed to that of BglII by substituting asparagine and serine for Asp154 and Arg155. However, random mutagenesis of these residues in BamHI fails to yield any viable mutants capable of binding the BglII AGATCT site in an in vivo transcriptional assay (I. Schildkraut, personal communication). The basis for this immutability appears to lie in how the DNA is contorted in the two complexes. When the BamHI and BglII DNA complexes are superimposed on the basis of equivalent protein residues (Fig. 8), the outer base pairs of the DNA are found to be displaced by as much as 2.5 Å as a result of a difference in DNA bending. Thus, it is clear that changing Asp154 and Arg155 in BamHI to match Asn140 and Ser141 in BglII would be insufficient to switch the specificity of BamHI, because the outer base pair of the BamHI DNA would be too far away to interact with the shorter serine residue.

Type II REases are characterized by a highly complementary protein–DNA interface, which has made them a paradigm for the study of events leading to protein–DNA recognition. The DNA in almost all cases is accommodated in a tight binding cleft (Aggarwal 1995; Pingoud and Jeltsch 1997, 2001), which prompts the question of what conformational changes the protein and the DNA undergo upon binding. To allow the entry of the DNA, BglII opens by an unprecedented scissor-like motion. This has the same effect of opening the

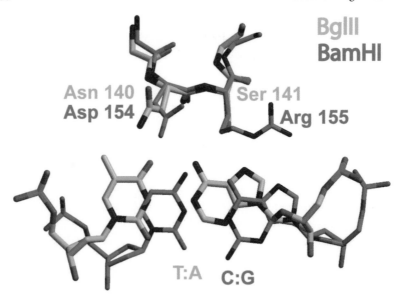

Fig. 8. Superimposition of the outer base-pair recognition residues Asn140 and Ser141 from BglII with Asn154 and Arg155 from BamHI. The superimposition reveals a 2.5 Å shift in the position of the outer bases. Produced with the programs MOLSCRIPT (Kraulis 1991) and RASTER3D (Merritt and Bacon 1997)

binding cleft as in free BamHI, but the motion of the subunits is in a direction parallel rather than perpendicular to the DNA axis. Because of the total encirclement of the DNA in the complex, the BglII monomers have to undergo a large motion to loosen their grip on the DNA. The 'tongs-like' motion seen in BamHI creates the problem of 'squeezing' the dimerization interface, which the scissor-like motion avoids by essentially allowing the dimerization helices to slide past each other. Interestingly, PvuII also completely encircles its DNA (Cheng et al. 1994), but it does not undergo the scissor-like motion seen in BglII. Instead, PvuII opens by the tongs-like motion, which is possible because each monomer is organized into dimerization and DNA binding subdomains (Athanasiadis et al. 1994; Cheng et al. 1994), and the hinge between them allows the DNA binding subdomains to move outward without any squeezing of the dimer interface. This is reminiscent of transcription factor NF-κB, which is also divided into dimerization and specificity subdomains (Ghosh et al. 1995; Muller et al. 1995), with the specificity subdomain changing configuration on different DNA sites (Muller et al. 1996). All in all, BglII is the first endonuclease to show alternative modes of dimerization, going from a parallel four-helix bundle in the complex to a more wedge-shaped arrangement of α-helices in the free enzyme. One of the key elements in the change from free to bound forms is a set of residues (Asn 69–Asp 84) termed the 'lever'. These residues, which make up parts of strands β3 and β4 and the loop connecting

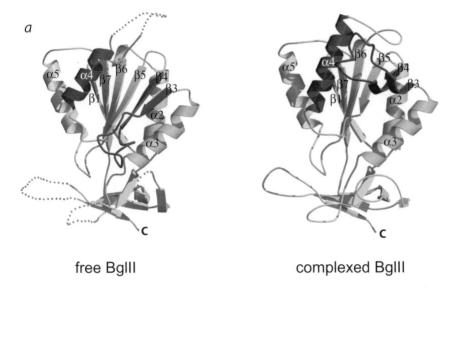

free BglII   complexed BglII

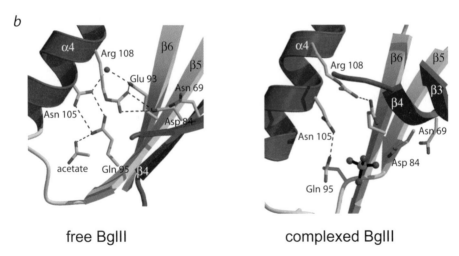

free BglII   complexed BglII

Fig. 9. Conformational changes in BglII. **a** The lever segment (*magenta*) and helix α4 (*blue*) undergo the largest local conformational change. The lever projects downward in the free enzyme (*left*) but upward in the complex (*right*). The length of helix α4 also changes in going from free to the complexed state. **b** A close up of the active site residues. In the free enzyme (*left*), the catalytic residues Asn69, Asp84, Glu93, and Gln95 are sequestered by extensive intramolecular hydrogen bonds with residues Asn105 and Arg108 from helix α4. In the complex (*right*), several of these hydrogen bonds are broken, and residues reorient to form the active site. The scissile phosphate is shown as a *ball-stick tetrahedron*. Produced with the programs MOLSCRIPT (Kraulis 1991) and RASTER3D (Merritt and Bacon 1997)

them, show the largest local conformational changes between the free and DNA-bound monomers. In the enzyme–DNA complex, this lever is in the 'up' position, forming rather curved strands that add significant twist to the large central β-sheet. When the lever is in the 'down' position in the free enzyme, the central β-sheet is larger but less twisted than in the complex (Fig. 9a). Remarkably, the active site residues are sequestered, and only become exposed for catalysis when the lever is raised to the 'up' position in the complex (Fig. 9b). This sequestering of active site residues is different from other Type II restriction endonucleases (for example, BamHI (Newman et al. 1994a), EcoRV (Winkler et al. 1993) and PvuII (Athanasiadis et al. 1994) as well as homing endonucleases, for example, I-CreI (Heath et al. 1997) and I-PpoI (Galburt et al. 2000), in which most of the active site residues face the solvent in the free proteins. The closest analogy in terms of sequestering of active site residues is with the Type IIs endonuclease FokI (Wah et al. 1997). FokI is composed of two separate domains, one for DNA recognition and one for nonspecific cleavage. The cleavage domain is sequestered by the recognition domain, but is thought to swing over to the distant cleavage sites on specific DNA binding. For FokI, it has been suggested that the sequestering of the cleavage domain is a mechanism to avoid hydrolysis at nonspecific DNA sites (Wah et al. 1997). It is likely that the sequestering of active site residues in the free BglII structure fulfills a similar function of preventing accidental cleavage. In the active site of BglII like in that of EcoRI (PDB code: 1QPS, unpublished data) only a single metal has been found. Whether this reflects a different mechanism of catalysis is unclear at present.

Conclusions

BamHI and BglII recognize sequences that differ only in the outer two base pairs. Although the elements carrying the DNA-recognition residues are superimposable in the two enzymes, a difference in how the DNA is contorted explains why the specificity of BamHI cannot be easily converted to that of BglII. The structures strongly reinforce a sense that it is the whole protein that contributes to the specificity. The lack of immutability in restriction enzymes, contrasts with transcription factors for which partial success has been achieved with proteins containing a similar DNA-binding motif. However, although most restriction endonucleases contain a similar α/β scaffold, each enzyme is found to decorate it with different secondary structural elements, loops and even subdomains for unique recognition. This immutability (and sequence diversity) of restriction enzymes may reflect an evolutionary pressure to not look too much alike. If the specificity of BamHI could be changed to that of BglII through a few point mutations, the host bacterium would become susceptible to lethal cleavage at unmethylated BglII sites on its genome. Restriction enzymes may be under strong selective pressure to evolve an interface that is not only complementary toward its recognition sequence, but also one which is not so easily changed to recognize another unprotected sequence.

Acknowledgements. This work was supported by National Institute of Health grants GM44006 (A.K.A.), and GM20015 (É.S.V.). H.V. was funded by a Fulbright/CONACYT Fellowship and C.M.L. was funded by a Cancer Research Fund of the Damon Runyon–Walter Winchell Foundation Fellowship.

References

Aggarwal AK (1995) Structure and function of restriction endonucleases. Curr Opin Struct Biol 5:11–19

Anton BP, Heiter DF, Benner JS, Hess EJ, Greenough L, Moran LS, Slatko BE, Brooks JE (1997) Cloning and characterization of the BglII restriction-modification system reveals a possible evolutionary footprint. Gene 187:19–27

Athanasiadis A, Vlassi M, Kotsifaki D, Tucker PA, Wilson KS, Kokkinidis M (1994) Crystal structure of PvuII endonuclease reveals extensive structural homologies to EcoRV. Nat Struct Biol 1:469–475

Ban C, Yang W (1998) Structural basis for MutH activation in E. coli mismatch repair and relationship of MutH to restriction endonucleases. EMBO J 17:1526–1534

Beese LS, Steitz TA (1991) Structural basis for the 3′-5′ exonuclease activity of Escherichia coli DNA polymerase I: a two metal ion mechanism. EMBO J 10:25–33

Bozic D, Grazulis S, Siksnys V, Huber R (1996) Crystal structure of Citrobacter freundii restriction endonuclease Cfr10I at 2.15 Å resolution. J Mol Biol 255:176–186

Brooks JE, Nathan PD, Landry D, Sznyter LA, Waite-Rees P, Ives CL, Moran LS, Slatko BE, Benner JS (1991) Characterization of the cloned BamHI restriction modification system: its nucleotide sequence, properties of the methylase, and expression in heterologous hosts. Nucleic Acids Res 19:841–850

Cheng X, Balendiran K, Schildkraut I, Anderson JE (1994) Structure of PvuII endonuclease with cognate DNA. EMBO J 13:3927–3935

Deibert M, Grazulis S, Janulaitis A, Siksnys V, Huber R (1999) Crystal structure of MunI restriction endonuclease in complex with cognate DNA at 1.7 Å resolution. EMBO J 18:5805–5816

Deibert M, Grazulis S, Sasnauskas G, Siksnys V, Huber R (2000) Structure of the tetrameric restriction endonuclease NgoMIV in complex with cleaved DNA. Nat Struct Biol 7:792–799

Dorner LF, Schildkraut I (1994) Direct selection of binding proficient/catalytic deficient variants of BamHI endonuclease. Nucleic Acids Res 22:1068–1074

Engler LE (1998) Specificity determinants in the BamHI endonuclease–DNA interaction.: 225. University of Pittsburgh, Pittsburgh

Fuxreiter M, Osman R (2001) Probing the general base catalysis in the first step of BamHI action by computer simulations. Biochemistry 40:15017–15023

Galburt EA, Chadsey MS, Jurica MS, Chevalier BS, Erho D, Tang W, Monnat RJ Jr, Stoddard BL (2000) Conformational changes and cleavage by the homing endonuclease I-PpoI: a critical role for a leucine residue in the active site. J Mol Biol 300:877–887

Ghosh G, van Duyne G, Ghosh S, Sigler PB (1995) Structure of NF-kappa B p50 homodimer bound to a kappa B site. Nature 373:303–310

Grabowski G, Jeltsch A, Wolfes H, Maass G, Alves J (1995) Site-directed mutagenesis in the catalytic center of the restriction endonuclease EcoRI. Gene 157:113–118

Grazulis S, Deibert M, Rimseliene R, Skirgaila R, Sasnauskas G, Lagunavicius A, Repin V, Urbanke C, Huber R, Siksnys V (2002) Crystal structure of the Bse634I restriction endonuclease: comparison of two enzymes recognizing the same DNA sequence. Nucleic Acids Res 30:876–885

Heath PJ, Stephens KM, Monnat RJ, Jr., Stoddard BL (1997) The structure of I-CreI, a group I intron-encoded homing endonuclease. Nat Struct Biol 4:468–476

Hickman AB, Li Y, Mathew SV, May EW, Craig NL, Dyda F (2000) Unexpected structural diversity in DNA recombination: the restriction endonuclease connection. Mol Cell 5:1025–1034

Horton JR, Cheng X (2000) PvuII endonuclease contains two calcium ions in active sites. J Mol Biol 300:1049–1056

Horton NC, Perona JJ (1998) Recognition of flanking DNA sequences by EcoRV endonuclease involves alternative patterns of water-mediated contacts. J Biol Chem 273:21721–21729

Horton NC, Dorner LF, Perona JJ (2002) Sequence selectivity and degeneracy of a restriction endonuclease mediated by DNA intercalation. Nat Struct Biol 9:42–47

Huai Q, Colandene JD, Chen Y, Luo F, Zhao Y, Topal MD, Ke H (2000) Crystal structure of NaeI-an evolutionary bridge between DNA endonuclease and topoisomerase. EMBO J 19:3110–3118

Huai Q, Colandene JD, Topal MD, Ke H (2001) Structure of NaeI-DNA complex reveals dual-mode DNA recognition and complete dimer rearrangement. Nat Struct Biol 8:665–669

Jen-Jacobson L, Engler LE, Jacobson LA (2000) Structural and thermodynamic strategies for site-specific DNA binding proteins. Structure Fold Des 8:1015–1023

Kim YC, Grable JC, Love R, Greene PJ, Rosenberg JM (1990) Refinement of EcoRI endonuclease crystal structure: a revised protein chain tracing. Science 249:1307–1309

Kostrewa D, Winkler FK (1995) Mg^{2+} binding to the active site of EcoRV endonuclease: a crystallographic study of complexes with substrate and product DNA at 2 Å resolution. Biochemistry 34:683–696

Kovall RA, Matthews BW (1997) Toroidal structure of lambda-exonuclease. Science 277:1824–1827

Kovall RA, Matthews BW (1998) Structural, functional, and evolutionary relationships between lambda-exonuclease and the Type II restriction endonucleases. Proc Natl Acad Sci USA 95:7893–7897

Kraulis P (1991) MOLSCRIPT: a program to produce both detailed and schematic plots of protein structures. Acta Crystallogr D Biol Crystallogr 24:946–950

Lesser DR, Kurpiewski MR, Jen-Jacobson L (1990) The energetic basis of specificity in the EcoRI endonuclease–DNA interaction. Science 250:776–786

Lukacs CM, Kucera R, Schildkraut I, Aggarwal AK (2000) Understanding the immutability of restriction enzymes: crystal structure of BglII and its DNA substrate at 1.5 Å resolution. Nat Struct Biol 7:134–140

Lukacs CM, Kucera R, Schildkraut I, Aggarwal AK (2001) Structure of free BglII reveals an unprecedented scissor-like motion for opening an endonuclease. Nat Struct Biol 8:126–130

McClarin JA, Frederick CA, Wang BC, Greene P, Boyer HW, Grable J, Rosenberg JM (1986) Structure of the DNA-EcoRI endonuclease recognition complex at 3 Å resolution. Science 234:1526–1541

Merritt EA, Bacon DJ (1997) Raster3D: photorealistic molecular graphics. Meth Enzymol 277:505–524

Mordasini T, Curioni A, Andreoni W (2003) Why do divalent metal ions either promote or inhibit enzymatic reactions? The case of BamHI restriction endonuclease from combined quantum-classical simulations. J Biol Chem 278:4381–4384

Muller CW, Rey FA, Sodeoka M, Verdine GL, Harrison SC (1995) Structure of the NF-kappa B p50 homodimer bound to DNA. Nature 373:311–317

Muller CW, Rey FA, Harrison SC (1996) Comparison of two different DNA-binding modes of the NF-kappa B p50 homodimer. Nat Struct Biol 3:224–227

Newman M, Strzelecka T, Dorner LF, Schildkraut I, Aggarwal AK (1994a) Structure of restriction endonuclease BamHI and its relationship to EcoRI. Nature 368:660–664

Newman M, Strzelecka T, Dorner LF, Schildkraut I, Aggarwal AK (1994b) Structure of restriction endonuclease BamHI phased at 1.95 Å resolution by MAD analysis. Structure 2:439–452

Newman M, Strzelecka T, Dorner LF, Schildkraut I, Aggarwal AK (1995) Structure of BamHI endonuclease bound to DNA: partial folding and unfolding on DNA binding. Science 269:656–663

Newman M, Lunnen K, Wilson G, Greci J, Schildkraut I, Phillips SE (1998) Crystal structure of restriction endonuclease BglI bound to its interrupted DNA recognition sequence. EMBO J 17:5466–5476

Perona JJ, Martin AM (1997) Conformational transitions and structural deformability of EcoRV endonuclease revealed by crystallographic analysis. J Mol Biol 273: 207–225

Pingoud A, Jeltsch A (1997) Recognition and cleavage of DNA by type-II restriction endonucleases. Eur J Biochem 246:1–22

Pingoud A, Jeltsch A (2001) Structure and function of Type II restriction endonucleases. Nucleic Acids Res 29:3705–3727

Roberts RJ, Halford SE (1993) Type II restriction endonucleases. In: SM Linn, RS Lloyd, RJ Roberts (eds) Nucleases. Cold Spring Harbor Laboratory Press, Cold Spring Harbor, NY

Roberts RJ, Vincze T, Posfai J, Macelis D (2003) REBASE: restriction enzymes and methyltransferases. Nucleic Acids Res 31:418–420

Selent U, Ruter T, Kohler E, Liedtke M, Thielking V, Alves J, Oelgeschlager T, Wolfes H, Peters F, Pingoud A (1992) A site-directed mutagenesis study to identify amino acid residues involved in the catalytic function of the restriction endonuclease EcoRV. Biochemistry 31:4808–4815

Sun J, Viadiu H, Aggarwal AK, Weinstein H (2003) Energetic and structural considerations for the mechanism of protein sliding along DNA in the nonspecific BamHI-DNA complex. Biophys J 84:3317–3325

Thielking V, Alves J, Fliess A, Maass G, Pingoud A (1990) Accuracy of the EcoRI restriction endonuclease: binding and cleavage studies with oligodeoxynucleotide substrates containing degenerate recognition sequences. Biochemistry 29:4682–4691

Thomas MP, Brady RL, Halford SE, Sessions RB, Baldwin GS (1999) Structural analysis of a mutational hot-spot in the EcoRV restriction endonuclease: a catalytic role for a main chain carbonyl group. Nucleic Acids Res 27:3438–3445

Tsutakawa SE, Jingami H, Morikawa K (1999a) Recognition of a TG mismatch: the crystal structure of very short patch repair endonuclease in complex with a DNA duplex. Cell 99:615–623

Tsutakawa SE, Muto T, Kawate T, Jingami H, Kunishima N, Ariyoshi M, Kohda D, Nakagawa M, Morikawa K (1999b) Crystallographic and functional studies of very short patch repair endonuclease. Mol Cell 3:621–628

van der Woerd MJ, Pelletier JJ, Xu S, Friedman AM (2001) Restriction enzyme BsoBI-DNA complex: a tunnel for recognition of degenerate DNA sequences and potential histidine catalysis. Structure (Camb) 9:133–144

Viadiu H, Aggarwal AK (1998) The role of metals in catalysis by the restriction endonuclease BamHI. Nat Struct Biol 5:910–916

Viadiu H, Aggarwal AK (2000) Structure of BamHI bound to nonspecific DNA: a model for DNA sliding. Mol Cell 5:889–895

Viadiu H, Kucera R, Schildkraut I, Aggarwal AK (2000) Crystallization of restriction endonuclease BamHI with nonspecific DNA. J Struct Biol 130:81–85

Wah DA, Hirsch JA, Dorner LF, Schildkraut I, Aggarwal AK (1997) Structure of the multimodular endonuclease FokI bound to DNA. Nature 388:97–100

Wah DA, Bitinaite J, Schildkraut I, Aggarwal AK (1998) Structure of FokI has implications for DNA cleavage. Proc Natl Acad Sci USA 95:10564–10569

Winkler FK, Banner DW, Oefner C, Tsernoglou D, Brown RS, Heathman SP, Bryan RK, Martin PD, Petratos K, Wilson KS (1993) The crystal structure of EcoRV endonuclease and of its complexes with cognate and non-cognate DNA fragments. EMBO J 12:1781–1795

Structure and Function of the Tetrameric Restriction Enzymes

V. Siksnys, S. Grazulis, R. Huber

1 Introduction

Type II restriction endonucleases recognize specific DNA sequences, typically 4–8 bp in length, and cleave phosphodiester bonds in the presence of Mg^{2+}, within or close to their recognition sites (Pingoud and Jeltsch 2001). Around 3500 species, from variety of bacteria with nearly 240 differing specificities, have now been characterized (Roberts et al. 2003). Most of the sequences recognized by Type II restriction endonucleases are palindromic, i.e., possess a twofold rotational axis of symmetry. On the basis of this observation, Kelly and Smith proposed the first model for the interaction of these restriction enzymes with DNA (Kelly and Smith 1970). According to their model (Fig. 1), recognition of the palindromic DNA sequence is achieved by two identical protein subunits related by a twofold axis of symmetry. Each subunit faces the same nucleotide sequence on the opposite DNA strand and contains one active site. Symmetrical nicking of opposite DNA strands by both monomers within a homodimer generates the double-strand break. Hence, the symmetry of the recognition sequence implies the oligomeric state of the restriction enzyme.

This simple strategy was subsequently confirmed by structural studies of Type II restriction endonuclease complexes with DNA (Pingoud and Jeltsch 2001). Despite individual structural peculiarities, differences in the recognition sites and cleavage patterns, these enzymes interacted with DNA as dimers, composed of two identical subunits. Since the majority of the sequences recognized by restriction endonucleases are palindromic, it is generally thought that most Type II restriction endonucleases interact with DNA as homodimers.

V. Siksnys, S. Grazulis
Institute of Biotechnology, Graiciuno 8, Vilnius 2028, Lithuania
R. Huber
Max Planck Institute of Biochemistry, Am Klopferspitz 18a, 8215 Martinsried, Germany

Nucleic Acids and Molecular Biology, Vol. 14
Alfred Pingoud (Ed.)
Restriction Endonucleases
© Springer-Verlag Berlin Heidelberg 2004

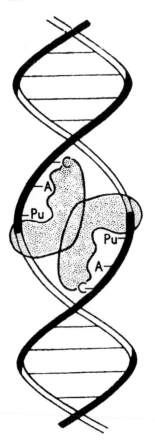

Fig. 1. Model of nucleotide sequence recognition by endonuclease R according to Kelly and Smith (1970). Two identical protein subunits related by a twofold axis of symmetry interact with CAPu sequences located on the opposite strands of the DNA duplex. (Modified from Kelly and Smith 1970)

However, studies of SfiI restriction endonuclease by Steve Halford and coworkers provided first biochemical evidence that higher oligomeric forms of restriction endonucleases might be of functional importance (Nobbs and Halford 1995; Wentzell et al. 1995). A tetramer has been reported to be the functionally active form of the SfiI restriction endonuclease that recognizes the interrupted palindromic sequence 5′-GGCCNNNN/NGGCC (cleavage point is indicated by /). Biochemical and kinetic studies revealed that SfiI interacts simultaneously with two copies of its recognition site before cleaving the DNA (Bilcock et al. 1999; Halford 1999). The oligomeric structure of SfiI and its mode of interaction with DNA were not dictated by its discontinuous recognition site. Further studies revealed that some restriction enzymes that recognize uninterrupted DNA sites, e.g., Cfr10I (Siksnys et al. 1999), are tetramers like SfiI. The architecture of the tetrameric restriction enzymes and their complexes with DNA, however, remained unknown.

In our efforts to elucidate structural and molecular mechanisms of the sequence discrimination by Type II restriction enzymes recognizing closely related DNA sites, we undertook the structure determination of Cfr10I,

Bse634I and NgoMIV specific for the 5′-Pu/CCGGPy (Cfr10I/Bse634I) and 5′-G/CCGGC (NgoMIV) sequences. The isoschizomeric restriction endonucleases Cfr10I (Janulaitis et al. 1983) and Bse634I (Repin et al. 1995) were originally isolated from the *Citrobacter freundii* and *Bacillus stearothermophilus* strains, respectively, and share ~30 % of identical amino acids in their protein sequences (Grazulis et al. 2002). Restriction endonuclease NgoMIV was originally identified in *Neisseria gonorrhoeae* (Stein et al. 1992).

The crystal structures of Cfr10I (Bozic et al. 1996) and Bse634I (Grazulis et al. 2002) were determined in the absence of DNA whereas the structure of NgoMIV (Deibert et al. 2000) was solved in the enzyme-product DNA complex. Crystallographic analysis revealed that Cfr10I, Bse634I, and NgoMIV are arranged as tetramers and provided us with the first structural models of the tetrameric restriction endonucleases. Solution experiments performed in parallel demonstrated that the tetrameric architecture of Cfr10I, Bse634I and NgoMIV is of functional importance (Deibert et al. 2000; Grazulis et al. 2002; Siksnys et al. 1999).

Here, we summarize results obtained by structural and functional studies of tetrameric restriction enzymes Cfr10I/Bse634I and NgoMIV which recognize closely related nucleotide sequences (see above).

2 Structural Anatomy of the Tetrameric Restriction Enzymes

2.1 Monomer Structure

The polypeptide chain of the Cfr10I, Bse634I and NgoMIV restriction endonucleases is folded into a compact α/β structure (Fig. 2). A twisted five-stranded mixed β-sheet sandwiched by α-helices makes up the core of the globule and is conserved within all three proteins (Fig. 2).

Spatially conserved pairs of helices $\alpha7$ and $\alpha8$ of Cfr10I, $\alpha6$ and $\alpha7$ of Bse634I, and $\alpha7$ and $\alpha8$ of NgoMIV, respectively, flank the convex side of the β-sheet. Structurally equivalent $\alpha3$ and $\alpha9$ helices of Cfr10I, $\alpha3$ and $\alpha8$ of Bse634I, and $\alpha3$ and $\alpha9$ of NgoMIV, respectively, pack on the concave side. Both Cfr10I/Bse634I and NgoMIV have an N-terminal substructure outside the conserved common core (Fig. 2). In addition, NgoMIV possesses a C-terminal substructure formed by $\alpha10$ and $\alpha11$ helices, and $\beta6$-strand, that is missing in Cfr10I/Bse634I (Fig. 2).

The structural analysis of Bse634I suggests (Grazulis et al. 2002) that the N-terminal substructure forms a domain that is connected to a C-terminal part by a hinge located between residues 70 and 90 in helix $\alpha3$ (Fig. 2). The N-terminal domain of Bse634I is comprised of two β-strands and two α-helices, whereas the C-terminal domain bears a conserved α/β core. Cfr10I exhibits similar domain organization that is also supported by limited proteolysis experiments. Indeed the N-terminal domain of Cfr10I that is structurally

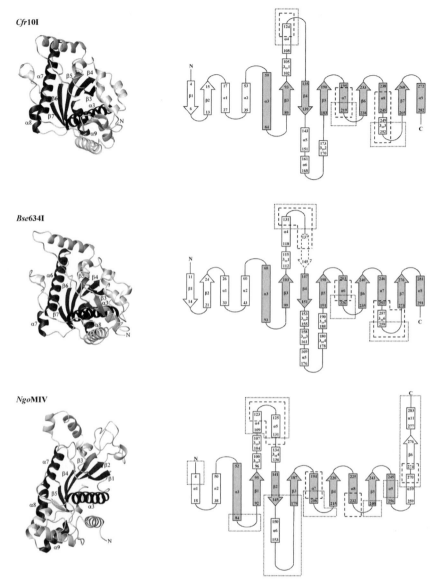

Fig. 2. Ribbon representation of the Cfr10I, Bse634I, and NgoMIV monomers and corresponding topological diagrams of the secondary structures. Conserved structural core is shown in *black* in ribbon structures and *shaded* in the diagrams. *Dashed* and *dotted boxes* indicate structural regions involved in dimerization and tetramerization, respectively

equivalent to that of Bse634I could be cleaved off by trypsin at K56 residue keeping the C-terminal subdomain intact (Skirgailiene A. and Siksnys V., unpublished data). The isoschisomeric restriction endonucleases Cfr10I and Bse634I share ~30 % of sequence identity. The crystal structures of both proteins are very similar and conserved structural cores overlap with RMSD of 1.0 Å for all atoms. To achieve superimposition of the N-terminal domains with a similar RMSD value, however, a nearly 10° rotation of the N-terminal domain of Cfr10I around the axis passing through the middle of the α3 helix was necessary (Grazulis et al. 2002). Thus, structural analysis suggests that the N-terminal subdomain of Cfr10I/Bse634I is connected to the conserved core by a relatively flexible hinge. The N-terminal substructure of NgoMIV is smaller and predominantly comprised of two α-helices. The general topology of Cfr10I, Bse634I, and NgoMIV monomers is similar to that of dimeric restriction enzymes (Pingoud and Jeltsch 2001). Hence, a similar conserved building block is used to construct both dimeric and tetrameric restriction endonucleases.

2.2 Dimer Arrangement

Two monomers of Cfr10I, Bse634I, and NgoMIV associate to build a primary dimer (Fig. 3) similar to that of the *bona fide* dimeric restriction enzymes, like EcoRI (Rosenberg 1991), that cleave hexanucleotide sequences producing four base pair staggered 5′-overhanging ends. In order to achieve such a cleavage pattern, restriction endonucleases should position their catalytic sites ~18–19 Å apart along the DNA axis (Aggarwal 1995). This constraint most likely is achieved through the conserved dimerization mode of restriction enzymes exhibiting the same cleavage pattern despite of the differences in their recognition sites. For example, EcoRV-like enzymes that recognize hexanucleotide sequences and cleave leaving blunt ends, exhibit a different dimerization mode (Aggarwal 1995). A common structural core and the active site structure, however, are conserved between EcoRI- and EcoRV-like enzymes (Venclovas et al. 1994). The similarities of dimer organization between Cfr10I, Bse634I, and NgoMIV and restriction enzymes producing four base-pair staggered 5′-overhanging ends further support the idea (Aggarwal 1995) that the cleavage pattern rather than recognition sequence plays a key role in the determination of the overall structure of the dimer.

Three conserved noncontiguous structural elements (Fig. 2) of the NgoMIV, Bse634I, and Cfr10I contribute to the dimer interface. First, amino acid residues located on the N-terminal parts of the conserved helices α7, α6, and α7 of NgoMIV, Bse634I, and Cfr10I, respectively, make a set of van der Waals and hydrogen bond contacts at the interface between the two monomers. Noteworthy, the N-termini of the latter α-helices are directed towards the major groove and carry aspartate and arginine residues involved

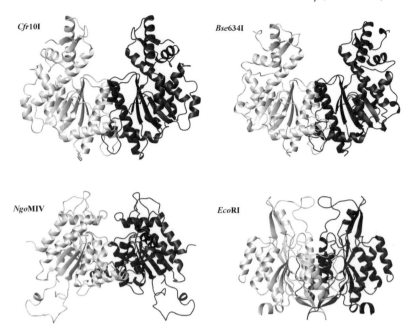

Fig. 3. Primary dimers of the tetrameric restriction enzymes Cfr10I, Bse634I, NgoMIV, and bona fide dimer of EcoRI. The proteins are represented as ribbons. Individual subunits are shown in *gray* and *black*

in DNA base contacts (see below). Hence, the dimerization and DNA recognition interfaces are intertwined. The second segment of the dimer interface is comprised of the amino acid residues located on the loops and α-helical stretches bridging the first two conserved β-strands (β3-β4 of Bse634I/Cfr10I and β1-β2-strands of NgoMIV, respectively). The third conserved module of the intersubunit interface spans across the C-terminal parts of the conserved helices α8, α7, and α8 and subsequent loops of NgoMIV, Bse634I, and Cfr10I, respectively. In NgoMIV, additional contacts at the dimer interface eminate from amino acid residues located at the C-terminus of the α10 helix which is missing in the Cfr10I and Bse634I.

Structure-based sequence alignment demonstrates that most of the 30 % of sequence identities between the Cfr10I and Bse634I correspond to the amino acid residues involved in the organization of the active site and DNA recognition and dimerization interfaces (Grazulis et al. 2002). Analysis of the intersubunit interface of the typical dimeric restriction enzymes (e.g., EcoRI, MunI, BamHI) producing four-base-pair staggered 5'-overhanging ends indicates that amino acid residues involved in dimer formation are located predominantly on the same conserved structural elements described for the Cfr10I, Bse634I, and NgoMIV (Pingoud and Jeltsch 2001).

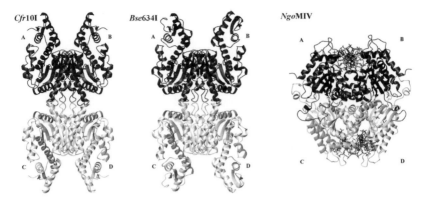

Fig. 4. Tetramers of the Cfr10I, Bse634I, and NgoMIV restriction endonucleases. Two primary dimers positioned back-to-back to each other are shown in *gray* and *black*. The monomers are labeled *A, B, C, D,* respectively. DNA molecules bound to the NgoMIV are shown in the *stick representation*

A U-shaped cleft large enough to accommodate a B-DNA molecule is formed between two monomers of Cfr10I/Bse634I (Fig. 3). In the NgoMIV a similar cleft is indeed occupied by the DNA (Fig. 4).

2.3 Tetramer Organization

Tetrameric restriction enzymes Cfr10I, Bse634I, and NgoMIV are arranged as dimers of primary dimers positioned back-to-back to each other with DNA binding clefts facing opposite directions (Fig. 4). The mutual orientation of the primary dimers, however, differs between Bse634I/Cfr10I and NgoMIV. In the NgoMIV tetramer the concavely shaped backsides of the dimers are slightly rotated around their local twofold axis, therefore the two bound DNA duplexes are positioned in an X-like fashion with their helical axes enclosing an angle of 60°.

The monomers in the Cfr10I and dimers of Bse634I crystal are related by a twofold crystallographic axis and form a symmetric tetramer with a similar back-to-back arrangement of dimers.

Two primary dimers become positioned at 90° angle in respect to each other. The differences in the mutual orientation of the two primary dimers of NgoMIV and Bse634I/Cfr10I are due to the distinct dimer-dimer interfaces. While the mode of the dimerization is mostly conserved between NgoMIV and Bse634I/Cfr10I, much less of the structural conservation is found at the dimer-dimer interfaces.

In the Bse634I/Cfr10I tetramers, structural elements (Fig. 2) involved in the dimerization also contribute to the interface between the two primary dimers.

For example, amino acid residues located on the N-termini of the structurally conserved α7/α6 helices of Bse634I/Cfr10I, respectively, make intersubunit contacts in the dimer, while residues located on the C–termini of the same helices are involved in tetramerization (Fig. 2). Of note is that some of the residues positioned on these helices contribute both to the dimerization and tetramerization interfaces. A single subunit of Cfr10I or Bse634I makes contacts to both subunits of the neighboring primary dimer. Totally, each subunit makes contacts with all three other protein subunits within the tetramer.

The tetrameric assembly of NgoMIV is fixed by side-by-side contacts (between subunits A/C and B/D) and cross-contacts (between subunits A x D and B x C) of primary dimers, respectively (Fig. 4). Cross-contacts are made exclusively by the helices α4 and α5 which also contribute to the dimerization interface. Similar contacts are conserved at the tetramerization interface of Bse634I/Cfr10I (Fig. 2). The most extensive contacts in the NgoMIV tetramer, between subunits A/C and B/D, are made by the tetramerization loop (residue 147 to 176 including helix α6) that spans across the neighboring monomer. The residues located on the α6 helix contacts the C-terminus of the α3 helix while other residues positioned on the loop make an intricate set of contacts to the residues with the N- and C-terminal residues of NgoMIV. Additional contacts at the A/C interface come from amino acid residues located on the C-termini of helices α7 and α8 and downstream regions. Similar contacts are made by structurally equivalent elements of Cfr10I/Bse634I. In total, the contacts between subunits at the tetramerization interface comprise eight salt-bridges and numerous hydrogen bonds and van der Waals interactions.

The tetramerization loop (residues 146–178) that contributes major contacts to the interdimer interface of NgoMIV (Fig. 2), is replaced by a short loop between two structurally equivalent β-strands in the dimeric restriction enzymes EcoRI, BamHI, and MunI (Pingoud and Jeltsch 2001). Interestingly, both Cfr10I and Bse634I tetramers possess loops (residues 140–182 and 152 to 190, respectively) that are topologically similar to the tetramerization loop of NgoMIV. However, in contrast to NgoMIV, loops bridging β4-β5-strands of Cfr10I/Bse634I are not involved in tetramerization but rather contact the N-terminal subdomain of the same subunit (Fig. 5).

The dual role performed by the topologically conserved loops in Cfr10I/Bse634I and NgoMIV remotely resembles a domain swapping mechanism for forming oligomeric proteins from their monomers (Bennett et al. 1995; Liu and Eisenberg 2002). Indeed, in the Cfr10I/Bse634I the structurally conserved loop makes contacts to the N-terminal domain of the same monomer (Fig. 5), while in NgoMIV it contacts the N-terminal domain of the neighboring subunit.

The structures of Bse634I/Cfr10I were solved in the apo-form, while the structure of the NgoMIV in complex with product DNA. One might speculate that DNA binding by Bse634I/Cfr10I might induce conformational changes and replace intermolecular contacts of the conserved loop by intersubunit

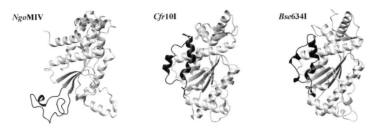

Fig. 5. Different conformations of the tetramerization loop of NgoMIV and topologically equivalent loops of Cfr10I and Bse634I. Loops are shown in *black*. In Cfr10I/Bse634I the tetramerization loop contact N-terminal domains of the same subunit, while in NgoMIV it contacts the N-terminal domain of the neighboring subunit (see Fig. 3)

contacts similar to those of NgoMIV. The requirement for the stringent conservation of the intra- and intersubunit loop contacts, however, might put a major constraint on the loop swapping in Bse634I/Cfr10I complexes with DNA. For example, fifty hydrogen bonds are made across the intersubunit interface by the residues located on the tetramerization loop of NgoMIV (Table 1). It is more likely that topologically equivalent loops in Bse634I/Cfr10I and NgoMIV were adopted to perform different structural roles.

Accessible surface area calculations indicate that in Cfr10I/Bse634I the largest accessible surface area of ~1350 and ~1550 Å^2, respectively, is buried at the primary dimer interface. Much less of the monomer protein surface is buried in contacts with both subunits of the neighboring primary dimer (Table 1). In NgoMIV, the contribution of individual intersubunit contacts to the total surface accessible area buried differs notably. The cross-contacts between monomers A × D comprise only around 400 Å^2. The contacts between subunits AB of the primary dimer hide ~1000 Å^2, while the largest surface area of nearly 2900 Å^2 is buried in side-by-side contacts between subunits A/C and B/D of the two primary dimers, respectively. Thus, the surface of each monomer shielded during tetramerization almost three times exceeds the surface buried at the dimer interface. The tetramerization interface of Cfr10I/Bse634I is much smaller in comparison to that of NgoMIV (Table 1).

Of note is that a single alanine replacement of the W220 residue of Cfr10I or structurally equivalent W228 residue in Bse634I located at the C-terminus of the conserved α7/α6-helices, respectively, disrupts the tetramerization interface generating a protein dimer (Siksnys et al. 1999). The tryptophan W220/W228 residue at the interface between two primary dimers most likely represents a hot spot that makes a major contribution to the binding energy (Bogan and Thorn 1998). If the loop would be involved in the stabilization of the tetramerization interface between the two primary dimers of Cfr10I/

Table 1. Protein interface parameters of the tetrameric restriction endonucleases[a]

Protein interface parameter	Cfr10I			Bse634I			NgoMIV		
	A/B	A/C	A/D	A/B	A/C	A/D	A/B	A/C	A/D
Interface ASA	1347	491	682	1544	411	504	1002	2886	409
Interface residue segments	3	3	2	4	2	2	4	8	2
Polar atoms in interface (%)	25.6	30.2	30.7	32.1	33.2	34.2	36.9	38.1	33.8
Nonpolar atoms in interface (%)	74.4	69.8	69.3	67.8	66.8	65.7	63.1	61.8	66.1
Hydrogen bonds	8	4	0	10	2	1	4	50	2

[a] Obtained using the Protein–Protein Interaction Server at
http://www.biochem.ucl.ac.uk/bsm/PP/server/

Bse634I similarly to NgoMIV, the disruption of the tetramers would be much less likely.

3 Active Sites of the Tetrameric Restriction Enzymes

Type II restriction endonucleases except of *Bfi*I (Sapranauskas et al. 2000) require cofactor Mg^{2+} for the catalysis to occur. The presence of two Mg^{2+} ions nearby the cleaved phosphate in the crystal structure of enzyme–product complex allowed us to identify unambiguously the catalytic/metal binding site of NgoMIV (Deibert et al. 2000). There are two Mg^{2+} ions in the active center (Fig. 6a).

Both Mg^{2+} ions exhibit octahedral coordination. The O2P of the 5'-phosphate, carboxylate of D140 and the acetate molecule (from the reservoir of the crystallization buffer) bridge both Mg^{2+}-ions. The remaining three ligands of the Mg^{2+} ion A are the backbone oxygen of C186 and water molecules 1 and 2. Water molecules 3, 4, and 5 complete the octahedral coordination of Mg^{2+} ion B. The ligand oxygen atoms are within 2.2–2.3 Å from Mg^{2+} ion A, and within 2.1– 2.4 Å from Mg^{2+} ion B. All Mg^{2+}-coordinated waters are fixed by either protein and/or DNA residues. Waters 1 and 2 are within the hydrogen bonding distance from NZ of K187, O1P of the adjacent Cyt1 and OE2 of E70 carboxylate, respectively. Water molecules 3 and 4 are hydrogen-bonded to the carboxylate group of E201 while water 5 is in contact with released O3' hydroxyl group of Gua3 (3.3 Å), the carboxylate of D140 (3.0 Å) and O5T oxygen of the 5'-phosphate (3.2 Å). The location of the metal ions of NgoMIV resembles the location of the metal ions at the active site of the BamHI restriction enzyme (Viadiu and Aggarwal 1998) and the 3'-5' exonuclease domain of DNA polymerase I (Beese and Steitz 1991). Indeed, Mg^{2+} ions at the active site of NgoMIV and Mn^{2+} ions at the active site of BamHI can be superimposed in their enzyme-product complexes (Deibert et al. 2000).

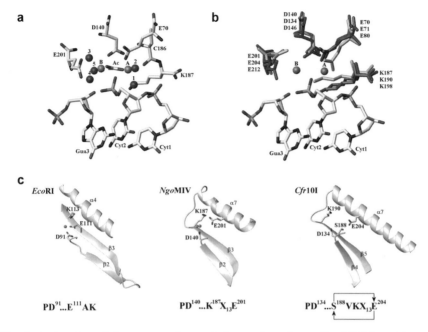

Fig. 6. Active sites of the tetrameric restriction endonucleases NgoMIV, Bse634I, and Cfr10I and the orthodox restriction enzyme EcoRI. **a** Coordination geometry of Mg^{2+} ions at the active site of NgoMIV in the enzyme–product complex. Two Mg^{2+} ions are shown as *gray spheres* and labeled *A* and *B*, respectively. Both Mg^{2+} exhibit octahedral coordination. The O2P of the 5′-phosphate, the carboxylate of D140 and the acetate molecule contribute to the coordination of both Mg^{2+} ions. The remaining three ligands of the Mg^{2+} ion A are backbone oxygen of C186 and water molecules 1 and 2 (shown as *black spheres*). Water molecules 3, 4, and 5 (shown as *black spheres*) complete the octahedral coordination of Mg^{2+} ion B. Coordination of glutamate residues E70 and E201 to the metal ions A and B, respectively, is mediated by water molecules. **b** Superimposed active sites of NgoMIV, Cfr10I, and Bse634I. Two Mg^{2+} ions present in NgoMIV structure are shown as *gray spheres* and labeled *A* and *B*, respectively. Only side chains of amino acid residues and a fragment of the DNA chain in the vicinity of the Mg^{2+} ions are shown. Water molecules and the acetate ion present at the NgoMIV active site were omitted for clarity. Stacked labels correspond to the conserved active site residues of NgoMIV (*top*), Cfr10I (*middle*) and Bse634I (*bottom*), respectively. **c** Structural localization of the active site residues of EcoRI, NgoMIV, and Cfr10I. Conserved structural elements bearing active site residues are shown as *ribbons*, active site residues are shown in *stick representation* and labeled. Mn^{2+}-ion present in the active site of EcoRI (PDB entry 1qps) is shown as a *gray sphere*. Two Mg^{2+} ions present in NgoMIV structure are shown as *grey spheres*. Active site motifs corresponding to the first metal ion binding site are shown below the each figure. *Arrows* indicate Cfr10I active site residues subjected to swapping (see text)

Structural comparisons reveal the conserved spatial location of the metal chelating residues of NgoMIV, Cfr10I, and Bse634I suggesting similar architecture of the catalytic/metal binding sites (Fig. 6b). Therefore, it is very likely that Cfr10I and Bse634I also coordinate two metal ions at their active sites. The metal ion B (Fig. 6b) in Cfr10I, Bse634I, and NgoMIV would be chelated by acidic residues of the conserved sequence motif $PDX_{46-55}KX_{13}E$. Aspartate residues belonging to the latter sequence motif and conserved glutamates E70/E71/E80 located on the α3-helix would contribute to the binding site of the metal ion A (Fig. 6b). Site-directed mutagenesis of the putative catalytic residues of Cfr10I (Fig. 6b) supports their active site function (Skirgaila et al. 1998).

Orthodox dimeric restriction endonucleases contain a single (EcoRI, BglII) or two metal ions (BamHI, PvuII, EcoRV) positioned at their active sites (Galburt and Stoddard 2002; Pingoud and Jeltsch 2001). One of the two-metal binding sites is homologous to the single site observed in EcoRI and BglII. Two acidic residues and a lysine defined by a weakly conserved sequence motif $PDX_{10-30}(D/E)XK$ are located at the ends of β-strands (see, EcoRI in the Fig. 6c) and form the first catalytic/metal binding site that is conserved within most restriction enzymes (Pingoud and Jeltsch 2001). Structural comparisons between the active sites of Cfr10I/Bse634I/NgoMIV and those of orthodox restriction enzymes reveal that the spatial location of the metal-chelating residues is conserved (Fig. 6), however, the order and position of the active site residues within the polypeptide sequence is different (Deibert et al. 2000; Grazulis et al. 2002; Skirgaila et al. 1998). This results in a sequence divergence in the second parts (shown in bold) of the catalytic motifs $PDX_{46-55}\mathbf{KX_{13}E}$ and $PDX_{19-30}\mathbf{(E/D)XK}$ characteristic for Cfr10I/Bse634I/NgoMIV and canonical restriction enzymes, respectively.

In EcoRI an acidic E111 residue from the EAK part of the conserved active site motif $PDX_{19}EAK$ is positioned at the end of the β-strand (Fig. 6c). In NgoMIV/Bse634I/Cfr10I an equivalent position is occupied either by glycine or serine residue. The conserved glutamate residue located at the N-terminus of the α7-helices of NgoMIV and Cfr10I, however, spatially substitutes the acidic residue located at the end of β-strand of EcoRI (Fig. 6c) and most other canonical restriction enzymes. Site-directed mutagenesis confirmed the functional significance of the E204 residue in Cfr10I catalysis (Skirgaila et al. 1998). Thus, the spatial arrangement of the active sites of NgoMIV/Bse634I/Cfr10I is similar to those of EcoRI or EcoRV, however, the order and position of the active site residues is not conserved. Such plasticity of the active sites has been reported for a number of different enzymes (Todd et al. 2002).

The primary importance of structure rather than sequence conservation of catalytic/metal binding residues is further supported by a residue swapping experiment (Skirgaila et al. 1998). Double mutation S188E/E204S in Cfr10I transposed the E204 residue from the α-helix to the end of the β-strand restoring a canonical active site structure (Fig. 6c). The Cfr10I mutant

S188E/E204S with a reengineered metal binding site retained significant catalytic activity (Skirgaila et al. 1998).

The sequence motif $PDX_{46-55}KX_{13}E$ first found in the NgoMIV/Bse634I/Cfr10I, however, does not seem to be limited only to the tetrameric restriction enzymes and might be more common than it was initially thought. Indeed, a similar catalytic motif (Fig. 7) was identified in the protein sequences of 11 restriction enzymes (and three putative restriction enzymes) which belong to the different subtypes (Type II, Type IIE, and Type IIF) but all have CCNGG, CCWGG or CCGG in their respective recognition sequences (Pingoud et al. 2002).

Mutational analysis of SsoII, Ecl18 kI (Pingoud et al. 2002; Tamulaitis et al. 2002) and PspGI (Pingoud, pers. comm.) confirmed the functional significance of the putative active site residues. Hence, the catalytic sequence motif first revealed for the tetrameric restriction enzymes most likely represents a different solution for coordination of metal ions at the active site of restriction enzymes.

According to the crystal structure D140 from $PDX_{46}KX_{13}E$ motif and E70 located on the α3-helix constitute a second metal ion binding site in NgoMIV (Deibert et al. 2000). Both E70 and D140 residues of NgoMIV are conserved within Cfr10I/Bse634I (Fig. 6b). Multiple sequence alignment (Fig. 7) reveals that E70 of NgoMIV is also conserved within a group of 11 restriction enzymes discussed above. Site-directed mutagenesis of the latter conserved residue supports its active site function in Cfr10I, SsoII, Ecl18 kI (Pingoud et al. 2002; Skirgaila et al. 1998; Tamulaitis et al. 2002) and PspGI (Pingoud, personal communication). The structurally equivalent glutamate residues E70/E71/E80 of NgoMIV/Cfr10I/Bse634I are involved in coordination of the

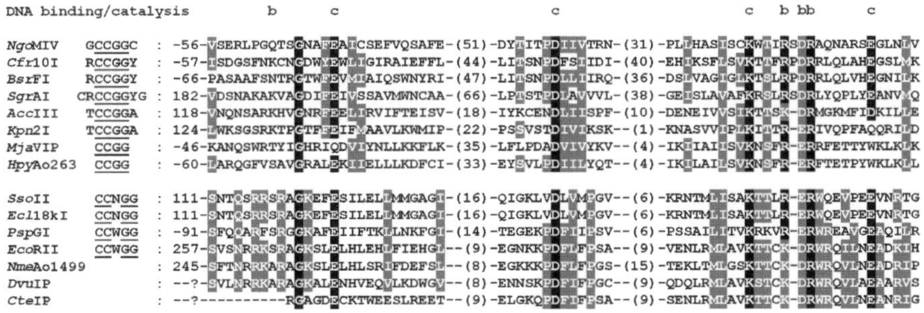

Fig. 7. Alignment of partial amino acid sequences of bona fide (recognition sequence indicated) or presumptive Type II restriction endonucleases (modified from Pingoud et al. 2002). The alignment is subdivided in two groups: the top group consists of restriction enzymes that recognize CCGG and the bottom group CCNGG or CCWGG. Similar or identical amino acid residues are indicated in *gray* or *black*. *Bold letters b* and *c* at *top* of the figure indicate NgoMIV amino acid residues involved in DNA binding or catalysis/metal ion binding

second metal ion are located on the helix α3 of the N-terminal subdomain whereas the rest of the active site residues are positioned in the conserved C-terminal protein core. Analysis of the subdomain motions in the Bse634I protein indicates that the C_α atom of the E80 moves 2.3 Å and C_α atom moves ~3 Å towards the active site residues located at the C-terminal subdomain (Grazulis et al. 2002). A similar "cantilever" α3 helix mediated movement of the N-terminal subdomain in the Bse634I (and probably Cfr10I) restriction enzyme during specific DNA binding might create an optimal geometry for the coordination of Mg^{2+} ions at the active site and couple catalysis and sequence recognition.

4 DNA Recognition by NgoMIV Restriction Endonuclease

The crystal structure of the NgoMIV restriction endonuclease in complex with product DNA provides a detailed picture of the protein–DNA recognition interface (Deibert et al. 2000). Two 10 bp oligonucleotide duplexes containing the recognition sequence 5′-GCCGGC-3′ are bound to the NgoMIV tetramer arranged as a dimer of dimers (Fig. 4). Two primary dimers are stacked back-to-back, placing the duplexes on the opposite sides of the tetramer. Each duplex is contacted by just two protein subunits of the primary dimer (Fig. 4). A single subunit of NgoMIV makes a set of hydrogen bonds in the major groove to the bases of the first 5′-GCC half site of the recognition sequence and a single minor groove contact to the C base of the second GGC-3′ half site (Fig. 8a). Thus, major groove contacts between the primary dimer of NgoMIV and palindromic recognition sequence follow the model suggested by Kelly and Smith (Fig. 1).

The recognition of the specific sequence 5′-GCCGGC-3′ by NgoMIV is achieved through a number of direct contacts between protein side chains and DNA bases. The structural elements of NgoMIV involved in the sequence recognition form two separate modules that are comprised by different segments of the protein chain. Since both half-sites of the recognition sequence make an identical set of contacts to the protein residues, only interactions with the 5′-GCC half-site are discussed.

Two central C:G base pairs are contacted by residues located on a short contiguous stretch of amino acids with the sequence 191RSDR just upstream and at the N-terminus of the α7 helix (Fig. 9a). Arginines R191 and R194 donate bidentate hydrogen bonds to the guanines while carboxylate oxygens of D193 bridge exocyclic amino groups of the neighboring cytosines (Fig. 8b). Thus, three residues R191, D193, and R194 of a single subunit unambiguously specify inner and middle C:G nucleotides within the first half-site of the NgoMIV recognition site 5′-GCC. Equivalent residues located on the symmetry related α7 helix make the same set of contacts to the central C:G base pairs of the second GGC-3′ half-site of the recognition sequence. Besides their key

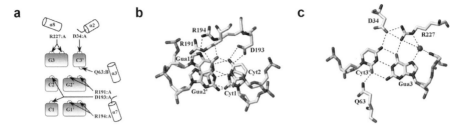

Fig. 8. NgoMIV–DNA interactions. **a** Schematic representation of the structural elements and amino acid residues involved in base recognition. Only contacts to one half-site of the recognition sequence 5′-GCCGGC are shown. Symmetrical contacts to the second half-site are not presented. Helices are represented as cylinders. Amino acid residues are labeled, letters *A* and *B* indicate the NgoMIV subunits bearing amino acids shown in the scheme. **b** Recognition of the inner C:G base pairs. NH2 and NE atoms of R191 and R194 make bidentate hydrogen bonds to Gua2′ and Gua1′, respectively. The carboxylate oxygens of D193 bridge the amino groups of the neighboring Cyt bases. **c** Recognition of the outer G:C base pair. D34 makes a hydrogen bond to Cyt3′. The NH2 atom of R227 makes a direct hydrogen-bond to the O6 oxygen of Gua3 while the NE of R227 in companion with the main chain nitrogen of S192 sandwiches a water molecule, which is hydrogen bonded to the N7 nitrogen of Gua3. The amino group of Q63 from the neighboring subunit, donates a hydrogen bond to the oxygen atom of Cyt3′ in the minor groove

role in the inner C:G recognition, D193 and R194 residues also contribute to the dimerization interface making two salt-bridges with R194 and D193 residues of neighboring subunit, respectively. Hence DNA recognition and dimerization interfaces in NgoMIV are intertwined.

The recognition of the outer G:C base pair by NgoMIV is achieved by a recognition module comprised of three noncontiguous structural elements: loop preceding helix α2, helix α8, and helix α3 of the neighboring subunit (Fig. 8b) The major groove contacts to the outer G:C base pair arise from the D34 and R227 residues located on loops upstream of helix α2, and helix α8, respectively (Fig. 8b). D34 makes a hydrogen-bond to the exocyclic amino group of Cyt3′ (Fig. 8c). The side chain of D34 is also buttressed by a salt bridge to R227 and a hydrogen bond to S36 (not shown in Fig. 8c). The NH2 atom of R227 makes a direct hydrogen bond to the O6 oxygen of Gua3, while the NE of R227 in companion with the main chain nitrogen of S192 sandwiches a water molecule, which is hydrogen bonded to the N7 nitrogen of Gua3. The amino group of Q63, located at the N-terminus of the α3 helix of the neighboring subunit, donates a hydrogen bond to the oxygen atom of Cyt3′ in the minor groove. The side chain conformation of Q63 residue is probably fixed through the interaction of its side chain carbonyl oxygen atom with the backbone nitrogen atoms of Q63 and G62 residues (not shown in the Fig. 8c). Thus both subunits of a primary dimer contribute to the recognition of the outer G:C pair of one half-site.

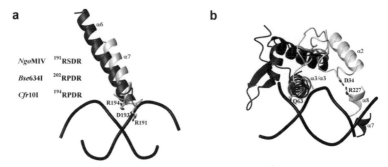

Fig. 9. Structural elements of NgoMIV involved in contacts with DNA bases and their structural equivalents in Bse634I. **a** Superimposed α7 and α6 helices of NgoMIV and Bse634I, respectively. Superimposition of Cα atoms of the active sites and conserved β-sheets of NgoMIV and Bse634I aligned α7 and α6 helices of NgoMIV and Bse634I. Amino acid residues of NgoMIV making contacts to the inner C:G base pairs at one half-site of the recognition sequence are shown in *stick representation*. Structurally equivalent amino acid residues of NgoMIV, Bse634I, and Cfr10I are shown on the left side of the figure. Only a backbone of a single DNA molecule (PDB entry 1fiu) is shown for the clarity. **b** Structural elements of the NgoMIV involved in the outer G:C base pair recognition and their comparison with topological equivalents of Bse634I. Superposition of the Cα atoms of the active sites and conserved β-sheets of the NgoMIV and Bse634I aligned α3 helices. The superimposition reveals a shift in the position of the α7 helix of Bse634I in respect to the structurally equivalent α8 helix of NgoMIV bearing R227 residue involved in contacts with outer Gua base (Fig. 6b). The N-terminal domain of Bse634I (shown in *black*) is also rotated nearly 90° in respect to the N-terminal domain of NgoMIV and faces the opposite major groove. Only a backbone of a single DNA molecule (PDB entry 1fiu) is shown for the clarity. Q63 residue of NgoMIV located on the α3 helix and involved in the minor groove contacts to the oxygen atom of the outer Cyt base is shown in stick representation. N73 and K64 of Bse634I and Cfr10I, respectively, are structural equivalents of Q63 residue of NgoMIV

Interestingly, there is one direct protein–base contact outside of the 6-bp recognition sequence which may explain the possible C:G preference for the bases flanking the NgoMIV recognition sequence. S36 of each subunit A, B and C makes a single hydrogen bond to Cyt4 outside of the recognition site of the DNA strands E, F, and H, respectively (not shown). The conformations of the S36 residues however slightly differ between subunits. Therefore S36 of monomer D makes a water-mediated contact to the backbone phosphate of the complementary Cyt3′ in strand H instead of direct contacts to the C base. Furthermore the backbone oxygens of S36 in subunits B and C are involved in water mediated contacts to the outermost guanine. The SgrAI restriction endonuclease recognizes 5′-CR/CCGGYG-3′ sequence (Tautz et al. 1990) that overlaps completely with the recognition sequence of Bse634I/Cfr10I. Protein sequence comparisons of SgrAI and Cfr10I/Bse634I restriction endonucleases reveal (Pingoud et al. 2002) that key catalytic and sequence recognition residues are conserved suggesting that recognition of the common 5′-R/CCGGY-3′ sequence is probably achieved by similar mechanisms. It would be interesting to see, how recognition of the extra C:G base pair is achieved by the SgrAI restriction endonuclease.

5 Possible Model of DNA Recognition by Bse634I/Cfr10I Restriction Endonucleases

Besides the common tetrameric architecture, NgoMIV and Bse634I/Cfr10I also share similarities in their recognition sites. Their recognition sequences, G/CCGGC and Pu/CCGGPy of NgoMIV and Bse634I/Cfr10I, respectively, differ only at the outer base pair flanking the common CCGG core. The crystal structures of both Bse634I and Cfr10I were solved in the absence of the DNA (Bozic et al. 1996; Grazulis et al. 2002). The superimposition of the active sites and common structural cores of NgoMIV and Bse634I/Cfr10I, however, allowed us to predict the putative DNA binding interface of Bse634I/Cfr10I (Grazulis et al. 2002).

The N-terminus of the $\alpha6$ helix of Bse634I overlaps with the helix $\alpha7$ of NgoMIV involved in the sequence-specific contacts to the inner C:G base pairs (Fig. 9a). Moreover, Bse634I residues R202, D204, and R205 (and equivalent residues R194, D196, and R197 in Cfr10I) superimpose with the R191, D193, and R194 residues of NgoMIV that make discriminating contacts to the inner C:G base pairs (Fig. 8b).

The conserved sequence motif R(P/S)DR corresponds to the structurally equivalent residues of Cfr10I, Bse634I and NgoMIV (Fig. 9a), suggesting that these enzymes use the same residues and structural mechanisms to interact with the central CCGG bases.

The recognition sequence 5′-Pu/CCGGPy-3′ of Bse634I/Cfr10I is less stringent in comparison to that of NgoMIV. Indeed, Bse634I/Cfr10I can tolerate two alternative purine nucleotides (A or G) at the first position of its recognition site while eliminating pyrimidines. The 5′-GCCGGC-3′ site of NgoMIV therefore overlaps only with one of the possible recognition sequences of Bse634I/Cfr10I. Does the structural comparison of NgoMIV and Bse634I/Cfr10I allow us to predict how the recognition of the Pu:Py base pair is achieved by Bse634I/Cfr10I?

The stringent discrimination of the outer G:C base pair by NgoMIV is achieved predominantly by three different residues (Fig. 8a, c). R227 located on the $\alpha8$ helix makes one direct (to the O6 oxygen) and one water-mediated (to the N7 nitrogen) hydrogen bonds to the outer Gua base (Fig. 9c). Such a hydrogen bond pattern is specific for the Gua base. Structural superposition of the Bse634I and NgoMIV indicates that structurally equivalent helix $\alpha7$ of Bse634I is shifted away from the DNA in respect of the $\alpha8$ helix of NgoMIV (Fig. 9b). The position of the helix $\alpha7$ of Bse634I therefore prevents direct interactions of amino acid residues with outer Ade/Gua bases of the recognition site.

The amino group of Q63 located at the N-terminus of α3 helix of NgoMIV donates a hydrogen bond to the oxygen atom of the outer Cyt base in the minor groove (Fig. 8 c). Noteworthy, in Bse634I N73 (or K64 in the case of Cfr10I) is a structural equivalent of Q63 in NgoMIV and in principle can make a hydrogen bond to the oxygen atom of the Thy or Cyt bases in the minor groove. Such a minor groove contact alone in the case of Bse634I/Cfr10I, however, is not sufficient to discriminate unambiguously the A:T and G:C base pairs against C:G and T:A base pairs, respectively.

The D34 of NgoMIV located at the N-terminal domain makes a hydrogen bond to the amino group of the outer Cyt base in the major groove. Superimposition of the crystal structures of Bse634I and NgoMIV reveals significant differences in the conformations of the N-terminal parts of both proteins (residues 1–89 in Bse634I and 1–79 in NgoMIV, respectively). Whereas the conserved structural cores (Fig. 2) of Bse634I and NgoMIV overlap, the N-terminal domain of Bse634I is rotated about 90° in respect to its equivalent in NgoMIV (Fig. 9b). Due to the different conformations of the N-terminal subdomain the loop bridging β1- and β2-strands in Bse634I becomes positioned close to the outer G:C base pair at the opposite recognition half-site than in the case of NgoMIV. Amino acid residues/backbone atoms situated in this loop or flanking regions might contact outer Pu:Py base pair of the PuCCG-GPy sequence.

The crystal structure analysis of the BsoBI restriction endonuclease (van der Woerd et al. 2001), specific for the C/PyCGPuG sequence, revealed that recognition of the degenerate base pair occurred in the major groove predominantly through the hydrogen bond of K81 to the N7 of the purine base (Ade in the crystal) and the water molecule sandwiched between the N6 of Ade and side chains of K81 and D246. The hydrogen bond of the lysine amino group to the N7 nitrogen atom is purine specific. In the crystal structure of HincII (Horton et al. 2002), specific for the GTPyPuAC sequence, a similar inner Gua N7 interaction with the amino group of N141 residue was the only hydrogen bond observed at the center C:G base pair (as seen in the crystal), consistent with the degenerate recognition of any Py–Pu dinucleotide at this position. Similar interactions with the N7 nitrogen atom of the outer purine base of the Bse634I/Cfr10I sequence in principle would allow to discriminate Gua/Ade bases against Cyt/Thy, consistent with the less stringent specificity of Bse634I/Cfr10I in comparison to the NgoMIV restriction enzyme.

Formation of the protein–DNA complexes, however, often evokes conformational rearrangements. One cannot exclude the possibility that DNA binding might induce conformational changes of the N-terminal domain of Bse634I leading to a different set of contacts with DNA. Therefore a proposal regarding the detailed mechanism of the external base pair recognition could only be made after the elucidation of the crystal structure of Bse634I(Cfr10I)-DNA complex.

Structural comparisons between NgoMIV and Bse634I/Cfr10I, however, suggest that recognition mechanisms of the common CCGG tetranucleotide are similar while the mechanism of outer base pair discrimination seems to be different.

A similar interaction pattern at the protein–DNA recognition interface was shown previously for the EcoRI and MunI restriction enzymes, specific for G/AATTC and C/AATTG sequences, respectively, (Deibert et al. 1999). The recognition of the AATT tetranucleotide common to both EcoRI and MunI recognition sites is achieved by a similar structural mechanism. In the case of EcoRI and MunI a short stretch of conserved residues (GNAXER) located on the α-helices structurally equivalent to the helix α7 of NgoMIV, are involved in similar hydrogen bonding and van der Waals interactions with AATT bases. The mechanism of the discrimination of the outer base pair, however differed between EcoRI and MunI (Deibert et al. 1999). Structural studies of EcoRI/MunI and NgoMIV/Bse634I/Cfr10I, however, demonstrate that enzymes recognizing overlapping nucleotide sequences can employ conserved mechanisms to interact with overlapping parts of their recognition sites. However, it seems not to be a general rule. The comparison of structural mechanisms of sequence recognition by BglII (A/GATCT) and BamHI (G/GATCC) indicate that both proteins display considerably different protein–DNA contacts at the common GATC sequence within their recognition sites (Lukacs and Aggarwal 2001; Lukacs et al. 2000).

It is worth to note that structural mechanism of the CCGG sequence recognition employed by NgoMIV/Bse634I/Cfr10I is probably shared by a wide group of restriction enzymes (Fig. 7) recognizing CCGG, CCNGG, or CCWGG in different sequence contexts (Pingoud et al. 2002). Moreover, these enzymes belong to different subtypes, viz. orthodox Type II enzymes (like SsoII or Ecl18KI), Type IIE enzymes (like EcoRII) and Type IIF enzymes (like NgoMIV). It was suggested (Pingoud et al. 2002) that these subfamilies share a not too distant common ancestor, which presumably was a homodimeric enzyme. By acquisition of an extra domain a Type IIE enzyme like EcoRII was derived. Alternatively, by development of a new interface for protein–protein interaction enabled tetramerization giving rise to representatives of Type IIF enzymes like Cfr10I and NgoMIV. Two lines of evidence come support this hypothesis. First, limited proteolysis experiments of EcoRII revealed that an extra N-terminal domain can be removed yielding a dimeric restriction enzyme that lost an absolute requirement for the second recognition site (Mucke et al. 2002). Second, the alanine replacement of a single tryptophane residue located at the tetramerization interface of Cfr10I (Siksnys et al. 1999) and Bse634I restriction enzymes (Zaremba and Siksnys, unpublished data) generates dimeric restriction enzymes that in the case of Bse634I W228A mutant retains a significant catalytic activity. These experiments provide first

experimental evidence supporting evolutionary relationships between different subtypes of restriction enzymes.

6 Functional Significance of the Tetrameric Architecture of Restriction Enzymes

The crystallographic evidence reveals that NgoMIV tetramer is composed of
two dimers with a separate DNA binding site in each (Fig. 4). Crystal structures of Bse634I (Grazulis et al. 2002) and Cfr10I (Bozic et al. 1996) show similar architecture. Therefore, tetrameric restriction enzymes might interact
with two recognition sites. Processing at these sites in principle may be independent or cooperative. Cleavage patterns of plasmids containing a single or
two recognition sites provide a general test whether a restriction enzyme acts
independently on the individual site or acts simultaneously at two copies of
the recognition site (Bath et al. 2002; Bilcock et al. 1999).

Representative data for NgoMIV cleavage of plasmids containing a single
or two recognition sites are shown in Fig. 10; Cfr10I and Bse634I exhibit qual-
itatively similar cleavage patterns (Deibert et al. 2000; Embleton et al. 2001;
Grazulis et al. 2002; Siksnys et al. 1999). Kinetic studies revealed that tetrameric enzymes NgoMIV, Bse634I, and Cfr10I all cleave plasmid DNA containing two recognition sites much faster than a single site plasmid. Moreover,
most of the supercoiled two-site DNA is directly converted into the final reaction product – linear DNA with two double-stranded breaks and only a small
fraction of open-circle DNA and linear DNA cut at one site is formed. We have
demonstrated that Cfr10I interacts simultaneously with both recognition
sites located *in cis* through DNA looping (Milsom et al. 2001; Siksnys et al.
1999). We suggest that a similar model could be applied for NgoMIV and
Bse634I. Hence, tetrameric restriction enzymes NgoMIV, Bse634I, and Cfr10I
simultaneously bind two copies of their recognition sequence and rapidly
cleave all four phosphodiester bonds at both sites during the lifetime of the
enzyme–DNA complex.

The cleavage rate of a plasmid containing a single recognition site by
tetrameric restriction enzymes NgoMIV, Bse634I, and Cfr10I was much
slower than that of two-site DNA. Addition of the oligodeoxynucleotide containing recognition sequence, however, increased the cleavage rate of the
supercoiled form of pUCGC1 and significantly reduced the amount of the
open circle DNA intermediate. In contrast, oligodeoxynucleotide lacking the
recognition sequence had no effect on the cleavage rate of plasmid containing
a single site (data not shown). We suggest that the activation of one-site plasmid DNA cleavage by cognate oligodeoxynucleotides is consistent with a
tetrameric architecture of NgoMIV, Bse634I, and Cfr10I. In the presence of the
specific oligodeoxynucleotide one of the dimers of the NgoMIV tetramer
interacts with recognition site on the plasmid while the other one interacts

Fig. 10. Cleavage of plasmid DNA by NgoMIV. **a** Cleavage of supercoiled plasmid pUCGC2 bearing two recognition sites of NgoMIV. Reaction mixtures contained 1.15 nM of pUCGC2 (2.3 nM of target sequence), 50 nM of NgoMIV (as a tetramer), 10 mM $(CH_3COO)_2Mg$, 33 mM Tris-acetate (pH 8.0, 25 °C) and 66 mM CH_3COOK at 25 °C. Amounts of supercoiled (*filled triangles*), open circle (*filled circles*), linear DNA with a single double-stranded break (*open squares*) and linear DNA with two double-stranded breaks (*inverted triangles*) forms are shown. **b** Cleavage of supercoiled plasmid pUCGC1 bearing a single recognition site of NgoMIV. Reaction mixtures contained 2.3 nM pUCGC1, 50 nM of NgoMIV (as tetramer) and other components as in **a**. Amounts of supercoiled (*filled triangles*), open circle (*filled circles*) and linear (*open squares*) forms of pUCGC1 are shown. **c** Cleavage of supercoiled plasmid pUCGC1 bearing a single recognition site of NgoMIV in the presence of cognate oligodeoxynucleotide. Reaction mixtures contained 2.3 nM pUCGC1, 200 nM of specific self-complementary 12 bp duplex 5′-GAG GCC GGC CTC-3′, 50 nM of NgoMIV (as tetramer) and other components as in **a**. Amounts of supercoiled (*filled triangles*), open circle (*filled circles*) and linear (*open squares*) forms of pUCGC1 are shown

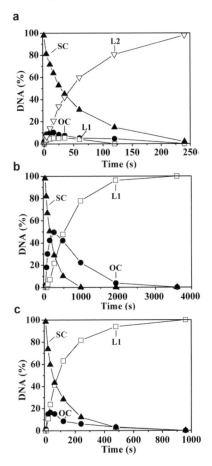

with a specific oligodeoxynucleotide and such interactions *in trans* increase the cleavage rate of pUCGC1. Transactivation by cognate oligonucleotides of plasmid DNA cleavage has been reported previously for tetrameric restriction enzyme SfiI (Nobbs and Halford 1995; Wentzell et al. 1995).

Hence, cleavage studies of plasmids containing a single and two recognition sites demonstrate that two recognition sites are required for effective DNA cleavage by NgoMIV, Bse634I, and Cfr10I and are consistent with a tetrameric architecture. Further mechanistic studies of tetrameric restriction enzymes and putative dimeric mutants are necessary to shed the light on the mechanism of cross-talking interactions between dimers.

Acknowledgements. Research support from Howard Hughes Medical Institute International Research Scholars program to V.S. is gratefully acknowledged. We thank Mindaugas Zaremba (IBT, Vilnius) for the help with figure preparation.

References

Aggarwal AK (1995) Structure and function of restriction endonucleases. Curr Opin Struct Biol 5:11–19

Bath AJ, Milsom SE, Gormley NA, Halford SE (2002) Many Type IIs restriction endonucleases interact with two recognition sites before cleaving DNA. J Biol Chem 277: 4024–4033

Beese LS, Steitz TA (1991) Structural basis for the 3'-5' exonuclease activity of *Escherichia coli* DNA polymerase I: a two metal ion mechanism. EMBO J 10:25–33

Bennett MJ, Schlunegger MP, Eisenberg D (1995) 3D domain swapping: a mechanism for oligomer assembly. Protein Sci 4:2455–2468

Bilcock DT, Daniels LE, Bath AJ, Halford SE (1999) Reactions of Type II restriction endonucleases with 8-base pair recognition sites. J Biol Chem 274:36379–36386

Bogan AA, Thorn KS (1998) Anatomy of hot spots in protein interfaces. J Mol Biol 280:1–9

Bozic D, Grazulis S, Siksnys V, Huber R (1996) Crystal structure of *Citrobacter freundii* restriction endonuclease Cfr10I at 2.15 Å resolution. J Mol Biol 255:176–186

Deibert M, Grazulis S, Janulaitis A, Siksnys V, Huber R (1999) Crystal structure of MunI restriction endonuclease in complex with cognate DNA at 1.7 Å resolution. EMBO J 18:5805–5816

Deibert M, Grazulis S, Sasnauskas G, Siksnys V, Huber R (2000) Structure of the tetrameric restriction endonuclease NgoMIV in complex with cleaved DNA. Nat Struct Biol 7:792–799

Embleton ML, Siksnys V, Halford SE (2001) DNA cleavage reactions by Type II restriction enzymes that require two copies of their recognition sites. J Mol Biol 311:503–514

Galburt EA, Stoddard BL (2002) Catalytic mechanisms of restriction and homing endonucleases. Biochemistry 41:13851–13860

Grazulis S, Deibert M, Rimseliene R, Skirgaila R, Sasnauskas G, Lagunavicius A, Repin V, Urbanke C, Huber R, Siksnys V (2002) Crystal structure of the Bse634I restriction endonuclease: comparison of two enzymes recognizing the same DNA sequence. Nucleic Acids Res 30:876–885

Halford SE (1999) Restriction enzymes that act simultaneously at two DNA sites. Biochem Soc Trans 27:A88

Horton NC, Dorner LF, Perona JJ (2002) Sequence selectivity and degeneracy of a restriction endonuclease mediated by DNA intercalation. Nat Struct Biol 9:42–47

Janulaitis AA, Stakenas PS, Berlin YA (1983) A new site-specific endodeoxyribonuclease from *Citrobacter freundii*. FEBS Lett 161:210–212

Kelly TJJ, Smith HO (1970) A restriction enzyme from *Hemophilus influenzae* II. Base sequence of the recognition site. J Mol Biol 51:393–409

Liu Y, Eisenberg D (2002) 3D domain swapping: as domains continue to swap. Protein Sci 11:1285–1299

Lukacs CM, Aggarwal AK (2001) BglII and MunI: what a difference a base makes. Curr Opin Struct Biol 11:14–18

Lukacs CM, Kucera R, Schildkraut I, Aggarwal AK (2000) Understanding the immutability of restriction enzymes: crystal structure of BglII and its DNA substrate at 1.5 Å resolution. Nat Struct Biol 7:134–140

Milsom SE, Halford SE, Embleton ML, Szczelkun MD (2001) Analysis of DNA looping interactions by Type II restriction enzymes that require two copies of their recognition sites. J Mol Biol 311:515–527

Mucke M, Grelle G, Behlke J, Kraft R, Kruger DH, Reuter M (2002) EcoRII: a restriction enzyme evolving recombination functions? EMBO J 21:5262–5268

Nobbs TJ, Halford SE (1995) DNA cleavage at two recognition sites by the SfiI restriction endonuclease: Salt dependence of *cis* and *trans* interactions between distant DNA sites. J Mol Biol 252:399–411

Pingoud A, Jeltsch A (2001) Structure and function of Type II restriction endonucleases. Nucleic Acids Res 29:3705–3727

Pingoud V, Kubareva E, Stengel G, Friedhoff P, Bujnicki JM, Urbanke C, Sudina A, Pingoud A (2002) Evolutionary relationship between different subgroups of restriction endonucleases. J Biol Chem 277:14306–14314

Repin VE, Lebedev LR, Puchkova L, Serov GD, Tereschenko T, Chizikov VE, Andreeva I (1995) New restriction endonucleases from thermophilic soil bacteria. Gene 157:321–322

Roberts RJ, Vincze T, Posfai J, Macelis D (2003) REBASE: restriction enzymes and methyltransferases. Nucleic Acids Res 31:418–420

Rosenberg JM (1991) Structure and function of restriction endonucleases. Curr Opin Struct Biol 1:104–113

Sapranauskas R, Sasnauskas G, Lagunavicius A, Vilkaitis G, Lubys A, Siksnys V (2000) Novel subtype of Type IIs restriction enzymes. J Biol Chem 275:30878–30885

Siksnys V, Skirgaila R, Sasnauskas G, Urbanke C, Cherny D, Grazulis S, Huber R (1999) The Cfr10I restriction enzyme is functional as a tetramer. J Mol Biol 291:1105–1118

Skirgaila R, Grazulis S, Bozic D, Huber R, Siksnys V (1998) Structure-based redesign of the catalytic/metal binding site of Cfr10I restriction endonuclease reveals importance of spatial rather than sequence conservation of active centre residues. J Mol Biol 279:473–481

Stein DC, Chien R, Seifert SH (1992) Construction of a *Neisseria gonorrhoeae* MS11 derivative deficient in NgoMI restriction and modification. J Bacteriol 174:4899–4906

Tamulaitis G, Solonin AS, Siksnys V (2002) Alternative arrangements of catalytic residues at the active sites of restriction enzymes. FEBS Lett 518:17–22

Tautz N, Kaluza K, Frey B, Jarsch M, Schmitz GG, Kessler C (1990) SgrAI, a novel class-II restriction endonuclease from *Streptomyces griseus* recognizing the octanucleotide sequence 5'-CR/CCGGYG-3'. Nucleic Acids Res 18:3087

Todd AE, Orengo CA, Thornton JM (2002) Plasticity of enzyme active sites. Trends Biochem Sci 27:419–426

van der Woerd MJ, Pelletier JJ, Xu S, Friedman AM (2001) Restriction enzyme *Bso*BI–DNA complex: a tunnel for recognition of degenerate DNA sequences and potential histidine catalysis. Structure (Camb) 9:133–144

Venclovas C, Timinskas A, Siksnys V (1994) Five-stranded β-sheet sandwiched with two α-helices: a structural link between restriction endonucleases EcoRI and EcoRV. Proteins 20:279–282

Viadiu H, Aggarwal AK (1998) The role of metals in catalysis by the restriction endonuclease BamHI. Nat Struct Biol 5:910–916

Wentzell LM, Nobbs TJ, Halford SE (1995) The SfiI restriction endonuclease makes a four-strand DNA break at two copies of its recognition sequence. J Mol Biol 248:581–595

Structure and Function of Type IIE Restriction Endonucleases – or:

From a Plasmid That Restricts Phage Replication to A New Molecular DNA Recognition Mechanism

M. REUTER, M. MÜCKE, D.H. KRÜGER

1 Introduction

Restriction endonucleases (REases) and DNA methyltransferases (MTases) are encoded by chromosomal, plasmid, or viral genes. In eubacteria as well as in archaea, they often form biologically active DNA restriction-modification (R-M) systems (for a review see Krüger and Reuter 1999). The first R-M systems, discovered by reversible growth reduction of bacterial viruses, were found to be encoded by chromosomal genes in *Escherichia coli* K-12, *Escherichia coli* B, and *Salmonella typhimurium* (for reviews: Arber 1974; Boyer 1971; Meselson et al. 1972). Besides the prophage P1 and the related plasmid p15 (Glover et al. 1963; Arber and Wauters-Willems 1970), also naturally occurring drug resistance plasmids ("R factors") of the *fi⁻* (fertility inhibition-minus) type have been shown to restrict and modify infecting bacteriophage or trans-conjugated plasmid DNAs in vivo. The first R-factor controlled R-M systems, called EcoRI and EcoRII today, have been defined by those phenotypic properties (Arber and Morse 1965; Bannister and Glover 1968, 1970; Watanabe et al. 1964). Endonucleolysis (Takano et al. 1968; Yoshimori et al. 1972) and cytosine methylation (Hattman 1972), respectively, have been proposed as the molecular mechanisms of EcoRII-specific restriction and modification. EcoRII was among the very first R-M enzymes for which the DNA substrate site was identified (Bigger et al. 1973; Boyer et al. 1973). The REase EcoRII recognizes the sequence 5′-CC(A/T)GG which exhibits a twofold rotational symmetry with a dyad axis corresponding to the central A–T pair. EcoRII cleaves the phosphodiester bond at the 5′ end of the first cytosine of the unmethylated sequence.

M. Reuter, M. Mücke, D.H. Krüger
Institute of Virology, Charité Medical School, Humboldt University, Schumannstr. 20/21, 10098 Berlin, Germany

Nucleic Acids and Molecular Biology, Vol. 14
Alfred Pingoud (Ed.)
Restriction Endonucleases
© Springer-Verlag Berlin Heidelberg 2004

The resulting DNA cleavage products exhibit cohesive ends with overhangs of five bases. By electron microscopy, Boyer et al. (1973) observed that EcoRII-generated cleavage products of SV40 DNA tend to re-associate at room temperature, which could be explained by the length and high CG-content of the "sticky" ends. Such re-association of cleaved EcoRII sites seemed to support experimental data that intracellular growth of bacteriophage λ is more efficiently restricted by EcoRI (efficiency of plating, eop=10^{-4}) than by EcoRII (eop=10^{-2}) even though there are 14 times as many EcoRII as EcoRI substrate sites per single genome (Boyer et al. 1973).

The EcoRII MTase converts the internal cytosine to 5-methylcytosine and thus, protects the DNA recognition sequence against endonucleolytic digestion (Bigger et al. 1973; Boyer et al. 1973; Buryanov et al. 1978). The chromosomally encoded cytosine methylase of *Escherichia coli* K-12, M.EcoKDcm, exhibits the same sequence specificity and cross-protects DNA substrates towards EcoRII digestion (Hattman 1977; Bogdarina et al. 1979). Genes coding for the EcoRII REase and MTase were cloned and their nucleotide sequence was determined. The two genes are convergently transcribed from separate promoters on opposite DNA strands (Kosykh et al. 1980; Som et al. 1987; Kosykh et al. 1989, Bhagwat et al. 1990; Reuter et al. 1999). The open reading frame of the EcoRII REase codes for a protein with a molecular mass of 44 and 45.6 kDa, respectively (Kosykh et al. 1982; Bhagwat et al. 1990), and that of the MTase gene codes for a protein of about 54.6 kDa (Som et al. 1987).

Investigations on the in vivo resistance of DNA towards the EcoRII R-M system have only been carried out to a limited extent. The single-stranded DNA phages, fd and M13, were found not to be subject of restriction in cells carrying EcoRII-encoding plasmids (Arber 1966) but they are in vivo substrates for EcoRII methylation (Hattman 1973). Vovis et al. (1975) noted that complete digestion of the replicative form (RF) DNA of the closely related phage f1 (2 sites per 6,407 bp) by EcoRII could not be obtained, and Hattman et al. (1979) reported that phage phiX174 RF DNA (two EcoRII sites per 5386 bp) could be completely digested only in the presence of unmodified heterologous DNA. The authors did not follow up this phenomenon.

2 The General Problem: Type IIE REases Need the Simultaneous Interaction with Two Copies of Their Recognition Sequence for Enzymatic Activity

Testing *E. coli* strains expressing different R-M systems regarding their ability to control growth of the virulent bacteriophages T7 and T3, it was found that these viruses are not restricted by EcoRII-producing cells though both phage genomes contain EcoRII-specific recognition sites (Krüger et al. 1985; Reuter 1985). In contrast to the isoschizomeric REases BstNI and MvaI, EcoRII can not even cleave the purified DNA genomes in vitro (Fig. 1). The reason for this

a

Phage · grown on E. coli strain	relative efficiency of plating on E. coli strain		
	0	0(NTP14)	0(R245)
	Res⁻ Mod⁻	R.EcoRI⁺ M.EcoRI⁺	R.EcoRII⁺ M.EcoRII⁺
$\lambda \cdot 0$	1	10^{-3}	10^{-2}
$\lambda \cdot 0(\text{NTP14})$	1	1	10^{-2}
$\lambda \cdot 0(\text{R245})$	1	10^{-3}	1
$T7 \cdot 0$	1	1	1
$T7 \cdot 0(\text{NTP14})$	1	1	1
$T7 \cdot 0(\text{R245})$	1	1	1

b

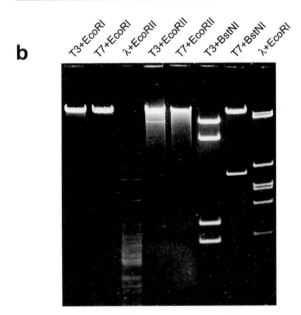

Fig. 1. DNA restriction of phage lambda by EcoRII in vivo as well as in vitro and missing effect on phage T7 and T3. **a** Bacteriophage growth efficiency in EcoRII-encoding host cells. Plating of phage T3 resulted in the same data as shown for T7. **b** Cleavage of DNAs isolated from phage lambda, T7 and T3 by EcoRII in vitro. BstNI is an EcoRII isoschizomer indicating the presence of specific recognition sites in T7 and T3 DNAs (data from Reuter 1985). In contrast to the resistance of T3 and T7 DNA to EcoRII, which is caused by the low frequency of DNA recognition sites (see text), resistance of the phage genomes towards EcoRI in vivo and in vitro is due to the lack of DNA recognition sites for this restriction endonuclease

cleavage resistance can not be DNA methylation, because at least T3 DNA is not methylated at all due to the activity of a virus-encoded S-adenosyl-L-methionine (SAM)-hydrolase (Krüger et al. 1985). Noticeable was the fact of strong avoidance of the pentameric EcoRII recognition sequence 5'-CC(A/T)GG in the virus genomes; T7 and T3 contain this nucleotide sequence only two and three times, respectively. However, on the basis of the occurrence of its subsequences, 5'-CC(A/T)G, 5'-C(A/T)GG, and 5'-C(A/T)G, the EcoRII site is suspected to occur 56 times at least in the T7 genome demonstrating a strong counterselection against this particular pentameric sequence in the evolution of the virus (Schroeder et al. 1986; Krüger et al. 1989).

The resistance of the rarely occurring EcoRII-specific recognition sites in the genomes of the closely related phages T3 and T7 against cleavage by EcoRII led to the question whether this phenomenon is due to the effect of neighboring bases. Upon cloning of an EcoRII recognition site including the 5'- and 3'- adjacent bases out of the T7 genome into the vector pUC19, which is itself cleavable by EcoRII, the T7-derived recognition site became susceptible to EcoRII cleavage. It was generalized from this result that the site is not refractory per se and base sequences immediately around the EcoRII recognition site are not the primary reason for cleavage resistance (Krüger et al. 1990).

Independently, it was observed that the recognition sites in T7 and T3 phage genomes become susceptible to EcoRII cleavage in the presence of DNAs, which contain an abundance of EcoRII-sensitive sites, e.g., pBR322 plasmid or phage lambda DNA (Krüger et al. 1988). Studies using fragments of pBR322 DNA containing different numbers of EcoRII sites showed that the susceptibility to EcoRII cleavage is proportional to the number of sites in the individual fragment. An isolated fragment containing only one site was demonstrated to be highly resistant to cleavage. In the presence of increasing amounts of pBR322 DNA and a constant amount of T3 DNA, cleavage only went to completion when the molar ratio of recognition sites in pBR322:T3 DNAs was at least 2:1. Lower pBR322 concentrations in the reaction mixture did not suffice. The direct correlation between the amount of activator DNA and the extent of T3 DNA cleavage is indicative of a stoichiometric relationship of the DNA molecules involved in EcoRII cleavage activation. It was postulated that EcoRII is the prototype of REases that require the presence of at least two simultaneously bound DNA sites for their activity (Krüger et al. 1988) – a feature that so far was only known for enzymes involved in DNA transposition, recombination or replication. Today, REases with similar reaction mechanisms and substrate requirements as EcoRII form a group of their own – named Type IIE – within the family of Type II restriction enzymes (Roberts et al. 2003).

3 Activation of Type IIE REases by Synthetic Oligonucleotide Duplexes

The inability of EcoRII to cleave non-methylated recognition sites in certain DNA molecules is due to an unusual reaction mechanism which is characterized by the simultaneous interaction of the protein with two DNA sites. What are the structural and functional prerequisites of an activator DNA molecule? Synthetic oligonucleotide duplexes of only 14-bp length with a single recognition site for EcoRII are sufficient to activate the cleavage of T3 DNA as well as the other high-molecular-weight DNAs mentioned above (Pein et al. 1989). In these reactions, EcoRII cleaves the short activator molecules themselves. A single-stranded oligonucleotide does not activate the cleavage of T3 DNA by EcoRII. Activation of EcoRII depends on the concentration of the oligonucleotide duplexes. The complete cleavage of T3 DNA requires a molar ratio of recognition sites on the oligonucleotide duplexes to recognition sites on T3 DNA of 140:1. This high molar excess of the oligonucleotide duplexes compared to pBR322 plasmid DNA (2:1) is probably due to differences in the stability of the formed enzyme–substrate complexes that correlate with the length of the DNA (Pein et al. 1989). The fact that the 14-bp oligonucleotide duplexes are cleaved themselves in the reaction implies that EcoRII can simultaneously bring together two of these molecules in an enzyme–substrate complex. This was the first example of a REase that can bind two copies of its DNA recognition sequence *in trans*.

Conrad and Topal (1989) described two classes of DNA recognition sites for NaeI, a Type IIE REase isolated from *Nocardia aerocolonigenes* – sites susceptible and sites resistant to cleavage. They showed by kinetic analyses that NaeI is activated by addition of DNA containing cleavable sites to cleave resistant sites. The activation relies on an increase of the rate of DNA cleavage but did not alter substrate binding. In addition, they observed that in the presence of 1 mM spermidine, resistant NaeI sites become cleavable although digestion of cleavable DNA is inhibited. Besides EcoRII, NaeI is the second Type IIE REase that has been comprehensively investigated.

4 Activation of Type IIE REases by Cleavage Products and by Non-Cleavable Oligonucleotide Duplexes

To understand the activation mechanism of Type IIE REases, the question had to be addressed whether the activator DNA molecules necessarily have to be cleavable. The six EcoRII recognition sites of pBR322 plasmid DNA, which can activate the cleavage of a priori resistant EcoRII DNA recognition sites in phage T3 DNA, were destroyed by EcoRII digestion. From the reaction sample, the enzyme was removed by phenol extraction and ethanol precipitation and the resulting pBR322 DNA fragments were incubated with T3 DNA and

EcoRII again. It became evident that the EcoRII-derived pBR322 cleavage products stimulate T3 cleavage (Pein et al. 1991). Remarkably, no enzyme activation was obtained using pBR322 cleavage products derived by digestion with the isoschizomers BstNI or MvaI. The data demonstrate that only the authentic EcoRII DNA cleavage products with their characteristic five-base 5' overhangs can activate EcoRII. BstNI and MvaI do recognize the same DNA recognition sequence as EcoRII, but release DNA fragments with only a single-base 5' overhang. Moreover, even oligonucleotides mimicking EcoRII cleavage products are efficient activators as long as they provide the 5' phosphate group (Pein et al. 1991). Another unconventional reaction mechanism that is similar to the EcoRII enzyme activation by cleavage products has recently been reported for the REase SgrAI (Bitinaite and Schildkraut 2002).

Moreover, in contrast to the sensitivity of EcoRII to the 5' phosphate group, oligonucleotide duplexes displaying intact recognition sequences but with a modified base (e.g., 6-methyl-2'-deoxyadenosine, N^4-methyl-2'-deoxycytidine, I-2'-deoxyinosine) are still capable of activating the enzyme without being cleaved themselves (Pein et al. 1991). Interestingly, though confirming that EcoRII does not cleave oligonucleotide duplexes containing 6-methyl-2'-deoxyadenosine in the recognition sequence, Petrauskene et al. (1995) reported that DNA molecules with this modified base in place of the central adenine are hydrolyzed in the presence of a canonical substrate. For NaeI, it was demonstrated that oligonucleotide duplexes, which contain a NaeI recognition sequence with an uncleavable phosphorothioate at the cleavage position, are similarly potent NaeI-activators as the respective oligonucleotide duplex with a scissile phosphodiester bond (Conrad and Topal 1992; Senesac and Allen 1995).

By investigating the cleavage of oligonucleotide duplexes of different lengths by EcoRII, an indirect correlation between their length and their cleavability by the enzyme was found. Oligonucleotide duplexes with a length of 14, 30, and 71 bp, respectively, are cleaved with decreasing efficiency (Pein et al. 1991). Though EcoRII has a two orders of magnitude higher affinity for a 30-bp substrate compared to a 14-bp substrate, the former substrate is hydrolyzed more slowly (Petrauskene et al. 1992, 1996). The isoschizomeric REase MvaI cleaves DNA substrates independent of their length (Pein et al. 1991). Moreover, EcoRII does not cleave DNA fragments of around 500–600 bp with only a single EcoRII recognition sequence, which were generated by pBR322 digestion with different REases. If there are two or more recognition sites on a DNA fragment, the distance between them plays a decisive role. The inter-site distance must not exceed about 1000 bp (Krüger et al. 1988; Pein et al. 1991). Furthermore, EcoRII simultaneously interacts with both strands of the substrate, but independently cuts them in a single binding event (Yolov et al. 1985; Petrauskene et al. 1998). When compared to the isoschizomeric MvaI REase, EcoRII is much more sensitive towards structural modifications in the DNA recognition sequence (Cech et al. 1988).

These data prove that it is not the cleavage of the activator molecule which activates Type IIE REases, because cleavage products as well as uncleavable substrate analogues are good activators. Although the activator DNA has not to be cleaved itself, it has to fulfil particular requirements: (1) EcoRII has to bind and to recognize the activator DNA, (2) for an effective intermolecular enzyme reaction (interaction of EcoRII with two recognition sites located *in trans*), the activator molecule has to be small enough to sterically allow the formation of the active enzyme–DNA complex, and (3) for an intramolecular reaction of EcoRII (interaction with two recognition sites located *in cis*), the distance between two recognition sites must be within a certain limit.

One can summarize these experimental details in the following model: in the active enzyme–substrate complex, Type IIE REases interact with two copies of their recognition sequence. One of them serves as an allosteric activator and does not have to be cleaved. The allosteric site (activator) and the site that has to be cleaved can be located on the same or on separate DNA molecule(s). The interaction of Type IIE REases with DNA sites on different DNA molecules depends on their length and concentration, whereas the interaction within the same DNA molecule is limited by the distance between two DNA sites.

5 The Enzymes' Reaction Mechanism – General Aspects and Details

5.1 Cooperative Interaction with Two Recognition Sites

The suggested two-site reaction model of Type IIE REases implies some kind of substrate cooperativity. Studying the DNA cleavage rate of numerous REases on various substrates, Oller et al. (1991) differentiated between resistant, slow, and cleavable substrate sites. Cleavage of the former two kinds of DNA sites could be significantly enhanced by the addition of cleavable DNA or spermidine. Based on the effect of the activator DNA on V_{max} and K_m of the reaction, they established two different action mechanisms. The REases NaeI and BspMI give hyperbolic substrate saturation curves with varying V_{max} and constant K_m. Cleavage was not significantly enhanced by increasing substrate concentrations. In contrast, substrate cleavage by the enzymes HpaII, NarI, and SacII showed sigmoidal dependence of the initial rate on the substrate concentration. This indicates that the substrate binding at the active site was cooperative. With increasing concentration of the respective activator DNA, K_m decreased, whereas V_{max} remained constant. Based on these results, it has been concluded that all five enzymes possess independent binding sites for the activator and the substrate DNA, but follow different positive allosteric enzyme activation mechanisms (Oller et al. 1991). Moreover, it has been proven that the two DNA-binding sites of NaeI are nonidentical and that they recognize different families of sequences flanking the recognition sequence. By measuring the relative binding affinities of NaeI for a 14-bp oligonucleotide duplex, which contained the NaeI recognition sequence 5'GCCGGC flanked by various sequences, Yang and Topal (1992) demonstrated that the activator binding site of NaeI prefers GC-rich flanking sequences, whereas the substrate binding site of NaeI most efficiently binds AT-rich flanking sequences.

Though detailed studies on how different neighboring bases influence the cleavage efficiency of EcoRII recognition sites have not been carried out, it is unlikely that EcoRII can strictly discriminate between activator and substrate molecules. Recognition site density, distance between recognition sites within a DNA molecule, and DNA concentration determine the catalytic activity of EcoRII. Gabbara and Bhagwat (1992) showed that, at low concentrations, short 14-bp activator oligonucleotide duplexes themselves are poorly cleaved by EcoRII. The reaction showed positive cooperativity, and at high concentrations oligonucleotide duplexes are likewise activators and good substrates. The turnover number for EcoRII was determined to be about 700 double-strand scissions per minute per dimer, a number that is significantly higher than that of other REases. Moreover, these authors showed that the amount of enzyme–substrate complexes is not the limiting factor for product formation at substrate concentrations at which poor yields of the product are obtained. The enzyme–substrate complexes are stable, even in the absence of Mg^{2+} ions (Gabbara and Bhagwat 1992). Cleavage of longer synthetic oligonucleotide duplexes (30 and 111 bp) with a single EcoRII recognition site confirmed the sigmoidal dependence of DNA cleavage efficiency on DNA site concentration. In contrast, a 111-bp oligonucleotide duplex with two EcoRII recognition sites was still cleaved significantly at the lowest DNA concentrations tested (Kupper et al. 1995); this is due to the intramolecular cooperation of EcoRII with the two recognition sites.

5.2 Stoichiometry of the Active Enzyme–Substrate Complexes

EcoRII and NaeI preferentially exist as dimeric molecules in solution (Kosykh et al. 1982; Baxter and Topal 1993). Nevertheless, the cooperative interaction of Type IIE REases with two DNA recognition sites can be achieved either by one dimer alone or by binding of one dimer per DNA recognition site and subsequent formation of an active tetrameric protein–DNA complex.

Both EcoRII and NaeI bind to their DNA recognition sequence in a symmetrical manner. Using a 259-bp fragment with a single NaeI recognition site, Baxter and Topal (1993) showed by DNaseI digestion that NaeI protects a DNA sequence of 24 bp, which is symmetrically located around the 6-bp recognition sequence. The DNA region protected by EcoRII has been estimated to be 20–22 bp based on the dissociation kinetics of the protein bound to synthetic concatemeric DNA substrates (Vinogradova et al. 1990). DNaseI footprinting analyses with a DNA substrate of 282-bp length containing two EcoRII recognition sites 82 bp apart reduced the DNA region specifically contacted by EcoRII to 16–18 bp. The DNaseI footprinting patterns were symmetic in the CCAGG as well as in the CCTGG strands of both recognition sites. However, EcoRII did not protect the DNA between the two recognition sites against DNaseI (Mücke et al. 2000).

In NaeI cleavage assays, activator DNA stimulated either DNA nicking or DNA cleavage depending on NaeI concentration (Baxter and Topal 1993). The transition from nicking to cleavage activity depends on the increasing NaeI concentrations and was interpreted to be due to two different structures of NaeI. The first structure is a NaeI monomer that is able to cleave one strand of a recognition site and the second structure is a dimer able to cleave both strands of a recognition site. Both, the nicking and the cleavage activity were postulated to require bound activator DNA. In gel electrophoretic mobility shift experiments, NaeI formed two different complexes with DNA – monomer–DNA and dimer–DNA complexes. It was stated that both, the NaeI monomer and the NaeI dimer possess two DNA-binding sites. Nevertheless, the completely active enzyme–substrate complex, which cleaves both strands of the recognition sequence, can only be formed by the NaeI dimer and when the activator binding site is occupied (Baxter and Topal 1993).

To determine the stoichiometry of the active EcoRII–substrate complex, Petrauskene et al. (1994) carried out single turnover experiments using increasing enzyme concentrations. The reaction rates increased with increasing enzyme concentration until the ratio of dimeric EcoRII molecules and DNA recognition sites of 1:2. Excess of EcoRII over DNA recognition sites inhibited the enzyme activity. This data indicated that the active EcoRII–DNA complex is formed when two subunits of the enzyme interact with two DNA recognition sites. Reuter et al. (1998) confirmed that the cooperative interaction with two recognition sites can be achieved by one EcoRII dimer. These authors demonstrated that the enzyme inhibition of EcoRII at high enzyme concentrations is reversible and that it is not due to high EcoRII concentration per se; a balanced increase of EcoRII recognition sites to the optimal molar ratio of one dimer per two recognition sites restored DNA cleavage to the normal extent. Moreover, the molecular weight of the EcoRII–DNA complex eluted from a gel filtration column corresponded to a dimeric enzyme structure bound to two DNA substrate sites (Reuter et al. 1998).

In contrast, studying EcoRII binding to various synthetic DNA duplexes, which contained either the modified or the canonical EcoRII recognition sequence, with native gel electrophoresis, Karpova et al. (1999a) claimed that the active complex is most likely formed by one monomeric EcoRII subunit and one DNA recognition site. This model implies that after occupation of both DNA-binding sites on the EcoRII dimer, the complex dissociates into two monomer–DNA complexes that are both catalytically active in the presence of

Mg^{2+}. After product release the monomeric EcoRII subunits are supposed to form again dimeric enzyme molecules.

Indeed, Mücke et al. (2000) demonstrated that a monomeric EcoRII subunit binds to the EcoRII recognition sequence. Single amino acid replacement of Val^{258} by Asn yielded an EcoRII mutant that was unaffected in substrate affinity and in its DNaseI footprinting pattern when compared to EcoRII wild type, but exhibited a profound decrease in cooperative DNA binding and cleavage activity. Analysis of the molecular mass of mutant V258N showed a high percentage of protein monomers in solution. The dissociation constant of mutant V258N confirmed a 350-fold decrease in the enzyme's dimerization capability. The electrophoretic mobility of the mutant enzyme–DNA complex was significantly higher than that of the wild-type EcoRII, and both can easily be distinguished in gel shift assays. Taking into account that EcoRII forms stable dimers with a K_D value for dimerization of 2.89 nM and that comparably fast-migrating enzyme–DNA complexes as in the case of the monomeric V258N mutant do not occur in the case of the wild-type EcoRII (Mücke et al. 2000), the data contradict the model proposed by Karpova et al. (1999a).

5.3 How do Type IIE REases Communicate Between Remote DNA Recognition Sites in a DNA Molecule?

In principle, different enzymatic processes can be discussed for the interaction of protein molecules with distant, *in cis* located DNA elements – ATP-dependent DNA translocation and ATP-independent DNA sliding and looping mechanisms (Wang and Giaever 1988; Dröge 1994). Because the enzyme activity of Type IIE REases does not depend on ATP, DNA translocation seems unlikely. Topal et al. (1991) demonstrated by electron microscopy that in the presence of Ca^{2+} or Mg^{2+} ions, NaeI simultaneously binds to multiple recognition sites in pBR322 DNA to form multiloop structures with NaeI bound to the loop's base. The maximum number of loops formed with a common base suggests four binding sites per enzyme molecule. NaeI binds also to other than the cognate DNA sites. These complexes probably represent enzyme–DNA intermediates trapped during the process of scanning through the DNA for specific binding sites (Topal et al. 1991).

In contrast, Milsom et al. (2001) did not detect NaeI–mediated DNA loops using plasmids with two NaeI recognition sites interspersed with two *res* sites for site-specific recombination by the Tn21 resolvase in the presence of Ca^{2+}. This assay evaluates the degree of inhibition of the resolvase from the extent to which the *res* sites were sequestered into separate loops by the REase. Recombination by the resolvase can only occur after the loop formed by the REase has dissociated. Because the resolvase inhibition test can only detect DNA loop structures if the rate of loop disassembly is slower than the rate of recombination by the resolvase, it can be assumed that DNA loops formed by NaeI had very short life-times under the used conditions (Milsom et al. 2001).

Several biochemical data provided the first hints that EcoRII also uses DNA looping to bridge the distance between two recognition sites *in cis* (Reuter et al. 1998): (1) a Lac repressor molecule as a "molecular barrier" bound between two cooperating EcoRII recognition sites, which were located 191 bp apart did not inhibit REase activity. This indicated that EcoRII communicates with two DNA sites by bending or looping of the intervening DNA stretch. In contrast, the bound Lac repressor blocks the essential cooperative interaction of the Type III REase EcoP15 with two inversely oriented DNA recognition sites. EcoP15–DNA complexes cooperate via an ATP-dependent DNA translocation mechanism (Meisel et al. 1995). Bound proteins and irregular DNA structures also hinder other REases, e.g., EcoRI that moves along the DNA helix by linear diffusion (Jeltsch et al. 1994). (2) Comparative cleavage of linear DNA substrates with differently spaced interacting EcoRII sites revealed an inverse correlation between site distance and cleavage efficiency starting from more than 80 % at a distance of 10 bp to 6 % at a distance of 952 bp. EcoRII probably bridges distances of more than 10 bp by distortions introduced into the enzyme–substrate complex, as by DNA bending or looping. Furthermore, this demonstrated that the probability of one DNA site to be occupied by an EcoRII dimer finding the second DNA site decreases with increasing distance. At least at the optimal distance of one helical turn, the EcoRII cleavage is independent of the orientation of the two recognition sites in the DNA double strand (Reuter et al. 1998). Comparable quantitative data on the correlation between site distance and cleavage efficiency as well as on optimal and limiting site distances are not available for NaeI.

Transmission electron microscopy provided evidence that EcoRII induces loop formation after binding of two EcoRII recognition sites 191 bp apart on linear DNA molecules in the presence and in the absence of Mg^{2+} ions (Mücke et al. 2000). The asymmetric location of the two EcoRII sites in the DNA molecule allowed to distinguish both DNA ends. In contrast, binding of EcoRII to a single-site substrate did not induce DNA loops. DNA loop formation specifically occurred between the two recognition sites. When comparing the frequency of loop formation of the dimeric wild-type EcoRII (20 %) and the monomeric mutant enzyme V258N (3 %), it was observed that induction of loop formation by EcoRII directly depends on an intact dimeric protein structure. These results again illustrate that the complex of the EcoRII dimer and two copies of the DNA recognition sequence represents the active form which is required for DNA hydrolysis.

6 Sequence-Specific DNA Recognition by Type IIE REases

By employing random mutagenesis of the NaeI-encoding gene with N-methyl-N'-nitrosoguanidine and selecting bacteria by their ability to survive in the presence of mutagenized NaeI and in the absence of a protecting methylase, Holtz and Topal (1994) isolated cleavage-deficient NaeI mutant enzymes. Analysis of the enzymatic properties of different classes of mutants assigned amino acids involved in catalysis and DNA recognition and predicted the existence of two independent protein domains; an N-terminal substrate-binding domain that also contains the catalytic center and a C-terminal domain that is only involved in DNA binding, probably the activator-binding domain.

To answer the question of how EcoRII achieves sequence-specific DNA binding Reuter et al. (1999) used matrix-bound peptide scans that represent the complete EcoRII amino acid sequence as overlapping peptides. Two separate DNA-binding regions emerged; amino acids 88–102 (binding site I, RHFGKTRNEKRITRW) and amino acids 256–273 (binding site II, NSVSNR-RKSRAGKSLELH), which share the consensus motif **KXRXXK**. Substitution analogs were obtained by exchanging every residue of both binding sites for all other amino acids. Results obtained with these substitution analogs demonstrated that the replacement of the basic amino acids within the consensus motif strongly affected DNA binding of the respective peptides. EcoRII mutant enzymes generated by replacing the consensus lysine residues for alanine or glutamic acid in one or both DNA-binding sites showed different effects – mutations in DNA-binding site I attenuated DNA binding, whereas corresponding mutations in DNA-binding site II inhibited DNA cleavage without changing the DNA dissociation constant when compared to the EcoRII wild-type. Apparently, the binding site I is important for DNA recognition by EcoRII, because introducing alanine instead of glutamic acid at position 96, nearly abolished DNA binding of EcoRII (Reuter et al. 1999). The motivation for this E96A mutant came from a search for putative catalytically relevant amino acids. Based on the catalytic consensus motif $P(D/E)X_n$ $(D/E)X(K/R)$ (Pingoud and Jeltsch 2001), E96 seemed to be part of a potential catalytic site.

To investigate whether the specific amino acid sequence or the overall amino acid composition of DNA-binding sites I and II is important for the DNA binding, "scrambled" peptides of both DNA-binding sites have been used. The data obtained indicated that in the C-terminal part of binding region II $(NH_2$-RKSRAGKSLELHLEH-COOH) the specific amino acid sequence mediates specific DNA binding (Fig. 2). In contrast to the original peptide sequence, peptides with an altered amino acid sequence did not bind DNA EcoRII-specifically. DNA binding was not affected by "scrambling" the amino acids in the DNA-binding site I of EcoRII (data not shown).

In 1993, Kosykh and coworkers described an amino acid sequence homology between the target recognition domains of three DNA MTases (M.EcoRII,

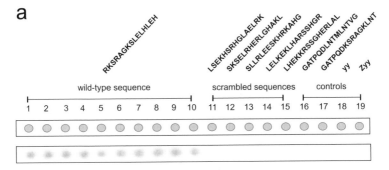

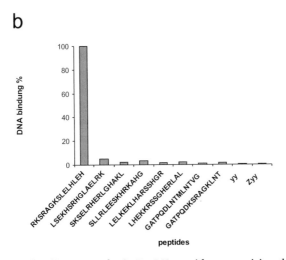

Fig. 2. DNA binding to synthetic EcoRII peptides comprising the 15 C-terminal amino acids of DNA binding region II arranged as wild-type or as scrambled sequence. a, upper part: Diagram of a cellulose strip with membrane-bound peptides and their sequences. *Spots 1–10* Amino acids 262–276 of the EcoRII wild-type sequence; *11–15* randomly scrambled sequences of amino acids 262–276 of EcoRII; *16* HIV-1 capsid epitope peptide; *17* HIV-1 capsid epitope peptide with the EcoRII consensus sequence KxRxxK; *18* β-alanine that is used as linker between peptide and membrane; *19* acetylated β-alanine residues; *lower part* scan of a cellulose strip after incubation with radioactively labeled specific oligonucleotide duplex in the presence of a twofold molar excess of unspecific oligonucleotide duplex over bound peptides. **b** Quantitation of bound radioactivity per peptide spot was performed as described in Reuter et al. (1999). The average value of bound radioactivity to the ten wild-type sequences was calculated and defined as 100 %

M.Dcm, M.SPR) and the REase EcoRII. Each of the enzymes recognizes the DNA sequence 5′CCWGG. The authors detected several regions of homology between these enzymes and speculated about their importance in DNA recognition. Interestingly, EcoRII DNA-binding site II, which was detected by synthetic peptide scans, is part of a homology region. Moreover, sequence

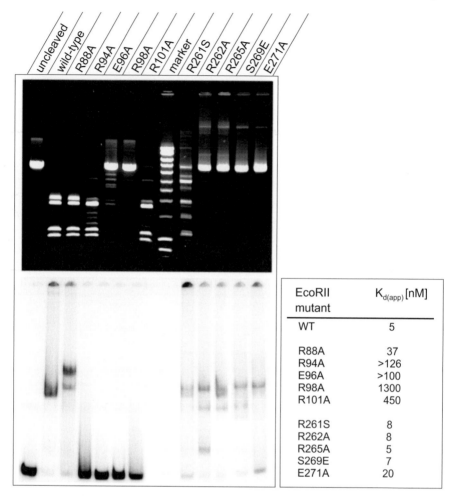

Fig. 3. DNA binding and cleavage efficiency of EcoRII mutants in DNA-binding site I and II, respectively. *Upper panel* BamHI-linearized pBR322 DNA (0.6 pmol EcoRII sites) was incubated with 0.3 pmol of EcoRII wild-type enzyme and the indicated enzyme mutants, respectively, in 20 µl for 1 h at 37 °C, and the products were analyzed in a 0.9 % agarose gel. Marker, 1-kb ladder. *Lower panel* DNA-binding affinity of EcoRII wild-type and enzyme mutants was evaluated by determining the apparent K_D values as described (Reuter et al. 1999). A constant amount of a 191-bp PCR fragment (1 nM) with a single EcoRII recognition site was incubated with increasing amounts of the respective protein. After polyacrylamide gel electrophoresis and quantification of the protein–DNA complexes by phosphorimaging, the K_D values were calculated as the protein concentration at which half of the DNA is shifted as complex. The polyacrylamide gel shows the complex formation obtained at 50 nM of the appropriate protein

alignments identified analogous motifs to EcoRII DNA-binding site II in various REases with terminal G:C or C:G base pairs in their recognition sequences (Reuter et al. 1999). Particularly, the great similarity of binding site II of EcoRII (/CCWGG) to a sequence in the EcoRII-homoisoschizomeric thermophilic restriction enzyme PspGI, to SsoII (/CCNGG) and its considerable resemblance to a ScrFI sequence (CC/NGG) lent independent support to a role of this site in sequence-specific DNA recognition (Morgan et al. 1998; Reuter et al. 1999). More comprehensive sequence- and structure-based alignments have predicted a third DNA-binding motif and a potential active site in the C-terminal part of EcoRII between amino acids Pro298 and Glu337 (Pingoud and Jeltsch 2001; Pingoud et al. 2002; Tamulaitis et al. 2002). Beside the basic lysine residues, these data also suggested a critical role for several arginine residues in the earlier described EcoRII binding sites I ([88]RHFGK-TRNEKRITRW[102]) and II ([256]NSVSNRRKSRAGKSLELH[273]). The generated EcoRII enzyme mutants confirmed the effect of the lysine substitutions (Reuter et al. 1999): (1) replacing the single arginine residues R88, R94, R98, and R101 by the neutral alanine in DNA-binding site I decreased the DNA-binding capability of the enzyme mutants R94A, R98A, and R101A, and indirectly their catalytic activity; (2) though amino acid substitutions at positions R262 and R265 in the DNA-binding site II did not affect the DNA binding, they yielded inactive enzyme mutants (Fig. 3). To further support the finding that the identified DNA-binding regions of EcoRII contribute to a different extent to DNA recognition and cleavage, data of a few other available EcoRII mutant enzymes were introduced into Fig. 3 (Reuter et al. 1999; Mücke et al. 2000; M.R. unpubl. data).

7 Domain Organization of Type IIE REases

Proteolytic cleavage of proteins provides information regarding protein folding units. Stably folded protein domains are expected to be better protected against proteolytic cleavage than linker or loop regions that connect protein domains. NaeI and EcoRII were subjected to limited proteolysis to map their domain organization, and the domains thus identified were purified after cloning of the respective coding DNA sequences. The trypsin and chymotrypsin digestion patterns were determined in the presence and in the absence of specific DNA (Colandene and Topal 1998; Mücke et al. 2002a). Both enzymes revealed two protease-resistant domains involving approximately the N- and C-terminal halves of the proteins. The experimental data showed that NaeI consists at least of one N-terminal domain (amino acids 1–120) and one C-terminal domain (amino acids 151–317). Both domains are linked by a protease-sensitive linker (amino acids 121–150). EcoRII consists of an N-terminal domain (amino acids 1–192) and a C-terminal domain (amino acids 173–404). Amino acids 173–192 probably represent a connecting linker region. DNA binding mainly protected the C-terminal domain of NaeI and the N-terminal domain of EcoRII, respectively, against proteolysis. This indicates that both domains present dominant DNA-binding sites. Photocrosslinking of EcoRII to an oligonucleotide duplex containing a single recognition site with 5-iododeoxcytidine substitutions for each C and 5-iododeoxyuridine substitutions for A, G, or T, strongly supported the predominant contribution of the N-terminal domain to DNA binding. Interestingly, EcoRII cross-linked specifically to the 5'C of the 5'CCAGG strand, but not to the 5'C of the 5'CCTGG strand yielding an asymmetrical cross-linking pattern. Mutational analyses of the electron-rich amino acids of the photocross-linked peptide in EcoRII revealed Tyr41 as the photocross-linking amino acid (Mücke et al. 2002b).

Characterization of the purified protein domains demonstrated that only the C-terminal domain of NaeI and the N-terminal domain of EcoRII specifically bind DNA. The apparent dissociation constants differed only by one order of magnitude from those of the complete enzymes. Mutational analysis of the N-terminal domain of NaeI implied that this domain mediates catalysis (Holtz and Topal 1994; Jo and Topal 1995). Nevertheless, the purified N-terminal domain of NaeI showed neither DNA binding nor catalytic activity (Colandene and Topal 1998). Although the N-terminus of EcoRII specifically bound to the DNA, it was not able to cleave DNA substrates. In contrast, the EcoRII C-terminal domain cleaved DNA specifically and much more efficiently than wild-type EcoRII. The fact that EcoRII-C specifically cleaved DNA implies that it also can specifically bind to DNA (Mücke et al. 2002a). Based on these data, it was concluded that the C-terminal EcoRII domain (termed EcoRII-C) contains all structural and functional components of an orthodox Type IIP REase that cleaves DNA at single recognition sites. Indeed, this truncated form of EcoRII no longer needs the simultaneous interaction with two copies of the recognition sequence and cleaves DNA substrates independent of the number and the distance of the DNA recognition sites in vitro and in vivo. DNA of phage T3 that resists cleavage by EcoRII wild-type enzyme is completely cleaved by EcoRII-C as efficiently as by the isoschizomer BstNI (Mücke et al. 2002a). These data obtained in vitro have been confirmed in vivo by plating phage T3 on E. coli cells that express EcoRII-C (Table 1). In contrast to EcoRII wild-type, EcoRII-C restricts growth of phage T3 by one and a half (1.5) orders of magnitude. Remarkably, the restriction of phage λ by EcoRII-C is less efficient as by the EcoRII wild-type.

Analytical ultracentrifugation experiments confirmed that EcoRII-C mainly forms dimers in solution (Mücke et al. 2002a). The dimeric structure allows specific cleavage of double-stranded DNA. The NaeI N-terminal domain was shown to form various aggregates in solution (Colandene and Topal 1998, 2000). In contrast, the N-terminal domain of EcoRII is monomeric in solution as is the C-terminal domain of NaeI (Colandene and Topal 1998; Mücke et al. 2002a). Studies using synthetic peptide scans to determine protein–protein contacts of EcoRII further support that mainly interactions in

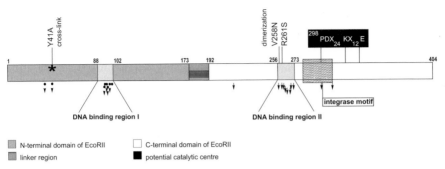

Fig. 4. Overview of the EcoRII domain organization, its functional important regions and amino acids. The bar corresponds to the 404 amino acids of the EcoRII monomer. The black box in the C-terminal domain shows the predicted active site (Pingoud et al. 2002; Tamulaitis et al. 2002). Y41 was photochemically cross-linked to the 5'C of the 5'CCAGG strand of the EcoRII recognition sequence, V258 and R261 apparently are involved in EcoRII dimerization. EcoRII mutants that are compromised in DNA binding and catalysis are indicated as dots and arrows, respectively. Mutants at several positions in the N-terminal domain show predominantly impaired DNA binding, but often also impaired catalysis. Because the latter is probably a secondary effect it is marked by arrow heads

Table 1. Efficiency of plating of unmodified phages T3 and λ on *E. coli* cells expressing EcoRII wild-type or EcoRII-C

E. coli strain	T3	λ
JM109	1	1
JM109(EcoRII)	1	10^{-3}
JM109(EcoRII-C)	5×10^{-2}	5×10^{-2}

the C-terminal part of the protein stabilize the EcoRII dimer (Petter 2001). However, these data do not exclude that the N-terminal domains of EcoRII dimerize on DNA binding.

In brief, NaeI and EcoRII show a similar two-domain organization with a linker structure that connects both domains. The C-terminal domain of NaeI appears to correspond functionally to the N-terminal domain of EcoRII in that both domains exhibit high DNA-binding affinity. The N-terminal domain of NaeI and the C-terminal domain of EcoRII apparently harbor the endonuclease catalytic sites, even though in contrast to EcoRII-C, the purified N-terminus of NaeI had no enzymatic activity. As an example, the EcoRII monomer is shown schematically together with the available amino acid substitutions and the corresponding loss of function supporting the predicted two-domain organization (Fig. 4).

8 Modular Architecture of NaeI and EcoRII and its Functional Implications

The crystal structure of NaeI was solved to 2.3 Å and showed that NaeI is a dimeric molecule with two domains per monomer (Huai et al. 2000). Each domain contains one potential DNA recognition motif. The N-terminal domain folds like a Type II REase, λ-exonuclease, and the DNA repair enzymes MutH and Vsr, and therefore was named Endo domain. The C-terminal domain contains the catabolite activator protein DNA-binding motif, a helix-turn-helix motif, which is present in a variety of DNA-binding proteins including topoisomerases. Because this structural homology matches the finding that replacement of Leu43 by lysine changed the NaeI endonuclease activity to a topoisomerase/recombinase activity (Jo and Topal 1995), the C-terminal domain was referred to as Topo domain. The Topo domain consists of a two-layer β-sandwich adjacent to four helices. In contrast to the well-ordered Endo domain, the Topo domain is loosely packed and appears to be of considerable flexibility. Both domains pack against one another by van der Waals interactions and hydrogen bonds (Huai et al. 2000). The assignment of the Endo domain to substrate binding implies that the Topo domain is responsible for the activator binding. Compared to the domains that were predicted from proteolysis studies, the domain location was revised to the following amino acid residues: Endo domain 1–162, hinge loop 163–171, and Topo domain 172–313. The structure of NaeI bound to a 17-bp cognate DNA substrate, which was determined to 2.5 Å, demonstrated that each of the two domains of NaeI recognizes one DNA duplex. DNA recognition induces structural rearrangements (Huai et al. 2001). The binding site of the Topo domain was narrowed and that of the Endo domain was widened to encircle and bend DNA by 45° for cleavage. The interface between the two NaeI monomers was completely restructured. The structure of the NaeI–DNA complex was the first example that two copies of the same DNA sequence are bound by two different amino acid sequences and two different structural motifs in one polypeptide chain (Huai et al. 2001). The structure of the DNA duplex is different depending on where it is bound; the DNA bound to the Endo domain shows B-DNA conformation, but with increased pitch and number of bases per helical turn. Furthermore, the DNA is bent approximately by 45° around the central CG base pairs of the recognition sequence. In contrast, the DNA bound to the Topo domain activator site is not bent and shows B-conformation with the expected values for the pitch and number of bases per helical turn. Moreover, it was shown that the Endo domain recognizes GCCGGC mainly through hydrogen bonds in the major and in the minor groove. In contrast, the Topo domains only contact the DNA via the major groove. In the complex with DNA, the ten hydrogen bonds that mediate the NaeI dimerization in the unligated form are replaced by five new hydrogen bonds. In addition, most of the van der Waals interactions between the NaeI monomers also rearrange. The NaeI dimer–DNA complex is symmetric, in contrast to the asymmetric monomer arrangement in the unligated NaeI dimer (Huai et al. 2001).

Structure-based alignments detected similarities of NaeI with eight REases (EcoRV, PvuII, BglI, BamHI, EcoRI, MunI, Cfr10I, FokI), with DNA repair endonucleases Vsr (very short patch repair) and MutH (mismatch repair), and with 5′-3′ λ-exonuclease (recombination and repair). These structural and topological similarities between REases and other DNA processing enzymes probably imply that DNA processing enzymes evolved from a common ancestor and share a similar catalytic mechanism (Huai et al. 2000).

For the EcoRII wild-type enzyme, two different crystals have been described that were obtained under varying experimental conditions. In the absence of Mg^{2+}, cubic EcoRII crystals were obtained which diffracted to about 4.0 Å (Karpova et al. 1999b). In another attempt using His_6-tagged EcoRII, monoclinic crystals appeared in the presence of Mg^{2+} and diffracted to 2.8 Å (Zhou et al. 2002). Both EcoRII wild-type crystals revealed two EcoRII monomers per asymmetric crystal unit. This is consistent with the fact that EcoRII is also a dimer in solution. The asymmetric unit of the crystals of a single amino acid substitution mutant of EcoRII, R88A, also contained two EcoRII monomers. The enzymatic properties of this mutant did not differ significantly from that of the EcoRII wild-type. The crystals were easier to grow and diffracted to 2.1 Å. Their structure was now determined (Zhou et al. 2003, 2004). The overall EcoRII structure consists of 18 α-helices and 13 β-strands and can be divided into two domains linked through a hinge loop. These are the N-terminal domain (residues 1–172), a linker (residues 173–193), and a C-terminal domain (residues 194–404). The N-terminal domain is built up from an eight-stranded β-sheet. The C-terminal domain has an endonuclease-like fold with a central five-stranded mixed β-sheet surrounded on both sides by α-helices. Its topology is similar to that of other Type II endonuclease structures with a common structural core (Pingoud and Jeltsch 2001). The search for structural homologues did not identify any known protein domain that is similar to the N-terminal domain of EcoRII and thus indicated that the N-domain has a novel fold (Zhou et al. 2004). The EcoRII N-terminal domain corresponds to the functional effector binding domain identified by limited proteolysis (Mücke et al. 2002a). It has a prominent cleft near its interface with the C-terminal domain. This cleft harbors numerous amino acids residues that were shown to be important for EcoRII DNA-binding earlier by different biochemical methods, e.g., Tyr41 that cross-linked to the 5′ of the CCAGG strand of the EcoRII recognition sequence (Mücke et al. 2002b) is right inside the cleft. In addition, the positively charged DNA-binding site I (residues 88–102) identified by membrane-bound peptide repertoires (Reuter et al. 1999) is also part of the cleft. Furthermore, several amino acid substitutions located in the cleft yielded mutant enzymes that were severely impaired in DNA binding (H36A, Y41A, K92E/K97E, R94A, E96A, R98A, and R101A).

Therefore, it is concluded that this cleft is involved in binding the effector DNA (Zhou et al. 2004).

The most similar structures to the C-terminal domain of EcoRII identified from structure databases are the REases NgoMIV (Deibert et al. 2000) and Cfr10I (Siksnys et al. 1999) followed by BsoBI (van der Woerd et al. 2001), FokI (Wah et al. 1997), and MunI (Deibert et al. 1999). This homology implies structural conservation and similar catalytic mechanisms among these endonucleases despite very low sequence identities. Structure-based alignments of the EcoRII C-terminal domain with NgoMIV and Cfr10I (Zhou et al. 2004) supported earlier predictions of the active center of EcoRII and putative catalytic residues E271, D299, K324, and E337 (Pingoud et al. 2002; Tamulaitis et al. 2002).

Based on the described structural alignment, Zhou et al. (2004) modelled a DNA duplex containing the EcoRII recognition sequence into the putative active center of EcoRII and showed that the catalytic site is spatially blocked by parts of the N-terminal domain and thus, is inaccessible for the substrate DNA. In this inactive state, specific DNA can only bind to the N-terminal effector domain. The active site only becomes accessible after the effector domain has moved away. This probably will be accomplished by effector DNA binding induced conformational rearrangements (Zhou et al. 2004).

9 Reaction Mechanism of Type IIE REases

The homodimeric Type IIE REases have to bind two copies of the recognition site to cleave the DNA. Therefore, the actual recognition sequence is $5'GCCGGC-(N)_x-GCCGGC$ for NaeI and $5'CCWGG-(N)_x-CCWGG$ for EcoRII. NaeI and EcoRII form loops with DNA molecules that possess two or more recognition sites in order to bring them together in the active complex. Though interaction with two recognition sites preferably occurs *in cis* (Schleif 1992), both REases can also interact with two recognition sites on different DNA molecules *in trans*. The cleavage efficiency *in trans* depends on DNA recognition site concentration and is inversely correlated to the length of the interacting DNA molecules.

The results of cleavage experiments with NaeI and EcoRII using plasmid DNAs with one and two recognition sites, respectively, emphasize that both enzymes require a second DNA recognition site to cleave efficiently. NaeI and EcoRII cleave supercoiled plasmid DNA with two recognition sites faster than plasmids with a single site (Embleton et al. 2001; G. Tamulaitis, M.R., M.M., V. Siksnys, in prep.). Furthermore, NaeI cleaves catenanes that consist of two plasmid rings each with a single recognition site faster than plasmids with single sites. At least under the experimental conditions tested, EcoRII and NaeI mainly convert a two-site plasmid into the full-length linear DNA that results from cleavage at only one of the two bound recognition sites (Embleton et al. 2001; G. Tamulaitis, M.R., M.M., V. Siksnys, in prep.). Cleavage at only one out of two recognition sites and the reduced cleavage efficiency of Type IIE REases with one-site substrates in general compared to orthodox Type IIP enzymes can be explained by the two-domain structure found for the Type IIE enzymes investigated so far.

NaeI and EcoRII consist of two stably folded domains flexibly connected by a linker structure (Huai et al. 2000, 2001; Zhou et al. 2004). These two domains are an endonuclease-like domain and a DNA-binding domain that both specifically interact with DNA (Huai et al. 2001; Mücke et al. 2002a). The DNA-binding domain binds one of the DNA recognition sites, the activator site. This binding activates the endonuclease-like domain in turn to bind and to cleave the second DNA recognition site, the substrate site. The essential positive cooperativity observed for DNA cleavage by EcoRII or NaeI is due to this two-site DNA-binding process. For NaeI, the crystal structure revealed a structural motif characteristic for the DNA-binding of the catabolite activator (CAP) protein of *E. coli* (Huai et al. 2000) in the DNA-binding domain. In the case of EcoRII, the DNA-binding domain shows a new fold (Zhou et al. 2004). Because the DNA-binding dissociation constants for the separately purified DNA-binding domains of NaeI and EcoRII differ only marginally from those of the complete enzymes, and because the isolated catalytic domains do not bind (NaeI) or bind only weakly (EcoRII) to DNA, it is supposed that in the complete enzymes the initial DNA-binding event is that to the activator domain (Colandene and Topal 1998; Mücke et al. 2002a). Structural comparison of the unligated and ligated NaeI showed that in the absence of DNA, the DNA-binding cleft at the catalytic site is too narrow to accommodate DNA. Only after DNA binding to the Topo domain, the DNA-binding cleft of the Endo domain widens by several Å and allows DNA substrate binding (Huai et al. 2001).

The situation in the EcoRII structure even seems more complex (Zhou et al. 2004). In the absence of DNA, EcoRII forms an inactive dimer structure. In the inactive structure the effector domain site is accessible for DNA, but the catalytic site is spatially blocked by the effector domains. It is assumed that in the presence of DNA, DNA binds first to the effector site. Effector binding induces domain rearrangements by moving the two effector domains away from the catalytic domains and by bringing the catalytic domains closer to each other to form the active dimer with the substrate DNA bound. This could mean that in the biological context, EcoRII is regulated by a typical autoinhibition mechanism. This kind of tight "on-site" repression of enzymatic activity was described for numerous transcription factors as well as for proteins involved in signal transduction in eukaryotic cells (for a review, see Pufall and Graves 2002), but this would be the first example observed among REases. This autoinhibition mechanism matches the observation that in limited proteolysis, initial DNA binding protects the N-terminal domains of EcoRII against proteases (Mücke et al. 2002a), whereas the C-terminal domains are accessible to proteolytic digestion. In contrast, in the absence of DNA the C-terminal domains are more resistant towards protease digestion, possibly because of their spatial protection by the N-terminal domains (Zhou et al. 2004). Moreover, it was shown that the N-terminal domain of EcoRII (EcoRII-N) specifically binds to the DNA recognition site but does not cleave it, while the C-terminal domain of EcoRII (EcoRII-C) alone acts as an orthodox Type IIP endonuclease, which specifically cleaves also at single 5'CCWGG sites (Mücke et al. 2002a). Apparently, cutting off the N-terminal domain relieved the autoinhibition of EcoRII and activated the catalytic C-terminal domain. Together with the modular EcoRII structure as a prerequisite, these experimental data perfectly match the definition of autoinhibition; activity enhancement of a particular protein domain in the absence of some other region of the protein indicates autoinhibition (Pufall and Graves 2002). Because autoinhibitory domains usually negatively regulate the function of another domain, a repression *in trans* can be expected. Cleaving pBR322 DNA with EcoRII-C is inhibited by increasing amounts of EcoRII-N from a molar ratio of EcoRII-N:EcoRII-C of 1:1 upward. DNA cleavage by EcoRII-C is completely abolished at a 20-fold molar excess of EcoRII-N over EcoRII-C (Fig. 5). The same inhibitory effect can be achieved by leaving the protein amount in the sample constant, but successively substituting EcoRII-N for EcoRII-C (data not shown). However, these experiments can not differentiate whether cleavage by EcoRII-C is inhibited by EcoRII-N due to inter-domain interactions or by competitive DNA-binding to EcoRII recognition sites on the DNA.

The structure-derived EcoRII activation model is different from that of NaeI whose effector-binding domains (Topo domains) have neither sequence

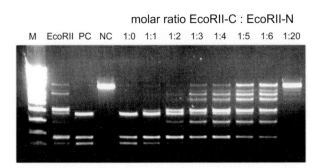

Fig. 5. Inhibition of the DNA cleavage function of EcoRII-C by EcoRII-N *in trans*. 100 ng HindIII-linearized pBR322 DNA (0.21 pmol EcoRII sites) were incubated in 20 µl samples with 0.105 pmol EcoRII-C and increasing molar ratios of EcoRII-N from 1:1 to1:20 for 1 h at 37 °C. BstNI cleavage was performed at 60 °C. DNA fragments were run in a 1 % agarose gel. *M* DNA molecular weight marker IV (Roche); *EcoRII* DNA cleaved with EcoRII wild-type; positive control), BstNI-cleaved DNA; negative control), uncleaved DNA

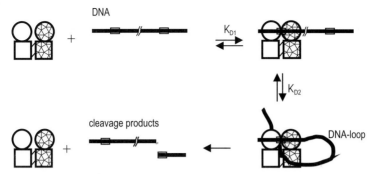

Fig. 6. Type IIE restriction endonuclease interaction with DNA. The two domains of a monomer are symbolized by a *circle* and a *rectangle*, respectively. *White* Monomer A, *black grid* monomer B of the homodimer. *Circles* DNA-binding domain; *rectangle* endonuclease domain. The *lines* between the *rectangles* symbolize dimerization contacts. *Black lines with boxes* DNA with specific DNA recognition sites. DNA first binds to the DNA-binding domains of the dimer. In a second event, the DNA-binding site in the endonuclease-domains becomes occupied. The enzyme forms a DNA loop with DNA containing two recognition sites. The DNA recognition site bound to the endonuclease-domains will be cleaved

nor structural homology with those of EcoRII and are physically distant from the substrate binding cleft (Huai et al. 2000, 2001).

In summary, the complex of the dimeric Type IIE REases with two DNA recognition sites is the active form of the enzymes. Only the DNA recognition site bound to the endonuclease-like domains of the dimer serves as substrate that is cleaved (Fig. 6).

10 Are NaeI and EcoRII Evolutionary Links Between REases, Topoisomerases and DNA Recombinases?

Although both enzymes, NaeI and EcoRII, possess an additional DNA-binding domain, EcoRII was the first REase that after dissection into functional domains revealed a still intact and even more active endonuclease unit that is regulated in the complete enzyme by an autoinhibition mechanism (Mücke et al. 2002a; Zhou et al. 2004). Based on these results, we like to suggest that during evolution the C-terminal domain of EcoRII acquired an additional DNA-binding domain and thus evolved a new protein function. Sequence homology studies among different subtypes of REases have also suggested this acquisition of an additional DNA-binding domain for EcoRII (Pingoud et al. 2002). The newly acquired function for EcoRII is the simultaneous interaction with two identical DNA recognition sites and, moreover, allows to control the "off" and the "on" state of the EcoRII enzyme activity, probably to prevent inappropriate function. Activity will only be achieved in the presence of an activating signal, namely by binding of DNA to the effector domain. In the activated EcoRII the autoinhibitory N-terminal domain fulfils a new function – mainly to achieve DNA binding of the enzyme.

The essential interaction of Type IIE enzymes with two recognition sites to form the catalytically active complex could be considered as a safety mechanism to avoid accidental cleavage of single unmethylated recognition sites in the host genome. Unmethylated sites can occur after incomplete DNA methylation or DNA repair (Bickle and Krüger 1993). Furthermore, the interaction of NaeI and EcoRII with their bipartite complex recognition sequences 5′GCCGGC-(N)$_x$-GCCGGC and 5′CCWGG-(N)$_x$-CCWGG, respectively, also provides the functional benefit of increased DNA specificity. The specificity of NaeI measured as k_{cat}/K_M for a noncognate recognition site was determined to be 10^{11}-fold lower than that for a cognate recognition site. This accuracy of NaeI cleavage is approximately 10^4-fold higher than that of restriction enzymes that require only one site (Yang et al. 1994).

The requirement for two DNA sites relates Type IIE REases to enzymes that participate in cellular processes such as DNA recombination and transposition in a wider sense. Evidence for a connection between EcoRII and these enzymes came from conserved amino acids between EcoRII and the integrase family of site-specific recombinases (Topal and Conrad 1993; Nunes-Düby et al. 1998). However, it was neither possible to isolate a covalent intermediate between EcoRII and DNA, the formation of which is a hallmark of integrase proteins (Topal and Conrad 1993) nor to demonstrate EcoRII-mediated DNA recombination so far.

Moreover, an evolutionary link between REases, topoisomerases, and recombinases has been suggested. The NaeI amino acid sequence uncovered similarity to the active site of human DNA ligase I, except that for Leu43 in NaeI a lysine is essential for the ligase activity. The replacement of Leu43 in NaeI by lysine produced a variant with NaeI-typical DNA topoisomerase activity, which includes DNA cleavage, strand passage, and religation (Jo and Topal 1995). The topoisomerase activity of mutant NaeI-L43K is NaeI-sequence dependent. Intercalating topoisomerase inhibitors completely inhibited DNA cleavage by NaeI-L43K, but only marginally inhibited NaeI wild-type (Jo and Topal 1996a). This single amino acid substitution significantly changed DNA recognition by NaeI. In contrast to NaeI, which recognizes double-stranded DNA, NaeI-L43K preferably interacts with single-stranded and mismatched DNA regions (Jo and Topal 1996b). Because NaeI-L43K changes the linking number of a single negatively supercoiled topoisomer of pBR322 by units of one and decatenated nicked DNA circles, it behaves like a prokaryotic Type I topoisomerase (Jo and Topal 1998). Recently, it has been shown that a positive charge at position Leu43 of NaeI dramatically changes the NaeI protein fold. The changes affect both, the Endo and the Topo domains (Carrick and Topal 2003). The discovery of a covalent

NaeI–DNA intermediate provided further direct evidence for an evolutionary link between Type IIE REases and the recombinase protein family (Jo and Topal 1995). The amino acid residue involved in the covalent contact to DNA has not yet been identified.

Interestingly, endonuclease-like structures have been found in DNA transposases, recombinases and DNA repair enzymes in prokaryota, eukaryota, and archaea, e.g., in the λ-exonuclease (Kovall and Matthews 1998), the MutH protein (Ban and Yang 1998), the Vsr (very short patch repair) endonuclease (Tsutakawa et al. 1999), and a transposase subunit coded by Tn7 (Hickman et al. 2000). Furthermore, sequence and structural alignments predicted evolutionary relationships between Type II REases and eukaryotic non-LTR retrotransposable elements (Yang et al. 1999; Volff et al. 2001), the Hjc resolvase, a recombination enzyme from archaea (Daiyasu et al 2000), the endonuclease VII, a repair enzyme of bacteriophage T4 (Aravind et al. 2000), and homing endonucleases (Bujnicki and Rychlewski 2001). The structural homologies could mean that these DNA–processing enzymes have evolved from a very distant ancestral nuclease fold, and that the function of this hypothetic nuclease was adapted to different fundamental cellular processes. This hypothesis is supported by an evolutionary tree of the endonuclease-like superfamily (Bujnicki 2000). Based on this phylogenetic relationship, it has been proposed that the nonspecific DNA cleavage domain of the Type IIS REase FokI is evolutionarily older than other endonucleases of the same branch. FokI might have acquired a separate domain for specific DNA binding (Bujnicki 2000). Because of the numerous hints for an evolutionary relationship between REases and DNA-processing enzymes, we suppose that acquiring an additional DNA-binding domain could be the first step on the evolutionary way to protein functions beyond phosphodiester bond cleavage.

11 Does Nature Construct Proteins with New Functions by Shuffling Protein Domains?

The described structural homology between protein folds apparently emphasizes the evolutionary link between enzymes involved in site-specific recombination and transposition on the one hand, and REases on the other – a link that might rely on the functions of R-M systems in nature. Besides the commonly accepted explanation that R-M systems counteract virus infections in bacterial cells, two other hypotheses have been discussed to explain the development and maintenance of the impressively high number and diversity of R-M systems during evolution (for reviews, see Arber 2000; Kobayashi 2001). Genetic stability for an individual or for a population at short term is as important as, genetic variability at long term. The endonucleolytic and methylating activities of DNA R-M systems likewise can accomplish both demands (Bickle and Krüger 1993). Arber (2000) referred to R-M genes as "evolution genes" and termed them together with repair genes "frequence modulators" of genetic variation; on the one hand, these gene products restrict the uptake of foreign genetic material into bacterial cells to a tolerable extent. On the other hand, they produce a pool of potential recombinogenic DNA fragments that can be incorporated into the bacterial genome. Comparison of the accumulated bacterial genomic sequences, their GC-content, and codon usage give evidence for the mobility of R-M genes within and between different genomes that often causes rearrangements of genomic sequences (Stein et al. 1998; Chinen et al. 2000; Xu et al. 2000; Kobayashi 2001; Lin et al. 2001). Further evidence for the frequent horizontal transfer of DNA R-M genes includes (1) that they are often flanked by insertion elements, (2) that DNA R-M genes themselves are often located on transposable elements and (3) that they were found in the vicinity of sequences encoding DNA transposases and recombinases (Gunn and Stein 1997; Rochepeau et al. 1997; Kobayashi 2001).

R-M genes were much more often found by genomic approaches than expected, though they are not always expressed. *Helicobacter pylori* strains, for instance, which can colonize the human stomach and are associated with the onset of stomach ulcer and cancer, contain 14 or 15 putative R-M systems corresponding to approximately 1 % of the genome (Tomb et al. 1997; Alm et al. 1999). In this regard, it seems exceptionally surprising that only one phage has been described so far that infects *Helicobacter pylori* strains (Kong et al. 2000). Otherwise, *Helicobacter* is naturally competent to accept the uptake of foreign DNA. Though interstrain-specific horizontal gene transfer is apparently a major source of genetic diversity in *Helicobacter*, the numerous R-M systems might indeed function as a barrier to control the DNA exchange (Ando et al. 2000). Moreover, in the genomes of *Neisseria gonorrhoeae* strains twenty or more putative R-M systems were identified (Roberts 1998). It is hard to believe that all these R-M systems serve as protectors against foreign DNA. Thus, it can be expected that right in these human pathogenic bacterial species, R-M systems influence further biological features, e.g., bacterial pathogenicity. Recently, an interrelation between the expression of the R-M system NmeSI and the hypervirulence of a *Neisseria meningitidis* strain has been described (Bart et al. 2001).

Another hypothesis postulates R-M systems to be selfish genetic elements like viruses or transposons. R-M systems ensure their maintainance within a population at the expense of their host cells, in that they kill cells that have lost the R-M genes. This observation was described for Type II restriction enzymes (Naito et al. 1995). In contrast, loss of Type I R-M systems does not seem to be lethal for the host cell (for a review, see Murray 2002). Rocha et al. (2001) analyzed to what extent small palindromic sequences occur in viral and bacterial genomes. The comparison demonstrated that in most cases bacteria much more counterselected palindromic sequences in their genomes than viruses. This result might support an invasive character of R-M systems and the resulting evolutionary pressure to the bacterial genomes to avoid palindromic sequences for their integration.

Taking into account that R-M genes presumably live like nomades and that the encoded enzymes can integrate DNA fragments into the host's genome, the shuffling of protein domains seems to be an efficient evolutionary strategy to generate proteins with new functions. There is not only evidence that genes have been passed between prokaryotes, but also from vertebrates into bacteria (reviewed by Ponting and Russell 2002). Comprehensive investigations on several protein superfamilies led to the conclusion that nature has re-engineered the same protein fold over and over to provide the manifold protein functions that are needed (for a review, see Babbitt and Gerlt 2001). In either case, Type IIE REases benefit from their additional DNA-binding domains in that they acquired the capability to interact simultaneously with two recognition sites. This is at the same time a striking feature of enzymes participating in cellular site-specific recombination and transposition processes. Achieving the capacity to transpose DNA could be regarded as a great evolutionary advantage for Type IIE REases because that way they could accomplish horizontal distribution of their encoding genes, like transposable elements.

12 Activation of Type IIE REases for Biotechnological Purposes and for Mapping Epigenetic DNA Modifications

Besides the prototypes EcoRII and NaeI, several other REases originating from different microorganisms were found to require the interaction with two recognition sites in the DNA for their endonucleolytic activity (Oller et al. 1991; Reuter et al. 1993). REases AtuBI, BspMI, Cfr9I, HpaII, NarI, SacII, SauBMKI, Eco57I, and Ksp632I can be stimulated by addition of synthetic oligonucleotide duplexes containing the specific recognition sequence to cleave all copies of the recognition sequence in a DNA molecule. In the case of BspMI, addition of cleavable sites *in trans* led to an increase of V_{max}, and in the case of HpaII, NarI, and SacII to an increase of K_m, respectively (Oller et al. 1991). It is conceivable that the enzymatic activity of these enzymes also significantly depends on the recognition site frequency and the distance between recognition sites. For the tetrameric Type IIS restriction endonuclease BspMI, it was reported that it acts concurrently with two DNA sites and cuts both DNA strands at both sites (Gormley et al. 2002). Nevertheless, detailed investigations on the other enzymes have to be done.

Inasmuch as these REases, as EcoRII and NaeI, are commonly used as tools in molecular biological research, the described method of enzyme activation is valuable to provide complete DNA digests. Exploiting the experiences with stimulation of EcoRII and NaeI by modified oligonucleotide duplexes (Pein et al. 1991; Conrad and Topal 1992), this method was commercialized for NaeI with an oligonucleotide duplex containing a modified NaeI recognition sequence that is not cleaved itself and offered as Turbo NaeI (Senesac and Allen 1995).

Genome mapping, determination of restriction fragment length polymorphisms, and DNA methylation pattern analyses rely on the complete cleavage of all unmethylated specific recognition sites in a DNA molecule. Using some of the described REases for these procedures may lead to misinterpretations of the DNA cleavage patterns, e.g., in DNA methylation analyses when uncleaved DNA recognition sites are stringently associated with an interfering DNA methylation. It was shown that such potentially misinterpreted results can appear during detection of Dcm methylation (5'Cm5CA/TGG) in bacterial DNA with the isoschizomeric REase pair EcoRII and BstNI as well as the detection of eukaryotic 5'-C5 mCGG methylation with the pair HpaII and MspI (Reuter et al. 1990; Kupper et al. 1997). Methylation-sensitive REases EcoRII and HpaII belong to the enzymes that only incompletely cleave DNA depending on the frequency and on the distance of the recognition sites and thus, they become sources of errors in determining DNA methylation patterns. This experimental problem was solved by adding site-specific oligonucleotide duplexes to the EcoRII and HpaII restriction digests, respectively. The DNA activator molecules stimulated DNA cleavage at all unmethylated recognition sites leaving the resistance of methylated DNA recognition sites unaffected (Reuter et al. 1990; Kupper et al. 1997).

Furthermore, the method of REase stimulation was adapted for cleavage of agarose-embedded high-molecular DNA (Kupper et al. 1999). Thus, activation of methylation-sensitive REases HpaII and EcoRII by cognate site-containing oligonucleotide duplexes improves the reliability of CpG and CpNpG cytosine methylation detection in small as well as in large DNA molecules. This approach is of some importance for the correct determination of epigenetic signals in current genomics.

13 Final Remarks

During the last years several other Type II REases with different oligomeric structures and reaction mechanisms, thus belonging to different subtypes of REases, were reported to require two DNA recognition sites for exertion of catalytic activity. Moreover, the functional dependence on two copies of the recognition sequence in the substrate DNA is also shared by REases of Types I, III, and IV, albeit the different enzymes follow different molecular mechanisms to interact with the DNA (reviewed by Roberts et al. 2003).

Acknowledgements. We gratefully acknowledge Carmen Bunn, Petra Mackeldanz, Elisabeth Möncke-Buchner, Cordula Petter, and Ursula Scherneck for performing experiments, the results of which are presented in this article. The work in the lab of the authors was supported by Deutsche Forschungsgemeinschaft (Re 879/2, Kr 1293/1), Bundesministerium für Forschung und Technik (Grant No. 0311014), Sonnenfeld-Stiftung, and Universitäre Forschungsförderung of the Humboldt University Medical School Charité.

References

Alm RA, Ling LS, Moir DT, King BL, Brown ED, Doig PC, Smith DR, Noonan B, Guild BC, de Jonge BL et al. (1999) Genomic-sequence comparison of two unrelated isolates of the human gastric pathogen *Helicobacter pylori*. Nature 397:176–180

Ando T, Xu Q, Torres M, Kusugami K, Israel DA, Blaser MJ (2000) Restriction-modification system differences in *Helicobacter pylori* are a barrier to interstrain plasmid transfer. Mol Microbiol 37:1052–1065

Aravind L, Makarova KS, Koonin EV (2000) Holliday junction resolvases and related nucleases: identification of new families, phylogenetic distribution and evolutionary trajectories. Nucleic Acids Res 28:3417–3432

Arber W (1966) Host-specificity of DNA produced by *Escherichia coli*. IX. Host-controlled modification of bacteriophage fd. J Mol Biol 20:483–496

Arber W (1974) DNA modification and restriction. Progr Nucl Acid Res Mol Biol 14:1–37

Arber W (2000) Genetic variation: molecular mechanisms and impact on microbial evolution. FEMS Microbiol Rev 24:1–7

Arber W, Morse (1965) Host specificity of DNA produced by *Escherichia coli*. VI. Effects on bacterial conjugation. Genetics 51:137–148

Arber W, Wauters-Willems D (1970) Host specificity of DNA produced by *Escherichia coli*. XII. The two restriction and modification systems of strain 15T-. Mol Gen Genet 108:203–217

Babbitt PC, Gerlt JA (2001) New functions from old scaffold: how nature reengineers enzymes for new functions. Adv Protein Chem 55:1–28

Ban C, Yang W (1998) Structural basis for MutH activation in *E. coli* mismatch repair and relationship of MutH to restriction endonucleases. EMBO J 17:1526–1534

Bannister D, Glover SW (1968) Restriction and modification of bacteriophage by R+ strains of *Escherichia coli* K-12. Biochem Biophys Res Commun 30:735–738

Bannister D, Glover SW (1970) The isolation and properties of non-restricting mutants of two different host specificities associated with drug resistance factors. J Gen Microbiol 61:63–71

Bart A, Pannekoek Y, Dankert J, van der Ende A (2001) NmeSI restriction-modification system identified by representational difference analysis of a hypervirulent *Neisseria meningitidis* strain. Infect Immun 69:1816–1820

Baxter BK, Topal MD (1993) Formation of a cleavasome: Enhancer DNA-2 stabilizes an active conformation of NaeI dimer. Biochemistry 32:8291–8298

Bickle TA, Krüger DH (1993) Biology of DNA restriction. Microbiol Rev 57:434–450

Bhagwat AS, Johnson B, Weule K, Roberts RJ (1990) Primary sequence of the EcoRII endonuclease and properties of its fusions with beta-galactosidase. J Biol Chem 265:767–773

Bigger CH, Murray K, Murray NE (1973) Recognition sequence of a restriction enzyme. Nat New Biol 244:7–10

Bitinaite J, Schildkraut I (2002) Self-generated DNA termini relax the specificity of SgrAI restriction endonuclease. Proc Natl Acad Sci USA 99:1164–1169

Bogdarina IG, Buryanov YI, Bayev AA (1979) Isolation and properties of DNA-cytosine methyltransferases EcoRII and E. coli K-12 (in Russian). Biochemistry (Moscow) 44:440–452

Boyer HW (1971) DNA restriction and modification mechanisms in bacteria. Annu Rev Microbiol 25:153–176

Boyer HW, Chow LT, Dugaiczyk A, Hedgpeth J, Goodman HM (1973) DNA substrate site for the EcoRII restriction endonuclease and modification methylase. Nat New Biol 244:40–43

Bujnicki JM (2000) Phylogeny of the restriction endonuclease-like superfamily inferred from comparison of protein structures. J Mol Evol 50:39–44

Bujnicki JM, Rychlewski L (2001) Unusual evolutionary history of the tRNA splicing endonuclease EndA: Relationship to the LAGLIDADG and PD-(D/E)XK deoxyribonucleases. Protein Sci 10:656–660

Buryanov YI, Bogdarina IG, Baev AA (1978) Site specificity and chromatographic properties of E. coli K12 and EcoRII DNA-cytosine methylases. FEBS Lett 88:251–254

Carrick KL, Topal MD (2003) Amino acid substitutions at position 43 of NaeI endonuclease – evidence for changes in NaeI structure. J Biol Chem 278:9733–9739

Cech D, Pein CD, Kubareva EA, Gromova ES, Oretskaya TS, Shabarova ZA (1988) The influence of modifications on the cleavage of oligonucleotide duplexes by EcoRII and MvaI endonucleases. Nucleosides Nucleotides 7:585–588

Chinen A, Uchiyama I, Kobayashi I (2000) Comparison between *Pyrococcus horikoshii* and *Pyrococcus abyssi* genome sequences reveals linkage of restriction-modification genes with large genome polymorphisms. Gene 259:109–121

Colandene JD, Topal MD (1998) The domain organization of NaeI endonuclease: Separation of binding and catalysis. Proc Natl Acad Sci USA 95:3531–3536

Colandene JD, Topal MD (2000) Evidence for mutations that break communication between the Endo and Topo domains in NaeI endonuclease/topoisomerase. Biochemistry 39:13703–13707

Conrad M, Topal MD (1989) DNA and spermidine provide a switch mechanism to regulate the activity of restriction enzyme NaeI. Proc Natl Acad Sci USA 86:9707–9711

Conrad M, Topal MD (1992) Modified DNA fragments activate NaeI cleavage of refractory DNA sites. Nucleic Acids Res 20:5127–5130

Daiyasu H, Komori K, Sakae S, Ishino Y, Toh H (2000) Hjc resolvase is a distantly related member of the Type II restriction endonuclease family. Nucleic Acids Res 28:4540–4543

Deibert M, Grazulis S, Janulaitis A, Siksnys V, Huber R (1999) Crystal structure of MunI restriction endonuclease in complex with cognate DNA at 1.7 Å resolution. EMBO J 18:5805–5816

Deibert M, Grazulis S, Sasnauskas G, Siksnys V, Huber R (2000) Structure of the tetrameric restriction endonuclease NgoMIV in complex with cleaved DNA. Nat Struct Biol 7:792–799

Dröge P (1994) Protein tracking-induced supercoiling of DNA: a tool to regulate DNA transactions in vivo? BioEssays 16:91–99

Embleton ML, Siksnys V, Halford SE (2001) DNA cleavage reactions by Type II restriction enzymes that require two copies of their recognition site. J Mol Biol 311:503–511

Gabbara S, Bhagwat AS (1992) Interaction of EcoRII endonuclease with DNA substrates containing single recognition sites. J Biol Chem 267:18623–18630

Glover SW, Schell J, Symonds N, Stacey KA (1963) The control of host-induced modification by phage P1. Genet Res 4:480–482

Gormley NA, Hilberg AL, Halford SE (2002) The Type IIs restriction endonuclease BspMI is a tetramer that acts concertedly at two copies of an asymmetric DNA sequence. J Biol Chem 277:4034–4041

Gunn JS, Stein DC (1997) The *Neisseria gonorrhoeae* S.NgoVIII restriction/modification system: a Type IIs system homologous to the *Haemophilus parahaemolyticus* HphI restriction/modification system. Nucleic Acids Res 25:4147–4152

Hattman S (1972) Plasmid-controlled variation in the content of methylated bases in bacteriophage lambda deoxyribonucleic acid. J Virol 10:356–361

Hattman S (1973) Plasmid-controlled variation in the content of methylated bases in single-stranded DNA phages M13 and fd. J Mol Biol 74:749–752

Hattman S (1977) Partial purification of the *E. coli* K-12 mec+ DNA-cytosine methylase: in vitro methylation completely protects bacteriophage lambda DNA against cleavage by R.EcoRII. J Bacteriol 129:1330–1334

Hattman S, Gribbin C, Hutchison CA (1979) In vivo methylation of bacteriophage phi XI74 DNA. J Virol 32:845–851

Hickman AB, Li Y, Mathew SV, May EW, Craig NL, Dyda F (2000) Unexpected structural diversity in DNA recombination: the restriction endonuclease connection. Mol Cell 5:1025–1034

Holtz JK, Topal MD (1994) Location of putative binding and catalytic sites of NaeI by random mutagenesis. J Biol Chem 269:27286–27290

Huai Q, Colandene JD, Chen, Y, Topal MD, Ke H (2000) Crystal structure of NaeI – an evolutionary bridge between DNA endonuclease and topoisomerase. EMBO J 19:3110–3118

Huai Q, Colandene JD, Topal MD, Ke H (2001) Structure of NaeI-DNA complex reveals dual-mode DNA recognition and complete dimer rearrangement. Nat Struct Biol 8:665–669

Jeltsch A, Alves J, Wolfes H, Maass G, Pingoud A (1994) Pausing of the restriction endonuclease EcoRI during linear diffusion on DNA. Biochemistry 33:10215–10219

Jo K, Topal MD (1995) DNA topoisomerase and recombinase activities in NaeI restriction endonuclease. Science 267:1817–1820

Jo K, Topal MD (1996a) Changing a leucine to a lysine residue makes NaeI endonuclease hypersensitive to DNA intercalative drugs. Biochemistry 35:10014–10018

Jo K, Topal MD (1996b) Effects on NaeI-DNA recognition of the leucine to lysine substitution that transforms restriction endonuclease NaeI to a topoisomerase: a model for restriction endonuclease evolution. Nucleic Acids Res 24:4171–4175

Jo K, Topal MD (1998) Step-wise DNA relaxation and decatenation by NaeI-L43K. Nucleic Acids Res 26:2380–2384

Karpova EA, Kubareva EA, Shabarova ZA (1999a) A model of EcoRII restriction endonuclease action: the active complex is most likely formed by one protein subunit and one DNA recognition site. IUBMB Life 48:91–98

Karpova EA, Meehan E, Pusey ML, Chen L (1999b) Crystallization and preliminary X-ray diffraction of restriction endonuclease EcoRII. Acta Cryst D 55:1604–1605

Kobayashi I (2001) Behavior of restriction-modification systems as selfish mobile elements and their impact on genome evolution. Nucleic Acids Res 29:3742–3756

Kong H, Lin LF, Porter N, Stickel S, Byrd D, Posfai J, Roberts RJ (2000) Functional analysis of putative restriction-modification system genes in the *Helicobacter pylori* J99 genome. Nucleic Acids Res 28:3216–3223

Kosykh VG, Buryanov YI, Bayev AA (1980) Molecular cloning of EcoRII endonuclease and methylase genes. Mol Gen Genet 178:717–718

Kosykh VG, Puntezhis SA, Buryanov YI, Bayev AA (1982) Isolation, purification, and characterization of restriction endonuclease EcoRII (in Russian). Biochemistry (Moscow) 47:619–625

Kosykh V, Repyk A, Kaliman A, Buryanov YI (1989) Nucleotide sequence of the EcoRII restriction endonuclease gene. Biochim Biophys Acta 1009:290–292

Kosykh VG, Repyk AV, Hattman S (1993) Sequence motifs common to the EcoRII restriction endonuclease and three DNA-(cytosine-C5) methyltransferases. Gene 125:65–68

Kovall RA, Matthews BW (1998) Structural, functional, and evolutionary relationships between λ-exonuclease and the Type II restriction endonucleases. Proc Natl Acad Sci USA 95:7893–7897

Krüger DH, Reuter M (1999) Host-controlled modification and restriction. In: Webster RG, Granoff A (eds) Encyclopedia of virology, 2nd edn. Academic Press, New York, pp 758–763

Krüger DH, Schroeder C, Reuter M., Bogdarina IG, Buryanov YI, Bickle TA (1985) DNA methylation of bacterial viruses T3 and T7 by different DNA methylases in *Escherichia coli* K12 cells. Eur J Biochem 150:323–330

Krüger DH, Barcak GJ, Reuter M, Smith HO (1988) EcoRII can be activated to cleave refractory DNA restriction sites. Nucleic Acids Res 16:3997–4008

Krüger DH, Schroeder C, Santibanez-Koref M, Reuter M (1989) Avoidance of DNA methylation: A virus-encoded methylase inhibitor and evidence for counterselection of methylase recognition sites in viral genomes. Cell Biophys 15:87–95

Krüger DH, Prösch S, Reuter M, Goebel W (1990) Cloning of the resistant EcoRII recognition site of phage T7 into an EcoRII-sensitive plasmid makes the site suceptible to the restriction enzyme. J Basic Microbiol 30:679–683

Kupper D, Reuter M, Mackeldanz P, Meisel A, Alves J, Schroeder C, Krüger DH (1995) Hyperexpressed EcoRII renatured from inclusion bodies and native enzyme both exhibit essential cooperativity with two DNA sites. Protein Expr Purific 6:1–9

Kupper D, Reuter M, Meisel A, Krüger DH (1997) Reliable detection of DNA CpG methylation profiles by the isoschizomers MspI/HpaII using oligonucleotide stimulators. BioTechniques 23:843–847

Kupper D, Möncke-Buchner E, Reuter M, Krüger DH (1999) Oligonucleotide stimulators allow complete cleavage of agarose-embedded DNA by particular Type II restriction endonucleases. Anal Biochem 272:275–277

Lin LF, Posfai J, Roberts RJ, Kong H (2001) Comparative genomics of the restriction-modification systems in *Helicobacter pylori*. Proc Natl Acad Sci USA 98:2740–2745

Meisel A, Mackeldanz P, Bickle TA, Krüger DH, Schroeder C (1995) Type III restriction endonucleases translocate DNA in a reaction driven by recognition site-specific ATP hydrolysis. EMBO J 14:2958–2966

Meselson M, Yuan R, Heywood J (1972) Restriction and modification of DNA. Annu Rev Biochem 41:447–466

Milsom SE, Halford SE, Embleton ML, Szczelkun MD (2001) Analysis of DNA looping interactions by Type II restriction enzymes that require two copies of their recognition sites. J Mol Biol 311:515–527

Morgan R, Xiao JP, Xu SY (1998) Characterization of an extremely thermostable restriction enzyme, PspGI, from a *Pyrococcus* strain and cloning of the PspGI restriction-modification system in *Escherichia coli*. Appl Environ Microbiol 64:3669–3673

Murray NE (2002) Immigration control of DNA in bacteria: self versus non-self. Microbiology 148:3–20

Mücke M, Lurz R, Mackeldanz P, Behlke J, Krüger DH, Reuter M (2000) Imaging DNA loops induced by EcoRII: A single amino acid substitution uncouples target recognition from cooperative DNA interaction and cleavage. J Biol Chem 275:30631–30637

Mücke M, Grelle G, Behlke J, Kraft R, Krüger DH, Reuter M (2002a) EcoRII: A restriction enzyme evolving recombination functions? EMBO J 21:5262–5268

Mücke M, Pingoud V, Grelle G, Behlke J, Kraft R, Krüger DH, Reuter M (2002b) Asymmetric photocross-linking pattern of restriction endonuclease EcoRII to the DNA recognition sequence. J Biol Chem 277:14288–14293

Naito T, Kusano K, Kobayashi I (1995) Selfish behavior of restriction-modification systems. Science 267:897–899

Nunes-Düby SE, Kwon HJ, Tirumalai RS, Ellenberger T, Landy A (1998) Similarities and differences among 105 members of the Int family of site-specific recombinases. Nucleic Acids Res 26:391–406

Oller AR, Van den Broek W, Conrad M, Topal MD (1991) Ability of DNA and spermidine to affect the activity of restriction endonucleases from several bacterial species. Biochemistry 30:2543–2549

Pein CD, Reuter M, Cech D, Krüger DH (1989) Oligonucleotide duplexes containing CC(A/T)GG stimulate cleavage of refractory DNA by restriction endonuclease EcoRII. FEBS Lett 245:141–144

Pein CD, Reuter M, Meisel A, Cech D, Krüger DH (1991) Activation of restriction endonuclease EcoRII does not depend on the cleavage of stimulator DNA. Nucleic Acids Res 19:5139–5142

Petrauskene OV, Kubareva EA, Gromova ES, Shabarova ZA (1992) Mechanism of the interaction of EcoRII restriction endonuclease with two recognition sites. Probing of modified DNA duplexes as activators of the enzyme. Eur J Biochem 208:617–622

Petrauskene OV, Karpova EA, Gromova ES, Guschlbauer W (1994) Two subunits of EcoRII restriction endonuclease interact with two DNA recognition sites. Biochem Biophys Res Comm 198:885–890

Petrauskene OV, Schmidt S, Karyagina AS, Nikolskaya II, Gromova ES, Cech D (1995) The interaction of DNA duplexes containing 2-aminopurine with restriction endonucleases EcoRII and SsoII. Nucleic Acids Res 23:2192–2197

Petrauskene OV, Tashlitsky VN, Brevnov MG, Backmann J, Gromova ES (1996) Kinetic modeling of the mechanism of allosteric interaction of restriction endonuclease EcoRII with two DNA sites (in Russian). Biochemistry (Moscow) 61:1257–1269

Petrauskene OV, Babkina OV, Tashlitsky VN, Kazankov GM, Gromova ES (1998) EcoRII endonuclease has two identical DNA-binding sites and cleaves one of two co-ordinated recognition sites in one catalytic event. FEBS Lett 425:29–34

Petter C (2001) Aufklärung der Struktur und Domänenorganisation der Restriktionsendonuklease EcoRII: Identifizierung von potentiellen Protein-Protein-Kontaktstellen. Diplomarbeit, Martin-Luther-Universität Halle-Wittenberg/Humboldt-Universität Berlin

Pingoud A, Jeltsch A (2001) Structure and function of Type II restriction endonucleases. Nucleic Acids Res 29:3705–3727

Pingoud V, Kubareva E, Stengel G, Friedhoff P, Bujnicki JM, Urbanke C, Sudina A, Pingoud A (2002) Evolutionary relationship between different subgroups of restriction endonucleases. J Biol Chem 277:4306–4314

Ponting CP, Russell RR (2002) The natural history of protein domains. Ann Rev Biophys Biomol Struct 31:45–71

Pufall MA, Graves BJ (2002) Autoinhibitory domains: Modular effectors of cellular regulation. Annu Rev Cell Dev Biol 18:421–462

Reuter M (1985) Wirtskontrollierte Modifikation und Restriktion von Bakterienviren auf DNA- und Proteinebene. Dissertation, Humboldt-Universität Berlin

Reuter M, Pein CD, Butkus V, Krüger DH (1990) An improved method for the detection of Dcm methylation in DNA molecules. Gene 95:161–162

Reuter M, Kupper D, Pein CD, Petrusyte M, Siksnys V, Frey B, Krüger DH (1993) Use of specific oligonucleotide duplexes to stimulate cleavage of refractory DNA sites by restriction endonucleases. Anal Biochem 209:232–237

Reuter M, Kupper D, Meisel A, Schroeder C, Krüger, DH (1998) Cooperative binding properties of restriction endonuclease EcoRII with DNA recognition sites. J Biol Chem 273:8294–8300

Reuter M, Schneider-Mergener J, Kupper D, Meisel A, Mackeldanz P, Krüger DH, Schroeder C (1999) Regions of endonuclease EcoRII involved in DNA target recognition identified by membrane-bound peptide repertoires. J Biol Chem 274:5213–5221

Roberts RJ (1998) Bioinformatics: A new world of restriction and modification enzymes. The NEB Transcript 9:1–4

Roberts RJ, Belfort M, Bestor T et al. (2003) A nomenclature for restriction enzymes, DNA methyltransferases, homing endonucleases and their genes. Nucleic Acids Res 31:1805–1812

Rocha EPC, Danchin A, Viari A (2001) Evolutionary role of restriction/modification systems as revealed by comparative genome analysis. Genome Res 11:946–958

Rochepeau P, Selinger LB, Hynes MF (1997) Transposon-like structure of a new plasmid-encoded restriction-modification system in *Rhizobium leguminosarum* VF39SM. Mol Gen Genet 256:387–396

Schleif R (1992) DNA looping. Annu Rev Biochem 61:199–223

Schroeder C, Jurkschat H, Meisel A, Reich JG, Krüger DH (1986) Unusual occurrence of EcoPl and EcoP15 recognition sites and counterselection of Type II methylation and restriction sequences in bacteriophage T7 DNA. Gene 45:77–86

Senesac JH, Allen JR (1995) Oligonucleotide activation of the Type IIe restriction enzyme NaeI for digestion of refractory sites. BioTechniques 19:990–993

Siksnys V, Skirgaila R, Sasnauskas G, Urbanke C, Cherny D, Grazulis S, Huber R (1999) The Cfr10I restriction enzyme is functional as a tetramer. J Mol Biol 291:1105–1118

Som S, Bhagwat AS, Friedman S (1987) Nucleotide sequence and expression of the gene encoding the EcoRII modification enzyme. Nucleic Acids Res 15:313–332

Stein DC, Gunn JS, Piekarowicz A (1998) Sequence similarities between the genes encoding the S.NgoI and HaeII restriction-modification systems. Biol Chem 379:575–578

Takano T, Watanabe T, Fukasawa T (1968) Mechanism of host-controlled restriction of bacteriophage lambda by R-factors in *Escherichia coli* K-12. Virology 34:290–302

Tamulaitis G, Solonin AS, Siksnys V (2002) Alternative arrangements of catalytic residues at the active sites of restriction enzymes. FEBS Lett 518:17–22

Tomb JF, White O, Kerlavage AR, Clayton RA, Sutton GG, Fleischmann RD, Ketchum KA, Klenk HP, Gill S, Dougherty BA et al. (1997) The complete genome sequence of the gastric pathogen *Helicobacter pylori*. Nature 388:539–547

Topal MD, Thresher RJ, Conrad M, Griffith J (1991) NaeI endonuclease binding to pBR322 DNA induces looping. Biochemistry 30:2006–2010

Topal MD, Conrad M (1993) Changing endonuclease EcoRII Tyr308 to Phe abolishes cleavage but not recognition: possible homology with the Int-family of recombinases. NucleicAcids Res 21:2599–2603

Tsutakawa SE, Jingami H, Morikawa K (1999) Recognition of a TG mismatch: the crystal structure of very short patch repair endonuclease in complex with a DNA duplex. Cell 99:615–623

van der Woerd MJ, Pelletier JJ, Xu S, Friedman AM (2001) Restriction enzyme BsoBI-DNA complex: a tunnel for recognition of degenerate DNA sequences and potential histidine catalysis. Structure (Camb). 9:133–144

Vinogradova MV, Gromova ES, Kosykh VG, Buryanov YI, Shabarova ZA (1990) Interaction of EcoRII restriction and modification enzymes with synthetic DNA fragments. Determination of the size of EcoRII binding site (in Russian). Mol Biol (Moscow) 24:847–850

Volff JN, Korting CV, Froschauer A, Sweeney K, Schartl M (2001) Non-LTR retrotransposons encoding a restriction enzyme-like endonuclease in vertebrates. J Mol Evol 52:351–360

Vovis GF, Horiuchi K, Zinder ND (1975) Endonuclease R-EcoRII restriction of bacteriophage f1 DNA in vitro: ordering of genes V and VII, location of an RNA promoter for gene VIII. J Virol 16:674–684

Wah DA, Hirsch JA, Dorner LF, Schildkraut I, Aggarwal AK (1997) Structure of the multimodular endonuclease FokI bound to DNA. Nature 388:97–100

Wang JC, Giaever GN (1988) Action at a distance along a DNA. Science 240:300–304

Watanabe T, Nishida H, Ogata C, Arai T, Sato S (1964) Episome-mediated transfer of drug resistance in Enterobacteriaceae. VII. Two types of naturally occurring R factors. J Bacteriol 88:716–726

Xu Q, Morgan RD, Roberts RJ, Blaser MJ (2000) Identification of Type II restriction and modification systems in *Helicobacter pylori* reveals their substantial diversity among strains. Proc Natl Acad Sci USA 97:9671–9676

Yang CC, Topal MD (1992) Non-identical DNA-binding sites of endonuclease NaeI recognize different families of sequences flanking the recognition site. Biochemistry 31:9657–9664

Yang CC, Baxter BK, Topal MD (1994) DNA cleavage by NaeI: Protein purification, rate-limiting step, and accuracy. Biochemistry 33:14918–14925

Yang J, Malik HS, Eickbush TH (1999) Identification of the endonuclease domain encoded by R2 and other site-specific, non-long terminal repeat retrotransposable elements. Proc Natl Acad Sci USA 96:7847–7852

Yolov AA, Gromova ES, Kubareva EA, Potapov VK, Shabarova ZA (1985) Interaction of EcoRII restriction and modification enzymes with synthetic DNA fragments. V. Study of single-strand cleavages. Nucleic Acids Res 13:8969–8981

Yoshimori R, Roulland-Dussoix D, Boyer HW (1972) R factor-controlled restriction and modification of deoxyribonucleic acid: restriction mutants. J Bacteriol 112:1275–1279

Zhou X, Reuter M, Meehan EJ, Chen L (2002) A new crystal form of restriction endonuclease EcoRII that diffracts to 2.8 Å resolution. Acta Crystallogr D 58:1343–1345

Zhou XE, Wang Y, Reuter M, Mackeldanz P, Krüger DH, Meehan EJ, Chen L (2003) A single mutation of restriction endonuclease EcoRII led to a new crystal form that diffracts to 2.1 Å resolution. Acta Crystallogr D 59:910–912

Zhou XE, Wang Y, Reuter M, Mücke M, Krüger DH, Meehan EJ, Chen L (2004) Crystal structure of Type IIE restriction endonuclease EcoRII reveals an autoinhibition mechanism by a novel effector fold. J Mol Biol 335:307–319

Analysis of Type II Restriction Endonucleases that Interact with Two Recognition Sites

A.J. WELSH, S.E. HALFORD, D.J. SCOTT

1 Introduction

Many students of molecular biology know for certain that the Type II restriction endonucleases are dimeric proteins that recognise a palindromic DNA sequence, 4–8 bp (base pairs) long, and cut this single sequence with exquisite specificity. This idea is also propagated by many text-books and commonly-used laboratory manuals in molecular biology (viz. Sambrook and Russell 2001). Like most generalisations, it is in fact wrong. The Type II restriction enzymes encompass not only dimeric proteins but also monomers, tetramers and higher assemblies. Their recognition sequences are not always unique palindromes but can instead be degenerate, discontinuous or asymmetric sequences (Roberts et al. 2003b). The sites at which they cleave the DNA are not necessarily within the recognition sequence. Many cleave the DNA at fixed positions several bp away from their recognition sequence. Most excitingly, many of the Type II enzymes have to interact with two copies of their recognition sites before they can cleave DNA (Halford 2001). When these enzymes bind two sites in the same DNA molecule, the intervening DNA is sequestered in a loop. DNA-looping interactions play central roles in many genetic processes (Schleif 1992; Rippe et al. 1995), such as replication, recombination and transcription, but these processes are often difficult to study. In contrast, the reactions of restriction enzymes can be monitored by a variety of simple techniques, so these enzymes make good tools for analysing the nature of long-range interactions between distant DNA sites. Indeed, the restriction enzymes that act at two DNA sites have become one of the principal paradigms for studying such interactions.

A.J. Welsh, S.E. Halford, D.J. Scott
Department of Biochemistry, School of Medical Sciences, University Walk, University of Bristol, Bristol BS8 1TD, UK

Nucleic Acids and Molecular Biology, Vol. 14
Alfred Pingoud (Ed.)
Restriction Endonucleases
© Springer-Verlag Berlin Heidelberg 2004

2 Different Classes of Interactions

The Type II restriction endonucleases have been divided into several subsets (Roberts et al. 2003a), a few of which are shown in Fig. 1. One subset, the Type IIP enzymes, contains the enzymes that recognise palindromic DNA and act at individual copies of their target sequence. These are normally dimeric proteins that bind symmetrically to their recognition sites, so that the active site from one subunit is positioned to cleave one strand of the DNA and likewise the second subunit with the complementary strand (Pingoud and Jeltsch 2001). However, the enzymes in this subset, such as EcoRI, EcoRV and BamHI, now seem increasingly to be the exception to the general rule for Type II restriction enzymes. The enzymes in all of the other subsets shown in Fig. 1 require two copies of their recognition sites for their DNA cleavage reactions (Halford 2001).

The Type IIS enzymes are grouped as a subset on the basis of their recognition sites, asymmetric sequences that are cleaved fixed distances away from the site (Roberts et al. 2003a). For example, FokI, the prototype of this group, recognises the sequence GATGG and cleaves the DNA downstream of this site, 9 and 13 bases away in top and bottom strands respectively. However, FokI cleaves DNA with one copy of this sequence at a very slow rate but it cleaves DNA with two copies at a much faster rate (Bath et al. 2002). Indeed, most

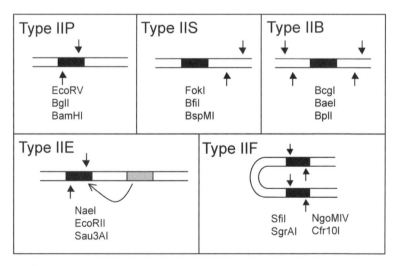

Fig. 1. Different subsets of Type II restriction endonucleases. Each *panel* indicates the mode of action of the Type II restriction enzymes within a particular subset, and lists examples of the enzymes within that group. In all panels, duplex DNA is represented as *parallel lines*, with the recognition sequence(s) as *filled (black)* segment(s); for the Type IIE subset, the *grey* segment marks the copy of the recognition sequence that functions as the activator. *Vertical arrows* mark the sites of DNA cleavage on each strand

Type IIS enzymes cleave DNA with two target sites more rapidly than DNA with one target, though they use a variety of different mechanisms (Soundararajan et al. 2002; Gormley et al. 2002; Lagunavicius et al. 2003).

In the case of FokI, the protein exists in solution as a monomer and the monomer recognises its asymmetric recognition site in its entirety, but this monomer has a single active site capable of cleaving at most one strand of the DNA (Wah et al. 1997). To cut both strands, the monomer associates with a second subunit to form a dimer with two active sites (Wah et al. 1998; Bitinate et al. 1998; Vanamee et al. 2001). However, FokI monomers bound to separate sites in a DNA with two FokI sites associate with each other much more readily than monomers in free solution or monomers bound to separate DNA molecules. Even so, the dimer formed on the DNA with two recognition sites cleaves the DNA at only one site, like the Type IIE enzymes noted below (Bath et al. 2002). MboII also acts in this manner (Soundararajan et al. 2002).

BfiI, on the other hand, another Type IIS enzyme that interacts with two sites, is a dimeric protein but this dimer has only one active site, at the interface between the subunits (Lagunavicius et al. 2003). Unlike other restriction enzymes, BfiI does not need divalent metal ions (Sapranauskas et al. 2000) and has a radically different active site from the Mg^{2+}-dependent enzymes. BfiI shows its maximal activity only when both subunits have bound separate copies of the recognition sequence. Hence, it is more active against DNA with two BfiI sites than DNA with one site but it cleaves the two-site DNA in four sequential steps, each cutting one strand (Lagunavicius et al. 2003).

BspMI is another example of a Type IIS enzyme that requires two copies of its recognition sequence. In contrast to FokI and to BfiI, monomeric and dimeric proteins respectively, BspMI is a tetramer of identical subunits (Gormley et al. 2002). Moreover, while the reactions of BfiI and FokI result in the cleavage of, respectively, one or two phosphodiester bonds, BspMI cuts four phosphodiester bonds per turnover. It converts a DNA with two copies of its recognition sequence directly to the final products cut in both strands at both sites. The tetrameric protein presumably contains four active sites that each cleave one phosphodiester bond, like the Type IIF enzymes noted below.

The Type IIB restriction enzymes (Fig. 1) are also grouped as a subset on the basis of their recognition sequences. They recognise interrupted asymmetric sequences and make double-stranded breaks on both sides of the target sequence, to liberate a small fragment of DNA of about 30 bp that contains the intact recognition site. For example, BcgI, the prototype of this group, recognises the sequence $CGA(N_6)TGC$ (where N is any base) and cuts the DNA in both strands on both the "left" and the "right" of this sequence, in both cases 10 bases away in one strand and 12 bases away in the other (Roberts et al. 2003b). Even though it cleaves four phosphodiester bonds at each copy of its recognition site, BcgI acts better still on DNA with two copies of its target sequence (Kong and Smith, 1998). Moreover, enhanced activity against DNA with two sites appears to be a feature of almost all of the Type IIB enzymes (DM Gowers and J.J.T. Marshall, pers. comm.). While most Type II enzymes are active as oligomers of identical subunits, BcgI is composed of different polypeptides and probably exists as an A_4B_2 hexamer, though its interactions with two DNA sites may involve higher-order oligomers (Kong 1998).

The Type IIE subset of restriction enzymes (Fig. 1) is, however, defined by a reaction mechanism rather the nature of the recognition site. The defining feature of the Type IIE enzymes is that they interact with two copies of their respective target sequence but cleave only one copy: the second copy acts as an enhancer for the DNA cleavage reaction at the first copy (Krüger et al. 1988; Oller et al. 1991) The best characterised of the Type IIE enzymes, EcoRII and NaeI, are dimeric proteins with two DNA-binding clefts (Yang and Topal 1992; Gabbara and Bhagwhat 1992; Reuter et al. 1998). The two clefts bind the same sequence but only one possesses the catalytic functions for phosphodiester hydrolysis (Huai et al. 2001), yet it seems to be active only when the enhancer cleft is also filled with cognate DNA. Consequently, the Type IIE enzymes cleave DNA with two target sites more rapidly than DNA with one site, but the two-site DNA is cleaved at only one site (Embleton et al. 2001). The Sau3AI restriction enzyme, on the other hand, acts like a Type IIE enzyme though it is a monomer in solution: two monomers probably associate with two Sau3AI sites to give an assembly that cleaves both strands at one site (Friedhoff et al. 2001).

The Type IIF subset (Fig. 1) is also defined by a reaction mechanism. The enzymes in this subset, such as SfiI and Cfr10I (Wentzell et al. 1995; Siksnys et al. 1999), bind simultaneously to two DNA sites but then, in contrast to the Type IIE enzymes, act concertedly at both sites. They thus convert a DNA with two target sites directly to the final product cut at both sites, largely bypassing the intermediates cut at a single site (Embleton et al. 2001). The majority of the Type IIF endonucleases are tetramers of identical subunits with two DNA-binding clefts, each cleft being made from two subunits (Deibert et al. 2000). Yet each pair of subunits must be able to sense whether cognate DNA is bound at the other cleft, since the Type IIF enzymes are generally inactive unless both clefts are filled with cognate DNA (Embleton et al. 1999; Siksnys et al. 1999; Williams and Halford 2002). However, the Type IIF enzymes are not all tetramers. The SgrAI enzyme is a dimer in free solution and when bound to a single copy of its recognition site, but, on a DNA with two SgrAI sites, the dimers bound to each site associate to a tetramer (Daniels et al. 2003). The tetramer bridging two sites is more active than the dimers at the solitary sites, and cleaves both sites concertedly.

These examples (Fig. 1) show that many Type II restriction endonucleases require two target sites for their DNA cleavage reactions. Many more such enzymes are currently being identified. However, the purpose of this article is not to review any one restriction enzyme acting at two sites or any subset of such enzymes, as these have been covered elsewhere in this volume (Siksnys et al., this Vol.; Reuter et al., this Vol.). Instead, this chapter will consider the various strategies that have been used to identify whether a restriction enzyme interacts with two DNA sites, and to characterise the nature of its communications between distant DNA sites.

3 Information from *cis* Reactions

Interactions between two DNA sites *in cis*, in the same DNA molecule, are usually favoured over interactions across sites *in trans*, on separate DNA molecules (Schleif 1992; Rippe et al. 1995). For example, the Type IIE enzyme EcoRII can cleave all 6 of its recognition sites in pBR322 but it fails to cleave its single site in phage T7 DNA, as the latter requires a *trans* interactions (Krüger et al. 1988).

The mean distance between two sites in the same DNA molecule is governed by the length of the intervening DNA while the mean distance between sites in separate molecules is a function of the concentration of the DNA and will usually be much longer than that between sites *in cis*. The effective concentration of one site in the DNA, in the local vicinity of another site, is thus much higher when both sites are in the same DNA molecule than when the sites are in separate DNA molecules (Rippe 2001). Hence, the energy needed to bring together sites *in cis* is usually much smaller than that for sites *in trans*. However, the energy required to span sites *in cis* varies with the length of the intervening DNA (Schleif 1992). At lengths <200 bp, bridging interactions become less favourable, due to the need to bend and/or twist the intervening DNA. Increasing lengths >400 bp also destabilise *cis* interactions, due to the entropic cost in tethering a long loop. Nevertheless, sites *in cis* are still almost always favoured over sites *in trans*.

3.1 One-Site/Two-Site Assays

The preference for *cis* over *trans* interactions yields a very simple diagnostic to see whether a restriction endonuclease requires two sites: by comparing the rates of its reactions on two DNA substrates that have, respectively, either one or two copies of its recognition site (Bilcock et al. 1999). If the enzyme has to interact with two sites before it can cleave DNA, it may cleave the DNA with two copies of the site more rapidly than the DNA with one site, as the former carries the sites *in cis*. In contrast, the cleavage of the one-site DNA will require an unfavourable interaction *in trans*, bridging sites in separate molecules of the DNA.

Two caveats need to be born in mind before using the comparison of reaction rates on one-site and two-site substrates as a diagnostic. First, a difference in the reaction rates on the one-site and the two-site DNA will be observed only under reaction conditions where the inefficiency of the *trans* interaction, relative to the *cis* interaction, impinges on the reaction velocity.

Under conditions where the enzyme has a high affinity for its recognition site, it may interact with sites *in trans* sufficiently well to give its maximal velocity. Its velocity on the two-site DNA would then be no faster than that on the one-site DNA. Hence, several enzymes that are active only after interacting with two sites cleave one-site and two-site substrates at the same rate at low ionic strength, where they bind tightly to their recognition sequence, but cleave the two-site DNA more rapidly than the one-site DNA at higher ionic strength (Nobbs and Halford 1995; Bilcock et al. 1999).

Second, it has long been known that the rate at which a restriction enzyme acts at a recognition site can vary with the sequence flanking that site (Thomas and Davis 1975; Halford et al. 1980). For enzymes that recognise discontinuous sites, such as SfiI at $GGCC(N_5)GGCC$, the non-specific bp in the middle of the sequence can further influence the rate (Williams and Halford 2001). For the Type IIS enzymes, the sequence between the recognition site and the downstream site of cleavage can also affect the reaction (Bath et al. 2002). Hence, it is essential to ensure that the two sites in the two-site substrate are identical to each other and to the site in the one-site DNA, in terms of their sequence context. Otherwise, a difference in rates on the one-site and the two-site DNA could be simply due to local sequence effects.

The assays are usually carried out by withdrawing samples from the reactions at various times, mixing the samples immediately with a quench to stop the reaction, and then analysing the DNA by electrophoresis through agarose to separate the DNA substrate from the various reaction products. EDTA is commonly used as the quench, as this chelates the Mg^{2+} ions needed by almost all Type II enzymes. To determine the concentrations of each form of the DNA at each time point sampled, it is best to use a radiolabelled substrate, so that the amount of each form of the DNA at each time point sampled can be measured from the radioactivity present in each band in the gel (Vipond et al. 1995). Alternatively, the gel can be stained with a fluorescent dye, such as ethidium bromide or SYBR green, and the amount of each form determined from a digital record of the fluorescence from the gel. The latter procedure is, however, more prone to errors, due to the fact that different forms of DNA bind ethidium to different extents. For the reasons noted above, the reactions then need to be repeated at different ionic strengths.

3.2 Determining Reaction Mechanism

The rates at which a restriction enzyme cleaves DNA substrates with one or with two copies of its recognition site can reveal directly whether or not the enzyme needs to interact with two sites. But to determine which particular reaction mechanism is followed by the enzyme, it is also necessary to examine the nature of the products from the one-site and the two-site substrates (Fig. 2). This analysis needs to be carried out on reactions containing much

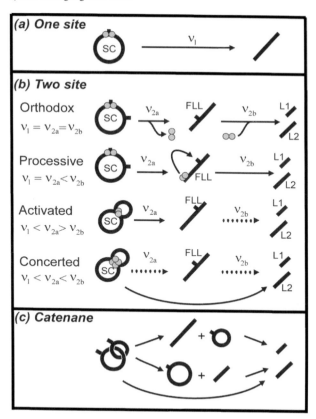

Fig. 2. a A supercoiled (*SC*) circle of DNA with one recognition site for the restriction enzyme under investigation is cleaved to a linear DNA, at a velocity v_1. **b** A supercoiled circle (*SC*) with two target sites is cleaved at one site to give the full-length linear (*FLL*) form of the DNA, at a velocity v_{2a}, and then at the residual site to give the two linear fragments L1 and L2, at a velocity v_{2b}. Four different modes of action of a Type II restriction enzyme are shown. **c** A catenane comprising two circles of DNA of unequal size, with one target site (*hatch mark*) in each circle, can be cleaved at either individual site, to give a large linear and a small circular DNA, or vice versa: the free circle is then linearised in a second reaction. Alternatively, in the lower pathway, concerted action at both sites converts the catenane directly to the two linear products

lower concentrations of enzyme than DNA, so that only a small fraction of the DNA is bound to the enzyme at any one time: the observed products are then the products liberated from the enzyme at the end of its reaction rather than intermediates still bound to the enzyme.

A supercoiled (SC) circle of DNA with one recognition site for the enzyme under investigation will be cleaved to a linear DNA, at a velocity v_1 (Fig. 2a). [The enzyme will in fact first cut one strand of the SC DNA, to give a nicked open-circle, and then the second strand, to give the linear product. Normally, both reactions are completed before the enzyme dissociates from the DNA and the nicked DNA is observed only in reactions with enzyme in excess of the DNA (Terry et al. 1987; Erskine et al. 1997).] A SC circle of DNA with two copies of the recognition sequence will be cleaved first at one site to give the full-length linear (FLL) form of the DNA, at a velocity v_{2a}, and then at the other site, to convert the FLL form to two linear fragments L1 and L2 at a velocity v_{2b} (Fig. 2b). However, the different reaction mechanisms that have been identified to date for Type II restriction enzymes lead to at least four different profiles of product formation from the two-site substrate, which can be distinguished from each other from the relative values of v_1, v_{2a} and v_{2b}.

For an orthodox restriction enzyme that catalyses independent reactions at each copy of its recognition site, the velocities on the one-site substrate and at each site in the two-site DNA are all equal to each other: i.e. $v_1=v_{2a}=v_{2b}$ (Fig. 2b). An indication of the relative magnitudes of v_{2a} and v_{2b} can be obtained from the maximal amount of FLL DNA liberated during the course of the reaction on the two-site substrate. If $v_{2a}=v_{2b}$, the FLL form builds up to a maximum at 40% of the total DNA before it is converted to the final products, L1 and L2 (Bilcock et al. 1999; Bath et al. 2002). If $v_{2a}>v_{2b}$, the peak amount of FLL DNA will be >40% of the total. Conversely, if $v_{2a}<v_{2b}$, the peak will be at <40% of the total.

A restriction enzyme that catalyses separate reactions at each copy of its recognition site can act processively on the DNA with two sites (Fig. 2b). Instead of departing from the DNA molecule after cutting one site before returning to cut the residual site, the enzyme cuts one site and then translocates directly to the second site, without departing from that molecule of DNA (Terry et al. 1987; Stanford et al. 2000). Though it is commonly thought that diffusional transfers of proteins along DNA occur by one-dimensional "sliding", the translocation actually proceeds by successive dissociation/re-association steps with the same molecule of DNA (Stanford et al. 2000; Gowers and Halford 2003). The hallmark of this behaviour is that the enzyme will still cleave its recognition sequence in the one-site and two-site substrates at the same rate ($v_1=v_{2a}$), but less of the FLL DNA will be produced than for the case of independent and equal reactions at each site (i.e. $v_{2a}<v_{2b}$).

The primary hallmark of a restriction enzyme that requires two copies of its recognition site is, as noted above, a faster velocity on the two-site substrate than the one-site DNA ($v_1<v_{2a}$). The Type IIE enzymes do not cut the second site but use it to activate the cleavage of the first site (Fig. 1). These activated enzymes thus rapidly cleave one site in the two-site substrate, to produce FLL DNA, but then cleave the residual site at a much slower rate ($v_{2a}>v_{2b}$; Fig. 2b). This is because the cleavage of one site is initially enhanced by the second site acting as an activator *in cis* but, once the first site is cleaved, the enzyme can cleave the second site only after interacting with an activator *in trans*. Examples of Type IIE endonucleases that display this behaviour include NaeI and Sau3AI (Embleton et al. 2001; Friedhoff et al. 2002). In contrast, after a Type IIS enzyme has cleaved the two-site DNA at one site, the DNA still possesses two intact copies of its recognition sequence, since these enzymes cleave DNA away from their sites. Hence, even after a Type IIS reaction at one site, the recognition sequence adjacent to the point of cleavage can still activate the reaction at the other site (Bath et al. 2002).

The Type IIF enzymes share with the Type IIE enzymes enhanced reaction velocities on the two-site substrate over the one-site substrates (i.e. v_1 is again $<v_{2a}$), but they differ from the IIE systems by acting concertedly at both sites (Fig. 2b). In the reaction of a Type IIF enzymes on a two-site substrate, the majority of the DNA is converted directly to the two final products L1 and L2, without liberating the FLL form cut at one site, so that in effect $v_{2a}<<v_{2b}$. In reality, the enzyme will occasionally dissociate from the DNA after cutting just one site, to liberate some of the FLL DNA. The yield of the FLL form increases under conditions that reduce the affinity of the protein for the DNA, such as high salt or high temperatures (Nobbs and Halford 1995; Nobbs et al. 1998).

This approach is illustrated here by comparing the reaction of the SfiI endonuclease on a plasmid with two SfiI sites (Fig. 3a) to that on a plasmid

Fig. 3. SfiI reactions. **a** Samples were removed at the indicated times from a reaction containing SfiI endonuclease (0.125 nM) and a circular plasmid (5 nM) with two SfiI sites, in SfiI assay buffer at 50 °C. The samples were analysed to determine the concentrations of the supercoiled DNA substrate (*SC, open squares*), the full-length linear from of the DNA cut at one site (*FLL, open circles*) and the mean of the two linear fragments from cutting both sites (L1/L2, *filled circles*). **b** The reactions containing SfiI endonuclease (0.125 nM) and a plasmid (5 nM), in SfiI assay buffer at 50 °C. In one case (*open circles*), the plasmid had two SfiI sites and the concentration of the final products cut at both sites was measured. In another reaction (*filled circles*), the plasmid had one SfiI site and the concentration of the product cut at this single site was measured. (Data from Wentzell et al. 1995: copyright 1995, with permission from Elsevier)

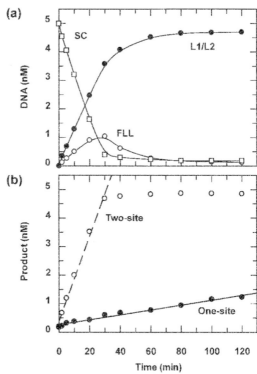

with one site (Fig. 3b). SfiI converts the majority of the two-site plasmid directly to the two linear fragments, the DNA cut at both sites: only a small amount of the singly-cut FLL DNA is liberated during the course of the reaction (Fig. 3a). By itself, this data shows that SfiI follows either the processive or the concerted scheme (Fig. 2b). However, SfiI generates its final products from the two-site plasmid much more rapidly than that from the one-site plasmid, even though the former requires two cleavage events and the latter just one (Fig. 3b). The lack of FLL DNA from the two-site substrate therefore cannot be due to processivity and must instead be due to concerted action.

3.3 The Effect of Supercoiling

The differences in bringing together sites *in cis*, relative to sites *in trans*, are further compounded when the DNA is supercoiled. A supercoiled DNA is more compact than its linear equivalent and occupies a smaller volume (Vologodskii and Cozzarelli 1994). The mean distance between two sites in a supercoiled molecule will always be smaller than that between the sites in relaxed DNA. The probability of the sites coming into close proximity of each other is therefore much higher in the supercoiled form of the DNA than in the relaxed form (Vologodskii and Cozzarelli 1996). Supercoiling thus enhances by a major factor the efficiency of *cis* interactions.

In contrast, supercoiling reduces the efficiency of *trans* interactions (Wentzell et al. 1995). An interaction between two sites in two separate molecules of DNA requires that one of the DNA chains penetrates the volume occupied by the other DNA. This interpenetration is greatly facilitated if at least one of the molecules is linear rather than circular, since linear DNA can worm its way into the volume occupied by another DNA in an end-on manner. Consequently, when comparing *cis* and *trans* reactions, the difference will be very much larger with supercoiled DNA than with relaxed DNA, on account of both the enhancement of the *cis* interaction and the inhibition of the *trans* interaction (Wentzell et al. 1995).

3.4 Catenane Substrates

In principle, interactions across distant DNA sites can occur by either a "tracking" mechanism following the 1-D path of the DNA, or by a "looping" mechanism bridging the sites through 3-D space. An unambiguous distinction between the 1-D and 3-D schemes can be obtained by testing the enzyme against a DNA catenane (Adzuma and Mizuuchi 1989), comprised of two interlinked rings of DNA with a copy of the target site in each ring (Fig. 2c). Such catenanes can be made in vitro by the site-specific recombination reaction of a resolvase on a plasmid with two *res* sites interspersed with two targets for the enzyme under study (see below, Fig. 6).

A protein that follows a tracking mechanism will never get from its site in one ring to the other ring: the Type I restriction enzymes (McCleland and Szczelkun, this Vol.) show this behaviour (Szczelkun et al. 1996). However, the mean distance between the sites in the interlinked rings of the catenane is similar to that between the sites *in cis*, in the plasmid from which the catenane was obtained (Levene et al. 1995). Hence, a looping enzyme ought to be able to bridge two sites in the separate rings of a catenane almost as readily as two sites *in cis*, in the same ring of DNA. Moreover, the interaction bridging the catenane rings should occur much more readily than that between the same two rings of DNA after they have been unlinked from each other. Several restriction enzymes have been found to act in this way, including SfiI, NaeI and NgoMIV (Szczelkun and Halford 1996; Embleton et al. 2001).

Further information about the mode of action of a restriction enzyme can be obtained by analysing the products emanating from its reaction on a catenane with a target site in each ring. The cleavage of either one of the sites in the catenane will lead to one linear and one circular product: either a large linear DNA and a small circle, or vice versa (Fig. 2 c). On the other hand, concerted action at both sites will convert the catenane directly to the two linear products, bypassing either of the free circles (Fig. 2 c). The Type IIF enzymes, SfiI, Cfr10I and NgoMIV all cleaved a catenane directly to two linear products (Szczelkun and Halford 1996). However, while the Type IIE enzyme NaeI cut the catenane more rapidly than the DNA with one site, it cleaved the catenane in just one ring: the site in one ring presumably activates the reaction at the NaeI site in the other ring, even though it is not cleaved itself (Embleton et al. 2001).

4 DNA Binding and Looping with Sites in cis

In addition to analysing the kinetics of their DNA cleavage reactions, several other approaches have been used to characterise DNA-looping interactions by the Type II restriction enzymes that interact with two copies of their recognition sequence. Perhaps the most direct is electron microscopy (EM). For several of these enzymes, DNA loops have been observed by EM after adding the enzyme to a DNA with two recognition sites in the absence of Mg^{2+} (Topal et al. 1991; Siksnys et al. 1999; Friedhoff et al. 2001). However, EM is a destructive technique: the staining and shadowing of DNA needed for EM will obliterate the enzyme bound to the DNA. It then remains an open question whether the looped complex observed by EM occurs during catalysis by the enzyme.

DNA-looping interactions were in fact first identified by analysing the effect of changing the length of DNA between two target sites (Dunn et al.

1984). If a protein is to bind concurrently to two DNA loci, both loci must present to the protein the appropriate surface of the DNA, commonly the major groove. This surface will repeat itself every helical turn of the DNA. Hence, alterations in the length of DNA between the sites by 10 or 11 bp, close to the helical repeat, will not alter significantly the orientation of the surface at one site relative to that at the other. However, 5- or 6-bp alterations will rotate one surface relative to the other by about 180° and the intervening DNA will then need to be under- or over-twisted to bring the sites back into register. This is energetically unfavourable. DNA-looping interactions should thus show a cyclical response to variations in site separation, with a periodicity corresponding to the helical repeat of the DNA (Schleif 1992). The amplitude of the cyclical pattern will, however, decrease at large separations, >300 bp, as the change in twist becomes distributed over a greater number of bp. In one application of this approach, the SfiI endonuclease was tested against a series of plasmids that had two SfiI sites separated by various lengths of DNA (Wentzell and Halford 1998). SfiI did indeed show the cyclical response characteristic of DNA looping (Fig. 4). Some plasmids were cleaved readily; for example, those with site separations of 154 or 164 bp. Others were cleaved less readily, for example the 160- and 170-bp separations. The pattern yields a periodicity of 11.4 bp, close to that expected for the helical repeat under the reaction conditions used here (Fig. 4).

One of the best methods for analysing the equilibrium binding of a DNA-looping protein to a DNA with two target sites is gel retardation, as was shown many years ago with the Lac repressor (Krämer et al. 1987). If a protein has two separate binding surfaces for the target DNA and the DNA two separate copies of the recognition sequence, there exist potentially a large number of alternative complexes: the DNA with one molecule of the protein at one site; the DNA carrying two molecules of the protein, one at each site; the looped complex, where the protein bridges the two sites *in cis*; a host of different *trans*

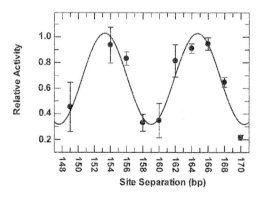

Fig. 4. Periodicity of SfiI. The activity of the SfiI endonuclease was measured against a series of ~6.7-kb plasmids that carry two SfiI sites: the separation of the sites in each plasmid (in bp) is shown on the x-axis and the relative activity on the y-axis. The line drawn is the optimal fit of the data to a sine function: the best fit was with a periodicity of 11.4 bp. (Data from Wentzell and Halford 1998: copyright 1998, with permission from Elsevier)

complexes, where the protein(s) bridge(s) one site in one DNA and one in the other DNA. The power of gel retardation is that, under appropriate conditions, these complexes can all be separated from each other, as they have different electrophoretic mobilities, and the amounts of each determined (Krämer et al. 1987). Moreover, even different forms of the looped complex can have different mobilities (Watson et al. 2000).

In the presence of Ca^{2+}, so as giving specific binding without cleavage (Embleton et al. 1999), the binding of SfiI to DNA fragments carrying two SfiI sites 149–170 bp apart (as in Fig. 4) yields two electrophoretically distinct forms of the looped complex: one predominates when the inter-site spacing lies between n and $n+1/2$ helical turns of DNA (where n is an integer) and the other at spacings between $n+1/2$ and $n+1$ turns (Watson et al. 2000). If the two DNA segments bound to SfiI overlay each other at an angle of ~90° (Fig. 5), as is likely to be the case (Siksnys et al., this Vol.), the looped complex will have its maximal stability when the inter-site spacing corresponds to $n+1/2$ helical turns rather than n turns (Watson et al. 2000). If the separation is slightly less than $n+1/2$ turns, the intervening DNA will need to be overwound, as in positively supercoiled DNA: if the separation is slightly more than $n+1/2$ turns, the intervening DNA will need to be underwound, as in negatively supercoiled DNA (Fig. 5). Gel-retardation analysis can thus reveal not only the existence of a looped complex but also the geometric structure of the loop.

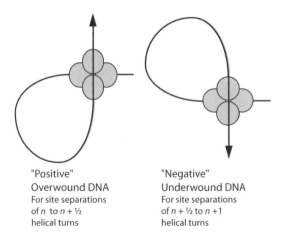

"Positive"
Overwound DNA
For site separations
of n to $n + ½$
helical turns

"Negative"
Underwound DNA
For site separations
of $n + ½$ to $n + 1$
helical turns

Fig. 5. Geometry of SfiI loops. A tetrameric protein is represented by four spheres: the two upper subunits constitute one DNA-binding surfaces and likewise the two lower subunits. Two schemes are shown for the complex of the tetramer bound to two DNA sites *in cis*. In both, the path of the DNA runs from the lower pair of subunits to the upper pair but the intervening DNA can have the same handedness as in either positively supercoiled DNA (shown on the *left*) or negatively supercoiled DNA (on the *right*). SfiI traps loops with the former geometry when the intervening DNA needs to be overwound and the latter geometry by underwinding. (Data from Watson et al. 2000: copyright 2000, with permission from Elsevier)

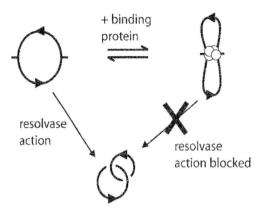

Fig. 6. Use of resolvase to detect DNA loop formation. A plasmid contains two target sequences for resolvase in direct repeat, the *res* sites (*black triangles*), and two recognition sites (*hatch marks*) for a protein that can bind concurrently to two sites. The binding of this protein (shown as four *spheres*) to both sites sequesters the *res* sites into separate loops. In the absence of this binding protein, resolvase acts on the plasmid to yield a DNA catenane. However, resolvase is blocked from its recombination reaction whilst the *res* sites are sequestered in the separate loops

Gel retardation is, however, limited in the main to relatively short DNA fragments, typically <400 bp. A strategy for detecting DNA-looping interactions in native DNA (i.e. supercoiled molecules of kb size) exploits the requirements of resolvase for its site-specific recombination reaction (Parker and Halford 1991; Szczelkun and Halford 1996). The resolvases from the Tn3-like transposons catalyse a recombination event between two copies of a specific DNA sequence called *res*, but only if the *res* sites are in direct repeat in the same contiguous molecule of supercoiled DNA (Fig. 6). If a plasmid carries two *res* sites interspersed with two target sites for a DNA-looping protein, the concurrent binding of that protein to both of its target sites will sequester the *res* sites into separate loops of DNA (Fig. 6). Resolvase cannot function on *res* sites in topologically separate loops so the recombination reaction is blocked for as long as the looping interaction persists. The amounts of the plasmid in looped and unlooped states can thus be measured from the fraction that is converted to the catenane on addition of resolvase. Moreover, after adding resolvase to the looped DNA, the rate of recombination may be slower than that in the absence of the looping protein, so the dissociation rate of the looped complex can also be measured. This strategy has been used to detect DNA looping by the SfiI and the Cfr10I endonucleases, to analyse the thermodynamics of the looping interactions and to measure the dissociation rates of the complexes (Szczelkun and Halford 1996; Milsom et al. 2001). Surprisingly large differences were observed in the breakdown rates: under optimal conditions in the presence of Ca^{2+}, the SfiI loop remained intact for many hours while the Cfr10I loop fell apart in ~90 s.

5 Information from trans Reactions

Although *cis* interactions are generally more favourable than *trans* interactions, the Type II restriction endonucleases that interact with two sites can under certain circumstances use two sites *in trans*, on separate molecules of DNA. Indeed, the first indication that a Type II restriction enzyme might need to interact with two copies of its recognition sequence before cutting DNA came from a *trans* reaction (Krüger et al. 1988): EcoRII is barely active against DNA with one cognate site but is activated against such DNA by oligonucleotide duplexes that carry its recognition sequence.

5.1 Kinetic Studies

Two factors commonly thwart proteins from interacting simultaneously with two sites in separate DNA molecules. One is the local concentration effect noted above. The effective concentration of one DNA site in the vicinity of another is generally much lower for sites *in trans* than for sites *in cis*, unless the DNA is present in solution at a very high concentration (Rippe 2001). The other, also noted above, is the steric difficulty in bringing together the site in one chain with a site in another chain, particularly when both DNA are supercoiled: at least one of the DNA ought to be linear (Wentzell et al. 1995). Both factors can be circumvented to a major extent by using as one of the substrates for a *trans* reaction a duplex made from synthetic oligodeoxyribonucleotides. Oligoduplexes can generally be employed at much higher concentrations than plasmids or phage DNA, and the duplex can readily encounter the recognition sequence in a large DNA molecule even when the latter is supercoiled.

To date, perhaps the most commonly used strategy to identify the restriction enzymes that need two site is "oligonucleotide activation" (for example, Oller et al. 1991; Reuter et al. 1993; Wentzell et al. 1995; Siksnys et al. 1999; Lagunavicius et al. 2003). In this procedure, the restriction enzyme is tested against a plasmid or similar DNA that has a single copy of the relevant recognition sequence, in both the absence and the presence of an oligonucleotide duplex that also has the recognition sequence, ideally at several different concentrations of the duplex (viz. Gormley et al. 2002). The extent of cleavage of the plasmid is then measured.

For a Type IIP restriction enzyme that acts at individual sites, such as EcoRV (Fig. 1), the duplex will function as a competitive inhibitor against the plasmid: increasing concentrations of the duplex will progressively reduce the extent of plasmid cleavage. In contrast, for enzymes that interact with two sites, the extent of plasmid cleavage is enhanced at relatively low concentrations of the duplex, but is reduced at higher concentrations. The enhancement is most likely due to the enzyme being activated by binding concurrently to its recognition sites on both the plasmid and the oligoduplex. The reduced cleavage of the plasmid at high concentrations of the duplex is most likely due to the enzyme then binding to two molecules of the duplex rather than one duplex and one plasmid. However, an exception to this behaviour exists: though the SgrAI enzyme acts as a tetramer to cleave DNA with two recognition sites concertedly, it remains a dimer when cleaving a DNA with one SgrAI site, so its reaction on a one-site plasmid is inhibited rather than activated by a cognate duplex (Daniels et al. 2003).

"Oligonucleotide activation" experiments monitor the cleavage of the plasmid rather than the oligonucleotide. Nevertheless, there exist many methods for following the cleavage of an oligonucleotide duplex by a restriction enzyme: the hyperchromic shift assay (Waters and Connolly 1992) is the most suitable for kinetic analyses as it provides a continuous record of the progress of the reaction. In many cases, the reaction velocities increase with increasing concentrations of the duplex in a sigmoidal manner, and not in the hyperbolic manner expected for systems following Michaelis-Menten kinetics (Yang and Topal 1992; Gabbara and Bhagwat 1992; Embleton et al. 1999). A sigmoidal relationship between v and [S] indicates a cooperative system where the enzyme displays its maximal rate only after binding two or more molecules of substrate, so this can provide yet further evidence for the interaction with two copies of the recognition sequence.

5.2 Binding Studies

Direct evidence for the concurrent binding of two oligonucleotide duplexes to a DNA-looping protein can be obtained by gel-retardation experiments using two duplexes of different lengths (Embleton et al. 1999; Lagunavicius et al. 2003). For example, the electrophoretic mobility of the complex formed between the SfiI endonuclease and a cognate duplex 17 bp long differs from its complex with a 30-bp duplex (Fig. 7). However, three complexes were observed when SfiI was mixed with both the 17- and 30-bp duplexes: two had the same mobilities as the complexes formed with, respectively, the 17- and 30-bp duplexes; the third had an intermediate mobility (Fig. 7). Moreover, the yields of the complexes varied with the ratio of the two duplexes in binomial fashion: equal concentrations of the two duplexes gave the 1:2:1 distribution. Hence, the complex with the highest mobility must be SfiI bound to two 17-bp duplexes: the slowest must be SfiI bound to two 30-bp duplexes; while the complex with intermediate mobility must be SfiI bound concurrently to one 17-bp and one 30-bp duplex (Embleton et al. 1999).

Further studies employed one duplex with the recognition sequence for SfiI but carrying a phosphorothioate instead of the phosphate at the scissile bond, to block cleavage by SfiI, and a second cleavable duplex with phosphate oxyesters throughout. As in Fig. 7, SfiI formed a hybrid complex with one

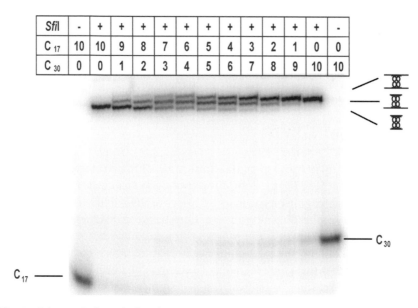

Sfil	-	+	+	+	+	+	+	+	+	+	+	+	-
C_{17}	10	10	9	8	7	6	5	4	3	2	1	0	0
C_{30}	0	0	1	2	3	4	5	6	7	8	9	10	10

Fig. 7. Gel-retardation of SfiI. The reactions, in Ca^{2+}-binding buffer, contained 10 nM ^{32}P-labelled DNA and either no SfI (*outside lanes*, marked *SfiI* -) or SfiI endonuclease at 5 nM (*inner lanes*, marked *SfiI* +). The DNA was: a 17-bp duplex with the recognition sequence for SfiI; a 30-bp duplex with the recognition sequence for SfiI; a mixture of the two. The samples were subjected to electrophoresis though polyacrylamide and the gel analysed by phosphorimager. The electrophoretic mobilities of the free 17-bp and the free 30-bp duplexes are on indicated as C_{17} and C_{30}, on the *left* and the *right* of the gel, respectively. The cartoons on the *right* denote the complexes of the SfiI tetramer (*four circles*) bound: to two long duplexes; to one long and one short duplex; to two short duplexes. (Data from Embleton et al. 1999: copyright 1999, with permission from Elsevier)

phosphorothioate duplex and one oxyester duplex but, when Mg^{2+} was added to this hybrid, it failed to cleave even the oxyester duplex (Williams and Halford 2001). Hence, the information that one of the two DNA-binding clefts contains a phosphorothioate moiety must be transmitted to the other cleft, to prevent the latter from cleaving the oxyester.

As an alternative to gel retardation, the concurrent binding of two DNA duplexes can also be measured by "pull-down" assays. These usually employ two duplexes, both with the requisite recognition sequence: one tagged with a biotin moiety and the other ^{32}P-labelled. The question posed here is then whether the ^{32}P-labelled duplex can be pulled down on a streptavidin-coated magnetic bead in the presence of both the enzyme and the biotinylated duplex. The biotin on the latter will bind strongly to the streptavidin-coated beads so, if the enzyme can bind simultaneously to the radiolabelled and the biotin-tagged duplexes, some of the radiolabel will become fixed to the bead.

When this procedure was applied to FokI and to BfiI, the radiolabelled duplex was in both cases pulled down with the beads, but, if either the enzyme or the biotinlyated duplex was omitted, no radiolabel became associated with the beads (Vanamee et al. 2001; Lagunavicius et al. 2003). Hence, these Type IIS both form synaptic complexes with two cognate sites.

6 Conclusion

The aim of this review was to give an indication of the strategies available to first identify whether a Type II restriction endonucleases needs to interact with two copies of its recognition site in order to cleave DNA, and second to characterise the nature of the long-range interactions between distant DNA sites. One strategy is, however, notably absent from this account: X-ray crystallography. Obviously, a wealth of information is available from the crystal structures of enzymes bound to two DNA segments (Deibert et al. 2000; Huai et al. 2001). Crystallographic studies of the restriction enzymes that bind two DNA sites are reviewed elsewhere in this volume (see Siksnys et al., this Vol.).

The biochemical approaches described in this chapter include both reactions *in cis*, spanning two sites in the same molecule of DNA, and reactions *in trans*, bridging two DNA molecules. Though *cis* interactions are inherently favoured over *trans* interactions (Schleif 1992; Rippe et al. 1995), the restriction enzymes that need two sites can still generally function in vitro with sites *in trans*, to at least some extent. Nevertheless, *trans* reactions are unlikely to play any role in vivo in the restriction of foreign DNA as it enters the bacterial cell. *Trans* reactions in vitro typically require high concentrations of DNA without supercoils, conditions that will not be encountered in vivo. The SfiI restriction enzyme is thus largely unable to restrict DNA with one SfiI site though it can restrict DNA with two or three SfiI sites, at efficiencies that equal, respectively, those for a single-site enzyme–EcoRI – on DNA with one and with two EcoRI sites (Bilcock and Halford 1999).

Despite their inefficiency at restricting DNA with one site in vivo, the reason why many restriction enzymes have to interact with two sites before they cut DNA is almost certainly because this provides a double check on their accuracy (Halford 2001). By recognising the nucleotide sequence at two separate locations in the DNA, and by proceeding into their DNA cleavage reactions only when both loci possess the correct sequence, they ensure that they cleave DNA only at the cognate sequence. For example, the SfiI endonuclease is active only in its synaptic complex with two DNA segments but, even though it can bind non-cognate sites, it binds only one such site at a time: it cannot even form a synaptic complex with one non-cognate and one cognate site (Embleton et al. 1999). Hence, it is precluded from cleaving DNA at any alternative sequence.

References

Adzuma K, Mizuuchi K (1989) Interaction of proteins located at a distance along DNA: mechanism of target immunity in the Mu DNA strand-transfer reaction. Cell 57:41–47

Bath AJ, Milsom SE, Gormley NA, Halford SE (2002) Many Type IIs restriction endonucleases interact with two recognition sites before cleaving DNA. J Biol Chem 277:4024–4033

Bilcock DT, Daniels LE, Bath AJ, Halford SE (1999) Reactions of Type II restriction endonucleases with 8-base pair recognition sites. J Biol Chem 274:36379–36386

Bilcock DT, Halford SE (1999) DNA restriction dependent on two recognition sites: activities of the SfiI restriction-modification system in *Escherichia coli*. Mol Microbiol 31:1243–1254

Bitinaite J, Wah DA, Aggarwal AK, Schildkraut I (1998) FokI dimerization is required for DNA cleavage. Proc Natl Acad Sci USA 95:10564–10569

Daniels LE, Wood KE, Scott DJ, Halford SE (2003) Subunit assembly for DNA cleavage by restriction endonuclease SgrAI. J Mol Biol 327:579–591

Deibert M, Grazulis S, Sasnauskas G, Siksnys V, Huber R (2000) Structure of the tetrameric restriction endonuclease NgoMIV in complex with cleaved DNA. Nat Struct Biol 7:792–799

Dunn TM, Hahn S, Ogden S, Schleif RF (1984). An operator at –280 base pairs that is required for repression of *araBAD* operon promotor: addition of helical turns between the operator and promoter cyclically hinders repression. Proc Natl Acad Sci USA 81:5017–5020

Embleton ML, Williams SA, Watson MA, Halford, SE (1999) Specificity from the synapsis of DNA elements by the SfiI endonuclease. J Mol Biol 289:785–797

Embleton ML, Siksnys V, Halford SE (2001) DNA cleavage reactions by Type II restriction enzymes that require two copies of their recognition sites. J Mol Biol 311:503–514

Erskine SG, Baldwin GS, Halford SE (1997) Rapid-reaction analysis of plasmid DNA cleavage by the EcoRV restriction endonuclease. Biochemistry 36:7567–7576

Friedhoff P, Lurz R, Lueder G, Pingoud A (2001) Sau3AI, a monomeric Type II restriction endonuclease that dimerizes on the DNA and thereby induces DNA loops. J Biol Chem 276:23581–588

Gabbara S, Bhagwat AS (1992) Interaction of EcoRII endonuclease with DNA substrates containing single recognition sites. J Biol Chem 267:18623–630

Gormley NA, Hillberg AL, Halford SE (2002) The Type IIs restriction endonuclease BspMI is a tetramer that acts concertedly at two copies of an asymmetric DNA sequence. J Biol Chem 277:4034–4041

Gowers DM, Halford SE (2003) Protein motion from non-specific to specific DNA by three-dimensional routes aided by supercoiling. EMBO J 22:1410–1418

Halford SE (2001) Hopping, jumping and looping by restriction enzymes. Biochem Soc Trans 29:363–373

Halford SE, Johnson NP, Grinsted J (1980) The EcoRI restriction endonuclease with bacteriophage lambda DNA: kinetic studies. Biochem J 191:581–592

Huai Q, Colandene JD, Topal MD, Ke H (2001) Structure of NaeI-DNA complex reveals dual-mode DNA recognition and complete dimer rearrangement. Nat Struct Biol 8:665–669

Kong H (1998) Analyzing the functional organization of a novel restriction modification system, the BcgI system. J Mol Biol 279:823–832

Kong H, Smith CL (1998) Does BcgI, a unique restriction endonuclease, require two sites for cleavage? Biol Chem 379:605–609

Krämer H, Niemöller M, Amouyal M, Revet B, von Wicklen-Bergmann B, Müller-Hill B (1987) *lac* repressor forms loops with linear DNA carrying two suitably spaced *lac* operators. EMBO J 6:1481–1491

Krüger DH, Barcak GJ, Reuter M, Smith HO (1988) EcoRII can be activated to cleave refractory DNA recognition sites. Nucleic Acids Res 11:3997–4008

Lagunavicius A, Sasnauskas G, Halford SE, Siksnys V (2003) The metal-independent Type IIs restriction enzyme BfiI is a dimer that binds two DNA sites but has only one catalytic centre. J Mol Biol 326:1051–1064

Levene SD, Donahue C, Boles TC, Cozzarelli NR (1995) Analysis of the structure of dimeric DNA catenanes by electron microscopy. Biophys J 69:1036–1045

Milsom SE, Halford SE, Embleton ML, Szczelkun MD (2001) Analysis of DNA looping interactions by Type II restriction enzymes that require two copies of their recognition sites. J Mol Biol 311:515–527

Nobbs TJ, Halford SE (1995) DNA cleavage at two recognition sites by the SfiI restriction-endonuclease – salt dependence of *cis* and *trans* interactions between distant DNA sites. J Mol Biol 252:399–411

Nobbs TJ, Szczelkun MD, Wentzell, Halford SE (1998) Excision by the SfiI restriction endonuclease. J Mol Biol 281:419–432

Oller AR, Van den Broek W, Conrad M, Topal MD (1991) Ability of DNA and spermidine to affect the activity of restriction endonucleases from several bacterial species. Biochemistry 30:2543–2549

Parker CN, Halford SE (1991) Dynamics of long-range interactions on DNA – the speed of synapsis during site-specific recombination by resolvase. Cell 66:781–791

Pingoud A, Jeltsch A (2001) Structure and function of Type II restriction endonucleases. Nucleic Acids Res 29:3705–3727

Reuter M, Kupper D, Pein CD, Petrusyte M, Siksnys V, Frey B, Krüger DH (1993) Use of specific oligonucleotide duplexes to stimulate cleavage of refractory DNA sites by restriction endonucleases. Anal Biochem 209:232–237

Reuter M, Kupper D, Meisel A, Schroeder C, Krüger DH (1998) Cooperative binding properties of restriction endonuclease EcoRII with DNA recognition sites. J Biol Chem 273:8294–8300

Rippe, K (2001) Making contacts on a nucleic acid polymer. Trends Biochem Sci 26:733–740

Rippe K, von Hippel PH, Langowski J (1995) Action at a distance: DNA-looping and initiation of transcription. Trends Biochem Sci 20:500–6

Roberts RJ, Belfort M, Bestor T, Bhagwat AS, Bickle TA, Bitinaite J, Blumenthal RM, Degtyarev SKh, Dryden DT, Dybvig K, Firman K, Gromova ES, Gumport RI, Halford SE, Hattman S, Heitman J, Hornby DP, Janulaitis A, Jeltsch A, Josephsen J, Kiss A, Klaenhammer TR, Kobayashi I, Kong H, Krüger DH, Lacks S, Marinus MG, Miyahara M, Morgan RD, Murray NE, Nagaraja V, Piekarowicz A, Pingoud A, Raleigh E, Rao DN, Reich N, Repin VE, Selker EU, Shaw PC, Stein DC, Stoddard BL, Szybalski W, Trautner TA, Van Etten JL, Vitor JM, Wilson GG, Xu SY (2003a) A nomenclature for restriction enzymes, DNA methyltransferases, homing endonucleases and their genes. Nucleic Acids Res 31:1805–1812

Roberts RJ, Vincze T, Posfai J, Macelis D (2003b) REBASE: restriction enzymes and methyltransferases. Nucleic Acids Res 31:418–420

Sambrook J, Russell DW (2001) Molecular cloning. A laboratory manual. Cold Spring Harbor Laboratory Press, Cold Spring Harbor

Sapranauskas R, Sasnauskas G, Lagunavicius A, Vilkaitis G, Lubys A, Siksnys V (2000) Novel subtype of Type IIs restriction enzymes BfiI endonuclease exhibits similarities to the EDTA-resistant nuclease Nuc of *Salmonella typhimurium*. J Biol Chem 275: 30878–885

Schleif R (1992) DNA looping. Annu Rev Biochem 61:199–223

Siksnys V, Skirgaila R, Sasnauskas G, Urbanke C, Cherny D, Grazulis S, Huber R (1999) The Cfr10I restriction enzyme is functional as a tetramer. J Mol Biol 291:1105–1118

Soundararajan M, Chang Z, Morgan RD, Heslop P, Connolly BA (2002) DNA binding and recognition by the IIs restriction endonuclease MboII. J Biol Chem 277:887–895

Stanford NP, Szczelkun MD, Marko JF, Halford SE (2000) One- and three-dimensional pathways for proteins to reach specific DNA sites. EMBO J 19:6546–6557

Szczelkun MD, Halford SE (1996) Recombination by resolvase to analyze DNA communications by the SfiI restriction-endonuclease. EMBO J 15 1460–1469

Szczelkun MD, Dillingham MS, Janscak P, Firman K. Halford SE (1996) Repercussions of DNA tracking by the Type IC restriction endonuclease EcoR124I on linear, circular and catenated substrates. EMBO J 15:6335–6347

Terry BJ, Jack WE, Modrich P (1987) Mechanism of specific site location and DNA cleavage by EcoRI endonuclease Gene Amplif Anal 5:103–118

Thomas M, Davis RW (1975) Studies on the cleavage of bacteriophage lambda DNA with EcoRI restriction endonuclease. J Mol Biol 91:315–328

Topal MD, Thrsher RJ, Conrad M, Griffth J (1991) NaeI endonuclease binding to pBR322 DNA induces looping. Biochemistry 30:2006–2010

Vanamee ES, Santaga S, Aggarwal AK (2001) FokI requires two specific DNA sites for cleavage. J Mol Biol 309:69–78

Vipond IB, Baldwin GS, Oram M, Erskine SG, Wentzell LM, Szczelkun MD, Nobbs TJ, Halford SE (1995) A general assay for restriction endonucleases and other DNA-modifying enzymes with plasmid substrates. Mol Biotech 4:259–268

Vologodskii AV, Cozzarelli NR (1994) Conformational and thermodynamic properties of supercoiled DNA. Annu Rev Biophys Biomol Struct 23:609–643

Vologodskii AV, Cozzarelli NR (1996) Effect of supercoiling on the juxtaposition and relative orientation of DNA sites. DNA. Biophys J 70:2548–2556

Wah DA, Bitinaite J, Schildkraut I, Aggarwal AK (1998) Structure of FokI has implications for DNA cleavage. Proc Natl Acad Sci USA 95:10570–10575

Waters TR, Connolly BA (1992) Continuous spectrophotometric assay for restriction endonucleases using synthetic oligodeoxynucleotides and based on the hyperchromic effect. Anal Biochem 204:204–209

Watson MA, Gowers DM, Halford SE (2000) Alternative geometries of DNA looping: an analysis using the SfiI endonuclease. J Mol Biol 298:461–475

Wentzell LM, Halford SE (1998) DNA looping by the SfiI restriction endonuclease. J Mol Biol 281:433–444

Wentzell LM, Nobbs TJ, Halford SE (1995) The SfiI restriction-endonuclease makes a 4-strand DNA break at 2 copies of its recognition sequence. J Mol Biol 248:581–595

Williams SA, Halford SE (2001) SfiI endonuclease activity is strongly influenced by the non-specific sequence in the middle of its recognition site. Nucleic Acids Res 29:1476–1483

Williams SA, Halford SE (2002) Communications between catalytic sites in the protein-DNA synapse by the SfiI endonuclease. J Mol Biol 318:387–394

Yang CC, Topal MD (1992) Nonidentical DNA-binding sites of endonuclease NaeI recognize different families of sequences flanking the recognition site Biochemistry 31:9657–9664

The Role of Water in the EcoRI–DNA Binding

N. Sidorova, D.C. Rau

1 Introduction

In many respects, Type II restriction endonucleases are prototypical DNA-binding proteins. In order to avoid catastrophic consequences for the cell, however, these enzymes must be far more stringent in recognition of their target sequences and subsequent DNA cleavage than other specific sequence recognition proteins that regulate gene activity. In contrast to *E. coli* Lac and λ Cro repressors, for example, that show gradually decreasing binding energies as the recognition sequence is changed (Frank et al. 1997; Takeda et al. 1992), many restriction nucleases are exquisitely specific. EcoRI will bind to its recognition sequence, GAATTC, with an association equilibrium constant $K_{a,sp}$ ~10^{11} M^{-1} and to a completely nonspecific sequence with $K_{a,nonsp}$ ~10^7 M^{-1}. A change of even a single base pair is sufficient to decrease the binding constant at least by 10^3, bringing it within a factor ~10 or less of nonspecific binding (Lesser et al. 1990).

To understand the physical basis of this specificity it is necessary both to know the structures of the complexes and to understand the energetics of molecular interactions. There has been an explosion of DNA–protein structures, in general, and of restriction nuclease–DNA complexes, in particular, solved by X-ray crystallography and NMR spectroscopy. There has not been a comparable increase in our ability to calculate binding energies from these structures. In particular, hydration energies are known to play important role in determining binding energies, but quantitating their contribution is still problematic. X-ray structures have uncovered many waters buried at protein–DNA interfaces that mediate interactions (Janin 1999), but energetic significance of these waters is unclear. The link between structure and the ener-

N. Sidorova, D.C. Rau
Laboratory of Physical and Structural Biology, National Institute of Child Health and Human Development, National Institutes of Health, Bld. 9, Room 1E108, Bethesda, Maryland 20892, USA

Nucleic Acids and Molecular Biology, Vol. 14
Alfred Pingoud (Ed.)
Restriction Endonucleases
© Springer-Verlag Berlin Heidelberg 2004

getics of intermolecular interactions is thermodynamics. The goal of this chapter is to begin correlating differences in binding energies to differences in hydration between complexes. The well-known linkage relationships that connect changes in binding energy to changes in salt concentration or pH through differences in ion binding or protonation will be extended to water activity or osmotic pressure.

We focus in this chapter on DNA complexes of the restriction endonuclease EcoRI, as a model system for delineating the role of water, in particular, in specific recognition. We compare directly specific and nonspecific binding properties, rather than simply investigating specific binding only, for two reasons. Binding specificity of protein only has meaning as a comparison of binding energies to different DNA sequences. On a more practical level, the pertinent equilibrium for specific sequence DNA-binding proteins within the DNA rich cellular environment is likely not between free and bound proteins, but between specifically and nonspecifically bound proteins. It should also be recognized that the water activity within the crowded cellular environment is not the same as in dilute aqueous solutions typically used to measure specific binding and that measuring the sensitivity of binding to water activity likely has practical applications for understanding in vivo action.

We show that nonspecific complexes of EcoRI sequester about 110 water molecules more than with the specific recognition sequence complex (Sidorova and Rau 1996). At low osmotic pressures this amount of water is seen even with complexes of EcoRI with DNA sequences that differ by a single base pair change ('star' sites) from the recognition sequence, consistent with the stringent binding specificity of this enzyme (Sidorova and Rau 1999). Much of the water sequestered by these 'star' sequence complexes, but not by other nonspecific sequence complexes, is removed at high osmotic pressures.

By combining equilibrium results with measurements of the dissociation rate of the specific sequence complex (Sidorova and Rau 2000, 2001) we are able to differentiate the effects of salt, pH, and water on the nonspecific–specific binding equilibrium and on the rate of dissociation of nonspecifically bound enzyme. The osmotic dependence of the dissociation rate constant of EcoRI from its recognition sequence is dominated by the 110 waters difference between specific and nonspecific binding modes. The dissociation of nonspecifically bound protein is accompanied by the uptake of much fewer waters. In contrast, there is very little salt and pH difference between specific and nonspecific modes of EcoRI binding. Nearly all of the dependence of the overall dissociation rate on salt and pH is coupled to the dissociation rate of the nonspecifically bound protein from the DNA.

The osmotic stress technique has now been used to measure the changes in hydration accompanying the DNA binding of several other proteins: *E. coli* Gal repressor (Garner and Rau 1995), *E. coli* CAP protein (Vossen et al. 1997), *E. coli* Lac repressor (Fried et al. 2002), Hin recombinase (Robinson and Sligar

1996), ultrabithorax and deformed homeodomains (Li and Mathews 1997), *E. coli* Tyr repressor (Poon et al. 1997), Sso7d protein (Lundback et al. 1998), TBP (Wu et al. 2001; Khrapunov and Brenowitz 2003).

2 Thermodynamics

It is by now standard practice when characterizing the binding of DNA recognition proteins to measure the sensitivity to salt concentration. The electrostatic interactions between the phosphate groups on the highly charged DNA backbone and basic amino acids of the protein in the complex are typically seen as the release of salt ions accompanying binding. The commonly used expression relating the association binding constant, K_a, salt concentration, [NaCl], and the difference in the number of associated ions between the complex and the free DNA and protein, ΔN_{NaCl}, is (e.g., Record et al. 1998),

$$\frac{d\ln(K_a)}{d\ln([NaCl])} = \Delta N_{NaCl} \qquad (1)$$

Since there may be other conformational changes in the DNA or protein coupled to binding that may bind or release additional salt ions, ΔN_{NaCl} is not necessarily a direct measure of the number of DNA–protein ion pairs formed in the complex. Linkage equations just like the above can be written for any solution component such that net changes in, for example, metal ion or ligand binding or protonation accompanying the formation of a protein–DNA complex can also be measured. The ΔN values extracted from these linkage relations are the differences in the binding of these solution components between products and reactants.

Since many waters hydrating both the protein and DNA surfaces are typically displaced in forming the recognition interface of a specific complex (cf. Fig. 1), water itself should also be considered an important solution component. As with salt, the numbers of waters coupled to a reaction can be determined from the sensitivity of the equilibrium constant to water 'concentration' or activity. Although it is generally thought that 'water concentrations' (~55.6 M in dilute solutions) or activities change too little in ordinary solutions to affect reactions, the numbers of waters associated with macromolecular binding reactions are typically large enough, however, to cause significant effects. In order to change water concentration or activity a solute (osmolyte) must be added to the solution. This other component, of course, must necessarily not itself bind to the DNA or protein. Rather than speaking of a 'water concentration' ($[H_2O]$), however, it is more appropriate to use osmotic pressures or osmolal concentrations of solutes as measured by a vapor pressure osmometer, for example. For many commonly used solutes, osmolal concentrations are nearly the same (to within ~20%) as osmolyte

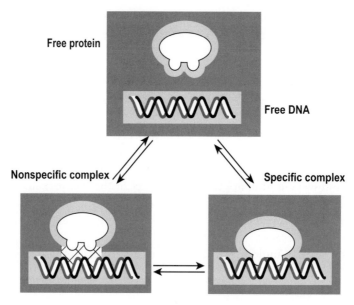

Fig. 1. Exclusion of water from protein, DNA, and their complexes. A schematic representation of a protein–DNA recognition reaction is shown to illustrate the two classes of protein- and DNA-associated water that exclude solutes and can affect binding constants as water chemical potential (osmotic stress) is changed. The protein is shown as globular with two lobes that represent helices that, e.g. specifically interact with DNA bases and are responsible for sequence recognition. Bulk solution (water and solute) is shown in *dark gray*. The free protein and DNA primarily exclude solutes from exposed surfaces through preferential hydration or crowding mechanisms. The *light gray* regions surrounding the protein and DNA surfaces represent a zone of osmolyte exclusion. The extent of solute exclusion (or water inclusion) from this zone will depend on the size and nature of the osmolyte probing the surface. In the specific complex (*lower right*), the DNA and protein come into direct contact, decreasing the amount of water that excludes solutes by preferential hydration. In addition to an exclusion by preferential hydration, the nonspecific EcoRI–DNA complex can also have a volume of water presumably in a cavity at the interface between surfaces, depicted by the *cross-hatched area* that sterically excludes solutes. Exclusion from this water will not depend on either solute size (after some minimum) or chemical nature

molal (mol/1 kg water) concentrations through ~1 molal. In terms of osmotic pressures expressed as osmolal concentrations,

$$\frac{d\ln(K_a)}{d[\text{osmolal}]} = -\frac{\Delta N_w}{55.6} \tag{2}$$

A more detailed discussion of linkage equations applied to water can be found in Parsegian et al. (1995, 2000).

There are generally two classes of waters illustrated for a DNA-binding protein (Fig. 1) that can be probed by adding osmolytes and changing water activity. Water sequestered in pockets, grooves, or cavities are often sterically inaccessible to solutes (the cross-hatched area of the nonspecific complex in Fig. 1). In this case, the size or chemical nature of the added solute does not matter. There is simply osmotic pressure acting on a volume of water.

For water hydrating protein and DNA surface areas that are exposed to the bulk solution (the light gray areas in Fig. 1), however, the number of waters probed depends on the competition between water and solute for interaction with the macromolecules and so varies with the size and chemical nature of the osmolyte. Two mechanisms are commonly considered. Crowding (Minton 1998) recognizes that there is a steric exclusion of large solutes from surfaces; osmolytes simply cannot approach as closely as water. Preferential hydration (Timasheff 1993, 1998) further recognizes that the interaction of water with groups on protein and nucleic acids surface may be energetically more favorable than with osmolyte. Many experiments measuring both exclusion of solutes from protein and DNA surfaces and the effect of this exclusion on macromolecular reactions show that the apparent number or change in the number of hydrating waters is constant over a wide range of solute concentrations for each osmolyte, but that this number is dependent on the particular solute probing surface hydration (e.g., Timasheff 1993; Courtenay et al. 2000; Davis-Searles et al., 2001). For the osmolytes we typically use, a total range of about three to five fold difference in exclusion is commonly observed. If a wide variety of osmolytes are examined, then the sensitivity to the solute nature can be used to distinguish changes in numbers of waters sequestered in pockets and cavities from changes in exposed surface area accompanying binding reactions.

Although we initially considered a simple binding reaction, linkage expressions such as Eqs. (1) and (2) can be written for any reaction and can even be applied to reaction rates. The competition reaction between specific and nonspecific DNA sequences for protein binding is $DNA_{nonsp} \cdot Protein + DNA_{sp} \Leftrightarrow DNA_{nonsp} + DNA_{sp} \cdot Protein$, with an equilibrium constant $K_{nonsp-sp}$. The osmotic pressure or salt concentration sensitivity, $d\ln(K_{nonsp-sp})]/d[osmolal]$ or $d[\ln(K_{nonsp-sp})]/d[\ln[NaCl]]$, gives the difference in 'bound' waters or salts: $(N_{nonspDNA} + N_{spComplex}) - (N_{spDNA} + N_{nonspComplex})$, where N is the total number of water molecules or salt ions associated with the reaction components. Since salt binding and solute exclusion from DNA is dominated by the sugar–phosphate backbone with very little contribution from the particular sequence, $N_{nonspDNA} \sim N_{spDNA}$ for both salt and water, leaving only the difference between specific and nonspecific complexes. It should be noted that free protein does not contribute to $K_{nonsp-sp}$. Competition experiments can either be done by measuring the change in specific binding constant as specific sequence DNA is titrated with protein with and without added competitor DNA or by measuring the loss of specific binding as specific DNA–protein complexes are titrated with competitor DNA.

Linkage relations can also be written for rate processes (Lohman 1985). If we consider the dissociation of a specifically bound protein, the osmotic or salt dependence of the rate, $d[\ln(k_d)]/d[osmolal]$ or $d[\ln(k_d)]/d[\ln[NaCl]]$, gives the difference in bound water or salt between the transition state and the specific complex. The transition state is defined for the rate-limiting kinetic step in dissociation as the high-energy structure from which it is about equally probable to rebind as to dissociate. Rate constants may also include a contribution from the osmolyte to the solution viscosity depending on whether the energy dissipation of the rate-limiting kinetic step is dominated by solvent friction or by the internal friction of the complex.

Many restriction nucleases, as EcoRI, are able to slide quite efficiently along DNA bound nonspecifically to facilitate finding recognition sequences (Wright et al. 1999; Stanford et al. 2000; Pingoud and Jeltsch 2001). The rate-limiting step for dissociation of EcoRI from DNA is the dissociation of the nonspecific complex. To a first order approximation, the overall dissociation rate, k_d, is the product of the dissociation rate of the nonspecific complex, $k_{d,nonsp}$, and the equilibrium constant for the specific to nonspecific complex reaction, $K_{nonsp-sp}$. The osmotic sensitivity (or salt or pH) of the dissociation can consequently be divided into these two steps,

$$\frac{d\ln(k_d)}{d[Osmolal]} = \frac{d\ln(k_{d,nonsp})}{d[Osmolal]} - \frac{d\ln(K_{nonsp-sp})}{d[Osmolal]} = -\frac{\Delta N_{w,d,nonsp}}{55.6} + \frac{\Delta N_{w,nonsp\text{-}sp}}{55.6} \quad (3)$$

If solution friction limits the dissociation rate, k_d can be corrected for viscosity. A number of waters linked to the dissociation rate of a nonspecifically bound protein, $k_{d,nonsp}$, can be calculated from the osmotic dependence of the specific-nonspecific binding equilibrium and of the overall dissociation rate, k_d.

Finally, it must be recognized that with added osmolytes sequestered waters are under a pressure. If there are alternate conformations of nonspecific complexes, for example, that sequester less water, then these states will be stabilized by osmotic stress. Loss of water in nonspecific complexes will be observed as a decrease in ΔN_w between specific and nonspecific complexes as the osmotic pressure increases. The energy of removing water from these nonspecific complexes can be calculated from pressure–volume ($\Pi\Delta V$) work accompanying this loss of water. It is important, however, that proper controls are performed to ensure that an apparent change in ΔN_w is actually due to water loss and not to the many other possible effects of osmolytes that may occur.

3 Experimental Applications

3.1 Equilibrium Competition

3.1.1 Osmotic Stress Dependence or $K_{nonsp-sp}$

Figure 2a shows a typical gel mobility-shift assay illustrating the competition for EcoRI binding between nonspecific poly(dI-dC) · poly(dI-dC) and a 322 bp DNA fragment containing the specific EcoRI recognition sequence, GAATTC, in the presence of different concentrations of triethylene glycol. In each series, EcoRI protein concentration is held constant and the amount of competitor DNA increased. The fraction of specific complex decreases with increasing competitor DNA concentration. Significantly more competitor DNA is required at 1 molal triethylene glycol to reach the same level of competition than in the absence of solute.

Fig. 2. Osmotic pressures favor specific sequence binding of EcoRI. **a** Poly(dI-dC)·poly(dI-dC) competes with a DNA fragment containing the specific recognition sequence for EcoRI binding. With increasing concentrations of neutral solutes, the ability of nonspecific polynucleotide to compete is significantly diminished. Competition experiments at three osmotic pressures are shown: no osmolyte added, 0.6 molal triethyleneglycol (0.62 osmolal), and 1 molal triethyleneglycol (1.05 osmolal). No divalent ion is present in order to avoid the enzymatic cleavage reaction. **b** The dependence of free energy difference between specific and nonspecific EcoRI–DNA complexes, $RT\ln(K_{nonsp-sp}/K^0_{nonsp-sp})$, on solute osmolal concentration is shown for several solutes. Competitive binding free energies scale linearly with osmotic pressure. The slope of the best fitting line translates into a difference of about 110±15 water molecules between specific and nonspecific EcoRI–DNA complexes

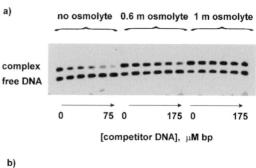

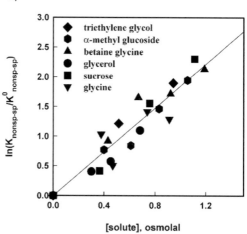

The ability of nonspecific polynucleotide to compete with specific DNA site for EcoRI binding depends of course on the ratio of specific and nonspecific binding constants ($K_{nonsp-sp}$). In the absence of osmolyte, $K^0_{nonsp-sp}$ was determined as $(2.8\pm0.4)\times10^4$. Compared with no added osmolyte, $K_{nonsp-sp}$ is about 4.5 times larger in 0.6 molal triethylene glycol and about 11 times larger in 1 molal triethylene glycol. The difference in the binding free energy of the EcoRI to specific and nonspecific DNA sequences is $RTln(K_{nonsp-sp})$. The dependence of free energy difference (divided by factor RT=0.6 kcal/mole) on concentration of added solute is shown in Fig. 2b as a function of osmolal concentration for several chemically distinct solutes. Two most important features in this graph to note are: (1) changes in competitive binding free energies scale linearly with changes in osmolal concentration or, equivalently, water chemical potential; (2) there is practically no difference among the several neutral solutes: betaine glycine, sucrose, glycerol, triethylene glycol, glycine, and α-methyl glucoside.

The difference in the number of solute excluding water molecules between specific and nonspecific EcoRI–DNA complexes, ΔN_w, can be determined from the slope of the best-fitting line to the data in Fig. 2b as specified by Eq. (2). The best fit to all data gives $\Delta N_w=-110\pm15$, indicating that the nonspecific complex sequesters about 110 waters more than the specific one. The observed insensitivity of ΔN_w to solute size and nature strongly suggests that observed difference in exclusion of solutes from 110 water molecules is strictly steric. Since the crystal structure of the specific EcoRI–DNA complex (McClarin et al. 1986; Kim et al. 1990) shows essentially anhydrous contact between DNA and protein surfaces, these 110 waters are likely sequestered at the protein–DNA interface of the nonspecific complex (see cartoon in Fig. 1).

3.1.2 pH and Salt Dependence of $K_{nonsp-sp}$

In contrast to the very strong dependence of specific binding of free EcoRI on salt concentration corresponding to the release of 10–12 ions (Jen-Jacobson 1997), only a small sensitivity to salt is seen for the equilibrium between specific and nonspecific binding of the EcoRI using the competitive binding assay. Between 90 and 160 mM NaCl the competitive binding constant of the EcoRI to specific versus nonspecific DNA sequences increases by only 50 % (Sidorova and Rau 2001). The salt dependence of $K_{nonsp-sp}$ on salt concentration can be analyzed either as a difference in ion binding reflecting a difference in DNA–protein charge interactions between specific and nonspecific complexes or as an indirect, osmotic effect of salt. The data over this limited salt concentration range are insufficiently precise to distinguish between these alternatives. If analyzed as a change in direct salt binding, the data are consistent with a release of an additional 0.6 ion in forming the nonspecific complex from the specific one. This small increase could be due to closer interactions of nonspecifically bound protein charge with DNA phosphate groups on the backbone. This dependence is, however, more likely due to the osmotic contribution of salt. As long as salt is excluded from the same water-filled cavities as the neutral solutes, an osmotic effect is required. If salt acts osmotically on the equilibrium between specific versus nonspecific binding with the same $\Delta N_w = -110$ molecules as for the neutral solutes, then the osmotic contribution would be equivalent to the release of ~ 0.5 ion over the salt concentration range examined, very close to the experimentally measured value of 0.6 ions. Several other DNA-binding proteins also show slightly more ions coupled to nonspecific binding than to specific binding (Lohman 1985; Record and Spolar 1990; Jen-Jacobson 1997). It seems probable that this general behavior is a reflection of a general osmotic action of salt on differences in water sequestered between specific and nonspecific complexes rather than an electrostatic effect as is commonly assumed.

Similarly, even though the equilibrium constant for specific binding of free EcoRI shows a strong dependence on pH (Jen-Jacobsen et al. 1983), there is practically no effect of pH on the competitive specific and nonspecific sequence binding equilibrium of EcoRI.

3.2 Dissociation Kinetics of EcoRI from its Specific Site

Although most thermodynamic work has focused on equilibrium constants for DNA–protein complex formation, studies of dissociation rates (k_d) of DNA–protein complexes are also extensive. Dissociation kinetics are important not only for understanding reaction rates of nucleases, ligases, polymerases, and repair enzymes, for example, but also because the binding of regulatory proteins to their target DNA sequences within a cell may be kinetically controlled and not an equilibrium reaction. Often the equilibrium constants and dissociation rates show similar dependences on salt concentration and pH (Lohman 1985). Differences, however, can lead to a better formulation of the detailed binding scheme. In particular, the dissociation of many DNA–protein complexes, including the specific EcoRI–DNA complex, occurs in two steps (Lohman 1985; Pingoud and Jeltsch 1997, 2001). There is a steady-state reaction between specific and nonspecific binding of the protein to the DNA. Nonspecifically bound protein can linearly diffuse along the DNA and can either return to the specific binding site or eventually dissociate from the DNA. Differences in the sensitivities of relative specific–nonspecific equilibrium binding constants and dissociation rates on solution conditions (salt concentration, pH, water activity, etc.) can distinguish between factors that are in common and that are different for specific and nonspecific binding. The total number of water molecules, salt ions or protons seen in the dissociation reaction of the EcoRI from its specific sequence can be separated into the contributions from the differences in the numbers of associated water molecules, ions or protons between the nonspecifically bound complex and the transition state for protein dissociation, and between EcoRI specifically and non-specifically bound complexes, respectively, as given in Eq. (3).

3.2.1 Osmotic Dependence of k_d

The dissociation of EcoRI from its specific site on DNA is well described by a single exponential and the rate of dissociation is sensitive to osmolyte concentration. Just as the dependence of an equilibrium binding constant on salt or water activity gives the difference in the number of ions or water molecules associated with products and reactants, the sensitivity of a rate constant to salt or water activities is determined from the difference in the number of ions or water molecules associated with the initial state and the transition state of the rate-limiting kinetic step. The sensitivity of the dissociation rate on water activity, for example, is determined by the difference in the numbers of water molecules associated with the initial specific complex of the enzyme with DNA and the transition state of the rate-limiting kinetic step of dissociation. Figure 3 shows plots of $\ln(k_d)$ versus solute osmolal concentration for a wide variety of solutes. Plots are linear for each solute (including the no osmolyte limit). All of the solutes are closely similar in their ability to slow down dissociation. There is, however, more solute-specific variation than we observed previously for $K_{nonsp-sp}$. The number of waters linked to the dissociation of the EcoRI from its specific site varies between 100 water molecules for sucrose and 155 for triethylene glycol and TMAO. These variations (50% at the most) are still much less than the factor differences of two to five expected for an osmotic effect based solely on exclusion of solutes from exposed surface

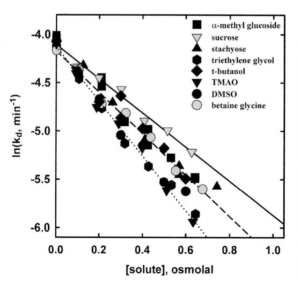

Fig. 3. Dependence of the dissociation rate on osmotic pressure. The dependence of $\ln(k_d)$ on solute osmolal concentration for the EcoRI for a wide variety of osmolytes. The slopes of the lines translate into 100 ± 6 waters for sucrose, 120 ± 8 for stachyose, 125 ± 6 for a-methyl glucoside, 120 ± 10 for t-butanol, 120 ± 6 for betaine glycine, 155 ± 8 for triethylene glycol, 145 ± 8 for dimethyl sulfoxide (DMSO), and 155 ± 4 for trimethylamine N-oxide (TMAO)

areas. Of these total waters coupled to EcoRI dissociation, 110 water molecules accompany the specific – nonspecific complex equilibrium. For the seven solutes other than sucrose the remaining number of waters accompanying the dissociation of nonspecifically bound protein is positive and varies from 10 to 45 extra water molecules. Given this dependence on solute identity, this reaction step is very likely characterized by a change in solute-accessible surface area. For sucrose this number was slightly negative (-10 ± 20) water molecules. This represents either no exclusion from or a small preferential inclusion of sucrose with the newly exposed surface area. Adjusting dissociation rates for the solution viscosity gave physically unrealistic results (Sidorova and Rau 2001), suggesting that the rate of dissociation of nonspecifically bound protein is limited by an internal friction rather than solution viscosity.

3.2.2 pH and Salt Dependence of k_d

The dissociation rate is coupled to a pH titration. Over the range of pH between 6.2 and 9.2, k_d increases by about 40-fold (Sidorova and Rau 2001). Since we observe virtually no pH-dependence of the equilibrium between specific and nonspecific binding measured by our competition assay, pH affects only the dissociation of the nonspecifically bound protein from the DNA. The pH-dependence can be fit adequately by assuming two titrating groups with identical pK values. The number of water molecules coupled to the rate of dissociation is insensitive to pH. Water activity and pH are acting independently to affect binding and dissociation.

Similarly, the dissociation rate is strongly dependent on salt concentration (Jen-Jacobson et al. 1986; Lesser et al. 1993; Sidorova and Rau 2001). We have measured binding of 5.8 ± 0.5 ions linked to the EcoRI dissociation from its specific site. Since the salt concentration dependence of the specific to nonspecific binding reaction is equivalent to the release of ~0.6 ions (see Sect. 3.1.2), ~6.4 ions are linked to the dissociation of nonspecifically bound EcoRI. Not surprisingly, pH and salt are coupled. The net binding of two extra Na^+ ions linked to the dissociation of nonspecifically bound enzyme accompanies the protonation of the two extra protein groups between pH 9.2 and 6.2. In contrast to the osmotic dependence, both salt and pH predominantly affect the dissociation of nonspecifically bound protein, not the equilibrium between specific and nonspecific binding.

These results clearly demonstrate the importance of water compared with salt and pH in distinguishing specific and nonspecific EcoRI binding. The osmotic dependence is primarily due to differences between specific and nonspecific binding of the EcoRI. Both salt and pH sensitivities are mostly due to the nonspecific binding of free protein and do not distinguish between specific and nonspecific binding. Our results also suggest that osmotic stress might be a convenient way to increase stability and lifetime of weak complexes for separation, chemical modification, or measurement of physical properties. Since the numbers of water molecules coupled with the dissociation of nonspecifically bound EcoRI are quite solute dependent then the ability to stabilize weak nonspecific complexes will also differ among osmolytes. Among all solutes used triethylene glycol, TMAO, and DMSO would be the most effective for stabilizing nonspecific EcoRI complexes. For example, half-life of a nonspecific complex should increase about 3 times between 0 and 1 molal triethylene glycol. In contrast, sucrose either would not influence the stability of a nonspecific complex or might even slightly destabilize it.

3.3 Removing Water from an EcoRI–Noncognate DNA Complex with Osmotic Stress

The equilibrium competition experiments discussed in Section 3.1.1 were restricted to comparatively low osmotic stresses and, therefore, small energy perturbations that are less than about 2RT or 1.2 kcal/mol. The linear dependence of the free energy difference on osmolality (Fig. 2b) shows that difference in the number of water molecules sequestered by nonspecific versus specific complex remains the same at least up to 1.2 osmolal. At high enough osmotic pressures, however, the pressure–volume or $\Pi\Delta V$ work gained by removing waters from a nonspecific complex will be comparable to the unfavorable interaction energy incurred in forcing closer contact between non-complementary surfaces. In principle, any sequestered water can be removed by applying high enough osmotic stress, but the work necessary to dehydrate complexes will obviously depend on the resulting DNA–protein contacts and complex structure.

It has long been known that EcoRI restriction endonuclease as well as many other Type II restriction endonucleases are capable of cleaving sequences that are similar to but not identical with the canonical recognition sequence, termed 'star' activity sites. The presence of neutral solutes, such as glycerol, dimethyl sulfoxide, ethanol, ethylene glycol, and sucrose, are among those solution conditions that promote 'star' activity. It has been shown by Robinson and Sligar (1993) that the increased digestion by EcoRI at 'star' sequences caused by neutral solutes is strictly correlated with water activity. One possible explanation for the effect of solutes on the 'star' activity is that osmotic stress modulates an equilibrium between a predominating nonspecific (water-mediated contact) and an energetically unfavorable, but enzymatically active, specific-like (direct protein–DNA contacts) modes of EcoRI binding to 'star' sequences.

3.3.1 Competitive Equilibrium at High Osmotic Stress

The general strategy is the same as described in Section 3.1.1. Comparative binding constants of EcoRI to different DNA sequences are measured by a competition assay. The loss of EcoRI binding to a DNA fragment containing its specific recognition sequence as the concentration of a competing oligonucleotide increases is measured using the gel mobility shift assay. Figure 4a shows the dependence on the osmolal concentration of betaine glycine of the relative binding free energies of EcoRI to two 30 bp oligonucleotides, differing only in that one contains a central 'star' sequence TAATTC and the other nonspecific oligonucleotide contains the inverted recognition sequence CTTAAG instead. The 'star' sequence oligonucleotide binds EcoRI only about twofold more strongly than the inverted sequence, nonspecific oligonucleotide. As seen from the slopes of the curves in Fig. 4a at low pressures, both oligonucleotide complexes sequester some 110 more waters than the specific sequence one. Even a single base pair change from the recognition sequence is sufficient to trigger the nonspecific-binding mode of EcoRI. Above ~2 osmolal the slope of the plot for the TAATTC 'star' sequence oligonucleotide complex is clearly smaller than at low pressures suggesting a loss of

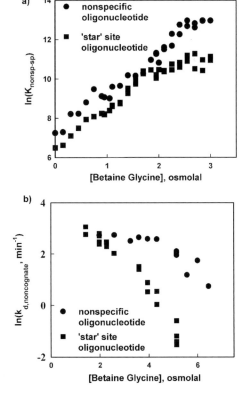

Fig. 4. Removing water at high osmotic pressures. **a** The binding energies, $\ln(K_{nonsp\text{-}sp})$, for two oligonucleotides relative to the specific complex is shown as a function of betaine glycine osmolal concentration. $K_{nonsp\text{-}sp}$ is the ratio between EcoRI binding constants to specific and competitor noncognate sequences measured in the experiments analogous to the one shown in Fig. 2a. The 'star' sequence competitor contains the sequence TAATTC. The nonspecific, inverted sequence oligonucleotide has CTTAAG instead. The two oligonucleotides are otherwise identical. The 'star' sequence complex seems to lose water at high osmotic pressures. **b** The dependence of $\ln(k_d)$ for the EcoRI dissociation from the two noncognate oligonucleotides, 'star' sequence and nonspecific, on betaine glycine osmolal pressure is shown. Significantly more water can be removed from the 'star' sequence complex compared with the nonspecific, inverted sequence one

water from the 'star' sequence complex with increasing osmotic stress. The average slope in the 2–3 osmolal range corresponds to 70 waters. No loss of water is seen for the complex with the nonspecific, inverted sequence oligonucleotide.

3.3.2 Dissociation Kinetics of the EcoRI from Noncognate Sites

The equilibrium experiments at high osmotic stress are difficult since dissociation rates are so very slow and the competitor DNA concentrations necessary to observe competition are very high. Alternatively, the osmotic dependence of the dissociation rate of 'star' and nonspecific DNA sequence complexes can be conveniently measured to very high osmotic stresses. The dissociation rate from noncognate DNA sequence complexes can be determined by adding EcoRI to a mixture of specific sequence DNA fragment and noncognate oligonucleotide and measuring the time course for specific binding. Under conditions of fast association, the appearance of specific complex depends on three factors: (1) the ratio of specific and nonspecific association rate constants, (2) the concentrations of specific sequence fragment and noncognate oligonucleotide, and (3) the nonspecific complex dissociation rate. The osmotic dependences of the EcoRI dissociation rate constant from nonspecific and 'star' sequence oligonucleotide complexes are shown in Fig. 4b.

At relatively low osmotic pressures there is only a small difference between two oligonucleotides as expected from equilibrium competition experiments (Fig. 4a). The slope at osmotic pressures <5 osm for the nonspecific sequence complex corresponds to ~10 waters, very close to the value inferred for the dissociation of nonspecifically bound protein from the specific complex dissociation rate measurements (Sect. 3.2.1). EcoRI dissociation from the 'star' site (TAATTC) oligonucleotide is obviously much more sensitive to osmotic stress. In the 2–3 osmolal range, the slope corresponds to ~40 waters or only ~80 waters left in the 'star' sequence complex, in reasonable agreement with the estimate from equilibrium experiments in the same pressure range (Sect. 3.3.1). At even higher pressures between 4 and 5 osmolal the osmotic sensitivity of $k_{d,nonsp}$ translates into uptake of about 100 waters for this noncognate complex, leaving only ~20 waters in the complex. Over the same high range of pressures the nonspecific, inverted sequence complex may also be losing some water, the slope giving ~25 waters or about 95 waters remaining in the complex. The precision of experiments, both equilibrium and kinetics, is not high enough to distinguish between a two state model, i.e., between a 'star' sequence complex with ~110 sequestered waters and a second discrete binding mode with much less associated water, or a continuum of states, i.e., a gradual loss of water from the 'star' complex. The approximate $\Pi\Delta V$ work for removing ~90 waters from the 'star' sequence complex is about 4 kcal/mol.

3.4 Other Applications of Osmotic Stress to Restriction Nucleases

Robinson and Sligar (1998) have measured the osmotic sensitivities of non-specific and specific EcoRI binding separately using a single osmolyte, ethylene glycol. Since these reactions are characterized by a large change in solution accessible surface area, apparent hydration numbers should depend on the solute size and nature. A difference of only ~70 waters (a release of 146 waters accompanying specific binding and 76 waters coupled to nonspecific complex formation) was observed rather than the 110 found using the competition assay and many more solutes (Sect. 3.1.1). Additionally, Lynch and Sligar (2000) report that the specific binding of BamHI was accompanied by the release of only about 20 waters. No osmotic dependence for nonspecific binding was reported.

The osmotic stress approach has been used to measure a number of waters coupled to enzymatic rates, k_{cat} or V_{max}. Robinson and Sligar (1998) find an uptake of ~28 coupled to the turnover rate of EcoRI; Lynch and Sligar (2000) report almost +90 waters linked to the BamHI turnover; and Wenner and Bloomfield (1999) measure 28 waters for EcoRV. Since these reaction rates were measured under conditions of multiple turnover it is not clear what step in the reaction osmotic stress is probing, cleavage or subsequent dissociation.

Robinson and Sligar (1993, 1995b) have reported that the 'star' activities not only of EcoRI (as already noted in Sect. 3.3), but also of PvuII and BamHI correlate well with osmotic stress for many different solutes. The 'star' activity of EcoRV, however, is not sensitive to osmolytes.

3.5 Application of Hydrostatic Pressure to Restriction Nucleases

Hydrostatic pressure is another probe of water structuring that has been applied to restriction endonuclease binding and cleavage kinetics, primarily by Sligar and coworkers (e.g., Lynch and Sligar 2002). Whereas osmotic pressure favors those species that have fewer numbers of waters that exclude solutes, hydrostatic pressure stabilizes those species that have larger densities or, equivalently, smaller volumes (e.g., Silva et al. 2001). If, for example, the hydrating water of DNA and protein that is released in forming a complex is more dense than bulk, then, all else remaining constant, hydrostatic pressure promotes complex dissociation, as observed for specific BamHI binding (Lynch et al. 2002). The stabilization energy is $P\Delta V$, where ΔV is the solution volume change. If the average molecular volume of bulk water is $\bar{v}_w^0$ and of the N_w released hydrating waters $\bar{v}_w$, then $\Delta V = N_w (\bar{v}_w - \bar{v}_w^0)$. Of course, protein and DNA conformational changes may also have accompanying density changes that will be sensitive to hydrostatic pressure. We have already noted that Sligar and coworkers found that the 'star' activity of a number of restriction nucleases (EcoRI, PvuII, and BamHI) is enhanced by osmotic stress.

Hydrostatic pressure acts to reverse the effect (Robinson and Sligar 1995a, b). Either hydrostatic pressure simply inhibits the turnover reaction or, more interestingly, alters the energetics of removing water from 'star' sequence complexes. The density of water in the DNA–protein cavity could be much different from bulk water.

4 Summary

No structure is available for the nonspecific EcoRI complex to compare with the specific sequence complex structure (McClarin et al. 1986; Kim et al. 1990) to confirm a water cavity at the protein–DNA interface. Structures of specific and nonspecific complexes of a closely similar restriction nuclease BamHI, however, have been reported (Newman et al. 1995, Viadiu and Aggarval 2000). The cavity at the protein–DNA interface of the nonspecific complex has a volume of 4763 $Å^3$, compared with only 282 $Å^3$ for the specific sequence complex. Assuming a typical volume of 30 $Å^3$ per water molecule, the difference in cavity sizes corresponds to 150 waters comparable to the 110 waters we find for EcoRI. This 'loose', 'water-lubricated' association of DNA within a binding cleft of the protein is a conformation that would easily allow linear diffusion of the protein along the DNA.

We see that ion release and proton binding are simply a result of nonspecific EcoRI binding. There is no further sensitivity to pH or salt (except for an apparent osmotic contribution from salt) between specific and nonspecific modes of binding. There is, however, a large difference in water. If indeed this water is at the DNA–protein interface of the nonspecific complex as suggested by the structure of the nonspecific BamHI complex, then 110 waters corresponds to ~1.5 hydration layers. The noncognate DNA and protein surfaces prefer to keep their hydration interactions, suggesting that water does play an important role in recognition. This is also consistent with heat capacity measurements. The formation of many specific DNA–protein complexes, including EcoRI (Ha et al. 1989), is accompanied by a large change in heat capacity. A large portion of this change seems to come from the release of hydration waters that are structured differently from bulk water (Spolar and Record 1994). In contrast, only very small heat capacity changes accompany the formation of several nonspecific complexes, consistent with retention of hydrating waters. The waters at the noncognate DNA–protein interface can only be removed with great difficulty. Indeed, we have only been able to remove water from EcoRI complexes with DNA sequences that differ by only one base pair from the recognition and even then the energy required is quite significant, ~4 kcal/mol.

Direct measurements of forces between macromolecules in condensed arrays indicates that water structuring forces do seem to dominate interactions at close spacings, the last 10–15 Å of separation between surfaces (Leikin et al. 1993). The key to understanding the binding strength and specificity of restriction endonuclease–DNA interactions is in understanding the energetics of the hydration interactions that must be replaced in forming a complex.

References

Courtenay ES, Capp MW, Anderson CF, Record MT Jr (2000) Vapor pressure osmometry studies of osmolyte-protein interactions: implications for the action of osmoprotectants in vivo and for the interpretation of "osmotic stress" experiments in vitro. Biochemistry 39:4455–4471

Davis-Searles PR, Saunders AJ, Erie DA, Winzor DJ, Pielak GJ (2001) Interpreting the effects of small uncharged solutes on protein-folding equilibria. Annu Rev Biophys Biomol Struct 30:271–306

Frank DE, Saecker RM, Bond JP, Capp MW, Tsodikov OV, Melcher SE, Levandovski MM, Record MT Jr (1997) Thermodynamic of the interactions of lac repressor with variants of the symmetric lac operator: effects of converting a consensus site to a nonspecific site. J Mol Biol 267:1186–1206

Fried MG, Stickle DF, Smirnakis KV, Adams C, MacDonald D, Lu P (2002) Role of hydration in the binding of lac repressor to DNA. J Biol Chem 277:50676–50682

Garner MM, Rau DC (1995) Water release associated with specific binding of gal repressor. EMBO J 14:1257–1263

Ha J-H, Spolar RS, Record MT Jr (1989) Role of the hydrophobic effect in stability of site-specific protein-DNA complexes. J Mol Biol 209:801–816

Janin J (1999) Wet and dry interfaces: the role of solvent in protein–protein and protein–DNA recognition. Structure 7:R277–R279

Jen-Jacobson L (1997) Protein–DNA recognition complexes: conservation of structure and binding energy in the transition state. Biopolymers 44:153–180

Jen-Jacobson L, Kurpiewski M, Lesser D, Grable J, Boyer HW, Rosenberg JM, Greene PJ (1983) Coordinate ion pair formation between EcoRI endonuclease and DNA. J Biol Chem 258:14638–14646

Jen-Jacobson L, Lesser D, Kupriewski M (1986) The enfolding arms of EcoRI endonuclease: role in DNA binding and cleavage. Cell 45:619–629

Khrapunov S, Brenowitz M (2004) Comparison of the effect of water release on the interaction of the *Saccharomyces cerevisiae* TATA binding protein (TBP) with "TATA box" sequences composed of adenosine or inosine. Biophys J 86:371–383

Kim YC, Grable JC, Love R, Greene PJ, Rosenberg JM (1990) Refinement of EcoRI endonuclease crystal structure: a revised protein change tracing. Science 249:1307–1309

Leikin S, Parsegian VA, Rau DC (1993) Hydration Forces. Annu Rev Phys Chem 44:369–395

Lesser DR, Kurpiewski MR, Jen-Jacobson L (1990) The energetic basis of specificity in the EcoRI endonuclease-DNA interaction. Science 250:776–786

Lesser D, Kurpiewski MR, Waters T, Connolly BA, Jen-Jacobson L (1993) Facilitated distortion of the DNA site enhances EcoRI endonuclease-DNA recognition. Proc Natl Acad Sci USA 90:7548–7552

Li L, Matthews KS (1997) Differences in water release with DNA binding by ultrabithorax and deformed homeodomains. Biochemistry 36:7003–7011

Lohman TM (1985) Kinetics of protein–nucleic acid interactions: use of salt effects to probe mechanisms of interaction. CRC Crit Rev Biochem 19:191–245

Lundback T, Hansson H, Knapp S, Ladenstein R, Hard T (1998) Thermodynamic characterization of non-sequence-specific DNA-binding by the Sso7d protein from *Sulfolobus solfataricus*. J Mol Biol 276:775–786

Lynch TW, Sligar SG (2000) Macromolecular hydratoion changes associated with BamHI binding and catalysis. J Biol Chem 275:30561–30565

Lynch TW, Sligar SG (2002) Experimental and theoretical high pressure strategies for investigating protein–nucleic acid assemblies. Biochim Biophys Acta 1595:277–282

Lynch TW, Kosztin D, McLean MA, Schulten K, Sligar SG (2002) Dissecting the molecular origins of specific protein–nucleic acid recognition: hydrostatic pressure and molecular dynamics. Biophys J 82:93–98

McClarin JA, Frederick CA, Wang BC, Greene P, Boyer HW, Grable J, Rosenberg JM (1986) Structure of the DNA–EcoRI endonuclease recognition complex at 3 Å resolution. Science 234 (4783):1526–1541

Minton AP (1998) Molecular crowding: analysis of effects of high concentrations of inert cosolutes on biochemical equilibria and rates in terms of volume exclusion. Methods Enzymol 295:127–149

Newman M., Strzelecka T., Dorner LF, Schildkraut I, Aggarwal AK (1995) Structure of BamHI endonuclease bound to DNA: partial folding and unfolding on DNA binding. Science 269:656–663

Parsegian VA, Rand RP, Rau DC (1995) Macromolecules and water: probing with osmotic stress. Methods Enzymol 259:43–94

Parsegian VA, Rand RP, Rau DC (2000) Osmotic stress, crowding, preferential hydration, and binding: a comparison of perspectives. Proc Natl Acad Sci USA 97:3987–3992

Pingoud A, Jeltsch A (1997) Recognition and cleavage of DNA by Type II restriction endonucleases. Eur J Biochem 246:1–22

Pingoud A, Jeltsch A (2001) Structure and function of Type II restriction endonucleases. Nucleic Acids Res 15:3705–3727

Poon J, Bailey M, Winzor DJ, Davidson BE, Sawyer WH (1997) Effects of molecular crowding on the interaction between DNA and the *Escherichia coli* regulatory protein TyrR. Biophys J 73:3257–3264

Record MT Jr, Spolar RS (1990) Some thermodynamic principles of nonspecific and site-specific protein–DNA interactions. In: Revzin A (ed) The biology of nonspecific DNA–protein interactions. CRC Press, Boca Raton, pp 33–69

Record MT Jr, Zhang W, Anderson CF (1998) Analysis of effects of salts and uncharged solutes on protein and nucleic acid equilibria and processes: a practical guide to recognizing and interpreting polyelectrolyte effects, Hofmeister effects, and osmotic effects of salts. Adv Prot Chem 51:281–353

Robinson CR, Sligar SG (1993) Molecular recognition mediated by bound water. A mechanism for star activity of the restriction endonuclease EcoRI. J Mol Biol 234:301–306

Robinson CR, Sligar SG (1995a) Hydrostatic pressure reverses osmotic pressure effects on the specificity of EcoRI–DNA interactions. Method Enzymol 259:395–427

Robinson CR, Sligar SG (1995b) Heterogenity in molecular recognition by restriction endonucleases: osmotic and hydrostatic pressure effects on BamHI, PvuII, and EcoRV specificity. Proc Natl Acad Sci USA 92:3444–3448

Robinson CR, Sligar SG (1996) Participation of water in Hin recombinase–DNA recognition. Protein Sci 5:2119–2124

Robinson CR, Sligar SG (1998) Changes in solvation during DNA binding and cleavage are critical to altered specificity of the EcoRI endonuclease. Proc Natl Acad Sci USA 95:2186–2191

Sidorova N, Rau DC (1996) Differences in water release for the binding of EcoRI to specific and nonspecific DNA sequences. Proc Natl Acad Sci USA 93:12272–12277

Sidorova N, Rau DC (1999) Removing water from and EcoRI–noncognate DNA complex with osmotic stress. J Biomol Struct Dynam 17:19–31

Sidorova N, Rau DC (2000) The dissociation rate of the EcoRI–DNA specific complex is linked to water activity. Biopolymers 53:363–368

Sidorova N, Rau DC (2001) Linkage of EcoRI dissociation from its specific DNA recognition site to water activity, salt concentration, and pH: separating their roles in specific and non-specific binding. J Mol Biol 310:801–816

Silva JL, Fougel D, Royer CA (2001) Presure provides new insights into protein folding, dynamics and structure. Trends Biochem Sci 26:612–618

Spolar RS, Record MT, Jr. (1994) Coupling of local folding to site-specific binding of proteins to DNA. Science 263:777–784

Stanford NP, Szczelkun MD, Marko JF, Halford SE (2000) One- and three-dimensional pathways for proteins to reach specific DNA sites. EMBO J 19:6546–6557

Takeda Y, Ross PD, Mudd CP (1992) Thermodynamics of Cro protein–DNA interactions. Proc Natl Acad Sci USA 89:8180–8184

Timasheff SN (1993) The control of protein stability and association by weak interactions with water: how do solvents affect these processes. Annu Rev Biophys Biomol Struct 22: 27–65

Timasheff SN (1998) Control of protein stability and reactions by weakly interacting cosolvents: the simplicity of the complicated. Adv Protein Chem 51:355–432

Viadiu H, Aggarwal AK (2000) Structure of BamHI bound to nonspecific DNA: a model for DNA sliding. Mol Cell 5:889–895

Vossen KM, Wolz R, Daugherty MA, Fried MG (1997) Role of macromolecular hydration in the binding of the *Escherichia coli* cyclic AMP receptor to DNA. Biochemistry 36:11640–647

Wenner JR, Bloomfield VA (1999) Osmotic pressure effects on EcoRV cleavage and binding. J Biomol Struct Dynam 17:461–471

Wright DJ, Jack WE, Modrich P (1999) The kinetic mechanism of EcoRI endonuclease. J Biol Chem 274:31896–31902

Wu J, Parhurst KM, Powell RM, Brenowitz M, Parkhurst LJ (2001) DNA bends in TATA-binding protein–TATA complexes in solution are DNA sequence-dependent.

Role of Metal Ions in Promoting DNA Binding and Cleavage by Restriction Endonucleases

J.A. Cowan

1 Introduction

While three major classes of restriction endonucleases have been identified (Types I, II, and III), Type II are the most straightforward inasmuch as they require divalent magnesium as an essential cofactor but have no need for ATP (Stasiak 1980; Bennett and Halford 1989; Bujnicki 2000; Murray 2000; Sapranauskas et al. 2000; Pingoud and Jeltsch 2001). A complete classification of Type II restriction nucleases has been presented elsewhere (Pingoud and Jeltsch 2001) and the family is noted for the remarkable specificity and simplicity of its function. These enzymes cleave both strands of double-strand DNA either at or near a recognition sequence that tends to be palindromic. Consequently most restriction endonucleases are dimeric and recognize symmetric DNA sequences. While showing many functional similarities in DNA recognition and catalytic cleavage, restriction endonucleases also display low sequence homology, and diversity in mechanisms of recognizing DNA target sequences and the positioning of metal cofactors (Aggarwal 1995; Wah et al. 1997; Viadiu and Aggarwal 1998). Such diversity results in subtle variations in both protein binding locations and potential functional roles for essential metal cofactors that are only now coming under investigation.

In spite of their low sequence homology, Type II restriction endonucleases display significant structural similarity. This is strongest when comparing endonucleases that share a similar cleavage pattern; for example, BamHI and EcoRI that result in a four base overhang, and EcoRV and PvuII that produce blunt ends. Nevertheless, all restriction endonucleases display a highly conserved structural core (Fig. 1; Aggarwal 1995; Pingoud and Jeltsch 2001), irre-

J.A. Cowan
Evans Laboratory of Chemistry, The Ohio State University, 100 West 18th Avenue, Columbus, Ohio 43210

Nucleic Acids and Molecular Biology, Vol. 14
Alfred Pingoud (Ed.)
Restriction Endonucleases
© Springer-Verlag Berlin Heidelberg 2004

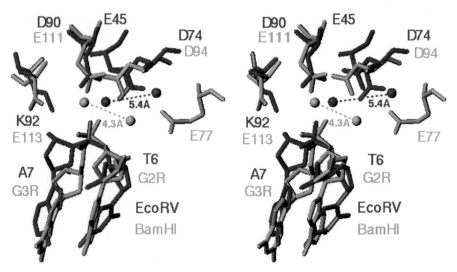

Fig. 1. A comparison of the BamHI (*blue*) and the EcoRV (*magenta*) active sites, prior to DNA cleavage. The bound metals differ both in separation and in their relative positions in the two enzymes. (Reprinted with permission from Viadiu and Aggarwal 1998)

spective of the nature of the cleavage chemistry and products. This core contains the catalytic center that includes two carboxylate residues and a lysine (Fig. 1), with the former implicated in binding divalent magnesium and the latter in stabilizing the increased negative charge built up in the transition state during attack by hydroxide ion. However, the role of Lys in transition state stabilization has been questioned inasmuch as the attacking hydroxide would be positioned on the same side of phosphate as the Lys residue. One structural study of EcoRV implicates H-bond formation with metal-bound water, and so Lys function may involve a positional influence on an essential metal cofactor (Horton et al. 1998). Consistent with this idea is the observation that Lys can never be replaced by the basic residue Arg, while BamHI is one of the few enzymes showing replacement of Lys (with Glu). The preservation of a structural core also belies the diversity of metal binding sites (comparing BamH1 and EcoRV) (Viadiu and Aggarwal 1998; Fig. 1) and the distinct quaternary structures of the metal-bound active dimers (comparing BglI and EcoRV) (Newman et al. 1998) that promote DNA recognition and cleavage. Clearly, the bound cofactors stabilize distinct quaternary structures and promote the orientation of active site residues in distinctive fashions, as dictated by tertiary structure refinement. Structural aspects of DNA recognition have been previously addressed (see Grigorescu et al.; Scheuring et al.; Winkler and Prota; Siksnys et al., this Vol.) and will not be considered further in this chapter, where the focus is exclusively on the role of essential metal cofactors in mediating both the interaction of restriction enzymes with cognate DNA sequences, and the catalytic hydrolysis of the phosphodiester backbone. Variations to be discussed include how the identity of the metal cofactor influences the reactivity (Mg^{2+}, Mn^{2+}, and Ca^{2+} being the most prominent choices), the role of the cofactor in promoting DNA binding, and whether the metal ions that promote DNA binding are distinct from cofactor(s) that mediate cleavage. Finally, a number of proposed functional roles for metal ions that promote cleavage of DNA will be summarized.

2 Selection of Metal Cofactors to Promote Endonuclease Activity

A comparison of the physicochemical properties of divalent magnesium and analogues has been made previously (Cowan 1992, 1995a, b, 1998). These works provide much insight on the selection of this metal cofactor for nuclease activity. In part this reflects the high natural abundance of this divalent cation, and a favorable combination of physical and chemical properties that include its redox inertness, small ionic radius and resulting high charge density, and a tendency to bind water molecules rather than bulkier ligands in the inner coordination shell, which leads to appropriate hydration states and slow solvent exchange rates. Consequently, magnesium usually maintains a limited ligation state with at most four, and more commonly two or three direct contacts with oxygen ligands on protein side-chains, and a preference for outer sphere contacts by hydrogen bond formation from waters of hydration to the nucleic acid backbone. In this regard the biological chemistry of divalent magnesium differs significantly from the other alkaline earth metals. Calcium in particular shows fewer bound water molecules (≤ 2) and an expanded coordination number of seven, reflecting the larger ionic radius of Ca^{2+} and reduced steric barriers to direct binding of larger protein-derived ligands. Primarily, it is the tendency of Mg^{2+} to maintain a moderate to high hydration state, and the involvement of these metal-bound H_2O in mediating the binding and catalytic chemistry of the magnesium cofactor that characterizes, and is the hallmark of the chemical interactions of this metal ion with nucleic acid substrates.

The fact that the essential magnesium divalent cofactor can be substituted by a number of other divalent metal ions (including Mn^{2+} and occasionally Fe^{2+}, Co^{2+}, Ni^{2+}, Zn^{2+}, and Cd^{2+}) but not Ca^{2+}, suggests that it is the size of the latter, and the likelihood of binding to other catalytically important carboxylate(s), that results in its inability to promote activity. Certainly Ca^{2+} binds to the catalytic site, where it acts as an inhibitor (Jose et al. 1999; Conlan and Dupureur 2002), however, it is most interesting to note that Mg^{2+}, Ca^{2+}, and Mn^{2+} do not necessarily bind to similar residues within the catalytic core (Pingoud and Jeltsch 2001), a result that has implications for extending mechanistic inference from crystallographic results obtained with the latter two metal ions. Inasmuch as Ca^{2+} promotes substrate binding, but not catalytic activity, this ion is a useful substitute for Mg^{2+} when one wishes to characterize the nuclease–substrate complex without concern over catalytic cleavage.

3 Magnesium Analogues

Divalent magnesium is essentially spectroscopically invisible (Cowan 1992, 1995a, 1998; Kim and Cowan 1992), and has such a low electron density that it is difficult to distinguish by most spectroscopic and crystallographic experiments. For this reason, the chemistry of magnesium-dependent enzymes has been studied with the use of transition metal probes and analogues (Cowan 1992, 1998; Kim and Cowan 1992). The underlying assumption in many studies of metal cofactor requirement in nucleic acid biochemistry is that magnesium analogues (especially manganese) show similar chemistry to that displayed by Mg^{2+}. In general this is not a good general assumption, and in fact, the stoichiometry and coordination mode of other metals may, and frequently do differ from Mg^{2+}. In particular, it is usually observed that Mn^{2+} (or other transition metals such as Co^{2+}, Fe^{2+}, or Zn^{2+}) confers higher levels of activity for both native enzyme, and Asp or Glu mutants, since these typically coordinate more tightly than the natural Mg^{2+} cofactor. These transition metal ions may also show changes in coordination geometry, or induce conformation perturbations in active site residues that influence substrate specificity. Consequently, a loss of specificity is commonly observed for restriction endonucleases with the use of metal cofactors other than Mg^{2+}.

It was noted earlier that Mn^{2+} will often promote higher levels of enzyme activity than Mg^{2+}. This should not be misconstrued as an indication that Mn^{2+} is in fact the natural cofactor, but simply reflects the distinct chemistry of this transition metal ion which promotes activity levels that are greater than the requirements for metabolism. While the use of analogs can provide insight on function, there are differences in coordination chemistry that can lead to changes in substrate selectivity and metal ion stoichiometry. Such problems need to be recognized and accounted for during data analysis.

4 Inhibitory Influence of Metal Cofactors

A variety of mechanistic models have been proposed and used to rationalize the metal dependence of nuclease activity (Jeltsch et al. 1992, 1995; Baldwin et al. 1995; Vipond et al. 1995; Vipond and Halford 1995; Horton et al. 1998). A general observation in almost all plots of enzyme activity as a function of metal cofactor concentration is an initial increase in activity with increasing $[Mg^{2+}]$, followed by a gradual decrease in activity. While direct enzyme inhibition is a possibility, the evidence for such is neither strong nor unequivocal. The author of this review is inclined to view this effect as a general example of substrate inhibition where the metal ions bind to the nucleic acid substrate, perturbing the charge density of the substrate and its interaction with the enzyme (Black and Cowan 1994). This may also be thought of as a form of competitive binding between the metal ions, and enzyme, to the substrate. Such a viewpoint is supported by the similarity of this response over a diverse group of enzymes (Demple et al. 1986; Black and Cowan 1994; Friedhoff et al. 1996), and the similarity of apparent binding constants for metal substrate complexation determined from these kinetic and thermodynamic measurements.

Significantly, this kind of inhibition effect is diminished by the presence of polyamines, which competitively displace Mg^{2+} ions, but have an overall reduced charge density relative to a number of divalent magnesium ions bearing an equivalent charge (Friedhoff et al. 1996). Such an observation may be taken as a diagnostic test for this kind of metal-mediated inhibition. The observation of sigmoidal behavior, or a lag-phase, in some kinetic profiles of activity versus metal cofactor concentration may be taken as evidence for the requirement of additional weakly bound metal cofactors that may either be required for catalytic activity, or to stimulate substrate binding.

5 Illustrative Examples and General Guidelines for Metal-Promoted Endonuclease Activity

General aspects of enzyme-mediated nucleic acid hydrolysis have been considered in detail in other recent reviews (Viadiu and Aggarwal 1998; Halford et al. 1999; Murray 2000; Rao et al. 2000; Halford 2001; Pingoud and Jeltsch 2001; Tsutakawa and Morikawa 2001). It appears likely that the general principles for promoting phosphodiester bond cleavage by the Type II endonucleases, and the specific roles for metal ion cofactors will show common themes for this family, and so any apparent differences in metal binding site preferences and related properties may actually provide a useful selection criterion for identifying the structural and mechanistic details that are likely to be common and of general importance.

Studies of phosphodiester bond cleavage by restriction enzymes all show inversion of the stereochemistry at phosphorus, consistent with direct attack by water and the absence of an intermediate enzyme-bound species (Grasby and Connolly 1992). However, this data does not allow one to distinguish between associative or dissociative-type mechanisms. Nevertheless, the importance of the metal cofactor is unquestioned, although the exact role and in particular the number of metal ions required to promote cleavage has been a topic of lively debate that this article is unlikely to resolve. As noted earlier, the propensity of various metal analogues to bind in a differing ways to the enzyme has muddied the waters. As illustrated in Figs. 2–4, one-, two-, and three-metal ion mechanisms have all been proposed (Lukacs et al. 2000;

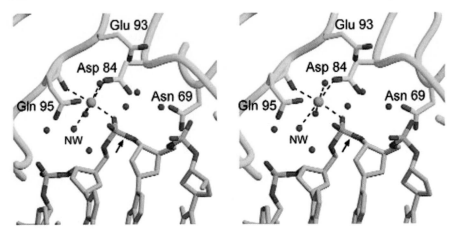

Fig. 2. The BglII active site around the scissile phosphodiester bond. Residues Asn 69, Asp 84, Glu 93, and Gln 95 correspond to the BamHI active site residues Glu 77, Asp 94, Glu 111, and Glu 113. The structure reveals numerous water molecules and an octahedrally coordinated cation (*orange sphere*) between the conserved residues and the DNA. Five of the six ligands have distances ranging from 2.2 to 2.5 Å; the sixth ligand (to the proposed nucleophilic water molecule) is 2.7 Å. One water molecule in the coordination sphere of the cation (labeled *NW*) makes a 155° angle with the scissile P–O3 bond, and appears to be the attacking nucleophile. (Reprinted with permission from Lukacs et al. 2000)

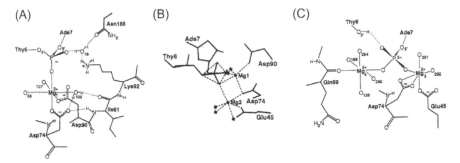

Fig. 3. **A** Schematic illustration of the catalytic pocket for the EcoRV-substrate complex. Note that only one metal cofactor is identified. **B** Proposed model for the transition state, showing recruitment of a second metal cofactor. However, in other work, the critical residue Glu45 was mutated to residues that do not bind magnesium and the mutants were found to be active. **C** Schematic illustration of the catalytic pocket for the EcoRV-product complex. Note that now two metal ions are observed, although metal ion concentrations in the range of 10–30 mM were used. (Reprinted with permission from Kostrewa and Winkler 1995)

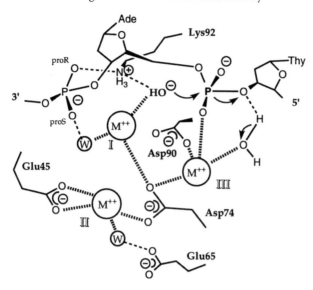

Fig. 4. Proposed transition state for the metal ion-mediated DNA cleavage reaction by EcoRV. The metal ion in site I generates the attacking nucleophile, which is stabilized and oriented by Lys-92. The metal in site III facilitates departure of the 3′-anion and compensates the developing negative charge in the transition state. The role of the metal in site II is structural. Metal-ligand inner-sphere contacts are shown as *hatched lines. Dotted lines* indicate hydrogen bonds. (Reprinted with permission from Horton et al. 1998)

Jeltsch et al. 1992; Vipond et al. 1995; Cowan 1998; Chevalier et al. 2001). Even more perplexing is the observation of distinct metal binding sites for the same metals enzyme partners in distinct crystallographic studies (also see references in Table 1; Kostrewa and Winkler 1995; Horton et al. 1998). Accordingly, much of the debate has focused on what can and cannot be inferred from structural data derived from crystals of enzyme co-complexed with metal and/or substrate DNA. In many cases it is not clear if the enzyme form under structural investigation is catalytically active, and so this concern must be factored into any analysis of mechanism. Common to these structural studies is the finding of a water molecule hydrogen-bonded to the 3′-phosphate and so it has been suggested that the substrate DNA may function as a catalytic base by promoting deprotonation of nucleophilic water (possibly a metal-bound water) (Jeltsch et al. 1993).

With regard to the "three-metal-ion" mechanism (Chevalier et al. 2001), it is important to keep in mind that no structure has been identified with three metal ions bound. Rather, these three sites have been identified from a variety of studies to be capable of binding metal cofactors, however, a maximum of two sites have been populated at any one time. Possibly, these sites represent snapshots of positions transiently occupied by one or more metal cofactors

Table 1. Summary of enzyme-bound metal ion ligands in restriction enzymes[a]. Adapted and expanded from (Jose et al. 1999)

Enzyme	Ligand1	Ligand2	Others			Ref^SDM	Ref^X·ray
EcoRI	E111	D91	D59			Wolfes et al. (1986); King et al. (1989); Grabowski et al. (1995, 1996)	
Mg^{2+}	x						Rosenberg (1991)
EcoRV	D90	D74	E45	Q69		Selent et al. (1992); Groll et al. (1997)	
Mg(II)	x	x			P, 3H$_2$O		Kostrewa and Winkler (1995)
1 Mg(II)[b]	x	x			P, 2H$_2$O		Kostrewa and Winkler (1995)
2 Mg(II)				x	P, 4H$_2$O		
BamHI	E111	D94	E77	E113	P, F112	Xu and Schildkraut (1991)	
1Ca(II)	x	x	x		P		Newman et al. (1994); Viadiu and Aggarwal (1998)
2Ca(II)	x						
PvuII	E68	D58	E55			Nastri et al. (1997)	Athanasiadis et al. (1994); Cheng et al. (1994); Horton et al. (1998)
Mg(II)	x	x			P, 3H$_2$O		Horton et al. (1998)
1Ca(II)	x	x		L69[c]	P, 2H$_2$O		Horton et al. (1998)
2Ca(II)	x	x		G56[c]	P, H$_2$O		Horton and Cheng (2000).
Cfr10I	E204	D134	E71	E113		Skirgaila et al. (1998)	Bozic et al. (1996)
BglII							
M	E93	D84		N95	P, V94,[c] 2H$_2$O		
	x			x			Lukacs et al. (2000)
BglI	D116	D142	E87[c]	E113	P, I142, 2H$_2$O		Newman et al. (1998)
1Ca(II)	x	x		x	P, 4H$_2$O		
2Ca(II)	x	x					

[a] x, Proposed ligand based on crystallographic analysis of wild-type enzyme soaked with metal ions. Mutation of underlined residues to Ala results in ≥1000-fold loss in specific activity. Other ligands are either proposed protein ligands, DNA, or H$_2$O. P, Coordination with phosphate of DNA. Ref^SDM and Ref^X·ray refer to relevant mutagenesis and crystallographic studies, respectively

[b] Product complex

[c] Indirect coordination of backbone carbonyl

during substrate cleavage. That is, all of the coordination sites may be occupied for some time depending on the temporal status of the cleavage chemistry. Figures 3 and 4 illustrate the mechanistic dichotomy that our understanding of reaction pathway faces, with two distinct mechanisms proposed for EcoRV that were devised from two distinct series of crystallographic studies (Kostrewa and Winkler 1995; Horton et al. 1998). Having said that, there is substantial evidence to suggest that the site defined by Glu45 (EcoRV numbering, Fig. 3) may serve a purely structural role (Haruki et al. 1994; Grabowski et al.; Groll et al. 1997; Nastri et al. 1997). To date, this site has not been found to be populated if either of the other two sites contain bound metal ion. Just as remarkable is the observation that deletion of this residue shows no significant change in the optimal activity of the enzyme, even with reduced metal content. The remaining two sites are defined by Asp74, Asp90, and Asp74, Asp90, Ile91, respectively. It seems plausible that these two sites may represent transient locations for the same metal cofactor as it moves to accommodate the dynamic coordination requirements of the cleavage reaction.

Another issue that pertains to metal ion stoichiometry stems from the optimum pH for endonuclease activity of ~8.5. This is clearly outside of the typical pK_a range for a magnesium-bound water (Burgess 1978), but not so if two metal ions resulted in a significant increase in positive charge density. However, with so many ionizable residues present in the catalytic pocket, and with a cluster of carboxylates that might increase the pK_a of an essential catalytic base, the nature of the chemistry controlling the pH dependence of the reaction has remained elusive.

5.1 EcoRV and EcoRI

Few magnesium-dependent nuclease enzymes have been studied in detail insofar as the role of the essential metal cofactor is concerned. Noteworthy exceptions are the restriction enzymes EcoRI, and especially EcoRV (Jeltsch et al. 1995; Kostrewa and Winkler 1995; Vipond et al. 1995; Vipond and Halford 1995; Halford et al. 1999) for which there is now an extensive body of structural and mechanistic information concerning the role of the metal cofactors in the function of this enzyme (Pingoud and Jeltsch 1997). EcoRI shows a high degree of structural homology with the active site of EcoRV (Fig. 5), although there is no general sequence homology (Pingoud and Jeltsch 2001). A mechanistic model has been suggested where the attacking water is activated by the phosphate placed 3′ to the scissile phosphodiester, and the leaving group is protonated by a water molecule associated with the Mg^{2+} cofactor (Jeltsch et al. 1993; Horton et al. 1998). This effect has been described as substrate-assisted catalysis, and has been supported by experiments that remove the phosphate group (substituting with an H-phosphonate) and by

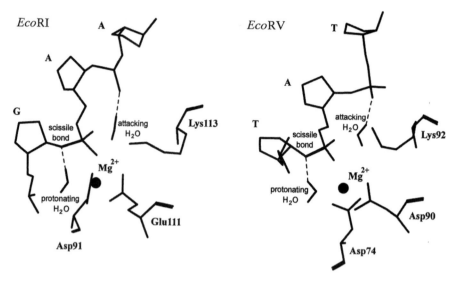

Fig. 5. A comparison of the metal cofactor binding domains, and other features of the catalytic pockets of EcoRI and EcoRV, illustrating the high degree of local homology for these two restriction endonucleases. (Reprinted with permission from Jeltsch et al. 1993)

phosphorothioate substitution. In general, activity is lost or severely diminished following these substitutions, and in the latter case activity is only observed if the negatively charged sulfur is in the R_p configuration. Nevertheless, there would appear to be a requirement for participation by a carboxylate residue to provide base catalysis, since the pK_a of a phosphodiester is too low ($pK_a < 2$) for this role.

Unlike EcoRV, EcoRI produces sticky-end products, since the homologous active site is offset relative to the DNA recognition sequence. In the presence of Mg^{2+}, association of EcoRI with palindromic tridecadeoxynucleotides containing the recognition site occurs at an almost diffusion controlled rate (Alves et al. 1989). Figure 5 illustrates how both enzymes show a similar structural arrangement in the vicinity of the scissile phosphate, and form a motif that is composed of four active site residues (Pro90, Asp91, Glu111, and Lys113 for EcoRI; and Pro73, Asp74, Asp90, and Lys92 for EcoRV) that appear to form a binding pocket for the essential divalent metal cofactor (Jeltsch et al. 1992, 1993). For EcoRI, both kinetic and structural studies are consistent with one essential metal cofactor in the active site (Jeltsch et al. 1992, 1993), however, the role of the metal cofactor has been most clearly elucidated for EcoRV and this is now taken up, although the mechanistic details of metal-promoted hydrolysis are most likely similar for each enzyme.

Unlike other restriction enzymes in this class, which demonstrate specific binding in the absence of Mg^{2+}, EcoRV binds nonspecifically in the absence of the divalent cofactor (Jeltsch et al. 1995). In 1993, on the basis of kinetic evidence, Pingoud and coworkers proposed a requirement for one metal ion to effect EcoRV catalysis (Jeltsch et al. 1992, 1993), but also pointed out that binding of an additional high affinity metal ion could not be excluded since an additional residue (Glu45) was available for binding and coordination of heavy metals (such as Pb^{2+}) had been recognized in crystallographic experiments. The 2 Å resolution structures of dimeric EcoRV co-complexed with both an undecamer substrate and product (Fig. 3; Kostrewa and Winkler 1995), published by Kostrewa and Winkler, showed one Mg^{2+} bound to the scissile phosphodiester group and two carboxylate oxygens from Asp 74 and Asp 90 in the substrate complex, while the product complex showed two Mg^{2+} bound to each 5'-phosphate oxygen (Fig. 3) (Mg_3^{2+} and Mg_4^{2+}). On the basis of this structure, and other kinetic data, Halford and Winkler and coworkers subsequently proposed a two-metal-ion transition-state model. Halford and coworkers also examined the metal requirement for EcoRI and EcoRV, and suggested a catalytic requirement of one metal cofactor for the former, and a two-metal-ion mechanism for the latter. It was observed that Ca^{2+} inhibits Mg^{2+} activation of EcoRV, but stimulates Mn^{2+} activation of the same enzyme, suggesting that Ca^{2+} can displace a metal from one of the two sites. Such a model was supported by stopped-flow fluorescence experiments that examined the influence of metal ions on DNA binding to EcoRV (Baldwin et al. 1995). Two Mg^{2+}-dependent transitions were observed and interpreted in terms of a model where one Mg^{2+} binds to the active site before phosphodiester hydrolysis, with a second binding subsequently to a preformed enzyme–DNA complex.

These facts notwithstanding, several problems nevertheless remained with this interpretation. First, the crystal complex with the substrate lacked activity, even in the presence of saturating magnesium, and the active site bound only one divalent magnesium ion (Fig. 3). Second, a Mg^{2+} concentration of ≥ 10 mM was required to populate both sites; however, the key issue is whether or not these sites can be populated under physiological conditions (~0.5 mM free $Mg^{2+}(aq)$). Third, and perhaps most importantly, the critical residues that define the binding site of the second metal ion (Mg_2^{2+} in Fig. 3), especially Glu45, have been mutated to nonbinding residues (such as Ala) and the mutants were found to be active (Thielking et al. 1992). It has therefore been suggested that a second, but spatially distinct and noncatalytic site might exist (Jeltsch et al. 1995), since the DNA binding specificity is increased by Mg^{2+} binding to a site that is distinct from that of the catalytic center (bound by Glu 45, Asp 74, Asp 90). Also, the Ala triple mutant, which cannot bind catalytic metal, was found to bind specifically to DNA in the presence of Mg^{2+}. Increasing the magnesium concentration was found to reduce nonspecific binding by a factor of 10^3 to 10^4, which was postulated to arise from the release of Mg^{2+} ions. Phosphorothioate derivatives provided further evidence for a Mg^{2+} site distinct from the catalytic center. Only one of these ions is required for catalysis; the other appears to promote binding of the substrate molecule. This most likely corresponds to the additional metal cofactor suggested by the stopped-flow kinetic studies of Halford and coworkers (Baldwin et al. 1995; Vipond et al. 1995).

Specific recognition of DNA by EcoRV can be promoted by Ca^{2+} ion. By analyzing gel shifts with a permuted set of DNA fragments, the degree of DNA bending was shown to be similar to that seen in the crystal structure of the cognate DNA–protein complex in the presence of Mg^{2+} (Vipond and Halford 1995). Calcium ion, therefore, mimics the ability of Mg^{2+} to generate a specific protein–metal–DNA complex, but is incapable of inducing the cleavage reaction. This may also reflect the tendency of Ca^{2+} to engage in inner sphere rather than outer sphere interactions, if the latter are required for catalysis. Outer sphere activation is also consistent with the absence of any significant influence on the relative cleavage rates for Mg^{2+}- vs. Mn^{2+}-induced activity with phosphorothioate substrates, since direct coordination to the metal cofactor should discriminate between a hard Mg^{2+} ion, and softer Mn^{2+} ion (Piccirilli et al. 1993).

Recent results for an Ile91Leu mutant show a greater than 1000-fold decrease in activity (Moon et al. 1996). A change in metal ion dependency was observed, with a preference for Mn^{2+} rather than Mg^{2+} as for the native enzyme. The mutant shows evidence of nonspecific DNA binding in gel-shift experiments, as does native. At noncognate sites that differ from the EcoRV site by one base pair, Mn^{2+} gives higher cleavage rates than Mg^{2+}, but the effect is reversed for the Ile 91 Leu mutant (Vipond et al. 1996). Since each mutant requires the same carboxylate residues for binding, compared to native, it is likely that the switch in metal preference arises from a structural perturbation of the metal binding pocket.

5.2 PvuII

Recent reports by Dupureur and coworkers on PvuII provide some of the most insightful evidence for the role of metal cofactors in promoting DNA binding (Conlan et al. 1999; Dupureur and Hallman 1999; Dupureur and Conlan 2000; Dominguez et al. 2001; Dupureur and Dominguez 2001; Conlan and Dupureur 2002a, b). These results are most likely relevant to understanding the general family of Type II endonucleases. The authors find evidence (Fig. 6; Conlan and Dupureur 2002b) for at least two metal ions participating in DNA binding and cleavage. By use of Ca^{2+} as a cofactor, binding could be monitored in the absence of cleavage. A 6000-fold increase in binding affinity to a cognate DNA sequence was observed in the presence of a metal cofactor (Conlan and Dupureur 2002a). This represents an energy contribution of only ~5.2 kcal/mol, which could easily be achieved either through direct metal-ligand contacts to the DNA substrate, or more likely in

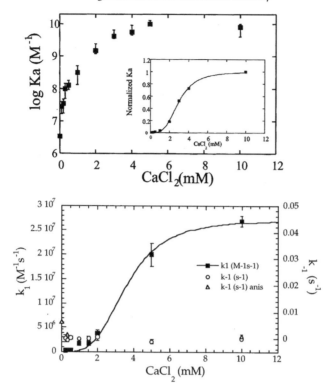

Fig. 6. *Top* Dependence of DNA binding affinity on metal ion concentration. The conditions were 50 mM Tris, pH 7.5, and 25 °C. The NaCl concentration was adjusted to a constant ionic strength of 107.5 mM. Below 1 mM $CaCl_2$, data were collected by fluorescence anisotropy at 25 °C with 2.5 nM HEX-labeled 14-mer duplex. Above 1 mM $CaCl_2$, binding constants were measured by nitrocellulose filter binding at 9 pM DNA duplex. The *inset* features a Hill analysis of the data normalized on a linear scale, yielding an n_H of 3.5±0.2 per enzyme dimer. *Below* Metal ion dependence of k_1 and k_{-1}. The conditions were 50 mM Tris, pH 7.5, and 25 °C. The NaCl concentration was adjusted to a constant ionic strength of 107.5 mM: *filled squares* k_1 obtained using nitrocellulose filter binding assays, *open circles* k_{-1} obtained via filter binding, and *triangles* k_{-1} obtained via fluorescence anisotropy. In general, errors were averaged over fits of multiple secondary plots. Hill analysis of k_1 data yielded an n_H of 3.6±0.2 per enzyme dimer. (Reprinted with permission from Conlan and Dupureur 2002b)

the case of magnesium ion through H-bonding contacts with metal-bound waters. Alternatively, the metal cofactor might perturb the enzyme structure to promote additional side-chain contacts with substrate DNA. In spite of the typically minor structural changes in the positions of active site residues that are induced by metal binding (Fig. 7), there is evidence to suggest that such minor perturbations do in fact result in global conformational change. Such tertiary structural change following metal ion binding was found in the case of PvuII through heteronuclear NMR experiments (Dupureur and

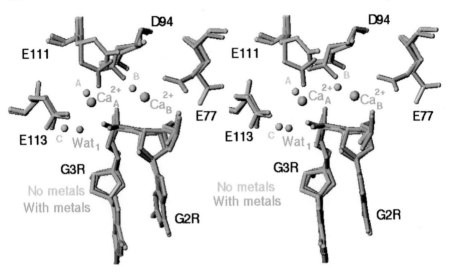

Fig. 7. A comparison of BamHI active sites in the presence (*blue*) and absence of metals (*yellow*). The structure in the absence of metals reveals three water molecules (A, B, and C) whose positions are compared to Ca^{2+}_A and Ca^{2+}_B and the attacking water molecule Wat1. (Reprinted with permission from Viadiu and Aggarwal 1998)

Dominguez 2001). Evidence for structural change is also manifest through the limited crystallographic data available, comparing apo- and holo-forms of restriction endonucleases (references in Table 1 and Fig. 7). The finding that DNA binding to the enzyme was only promoted by metal ions for the cognate sequence (Conlan and Dupureur 2002a, b) suggests that the placement of the metal cofactor is optimal to promote specific contacts with the cognate sequence, either through direct binding interactions or an indirect influence on enzyme structure.

Evidence for the requirement for more than one metal cofactor to promote activity is provided by the cooperativity displayed by metal binding to the enzyme, and by the sigmoidal behavior observed for the metal concentration dependence of the catalytic rate constant with Hill coefficients (n_H) of 3.5 and 3.6, respectively (Fig. 6; Conlan and Dupureur 2002a, b). While the possibility of two metals being required for activity has been described in support of a "two-metal-ion mechanism", care must be taken in the interpretation of what this term means. The paradigm of the two-metal-ion model is perhaps best defined as the requirement for two metal ions, located in close proximity (<4 Å), that are bridged by a common substrate (Beese and Steitz 1991; Cowan 1998). This allows a clear distinction from other enzymes that may bind two metal cofactors in the same catalytic domain, but do not function as a coherent catalytic unit. For example, crystallographic analysis of T5 5′-exonuclease shows two Mn^{2+} sites for data collected in the presence of 25 mm Mn^{2+} (Ceska et al. 1996). These sites are, however, separated by 8.1 Å (10 Å in the case of Taq

5'-exonuclease), which is significantly greater than the <4 Å separation observed in the putative two-metal nucleases. Two Mg^{2+} sites are separated by 5.4 Å in EcoRV (Fig. 3; Kostrewa and Winkler 1995). Accordingly, one would expect a difference in mechanism from the two-metal variety for these, and related enzymes. In fact, restriction endonucleases appear to display significant variations in their use of metal cofactors despite having remarkably similar folds for their catalytic regions. Notably the structural study of the calcium complex of PvuII demonstrated one Ca^{2+} in a position equivalent to that found in all structurally characterized endonucleases, whereas the second position was similar to that of endonuclease EcoRV but was distinct from that of endonucleases BamHI and BglI. The location of the second metal cation in PvuII, unlike that in either BamHI or BglI, did not allow for a direct interaction between the second metal and the O3' leaving group. However, the interactions between the scissile phosphate and the "catalytic metal ion" and the nucleophilic H_2O were the same. Nevertheless, this catalytic ion in turn was distinct from the coordination details in the proposed three-metal or the two-metal cation models of EcoRV. Also remarkable is the fact that the crystallographically characterized Mg^{2+} complex (Horton et al. 1998) shows only one bound Mg^{2+} ion in the catalytic domain. Again, the question remains as to whether the structural cation, represented by the calcium data (structural and functional) is relevant to the physiological magnesium-promoted reaction.

Overall the accumulated data suggest distinct mechanistic roles for the two metal cofactors that have been implicated in "two-metal" mechanisms, with one predominantly involved in promoting cleavage chemistry while the other (or both) serve structural roles and/or influence substrate binding. Clearly, the cooperativity observed for the metal-dependence of substrate binding supports the idea that each metal influences the binding of the other, either electrostatically and/or by conformational tuning of side chains. Whether the same holds true for the natural magnesium cofactor remains to be established.

5.3 BamHI

By use of Mn^{2+} and Ca^{2+} as Mg^{2+} analogs BamHI has been structurally investigated in its pre- and post-reaction states (Fig. 8; Viadiu and Aggarwal 1998) and a comparison is made with EcoRV which has been studied in a similar fashion (Fig. 3; Kostrewa and Winkler 1995), although a distinct use of metal ions cofactors is suggested relative to EcoRV. Importantly, these crystals were shown to be active with DNA cleavage directly in the crystals. It is possible that this simple fact of maintaining an active conformational state underlies the distinct results obtained with EcoRV since the latter enzyme was not active in the crystals (Kostrewa and Winkler 1995). Accordingly any claim to be mechanistically distinct should be treated with caution. For both EcoRV

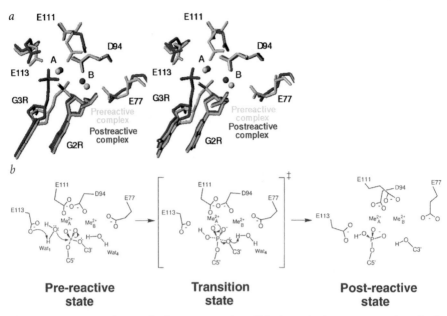

Pre-reactive   **Transition**   **Post-reactive**
state   **state**   **state**

Fig. 8. a A comparison of the pre-reactive (*blue*) and the post-reactive (*red*)
BamHI–DNA complexes. The calcium ions (*A* and *B*) bound to the pre-reactive complex
are shown as *blue spheres*, while the manganese ions (*A* and *B*) bound to the post-reactive complex are shown as *red spheres*. The phosphate group moves towards Glu 113 after
cleavage. **b** The BamHI reaction mechanism based on the structures of the pre- and
post-reactive complexes. Me^{2+}_A and Me^{2+}_B denote the metal sites; Wat_1 and Wat_4 refer to
water molecules. (Reprinted with permission from Viadiu and Aggarwal 1998)

and BamHI, it was observed that metals were only located in one of the subunits (the R active site) of the dimer (Viadiu and Aggarwal 1998) with a shift
of carboxylates to accommodate metal ions. Consequently, there remains the
question of whether two metal ions can bind simultaneously to each subunit,
which would be required for function.

In common with other Type II nucleases, the metal cofactor(s) mediates
an in-line attack by hydroxide that results in displacement of the 3′-OH and
inversion of configuration at the 5′-phosphate (Figs. 3 and 8). A combination
of activation of a water molecule and transition state stabilization is
undoubtedly required. While retaining a structurally similar active site as
other endonucleases (EcoRI, EcoRV, PvuII, etc.), BamHI differs inasmuch as
the conserved Lys residue is replaced by Glu (Fig. 1; Viadiu and Aggarwal
1998). As noted earlier, the proposed role for this Lys in transition state stabilization has been refined to that of a positional influence on an essential
metal cofactor (Horton et al. 1998). Nevertheless, it is very possible that the
Lys may in fact serve to stabilize the transition state in one enzyme while a
distinct residue at the analogous position in another enzyme might serve an

entirely distinct function. This is clearly evidenced by the diversity of roles of active site residues (Figs. 1–4) in spite of the apparent conservation of the structural core. As noted earlier in this review, there exists considerable diversity of both metal binding sites and the quaternary structures of the metal-bound active dimers (comparing BglI and EcoRV) (Newman et al. 1998) that promote DNA recognition and cleavage. The bound cofactors obviously stabilize distinct quaternary structures and promote the orientation of active site residues in distinctive fashions, as dictated by tertiary structure refinement.

Similarity has been noted with the 3'-5' exonuclease domain of *E. coli* DNA polymerase I (Beese and Steitz 1991) with two metal ions located ~4 Å apart in a fashion that optimizes them for interactions with the apical oxygens. One of these metal ions appears to be principally involved in stabilizing the transition state, while the other activates a water molecule. This contrasts with the situation for EcoRV where the inter-metal distance is ~5.4 Å and occupy a position in the equatorial plane of the trigonal bipyramidal transition state. For BamHI substantial movement of the scissile phosphate was observed following cleavage. Also, evidence was noted for significant loss of one of the metal cofactors in the complex with the cleaved product (Viadiu and Aggarwal 1998), possibly reflecting the fact that the free enzyme (following product release) has a low propensity for binding a second metal cofactor.

The finding that Ca^{2+} binds in a similar location to the Mn^{2+} ions supports the idea that Ca^{2+} inhibition stems from direct blocking of catalytically competent Mg^{2+} or Mn^{2+} ions, and is itself incapable of promoting catalysis. The latter most likely reflects the distinct coordination geometries and numbers preferred by Ca^{2+} and its more modest Lewis acidity.

6 Other Restriction Endonucleases

There is a paucity of structural and mechanistic information concerning other restriction enzymes, especially with regard to the chemistry of the metal cofactor. Nevertheless, the constellations of active site metal-binding residues in other structurally characterized endonucleases are similar to the families illustrated herein, suggesting that such enzymes proceed by mechanisms that show many of the features highlighted in this review.

The case of BglII (Fig. 2) is of considerable interest. It appears to show a single bound metal cofactor. By comparison with BamHI, which uniquely shows a glutamate residue replacing a catalytically essential lysine, BglII provides a unique example of a glutamine residue at this site. As noted earlier, the roles for the active site residues are significantly distinct from those of related enzymes (BamHI and EcoRV). For example, Gln95 in BglII H-bonds to the metal-bound (and putative) nucleophilic water, while the corresponding Glu113 in BamHI H-bonds to the phosphate oxygen.

FokI is a Type II restriction endonuclease that cleaves DNA at a site of 9 and 13 nucleotides downstream of a non-palindromic recognition site (5'-GGAT9–3'). Also in contrast to other restriction enzymes, the enzyme forms a dimeric state only in the presence of divalent metal cofactors and does so by binding two DNA substrates (Vanamee et al. 2001).

It has been observed that introduction of phosphoramide bonds at the cleavage position of one strand blocks cleavage of that strand for the endonucleases EcoRII and SsoII, and in the case of EcoRII reduces the rate of cleavage of the other natural strand (Gromova et al. 1987). In the presence of the Mg^{2+} cofactor, association rates are reduced threefold and dissociation rates are increased 1.5-fold, and so recognition and specific binding is strongly influenced by the metal cofactor. This is consistent with the breakdown in recognition specificity of Eco RI and other endonucleases (such as ApoI, AseI, BamHI, EcoRI, EcoRV, HindIII, KpnI, PstI, PvuII, and TaqI, among others) with magnesium analogues.

7 Concluding Remarks

Clearly the existing structural and thermodynamic results, while often differing in both minor and major aspects of metal ion binding, and the mechanistic import that follows, do provide a number of mechanistically reasonable pathways that should be considered. However, rather than being seen as an end in themselves, such models should form the basis for further studies. In particular, what is especially absent from current literature is a series of carefully planned kinetic studies under solution turnover conditions, using the natural divalent magnesium cofactor, that would allow these various mechanistic possibilities to be distinguished. While most likely extremely difficult to perform, such studies are vital to resolving the issues that the structural studies have framed, and are by themselves incapable of answering.

Acknowledgements. This work was supported by a grant from the Petroleum Research Fund, administered by the American Chemical Society (J.A.C.), the National Science Foundation, CHE-0111161 (J.A.C.).

References

Aggarwal AK (1995) Structure and function of restriction endonucleases. Curr Opin Struct Biol 5:11–19

Alves J, Urbanke C, Fliess A, Maass G, Pingoud A (1989) Fluorescence stopped-flow kinetics of the cleavage of synthetic oligodeoxynucleotides by the EcoRI restriction endonuclease. Biochemistry 28:7879–7888

Athanasiadis A, Vlassi M, Kotsifaki D, Tucker PA, Wilson KS, Kokkinidis M (1994) Crystal structure of PvuII endonuclease reveals extensive structural homologies to EcoRV. Nat Struct Biol 1:469–475

Baldwin GS, Vipond IB, Halford SE (1995) Rapid reaction anaylsis of the catalytic cycle of the EcoRV restriction endonuclease. Biochemistry 34:705–714

Beese LS, Steitz TA (1991) Structural basis for the 3′-5′ exonuclease activity of *Escherichia coli* DNA polymerase I: a two metal ion mechanism. EMBO J 10:25–33

Bennett P, Halford SE (1989) Recognition of DNA by Type II restriction enzymes. Curr Topics Cell Regul 30:57–104

Black CB, Cowan JA (1994) Inorg Chem 33:5805–5808

Bozic D, Grazulis S, Siksnys V, Huber R (1996) Crystal structure of *Citrobacter freundii* restriction endonuclease Cfr10I at 2.15 Å resolution. J Mol Biol 255:176–186

Bujnicki JM (2000) Phylogeny of the restriction endonuclease-like superfamily inferred from comparison of protein structures. J Mol Evol 50:39–44

Burgess J (1978) Metals in solution. Ellis Horwood, New York

Ceska TA, Sayers JR, Stier G, Suck D (1996) A helical arch allowing single-stranded DNA to thread through T5 5′-exonuclease. Nature 382:90–93

Cheng XK, Balendiran K, Schildkraut I, Anderson JE. (1994) Structure of PvuII endonuclease with cognate DNA. EMBO J 13(17):3927–3935

Chevalier BS, Monnat RJ, Stoddard BL (2001) The homing endonuclease I-CreI uses three metals one of which is shared between the two active sites. Nat Struct Biol 8:312–316

Conlan LH, Dupureur CM (2002a) Dissecting the metal ion dependence of DNA binding by PvuII endonuclease. Biochemistry 41:1335–1342

Conlan LH, Dupureur CM (2002b) Multiple metal ions drive DNA association by PvuII endonuclease. Biochemistry 41:14848–14855

Conlan LH, Jose TJ, Thornton KC, Dupureur CM (1999) Modulating restriction endonuclease activities and specificities using neutral detergents. Bio Techn 27:955–960

Cowan JA (1992) Transition metals as probes of metal cofactors in nucleic acid biochemistry. Comm Inorg Chem 13:293–312

Cowan JA (1995a) Biological chemistry of magnesium. VCH, New York

Cowan JA (1995b) Inorganic biochemistry. An introduction. VCH, New York

Cowan JA (1998) Metal activation of enzymes in nucleic acid biochemistry. Chem Rev (Washington, DC) 98(3):1067–1087

Demple B, Johnson A, Fung D (1986) Exonuclease-III and exonuclease-IV remove 3′ blocks from DNA-synthesis primers in H_2O_2-damaged *E. coli*. Proc Natl Acad Sci USA 83:7731–7735

Dominguez MA, Thornton KC, Melendez MG, Dupureur CM (2001) Differential effects of isomeric incorporation of fluorophenylalanines into PvuII endonuclease. Prot Struct Func Gen 45:55–61

Dupureur CM, Conlan LH (2000) A catalytically deficient active site variant of PvuII endonuclease binds Mg(II) ions. Biochemistry 39:10921–10927

Dupureur CM, Dominguez MA (2001) The PD...(D/E)XK motif in restriction enzymes: a link between function and conformation. Biochemistry 40:387–394

Dupureur CM, Hallman LM (1999) Effects of divalent metal ions on the activity and conformation of native and 3-fluorotyrosine-PvuII endonucleases. Eur J Biochem 261:261–268

Friedhoff P, Kolmes B, Gimadutdinow O, Wende W, Krause KL, Pingoud A (1996) Analysis of the mechanism of the *Serratia* nuclease using site-directed mutagenesis. Nucleic Acids Res 24:2632–2639

Grabowski G, Jeltsch A, Wolfes H, Maass G, Alves J (1995) Site-directed mutagenesis in the catalytic center of the restriction endonuclease EcoRI. Gene 157:113–118

Grabowski G, Maass G, Alves J (1996) Asp-59 is not important for the catalytic activity of the restriction endonuclease EcoRI. FEBS Lett 381:106–110

Grasby JA, Connolly BA (1992) Stereochemical outcome of the hydrolysis reaction catalyzed by EcoRV restriction endonuclease. Biochemistry 31:7855–7861

Groll DH, Jeltsch A, Selent U, Pingoud A (1997) Does the restriction endonuclease EcoRV employ a two-metal-Ion mechanism for DNA cleavage? Biochemistry 36:11389–11401

Gromova ES, Vinogradova MN, Uporova TM, Gryaznova OI, Isagulyant MG, Kosykh VG, Nikol'skaya II, Shabarova ZA (1987) DNA duplexes with phosphoamide bonds: the interaction with restriction endonucleases EcoRII and SsoII. Bioorg Khim 13:269–272

Halford SE (2001) Hopping jumping and looping by restriction enzymes. Biochem Soc Trans 29:363–373

Halford SE, Bilcock DT, Stanford NP, Williams SA, Milsom SE, Gormley NA, Watson MA, Bath AJ, Embleton ML, Gowers DM, Daniels LE, Parry SH, Szczelkun MD (1999) Restriction endonuclease reactions requiring two recognition sites. Biochem Soc Trans 27:696–699

Haruki M, Noguchi E, Nakai C, Liu YY, Oobatake M, Itaya M, Kanaya S (1994) Investigating the role of conserved residue Asp134 in *Escherichia coli* ribonuclease HI by site-directed random mutagenesis. Eur J Biochem 220:623–631

Horton JR, Cheng X (2000) PvuII endonuclease contains two calcium ions in active sites. J Mol Biol 300:1051–1058

Horton JR, Nastri HG, Riggs PD, Cheng X (1998) Asp34 of PvuII endonuclease is directly involved in DNA minor groove recognition and indirectly involved in catalysis. J Mol Biol 284:1491–1504

Horton NC, Newberry KJ, Perona JJ (1998) Metal-ion -mediated substrate-assisted catalysis in Type II restriction endonucleases. Proc Natl Acad Sci USA 95:13489–13494

Jeltsch A, Alves J, Maass G, Pingoud A (1992) On the catalytic mechanism of EcoRI and EcoRV A detailed proposal based on biochemical results structural data and molecular modeling. FEBS Lett 304:4–8

Jeltsch A, Alves J, Wolfes H, Maass G, Pingoud A (1993) Substrate-assisted catalysis in the cleavage of DNA by the EcoRI and EcoRV restriction enzymes. Proc Natl Acad Sci USA 90:8499–8503

Jeltsch A, Maschke H, Selent U, Wenz C, Koehler E, Connolly BA, Thorogood H, Pingoud A (1995) DNA binding specificity of the EcoRV restriction endonuclease is increased by Mg^{2+} binding to a metal ion binding site distinct from the catalytic center of the enzyme. Biochemistry 34:6239–6246

Jose TL, Conlan LH, Dupureur CM (1999) Quantitative evaluation of metal ion binding to PvuII restriction endonuclease. (JBIC) J Biol Inorg Chem 4:814–823

Kim S, Cowan JA (1992) Inert cobalt complexes as mechanistic probes of the biochemistry of magnesium cofactors. Application to topoisomerase I. Inorg Chem 31:3495–3496

King KS, Benkovic SJ, Modrich P (1989) Glu-111 is required for activation of the DNA cleavage center of EcoRI endonuclease. J Biol Chem 264:11807–11815

Kostrewa D, Winkler FK (1995) Mg^{2+} binding to the active site of EcoRV endonuclease: a crystallographic study of complexes with substrate and product DNA at 2 Å resolution. Biochemistry 34:683–696

Lukacs. CM, Kucera R, Schildkraut I, Aggarwal AK (2000) Understanding the immutability of restriction enzymes: crystal structure of BglII and its DNA substrate at 1.5 Å resolution. Nat Struct Biol 2:134–140

Moon BJ, Vipond IB, Halford SE (1996) Site-directed mutagenesis of Ile91 of restriction endonuclease EcoRV: dramatic consequences on the activity and the properties of the enzyme. J Biochem Mol Biol 29:17–21

Murray NE (2000) Type I restriction systems: sophisticated molecular machines (a legacy of Bertani and Weigle). Microbiol Mol Biol Rev 64:412–434

Nastri HG, Evans PD, Walker IH, Riggs PD (1997) Catalytic and DNA binding properties of PvuII restriction endonuclease mutants. J Biol Chem 272:25761–25767

Newman MK, Lunnen K, Lunnen K, Wilson G, Greci J, Schildkraut I, Phillips SEV (1998) Crystal structure of restriction endonuclease BglI bound to its interrupted DNA recognition sequence. EMBO J 17:5466–5476

Newman MT, Strzelecka T, Dorner LF, Schildkraut I, Aggarwal AK, (1994). Structure of restriction endonuclease BamHI and its relationship to EcoRI. Nature 368:660–664

Newman MT, Strzelecka T, Dorner LF, Schildkraut I, Aggarwal AK (1995) Structure of Bam HI endonuclease bound to DNA: partial folding and unfolding on DNA binding. Science 269:656–663

Piccirilli JA, Vyle JS, Caruthers MH, Cech TR (1993) Metal ion catalysis in the *Tetrahymena* ribozyme reaction. Nature 361:85

Pingoud A, Jeltsch A (1997) Recognition and cleavage of DNA by Type-II restriction endonucleases. Eur J Biochem 246:1–22

Pingoud A, Jeltsch A (2001) Structure and function of Type II restriction endonucleases. Nucleic Acids Res 29:3705–3727

Rao D, Saha S, Krishnamurthy V (2000) ATP-dependent restriction enzymes. Prog Nucl Acid Res Mol Biol 64:1–63

Rosenberg JM (1991) Structure and function of restriction endonucleases. Curr Opin Struct Biol 1:104–113

Sapranauskas RG, Sasnauskas G, Lagunavicius A, Vilkaitis G, Lubys A, Siksnys V (2000) Novel subtype of Type IIs restriction enzymes. BfiI endonuclease exhibits similarities to the EDTA-resistant nuclease Nuc of *Salmonella typhimurium*. J Biol Chem 275(40):30878–30885

Selent UT, Ruter T, Kohler E, Liedtke M, Thielking V, Alves J, Oelgeschlager T, Wolfes H, Peters F, Pingoud A (1992) A site-directed mutagenesis study to identify amino acid residues involved in the catalytic function of the restriction endonuclease EcoRV. Biochemistry 31:4808–4815

Skirgaila RS, Grazulis S, Bozic D, Huber R, Siksnys V (1998) Structure-based redesign of the catalytic/metal binding site of Cfr10i restriction endonuclease reveals importance of spatial rather than sequence conservation of active center residues. J Mol Biol 279:473–481

Stasiak A (1980) Restriction enzymes. I. Mechanisms of action of Type II restriction-modification systems. Postepy Biochem 26:343–367

Thielking VU, Selent U, Kohler E, Landgraf A, Wolfes H, Alves J, Pingoud A (1992) Magnesium(2+) confers DNA binding specificity to the EcoRV restriction endonuclease. Biochemistry 31:3727–3732

Tsutakawa SE, Morikawa K (2001) The structural basis of damaged DNA recognition and endonucleolytic cleavage for very short patch repair endonuclease. Nucleic Acids Res 29:3775–3783

Vanamee E, Santagata S, Aggarwal AK (2001) FokI requires two specific DNA sites for cleavage. J Mol Biol 309:69–78

Viadiu H, Aggarwal AK (1998) The role of metals in catalysis by the restriction endonu-
    clease BamHI. Nat Struct Biol 5:910–916
Vipond IB, Baldwin GS, Halford SE (1995) Divalent metal ions at the active site of the
    EcoRV and EcoRI restriction endonucleases. Biochemistry 34:697–704
Vipond IB, Halford SE (1995) Specific DNA recognition by EcoRV restriction endonu-
    clease induced by calcium ions. Biochemistry 34:1113–1119
Vipond IB, Moon BJ, Halford SE (1996) An isoleucine to leucine mutation that switches
    the cofactor requirement of the EcoRV restriction endonuclease from magnesium to
    manganese. Biochemistry 35:1712–1721
Wah DA, Hirsch JA, Dorner LF, Schildkraut I, Aggarwal AK (1997) Structure of the mul-
    timodular endonuclease FokI bound to DNA. Nature 388:97–100
Wolfes HJ, Alves J, Fliess A, Geiger R, Pingoud A (1986) Site directed mutagenesis exper-
    iments suggest that Glu 111 Glu 144 and Arg 145 are essential for endonucleolytic
    activity of EcoRI. Nucleic Acids Res 14:9063–9080
Xu SY, Schildkraut I (1991) Isolation of BamHI variants with reduced cleavage activities.
    J Biol Chem 266:4425–4429

Restriction Endonucleases: Structure of the Conserved Catalytic Core and the Role of Metal Ions in DNA Cleavage

J.R. Horton, R.M. Blumenthal, X. Cheng

1 Introduction

Type II restriction endonucleases (REases) are a fascinating group of proteins. With the REBase database currently listing ~3500 Type II REases having nearly 240 distinct DNA sequence specificities, they constitute one of the larger known families of enzymes (Roberts et al. 2003). These DNA-cleaving enzymes combine very high catalytic efficiencies ($k_{cat}/k_{uncat}=\sim10^{16}$) with exquisite DNA sequence selectivity. Restriction enzymes that are classified Type II (Welsh et al., this Vol.) cleave specifically within or close to their recognition sites, and do not require ATP hydrolysis for their nucleolytic activity. DNA cleavage by these enzymes can result in DNA with either 5′ or 3′ overhangs or blunt ends.

Most Type II REases function as homodimers. Two nonoverlapping catalytic sites, one per monomer, act cooperatively: each subunit acts on one strand of the duplex DNA substrate such that normally both strands are sequentially cleaved in a single binding event. Each monomer must contain structural elements corresponding to the three principal functions of these enzymes: monomer interaction (primarily dimerization, although some Type II enzymes form tetramers and require a greater oligomerization interface), DNA sequence recognition and catalyzing phosphodiester hydrolysis. Intra- and inter-monomer communication is necessary, because each subunit cooperates in the recognition process and in the catalytic reaction pathway. X-ray crystallography has profoundly affected our understanding of how Type II

J.R. Horton, X. Cheng
Department of Biochemistry, Emory University School of Medicine, 1510 Clifton, Atlanta, Georgia 30322, USA
R.M. Blumenthal
Department of Microbiology & Immunology, and Program in Bioinformatics & Proteomics/Genomics, Medical College of Ohio, Toledo, Ohio 43614–5806, USA

Nucleic Acids and Molecular Biology, Vol. 14
Alfred Pingoud (Ed.)
Restriction Endonucleases
© Springer-Verlag Berlin Heidelberg 2004

Table 1. Crystal structures of Type II endonucleases and related enzymes

Enzyme	Subtype	Recognition/cleavage	Catalytic residues	PDB (reference)
EcoRI-like subclass				
EcoRI	Orthodox	G↓AATTC	Asp91/Glu111/Lys113	1R1E, 1ERI, 1QPS (Mn²⁺) (Kim et al. 1990)
MunI	Orthodox	C↓AATTG	Asp83/Glu98/Lys100	1D02 (D83A) (Deibert et al. 1999)
BglII	Orthodox	A↓GATCT	Asp84/Glu93/Gln95 (Asn69)	1ES8, 1DFM (Ca²⁺), 1D2I (Mg²⁺) (Lukacs et al. 2000)
BamHI	Orthodox	G↓GATCC	Asp94/Glu111/Glu113	1BAM (Newman et al. 1994); 2BAM (Ca²⁺), 3BAM (Mn²⁺) (Viadiu and Aggarwal 1998)
BsoBI	Orthodox	C↓PyCGPuG	Asp212/Glu240/Lys242 (Glu71)	1DC1 (van der Woerd et al. 2001)
Cfr10I	Orthodox	Pu↓CCGGPy	Asp134/Glu204/Lys190 (Glu80)	1CFR (Bozic et al. 1996)
Bse634I	Orthodox	Pu↓CCGGPy	Asp146/Glu212/Lys198 (Glu70)	1KNV (Grazulis et al. 2002)
NgoMIV	IIF	G↓CCGGC	Asp140/Glu201/Lys187	1FIU (Deibert et al. 2000)
FokI	IIS	GGATGN₉↓NNNN↑	Asp450/Asp467/Lys469	2FOK (Wah et al. 1998) 1FOK (Wah et al. 1997)
TnsA (Tn7 transposase)		-	Glu63/Asp114/Lys132	1F1Z (Mg²⁺) (Hickman et al. 2000)
EcoRV-like subclass				
EcoRV	Orthodox	GAT↓ATC	Asp74/Asp90/Lys92 (Glu45)	1RVE (Winkler et al. 1993); 1AZ0 (Ca²⁺)(Perona and Martin 1997); 2RVE, 4RVE (Winkler et al. 1993); 1RVB (2 Mg²⁺) (Kostrewa and Winkler 1995); 1BSU (Ca²⁺)(Martin et al. 1999b); 1BSS (T93A) (2Ca²⁺) (Horton et al. 1998)
HincII	Orthodox	GTPy↓PuAC	-	1KC6 (Horton et al. 2002a)

Enzyme	Subtype	Recognition/cleavage	Catalytic residues	PDB (reference)
PvuII	Orthodox	CAG↓CTG	Asp58/Glu68/Lys70	1PVI (Cheng et al. 1994) 1PVU (Athanasiadis et al. 1994) 1EYU (Horton and Cheng 2000) 1F0O (Ca^{2+}) (Horton and Cheng 2000)
NaeI	IIE	GCC↓GGC	Asp86/Asp95/Lys97	1EV7 (Huai et al. 2000) 1IAW (Huai et al. 2001)
BglI	Orthodox	GCCNNNN↓NGGC	(Glu87) Asp116/Asp142/Lys144	1DMU (Ca^{2+}) (Newman et al. 1998)
Others				
λ-exonuclease	–		Asp119/Glu129/Lys131	1AVQ (Kovall and Matthews 1997)
MutH	–		Asp70/Glu77/Lys79	1AZO, 2AZO (Ban and Yang 1998)
VSR endonuclease	–		(Glu21) Asp51	1VSR (Tsutakawa et al. 1999b) 1CW0 (Tsutakawa et al. 1999a) 1OGD (Bunting et al., 2003)
T7 endonuclease I	–		(Glu20) Asp55/Glu65/Lys67	1FZR (E65 K) (Hadden et al. 2001) 1M0I (Hadden et al. 2002) 1M0D (Mn^{2+}) (Hadden et al. 2002) 1HH1 (Bond et al. 2001)
S.solfataricus				
Hjc Resolvase	–		Asp42/Glu55/Lys57	1GEF (Nishino et al. 2001b)
P.furiosus Hjc Resolvase	–		(Glu9) Asp33/Glu46/Lys48	1IPI (Nishino et al. 2001a)

REases accomplish these three functions and how they are linked to achieve concerted and specific DNA cleavage. Regarding this linkage, it is worth emphasizing that the selective pressure favoring REase specificity is quite strong, as cleavages at sites not protected by the cognate methyltransferase can kill the host cell (Heitman et al. 1999). Thus far, 14 REases have been structurally characterized along with six other evolutionarily related enzymes (Table 1). These include structures for the enzymes alone and in binary and ternary complexes with specific and nonspecific DNA and with divalent metal cations.

Only the catalytic core shares a common architecture among structurally characterized Type II REases. The remainder of these proteins, partially or entirely responsible for DNA binding and multimerization, exhibits great structural diversity even among some enzymes that recognize and cut the same DNA sequence (Grazulis et al. 2002; Lukacs et al. 2000; Pingoud and Jeltsch 2001). Surprisingly, the structurally conserved core is not specified by a contiguous gene segment, but is interspersed among coding regions for the nonconserved regions. The common architecture of the catalytic core has also been observed in other nucleases as well (Ban and Yang 1998; Bond et al. 2001; Hadden et al. 2001; Hickman et al. 2000; Kovall and Matthews 1998; Nishino et al. 2001b; Tsutakawa et al. 1999b; Bunting et al. 2003). This structural conservation has facilitated molecular phylogenetic analysis of the REase superfamily. [Members of this superfamily can be seen at the following URL for the SCOP database: http://scop.mrc-lmb.cam.ac.uk/scop-1.61/data/scop.b.d.gd.b.html].

However, despite the wealth of biochemical and structural data, summarized in recent reviews (Galburt and Stoddard 2002; Kovall and Matthews 1999; Pingoud and Jeltsch 2001), there is still controversy over the exact catalytic mechanism(s) used by Type II REases and whether one, two or three divalent metal ions are required. All Type II restriction enzymes, with the apparent exception of *Bfi*I (Sapranauskas et al. 2000), require Mg^{2+} or a similar divalent metal cation to cleave DNA. In some cases, these cations also play critical roles in specificity-determining contacts between the REase and its cognate DNA (Conlan and Dupureur 2002a; Lagunavicius et al. 1997; Skirgaila and Siksnys 1998; Soundararajan et al. 2002; Vipond and Halford 1995). Elucidating the detailed mechanism(s) of DNA cleavage depends critically on knowing how many Mg^{2+} ions are directly involved in catalysis.

In the rest of this chapter, we first discuss the structure and requirements of the conserved REase active site. We will summarize our current understanding of evolutionary relationships based on structure-based sequence alignments between REases and other enzymes associated with DNA phosphodiester hydrolysis. The one-, two-, and three-metal ion catalytic mechanisms will be discussed in the context of the structures of binary and ternary complexes of the enzymes. Finally, we will summarize our current (spring 2003) understanding of REase structure and function and the critical questions that remain to be answered.

2 Common Structural Attributes of Type II REases

Type II REases do not share appreciable amino acid sequence similarity and are thus not amenable to standard alignment procedures, except for a few obvious exceptions of closely related isoschizomers (Bujnicki 2000; Deva and Krishnaswamy 2001; Jeltsch et al. 1995a; Pingoud and Jeltsch 2001). Nevertheless, the crystal structure of EcoRV (Winkler et al. 1993) compared to that of EcoRI revealed a common catalytic core between these otherwise very different and seemingly unrelated enzymes (Rosenberg 1991; Venclovas et al. 1994; Winkler 1992).

The Type II enzymes in Table 1 have been divided into two endonuclease subgroups: EcoRI-like and EcoRV-like. This division is based on both structural and phylogenetic rationales. First, the EcoRI family binds DNA from the major groove side and produces sticky ends with 5′-overhangs, whereas the EcoRV family approaches DNA from the minor groove side and produces blunt ends (Aggarwal 1995). This has consequences for the positioning of the two active sites and subunit arrangement in the homodimers. Second, these families were found to correspond to the two main branches in a proposed phylogeny for the REase superfamily based on structure-guided sequence alignment of these proteins (Bujnicki 2000).

3 A Common Catalytic Core with Key Catalytic Side Chains

The REase core consists of a five-stranded mixed β-sheet flanked by α-helices that carries the catalytic center at one end, facing the reactive phosphodiester linkage (Fig. 1). Only four β-strands are absolutely conserved within the catalytic core; two of these β-strands contain the three amino acids directly involved in catalysis (see below). Of the four, three are shared with non-REase nucleases; in fact, superposing eight REases, the topoisomerase-related REase NaeI (see below), the repair nucleases MutH and Vsr, and the bacteriophage λ exonuclease revealed rms deviations of only ~2 Å for the Cα atoms in the three shared strands (Huai et al. 2000). The other strands could be critical for forming or stabilizing the β-sheet and may have faced fewer constraints during subsequent divergence.

The conserved core brings together two carboxylates, one from Asp and the second from either Glu or Asp, and one Lys. These belong to a moderately conserved sequence motif, $(P)DX_{10-30}(D/E)\Phi K$, where X is any residue and Φ is usually a hydrophobic one. Though the separation between the two carboxylates in REase primary sequences can vary considerably, their location on consecutive and structurally adjacent β-strands is highly conserved. In fact, the distance between the two carboxylate carbons ranges from 3.6–5.1 Å. The first Asp always appears at the N-terminus of one β-strand, adjacent to the Asp or Glu on the next strand. Sequence variation between the two carboxy-

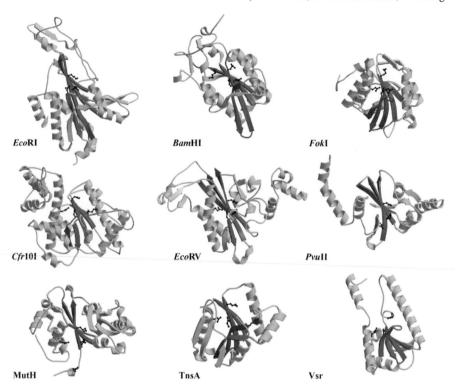

Fig. 1. Comparison of the REase folds in some members of the REase superfamily. The conserved central β-sheet fold is highlighted in *dark gray*. The catalytic residues are in ball-and-stick and colored *black*. Single protomers of BamHI, EcoRI, EcoRV, PvuII and Cfr10I and only the N-terminal catalytic domains of FokI and TnsA are shown

lates only affects the length of the strands. This sequence motif is somewhat relaxed, which makes it impossible in some cases to locate by simply inspecting the primary sequence. The first Asp is the only residue to be absolutely conserved among the structurally-characterized REases. The second acidic residue (D/E; more often a Glu than an Asp) is Ser in Cfr10I and NgoMIV. The hydrophobic residue (Φ) is a Cys in NaeI and NgoMIV, the conserved Lys (K) is replaced by Glu in BamHI and by Gln in BglII, and the Pro in some REases has been replaced by Ile, Val, Asp, and Thr. The roles of these residues are discussed below, but their naturally occurring substitutions suggest either that the REases can achieve phosphodiester hydrolysis through varied mechanisms, or that they are functionally flexible enough to have the replaced residues contributed by other, spatially equivalent portions of the protein.

4 Phylogenetic Analysis of REases

There is essentially no overall conservation of REase sequences, and the portions of REase genes that specify the structurally conserved catalytic region are noncontiguous and variably spaced. Nevertheless, the increasing number of crystal structures and methods for sequence comparison has allowed refined structure-guided alignments. These have revealed divergent evolutionary relationships (Bujnicki and Rychlewski 2001a, b; Deva and Krishnaswamy 2001; Jeltsch et al. 1995a). The releationships between REases also exhibit reconvergence of functionally divergent enzymes: a single amino acid change transforms restriction enzyme NaeI to a topoisomerase and recombinase (Jo and Topal 1995). NaeI shows no sequence similarity to other REases, yet its facile conversion into a REase and pairing with a modification methyltransferase provides a telling view of the evolutionary dynamics of REases.

Because catalysis is at least partially divorced from substrate binding, in the sense that distinct regions of the protein are responsible for each, REase activity appears to depend on a small number of spatially conserved side chains. Supported by mutational analysis (Skirgaila et al. 1998), the structures of Cfr10I (Bozic et al. 1996), NgoMIV (Deibert et al. 2000) and Bse634I (Grazulis et al. 2002) indicate that the sequence motif $PDX_{46-55}KX_{13}E$ corresponds to the conserved active sites of these three REases. Additionally, in these REases, the distance between the two carboxylate carbons has increased compared to those with the common motif to ~6.9 Å which may be specially required for their activity. SsoII and Ecl18 kI appear to have a similar arrangement based on sequence alignments and structural comparisons and verified by mutational analysis (Pingoud et al. 2002; Tamulaitis et al. 2002).

Divergent evolution within the Type II REase superfamily has led to enzymes that contain the protein fold of the common catalytic core but only have remnants of the catalytic sequence motif (Tsutakawa and Morikawa 2001; Tsutakawa et al. 1999b; Bunting et al. 2003). Very short patch repair (Vsr) endonuclease is a protein involved in DNA repair that recognizes a TG mismatch in the context of a specific sequence and cleaves the phosphate backbone on the 5′ side of the thymidine (Bhagwat and Lieb 2002). As noted earlier, the central part of the β-sheet core of Vsr superimposes well onto the catalytic domain β-strands of Type II REases. However, only Vsr Asp51, a catalytically essential residue, superimposes directly onto the first conserved Asp residue of the sequence motif in REases.

5 Possible Roles for Divalent Metal Ions

To catalyze phosphodiester bond cleavage, four main active-site components are required: first, a nucleophile to which the phosphoryl group can be transferred; second, a general base to activate and position the nucleophile; third, one or more Lewis acids to stabilize the negative charge that develops in the transition state; and fourth, a general acid that can protonate the leaving group oxygen. Due to the chemical diversity of amino acids and local structural diversity of proteins, there are many ways to form an active site with the necessary spatial distribution of appropriate functional groups. One way is to use divalent metal cations as cofactors. These metal ions activate and position the attacking water molecule in line with the scissile phosphoryl group.

Divalent metal ions are essential cofactors in many non-REase phosphoryl-transfer reactions. These include DNA and RNA polymerases (Brautigam and Steitz 1998; Cheetham and Steitz 2000; Steitz 1998), ribozymes (Steitz and Steitz 1993), transposases and many recombination enzymes (Declais et al. 2001; Lins et al. 2000; Lovell et al. 2002; Rice et al. 1996; Wlodawer 1999), and many exo- and endonucleases (Beernink et al. 2001; Beese and Steitz 1991).

Magnesium is the divalent metal ion of choice as a cofactor in most (if not all) Type II REases. In vitro, Mg^{2+} can be replaced by Mn^{2+} and, to varying extents depending on the enzyme, the divalent ions of Fe, Co, Ni, Zn, and Cd. However, Ca^{2+} generally inhibits cleavage by Type II REases while allowing formation of stable, inactive specific protein–metal–DNA complexes (Conlan and Dupureur 2002b; Vipond and Halford 1995). Thus, Ca^{2+} is a near-perfect analog for Mg^{2+} in terms of effects on DNA binding (Martin et al. 1999a).

It is not obvious from structures of pre-reactive complexes of the common catalytic core why Mg^{2+} and Ca^{2+} have such dramatically different behavior. However, NMR data on the PvuII REase show that Ca^{2+} binding perturbs a larger number of resonances than either Mn^{2+} or Mg^{2+} (Dupureur and Hallman 1999). Subtle differences between Mg^{2+} and Ca^{2+} could explain these results. First, the ionic radius of Ca^{2+} is 0.34 Å larger than that for Mg^{2+}. Second, Ca^{2+} favors a different coordination geometry (six to eight coordination vs. six for Mg^{2+}) and longer coordination distances (2.5 vs. 2.0 Å for Mg^{2+}) (Cowen, 1998). Third, the pK_a of Ca^{2+}-bound water (12.9) is higher than that for water bound by Mg^{2+} (11.4) or Mn^{2+} (10.6), which could lower the concentration of metal-hydroxyl ions for the nucleophilic attack when calcium is present. Furthermore, even minor deviation from the precise positioning of substrate and reactants with Ca^{2+} compared to Mg^{2+} or Mn^{2+} takes the reaction off the direct thermodynamic path to the transition state (Jen-Jacobson 1997). The inhibitory effects of Ca^{2+} on endonuclease activity are probably more significant in the transition state rather than in the initial binding in the pre-reactive ground-state complex. Computational modeling of endonuclease catalysis is beginning to support this view (Fuxreiter and Osman 2001; Mordasini et al. 2002).

6 The One-Two-Three Metal Debate

One- and two-metal ions have been observed at different sites via X-ray crystallography of ternary complexes of REase, cognate DNA, and divalent metal ion(s). These observations have led to proposals of one-, two-, or three-metal reaction mechanisms. The various proposed mechanisms also differ in which component of the active site functions as the general base that abstracts a proton from the attacking water. The general base could be (1) a divalent metal ion, (2) the Lys of the $(P)DX_{10-30}(D/E)\Phi\underline{K}$ motif and/or (3) the scissile phosphate itself (substrate-assisted catalysis). All three proposed general bases would require a considerable shift in pK from their unperturbed value in the free state. Whether additional amino acid residues are recruited for this purpose could differ from REase to REase.

The situation seems most complicated in the case of EcoRV, for which over twenty structures have been determined. Thirteen of these structures are of ternary complexes containing various divalent metal ions, different substrates, and native and mutants forms of EcoRV (Pingoud and Jeltsch 2001). Various configurations of metal ion binding in the catalytic sites include one Ca^{2+} ion (site II in Fig. 2) in wild-type enzyme ternary complex (Perona and

Fig. 2. Schematic of catalytic site and reaction. *Top* DNA backbone near the cleavage site with the phosphate drawn in the pentavalent transition state. The attacking and protonating water molecules are shown as "\o/". *Curved gray arrows* show the movement of electrons during the course of the REase reaction. Generalized divalent metal sites are labeled I, II, and III. Many REase family enzymes contain two divalent metal ions at sites I and II. At *bottom*, the moderately-conserved sequence motif is shown with additional acidic residues found in some REases upstream from, and sometimes included in, the motif (Galburt and Stoddard 2002). *Arrows* point to where key residues in the motif are located in the active site

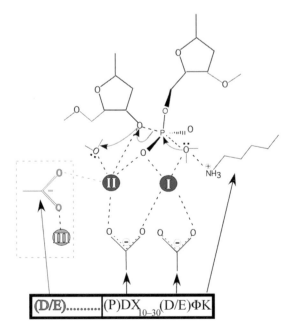

Martin 1997), two Ca^{2+} ions (sites I and III in Fig. 2) in a T93A mutant complex (Horton et al. 1998), and two Mg^{2+} ions (sites II and III in Fig. 2) in an unproductive complex (Kostrewa and Winkler 1995). Consequently, at least three different mechanisms have been proposed for EcoRV catalysis, all based on structural information but differing in the numbers and sites of Mg^{2+} ions involved (Groll et al. 1997; Horton et al. 1998; Jeltsch et al. 1992; Sam and Perona 1999; Vipond et al. 1995). However, the confusion in relating ternary complex crystal structures of EcoRV to mechanism pervades our understanding of REase catalysis in general.

While structural observations have often been used to support a particular catalytic mechanism, structural interpretation in the absence of functional analysis can be dangerous. For example, in several instances only one binding site per subunit is occupied and/or that only one subunit has divalent metal ions bound in a crystal structure, in spite of the fact that very high concentrations of divalent metal ions were used for the soaking, but also for the cocrystallization experiments (e.g., see Table 5 of Pingoud and Jeltsch 2001). The fact that high concentrations are being used for these experiments is also a point of concern, as they might lead to occupation of nonphysiological binding sites (Cowan 1998). Another problem is that biochemical experiments suggest that there must be additional divalent metal ion binding sites in the EcoRV–DNA complex, not yet seen in the cocrystals, which are involved in recognition (Jeltsch et al. 1995b).

7 Generalizations Regarding REase Catalysis

Structural observations of many Type II enzyme–DNA–metal complexes (Table 1) allow some generalizations to be made about the catalytic machinery and mechanism. These generalizations help to distinguish between the proposed one-, two-, and three-metal mechanisms. Comparing the schematic diagrams of protein–DNA–metal interactions, in both pre-reactive complexes containing Ca^{2+} and post-reaction complexes with either Mg^{2+} or Mn^{2+}, reveals some intriguing similarities (Figs. 3–7).

First, a divalent metal at site I (Fig. 2) is common in all structurally characterized REases and related enzymes, and is included in all proposed mechanisms. Metal-ion binding site I in pre-reactive ternary complexes is formed by the generally conserved two acidic residues of the $(P)\underline{D}X_{10-30}(\underline{D/E})\Phi K$ motif, by a main chain carbonyl of the hydrophobic amino acid Φ of the motif, and a non-bridging oxygen of the scissile phosphate.

Second, the divalent metal-ion-occupying binding site I is coordinated to a water molecule which is directly or indirectly (via a second water molecule) associated with the $proR_p$-oxygen of the phosphate 3' to the scissile phosphodiester bond. This water molecule is in position for an in-line attack on the bond to be cleaved, and is held there by the last residue of the

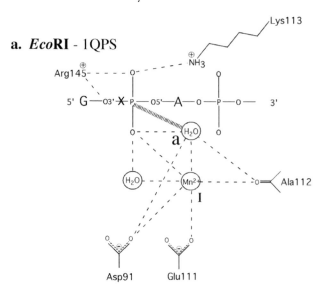

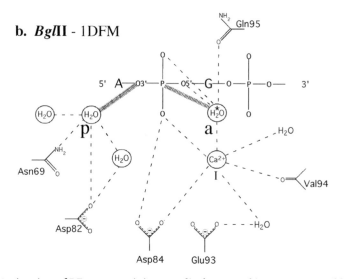

Fig. 3. Active sites of REases containing one divalent metal ion. **a** EcoRI, and **b** BglII. The figure was deduced from cocrystal structures of specific protein–DNA substrate complexes obtained in the presence of divalent metal ions (PDB code of structure is shown). The *broken lines* indicate metal coordination or hydrogen bonds; the *striped lines* indicate the contacts between a potential attacking water molecule (labeled *a*) and the target phosphate; and between the protonating water (labeled *p*) acting as a general acid and the leaving O3′ oxygen atom. Some water molecules (labeled *) have as many as six or seven coordination and, for clarity, not all of them are shown. Metals are labeled *I, II,* or *III* according to the metal site bound as shown in Fig. 2

a. *Bam*HI - 2BAM

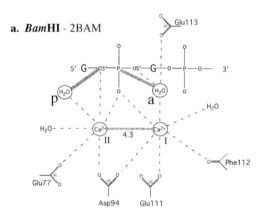

Fig. 4. Active sites of REases containing two divalent metal ions in a pre-reactive complex (uncleaved DNA). **a** BamHI, **b** PvuII, and **c** BglI

b. *Pvu*II - 1F0O

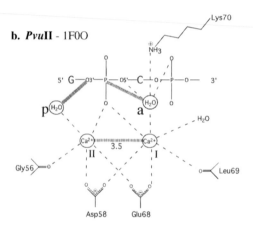

c. *Bgl*I - 1DMU

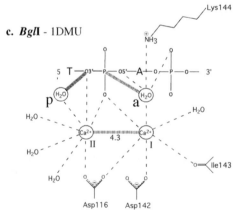

Fig. 5. Active sites of enzyme–metal complexes of other superfamily members.
a TnsA, and b T7 endonuclease I

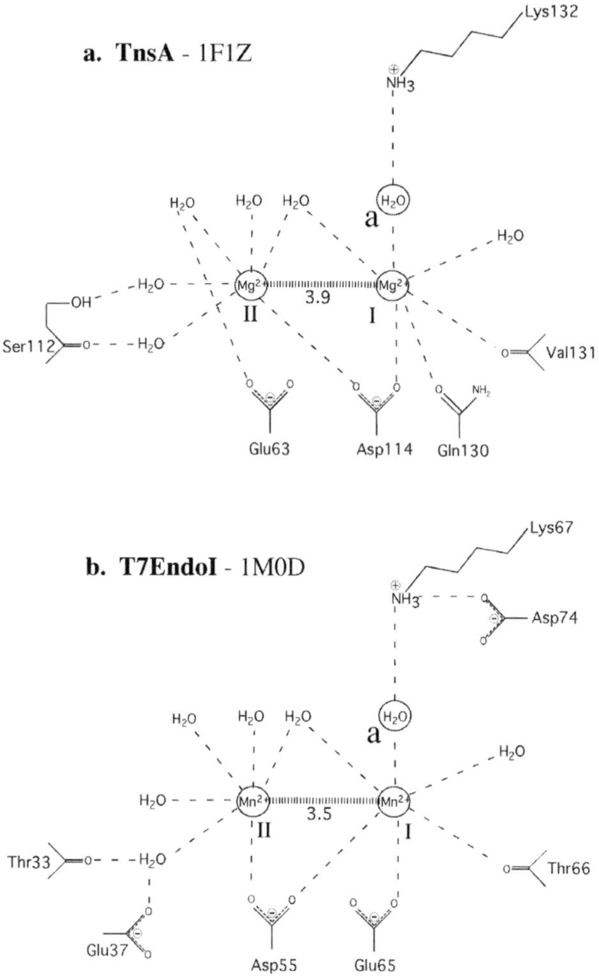

a. TnsA - 1F1Z

b. T7EndoI - 1M0D

(P)DX$_{10-30}$(D/E)ΦK motif. This is usually Lys, but is Glu in BamHI and Gln in BglII; all three are capable of forming a hydrogen bond to the attacking water molecule. Mg^{2+}-bound water from the inner hydration sphere is more acidic (pK$_a$=11.4) than bulk water, and may protonate the leaving group (Jeltsch et al. 1992). In this case, then even though basic, acidic and neutral side chains are located at structurally equivalent positions in different active sites, mechanistic divergence is not necessary implied. However, the roles of these side chains in the activation/positioning of the attacking nucleophile still need to be elucidated.

Third, in many pre-reactive complexes, a second divalent metal-ion binding site (site II in Fig. 2) is formed by the first and in some cases also by the second acidic residue of the (P)DX$_{10-30}$(D/E)ΦK motif. A divalent metal ion at

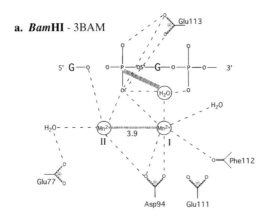

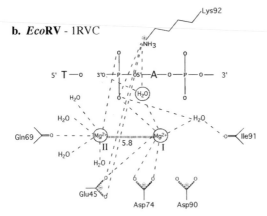

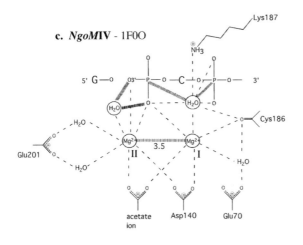

Fig. 6. Active sites of REases containing two divalent metal ions in a post-reactive complex (cleaving DNA). **a** BamHI, **b** EcoRI, and **c** NgoMIV. Although the metal ions are still labeled as at sites I and II, note that their coordination has changed from the pre-reactive complexes in Fig. 4. The metal at site I has taken on some of the coordination of the metal at site II in the pre-reactive complex, and the metal at site II has some coordination as a pre-reactive metal site III

Fig. 7. Active site of EcoRV containing three divalent metal ions. The interactions were extracted from two cocrystal structures containing metal ions. Sites I and III are from PDB 1BSS (Horton et al. 1998) and site II is from PDB 1BSU. (Martin et al. 1999b)

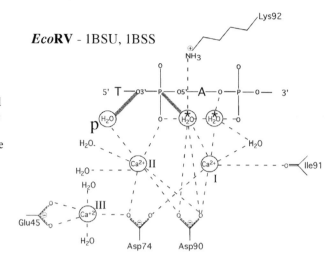

this site could serve to polarize a P–O bond and thereby to make the phosphorus more susceptible to nucleophilic attack. It could also help to coordinate a water molecule, which could be used to protonate or directly associate with the leaving group (3′-O-).

Fourth, comparison of pre- and post-reaction complexes suggest that the second acidic residue of the (P)DX$_{10-30}$(D/E)YK loses direct coordination to the divalent metal ions during the course of the reaction. Thus, the first acidic residue of the motif may be more important and play multiple roles during the course of the reaction. The metal ions themselves could play various roles in the mechanism, as they are required in some cases for DNA association or sequence recognition (Conlan and Dupureur 2002b; Erskine and Halford 1998; Martin et al. 1999a; Vipond and Halford 1995).

Clearly, the precise mechanism by which Type II enzymes catalyze DNA cleavage depends on the minimal number and positions of metal ions in the active site. Proposals for Type II REase catalytic mechanisms involve 1–3 metal ions – which has led to a "One-Two-Three-Metal Ion" debate (Kovall and Matthews 1999). Arguments can be made for generalized mechanisms requiring each of these numbers of metal ions. Discussion of additional aspects of the Type II REase catalytic mechanism can be found in Pingoud and Jeltsch (2001).

8 One-Metal Catalytic Mechanism: EcoRI, BglII

The first one-metal ion mechanism was based on a comparison of the enzyme–DNA cocrystal structures of EcoRI and EcoRV as well as on molecular modeling (Jeltsch et al. 1992, 1993). Later, a single metal was observed in

EcoRI structures (pdb code 1QPS) in both pre-transition and post-reaction ternary complexes of EcoRI–DNA–Mn^{2+}. In addition, a ternary complex of BglII revealed only one Mn^{2+} binding site per subunit (Lukacs et al. 2000).

In EcoRI and BglII, the single metal ion is bound to site I, and a metal-coordinated water molecule is positioned and activated for in-line nucleophilic attack on the electrophilic phosphorus atom (Figs. 2 and 3). In BglII, the probable nucleophilic water molecule also makes hydrogen bonds to Gln95 (which is in the position equivalent to the last residue of the (P)DX$_{10-30}$(D/E)ΦK motif), to both bridging and non-bridging oxygen atoms from the scissile phosphate, and to another water molecule bound to a non-bridging oxygen atom from the 3'-adjacent phosphate (Lukacs et al. 2000). A second metal-bound water molecule is positioned by Asn69, Asp82, and Asp84 (all in the X$_{10-30}$ motif spacer region), and makes a hydrogen bond to the leaving group oxygen. This second water molecule appears to be well positioned to donate a proton to, and thereby stabilize, the leaving group.

In EcoRI, the nucleophilic water is not contacted by a protein residue, but does form hydrogen bonds to both a 3'-phosphate-bound water and a non-bridging oxygen from the scissile phosphate. An unobserved, hypothetical water molecule positioned by Arg145 could protonate the leaving group (Pingoud and Jeltsch 2001). Thus, a single-metal mechanism for this enzyme could utilize an in-line, metal-bound nucleophile with a pKa that is lowered through metal association.

However, neither the EcoRI nor the BglII structure reveals an obvious general base, which would be required to abstract the proton from the nucleophilic water molecule in an associative nucleophilic attack. There are three possible ways to obviate the requirement for a protein-supplied general base. These include involvement of a bulk solvent-supplied metal-bound hydroxide ion that does not need to be deprotonated, existence of a dissociative transition state, and substrate-assisted catalysis wherein the general base is a non-bridging DNA phosphate oxygen atom that activates a bound water molecule.

Substrate-assisted catalysis by REases was first proposed by Jeltsch et al. (1992). In this model, the conserved acidic amino acids at the catalytic site bind a Mg^{2+} ion and position it near the scissile phosphate as in the previous model. Hydrolysis begins with in-line nucleophilic attack by an activated water molecule. The phosphate 3' of the scissile bond activates the water as a general base. The conserved Lys and/or a Mg^{2+} could also participate in activating the water and stabilizing the pentavalent transition state produced at the scissile phosphorus (Sam and Perona 1999). Inversion occurs as the 3'-OH leaving group is protonated by a Mg^{2+}-bound water upon exit.

Later cocrystal structures of enzymes bound to their cognate sequences and substitution experiments in the enzyme or DNA supported this model, and suggested some sort of substrate assistance to catalysis by many other members of the Type II REase superfamily (Jeltsch et al. 1993, 1995c; Pingoud and Jeltsch 1997). For example, the cleavage rates catalyzed by 17 enzymes were measured using modified and unmodified oligodeoxyribonucleotides, in which the phosphate group 3′ to the scissile bond was substituted by methyl phosphonate (Jeltsch et al. 1995 c). Five REases turned out not to be inhibited by this modification (BglII, Cfr10I, MunI, BstI, and BstYI). Twelve other REases, including BamHI and PvuII, cleaved the modified substrate at a strongly reduced rate or not at all, suggesting substrate assistance of their catalytic mechanism. In other experiments (Jeltsch et al. 1993) when the 3′ phosphates were absent or replaced, the DNA substrate was bound but not cleaved by EcoRI and EcoRV, again consistent with substrate assistance.

9 Two-Metal Catalytic Mechanisms

On the basis of a crystal structure containing two ions (Freemont et al. 1988) and later supported by the structure of an enzyme–product complex (Beese and Steitz 1991), the first two-metal ion phosphodiester cleavage mechanism was proposed for the 3′-5′ exonuclease domain of *E. coli* DNA polymerase I. Elements of the 3′-5′ exonuclease mechanism appear in proposed two-metal mechanisms for enzymes in the Type II REase superfamily as well as for other DNA-cutting enzymes (Fedor 2000; Nishino and Morikawa 2002).

Three REases have been crystallized in the pre-reactive, "substrate" state with Ca^{2+}, enzyme and cognate DNA: BamHI (Viadiu and Aggarwal 1998), PvuII (Horton and Cheng 2000), and BglI (Newman et al. 1998) (Fig. 4). Three post-reaction complexes, the result of catalysis before diffusional loss of the products, show two metal ions (Mg^{2+} or Mn^{2+}) near the cleaved cognate DNA structures: EcoRV (Kostrewa and Winkler 1995), BamHI (Viadiu and Aggarwal 1998) and NgoMIV (Deibert et al. 2000) (Fig. 6).

Methods other than X-ray crystallography point to a two-metal mechanism for Type II REases, and especially for PvuII. Isothermal titration calorimetry (ITC) and ^{19}F-NMR spectroscopy show that two Mn^{2+} or Ca^{2+} ions are bound per monomer (Dupureur and Conlan 2000; Jose et al. 1999). ^{25}Mg-NMR spectroscopy also demonstrates that more than one Mg^{2+} ion binds in each PvuII active site with affinities that are reasonable for Mg^{2+} under physiological conditions (Dupureur and Conlan 2000). Additionally, a study measuring association and dissociation rates as a function of metal ion concentration (Conlan and Dupureur 2002a) also suggests two metal ions per PvuII monomeric active site.

The two-metal mechanism of BamHI, PvuII, and NgoMIV shares features with the one-metal mechanism in the absence of substrate assistance, and resembles the mechanism proposed for 3′-5′ exonuclease. The metal ion at site I is again held by the two conserved acidic residues and the backbone carbonyl of the hydrophobic residue of the (P)$\underline{D}$X$_{10-30}$($\underline{D/E}$)ΦK motif, and by a non-bridging oxygen from the scissile phosphate. This metal ion is again responsible for positioning the nucleophilic water for in-line attack. Finally, as with the one-metal mechanisms, it is not clear for the two-metal mechanisms what plays the role of a general base for the reaction.

However, in these two-metal mechanisms a second metal ion is located at site II, on the other side of the scissile phosphate from the nucleophilic water molecule. Site II is created by the first conserved Asp of the motif, a substrate non-bridging oxygen, and a complement of other acidic residues, backbone carbonyls and water molecules depending on the specific enzyme. A divalent metal ion at site II is well positioned to activate a water molecule for protonation of the leaving group 3'-phosphate oxygen and can help to stabilize the negative charge that develops in an associative transition state. The oxygen of the scissile phosphate interacts with both metals in the three pre-reactive structures of PvuII, BglI, and BamHI (Fig. 4), suggesting that the two metals act together as Lewis acids to neutralize the negative charge created by the pentavalent transition state. The attacking water coordination is particularly similar between PvuII and BglI (Fig. 4). However, the slightly different locations of the second metal (Fig. 4) explain why the interactions between the leaving group O3' and metal II, observed in BamHI and BglI (Fig. 4), are not seen in PvuII. The metal at site II in BamHI and BglI is believed to stabilize the negative charge on the 3' phosphate oxyanion leaving group (Kovall and Matthews 1999). However, a direct interaction with metal ion may not be necessary. Stabilization of the negative charge could also be achieved by a water molecule in the first hydration sphere of the metal ion at site II, as may be the situation for PvuII.

9.1 EcoRV

Three alternative two-metal mechanisms have been presented for EcoRV, each differing in the identity of the general base. The first proposal suggests that a water molecule in the hydration sphere of Mg^{2+} bound to Asp74 and Asp90 (site II in Figs. 2 and 7) provides the attacking nucleophile, while a water molecule in the hydration sphere of Mg^{2+} ion bound to Glu45 and Asp74 (site III in Figs. 2 and 7) protonates the leaving group (Kostrewa and Winkler 1995). In contrast, the second proposal suggests that Mg^{2+} in site III functions as the general base (Vipond et al. 1995). The third mechanism is a modification of the second: the water molecule in the hydration sphere of Mg^{2+} in site III is deprotonated with the help of Glu45 from one subunit and Asp36 from the other subunit of the homodimer (Baldwin et al. 1999; Stanford et al. 1999). Unfortunately, each of these model mechanisms cannot explain observations from all biochemical experiments conducted with this REase – which has led to a three-ion proposal (see below).

9.2 BamHI

As noted above, two-metal REase mechanisms have not yet clearly identified the general base. Intriguingly, a pre-reactive complex of BamHI containing Ca^{+2} indicates that Glu113 hydrogen bonds the nucleophilic water molecule (2.9 Å) and represents a possible general base (Viadiu and Aggarwal 1998). This Glu is at the Lys position of the conserved motif, and appears to be critical to the BamHI mechanism. In fact, replacing this Glu113 with Lys results in a severe loss of activity (Xu and Schildkraut 1991). Others have concluded that Glu113 has an electrostatic effect on BamHI action rather than a direct involvement in catalysis as a general base; the replacement of Glu by Lys could certainly disrupt site II metal binding. However, Lys in the last position of the motif appears to be critical for other Type II REases, as a Lys92Glu mutation in EcoRV or Lys113Glu in EcoRI abolish their activity (Grabowski et al. 1995; Selent et al. 1992). Again, in these mutant proteins electrostatic changes could affect metal binding (Engler et al. 2001; Horton et al. 2002b). Also, neither PvuII nor BglI ternary complexes have a potential general base positioned close to the nucleophilic water. In summary, even though these REases bind two metal ions, the divergence in catalytic residues in BamHI and other REases suggest that these enzymes differ in their catalytic mechanism and reaction coordinates. BamHI, with a putative general base, would appear to proceed with an associative (S_N2) type reaction while PvuII, lacking a putative general base, may follow a more dissociative (S_N1) type reaction.

Using pre- and post-reactive ternary BamHI structures (Figs. 4a and 6a), a combination of quantum-mechanical and classical simulations is beginning to shed light on the functional role of metal ions and add credence to two-metal catalysis models (Mordasini et al. 2002). Several observations from these simulations involve inactive BamHI–DNA–Ca^{2+} and active BamHI–DNA–Mg^{2+} complexes. First, the site I Ca^{2+} ion appears to increase its coordination in the transition state, resulting in a strained rearrangement of the atoms around it, while a site I Mg^{2+} ion maintains a quasi-symmetrical octahedral structure throughout the reaction. Second, the role of the ion at site II appears to be crucial for stabilizing a pentacoordinate transition state for the phosphate. To accomplish this, the site II ion has to reduce its coordination with oxygen from six to four. This would allow a water molecule to exit the coordination sphere and position itself to stabilize, via hydrogen bonding, another water molecule which in turn drives protonation of the O3′ leaving group of phosphate. Coordination of Mg^{2+} by four oxygen atoms is not atypical. This is not true of Ca^{2+}, which tends toward overcoordination (relative to sixfold coordination) rather than undercoordination. In fact, when the BamHI–DNA–Mg^{2+} system is forced to accommodate the pentacoordinate phosphate, the bonding between the phosphate and the metal ion relaxes, but the BamHI–DNA–Ca^{2+} system does not readjust its coordination and attempts to retain all possible oxygen partners. Thus, these combined simulations reveal that a site II Ca^{2+} ion fails to stabilize a favorable transition state for the Type II enzyme reaction.

9.3 Tn7 Transposase

Two-metal 3'-5' exonuclease-like binding sites have been visualized and catalytic mechanisms have been proposed for Tn7 transposase (Hickman et al. 2000) and T7 endonuclease I (Hadden et al. 2002) (Fig. 5). The Tn7 transposase is a heterodimer of the polypeptides TnsA and TnsB, and introduces double-strand breaks at the transposon ends during transposition. TnsA is responsible for cleavage at the 5' ends of the transposon, while TnsB cuts the 3' ends and catalyzes subsequent strand transfer using the newly created free 3' OH groups as the nucleophiles. Contrary to expectation, the structure of TnsA at 2.4 Å resolution showed that the majority of the fold is that of a Type II REase common core (Hickman et al. 2000). On the basis of number of aligned residues, the most similar REases are FokI, and Cfr101. Catalytically important residues Glu63, Asp114, Lys132, and Glu149 correspond structurally to Glu71, Asp134, Lys190, and Glu204 in Cfr10I, and the Cfr10I residues are known to be important for activity (Skirgaila et al, 1998). In TnsA, two Mg^{2+} ions are in distorted octahedral coordination environments and located 3.9 Å from each other. These two metal ions are bidentately liganded by the carboxylate group of Asp114 (Fig. 5a). Two bound Mg^{2+} ions had not previously been observed in other REase-like active sites in the absence of substrates or analogs, suggesting that a two-metal nuclease active site could be assembled without the DNA substrate.

9.4 T7 Endonuclease I

Bacteriophage T7 endonuclease I (T7 Endo I) is a junction-resolving enzyme that selectively binds and cleaves four-way DNA junctions. The structure reveals a symmetrical homodimer arranged in two well-separated domains. Each domain, however, is composed of elements from both subunits, and amino acid side chains from both protomers contribute to the active site. While no significant structural similarity could be detected with any other junction-resolving enzyme, the active site was similar to that found in several REases and most like BglI. T7 Endo I active site residues were suggested by superimposition with Glu20, Asp55, Glu65, and Lys67 of BglI, with a root-mean-square deviation of just 0.33 Å (Hadden et al. 2001). Furthermore, Mn^{2+} ions were situated in T7 Endo I in similar positions to the calcium ions found in the BglI structure (Fig. 5b). The Mn^{2+} ion located at site I is coordinated by three protein ligands, while that located at site II is coordinated by one protein ligand. A two-metal mechanism has been proposed based on this observation and calorimetry experiments showing the enzyme binds two Mn^{2+} ions in a sequential process (Hadden et al. 2002). Involvement of substrate assistance is implied by the observation of a sulfate ion (a substitute for the phosphate of the DNA substate) hydrogen-bonded to a candidate attacking water molecule near the 3′ phosphate to the cleavage site.

10 Three-Metal Catalytic Mechanism

On the basis of the structure of the EcoRV mutant T93A and a crystal structure of the wild-type EcoRV–DNA complex, in which the activity of the protein is retained within the crystal lattice, molecular modeling and energy minimization were used to construct a model for the pre-transition state and to propose a three-metal mechanism of catalysis (Horton et al. 1998). This mechanism was called a metal ion-mediated substrate-assisted catalysis mechanism, to account for the fact that the metal ion in site I is coordinated to two water molecules, one that bridges to the $proS_p$-oxygen of the 3′-phosphate, and another that is positioned to attack the scissile phosphate. The positions of metals at sites I and II in this mechanism are similar to those observed in other Type II enzymes and in the two-metal mechanisms.

In the three-metal mechanistic scheme the metal at site I, which had not been observed in earlier structures of EcoRV, positions and activates a water molecule (or hydroxide ion) for nucleophilic attack on the scissile phosphate (Fig. 7). Lys92, a catalytically essential residue (Selent et al. 1992), also participates in orienting and stabilizing the hydroxide ion for attack. This is in contrast to previous schemes in which it was suggested that the function of Lys92 is to stabilize the negative charge of the transition state (Jeltsch et al. 1993; Kostrewa and Winkler 1995; Vipond et al. 1995). The metal at site II (Figs. 2 and 7) is assumed to both stabilize the negative charge of the transition state and to coordinate an inner sphere water molecule that acts as a general acid by protonating the 3′-oxygen of the leaving group. The role of the metal at site III (Figs. 2 and 7) is presumed to be primarily structural in nature, but necessary in orienting Asp74 for its interaction with the other two metals.

A major drawback of the three-metal ion model is that it is the result of combining information from several crystal structures. Ideally, the role of all the catalytic machinery may be observed in the course of the reaction as in a time-resolved Laue X-ray crystallography experiment (Stoddard 1998). However, this has not yet been attempted for any Type II REase or related enzymes. Nevertheless, it is interesting to compare the structure of the proposed three-metal ion mechanism of EcoRV (Fig. 7) with changes in the structure of the catalytic site observed between the pre- and post-reaction complexes (Figs. 4 and 6). While the catalytic site in the pre-reactive complexes contain metal binding sites I and II, that in the post-reaction complexes appears to have metal ions in locations more like sites II and III. The three-metal catalytic model seems then to be an average of pre- and post-reactive states and a cumulative illustration of metal ion movement during the reaction pathway. It is possible that the available crystal structures for EcoRV are just snapshots along the reaction pathway that cluster near pre- and post-reactive states, or that crystal conditions affect the electrostatics within the active sites.

11 "Hypothetical" Active Sites of REase Superfamily Members

Several cocrystal structures of specific REase–DNA complexes have been determined in the absence of divalent metal ions. In addition, several other nucleolytic enzymes listed in Table 1 share the Type II REase catalytic core that unites them in the same superfamily. The similarity of the common structural core in primary, secondary, and tertiary structure suggests divergent evolution from a common evolutionary origin, implying a common ancestral catalytic mechanism. The common core fold and its associated catalytic motif $(P)\underline{D}X_{10-30}(D/E)\Phi K$ identify the active sites in these metal-free structures. Many of the catalytic residues can be determined by superposition of common core structures and/or sequence alignment of the motif, and then tested by mutagenesis.

For example, catalytic and metal-binding residues were proposed for T7 Endo I based on comparison with active sites of BglI, EcoRV, EcoRI, and FokI (Hadden et al. 2001). Later, a two-metal ion mechanism for the hydrolytic cleavage of DNA junctions was proposed based upon the similarity of these active sites and mutagenesis experiments of T7 Endo I (Declais et al. 2001). Afterward, direct observation of metal ion binding to the T7 Endo I active site was shown to be consistent with the proposed mechanism. Catalytic residues, and sometimes mechanism, are implied in the following enzymes but metal binding has not been observed by X-ray crystallography. In addition, many of the biochemical and other structural approaches to characterization of the multiple crucial roles of divalent metal ions (outlined by Perona 2002) have yet to be used. Experimentally testing predictions from comparisons of REase active sites will provide an important test of our understanding of these enzymes.

11.1 MunI

The ability of MunI to form tight complexes is highly dependent on pH but not on the presence of metal ions (Haq et al. 2001). Asp83 of MunI is the first acidic residue of the $(P)\underline{D}X_{10-30}(D/E)\Phi K$ motif. An Asp83Ala mutant of MunI was cocrystallized with a decameric oligonucleotide. Although these crystals were grown in the presence of 50 mM $CaCl_2$, no Ca^{2+} ions were present in the active site probably because both acidic groups of the motif are required to bind a divalent ion (Deibert et al. 1999). Glu98 and Lys100 of the conserved motif (^{82}PDX$_{14}$EIK) are close to the scissile phosphate, and superimpose onto the catalytic residues of EcoRI. Site-directed mutagenesis experiments support this putative active site (Lagunavicius and Siksnys 1997). Thus, it is theorized MunI probably follows a one-metal catalytic mechanism similar to that proposed for EcoRl.

11.2 BsoBI

BsoBl is the only solved REase structure that has an obvious protein-supplied general base close to the active center (van der Woerd et al. 2001). Two strands from one monomer contain Asp212, Glu240, and Lys242 that together form a normal motif. BsoBI shares with other Type II REase structures the position of the attacking water molecule, which is located 3.7 Å from the phosphorus atom of the scissile phosphodiester bond and within hydrogen bonding distance of the proR$_p$ oxygen of the phosphate 3′ to the scissile bond. Furthermore, this water molecule is only 3.8 Å from Lys243, which in turn contacts the proS$_p$ oxygen of the phosphorus atom to be attacked.

However, BsoBI, like Vsr endonuclease, has a His in the active site that appears poised to deprotonate the attacking water molecule. In addition to this idiosyncrasy, residues from the other monomer – Asp251 and Glu252 – may participate in forming a divalent metal binding site. In a two-metal mechanism proposed from these observations, one Mg^{2+} is suggested to bind to Asp212 and Glu240 as well as to the pro-S$_p$ oxygen of the scissile phosphodiester bond, while the second Mg^{2+} binds to the proR$_p$ oxygen and Glu252 of the other subunit. His253 could be involved in recognition rather than catalysis, as it is involved in a hydrogen bond to N7 of the inner G of its cognate sequence. It is not clear, however, how the actual presence of Mg^{2+} will affect these relationships in a ternary complex.

11.3 NaeI

NaeI appears to be an evolutionary bridge between "simple" endonucleases and the more complex activities of topoisomerases and recombinases. Unlike the structures of most REases, NaeI contains two separate domains, both of which bind DNA (Huai et al. 2001). The N-terminal domain core folds like other enzymes in the REase superfamily, while a C-terminal domain contains a catabolite activator protein motif present in many DNA-binding proteins, including the Type IA and Type II topoisomerases.

Interestingly a single substitution in NaeI, Leu43Lys, abolishes REase activity and in its place yields topoisomerase and recombinase activities (Carrick and Topal 2003; Jo and Topal 1995). NaeI Leu43 is located near the C-terminus of an α-helix that forms part of the central hydrophobic core of the REase domain (Huai et al. 2000). The Leu43Lys substitution also results in a preference for binding single-stranded DNA and a sensitivity to salt and intercalative topoisomerase-inhibiting drugs that are lacking in REases (including wild-type NaeI) but are characteristic of topoisomerases (Jo and Topal 1996a, b).

A pairwise least-squares superposition of backbone atoms of NaeI and many Type II REases yielded an rms difference of ~2 Å over 21–71 structurally similar residues. This superposition and a subsequent structure-based sequence alignment implicate the side chains of Glu70, Asp86, Asp95, and Lys97, as the catalytic motif residues which are near the scissile phosphate. Based on this superposition with many other REases with two metals experimentally observed in the active site, it was suggested that two divalent metals are required for catalysis.

11.4 MutH

MutH is a nuclease that plays a key role in mismatch repair (Bhagwat and Lieb 2002). Identification of its active site was apparently based on structural similarity to those of the Type II REases and the conservation of amino acids in this region among MutH orthologs (Ban and Yang 1998). The structure of MutH is strikingly similar in overall fold to that of PvuII, except that MutH has insertions that allow it to function as a monomer. Excluding the first 33 amino acids, that form the dimer interface in PvuII, MutH and PvuII can be superimposed with an rms deviation of 2.3 Å over 83 pairs of Cα atoms. The catalytic sequence motif (D70...E77YK79) is present in MutH and resides on the consecutive β-strands of the core β-sheet at the bottom of a DNA-binding cleft. Although Mg^{2+} ions were included in the crystallization buffer, no Mg^{2+} was found in the refined structure even at 1.7 Å resolution. This may have been due to the flexibility of the active site and a lack of substrate. A two-metal binding active site is implied by its similarity to PvuII. Most site-directed mutagenesis has been directed toward hypothetical DNA-binding residues (based on known PvuII–DNA binding) instead where a lysine outside the common motif, but important for catalysis, has been detected (Loh et al. 2001). Recently, broader in vitro and in vivo mutagenesis studies of DNA mismatch repair in E. coli has been conducted (Junop et al. 2003). Indeed, alteration of residues in the catalytic sequence motif and the aforementioned lysine results in no detectable endonuclease activity of MutH as well as reduced DNA binding. In addition, Tyr212 of MutH is involved in DNA strand discrimination (Friedhoff et al. 2003). *Haemophilus influenzae* and *Vibrio cholerae* genes for *mutH* are able to fully complement a *mutH* defect in E. coli (Friedhoff et al. 2002).

11.5 Holliday Junction Resolvases, Hjc

The crystal structures of Holliday junction resolvases (Hjc) from *Sulfolobus solfataricus* (motif PD42...E55MK) and *Pyrococcus furiosus* (motif VD36...E46VK) have been determined. A pairwise least-squares superposition with Cα atoms of EcoRV yielded an rms difference of 2.1 Å over 72 structurally similar residues in the comparison to *S. solfataricus* Hjc, and 2.8 Å for over 80 residues with the *P. furiosus* Hjc. Additional superposition of *P. furiosus* Hjc with 83 residues of FokI and 87 residues of BglI yielded average rms values of 2.5 and 2.9 Å, respectively. In fact, this superposition of the two Hjcs onto the FokI, EcoRV, and BglI REases reveals a striking spatial coincidence of their catalytic motif residues to Asp33, Glu46, and Lys48 in *P. furiosus* Hjc and to Asp42, Gly55, and Lys57 in *S. solfataticus* Hjc. Metal-binding sites were implicated by comparison to a EcoRV–DNA structure. The putative metal-binding residues when mutated in *P. furiosus* Hjc have been shown to impair function (Komori et al. 2000).

12 Prospects for the Future

As noted above, elucidating the mechanism(s) of DNA cleavage by Type II REases depends critically on knowing how many Mg^{2+} ions are directly involved in catalysis. At present, the mechanisms of DNA cleavage ist not yet clear for any REase, and it is not even known whether all Type II REases with a legitimate $(P)DX_{10-30}(D/E)YK$ motif follow the same mechanism. Thus, the exact number of metal ions involved in nucleic acid phosphodiester hydrolysis is one of the most hotly debated issues in enzyme mechanisms.

Several Type II REases are currently (spring 2003) en route to having their structures completed either alone or in a complex (Table 2). To provide the best possible description of endonuclease catalytic mechanisms, each of these enzymes must be studied with a combination of crystallographic and biochemical techniques (Perona 2002). However, even after a variety of experimental results have been gathered, the exact molecular mechanism of catalysis still remains elusive. A way to test mechanistic proposals that is beginning to come of age is computational modeling of reaction pathways. New techniques under development, such as those used for BamHI (see above), may eventually reveal the best possible reaction coordinates for a catalyzed reaction given the crystallographically determined substrate and product complex structures (Mordasini et al. 2002). Understanding the enzymology of this Type II REase superfamily of enzymes will be advantageous to those who use these enzymes, and essential for those who are devoted to the ambitious goal of changing the properties of these enzymes to make them even more useful. Further structural and biochemical investigations of this superfamily of

Table 2. Crystals reported

Enzyme	Recognition/cleavage	Reference	PDB
DdeI	C↓TNAG	L. Dorner, S. Cook, I. Schildkraut	–
EcoRII	↓CCWGG	Zhou et al. (2002)	1NA6
HaeIII	GG↓CC	S. Cook, L. Dorner, I. Schildkraut	–
HhaII	G↓ANTC	Chandrasegaran et al. (1986)	–
HindIII	A↓AGCTT	I. Schildkraut, L. Dorner	–
MspI	C↓CGG	O'Loughlin et al. (2000); Xu et al. (2001)	
NdeI	CA↓TATG	L. Dorner, I. Schildkraut	–
PspGI	CA↓TATG	A. Foster, A. Friedman, J. Pelletier, K. Le Coz, S-Y Xu	–
SfiI	GGCCNNNN↓NGGCC	L. Dorner, I. Schildkraut	

enzymes will be relevant to those interested in the mechanism of DNA recognition and cleavage in general.

Note added in proof: There is now strong evidence for REases using "zero-metal" mechanisms. The best studied such REase is BfiI, for which limited proteolysis reveals two distinct domains [Zaremba M, Urbanke C, Halford SE, Siksnys V (2004) Generation of the BfiI restriction endonuclease from the fusion of a DNA recognition domain to a non-specific nuclease from the phospholipase D superfamily. J Mol Biol 336:81–92]. The N-terminal domain has dimerization and catalytic functions, and belongs to the phosholipase D superfamily. The C-terminal domain is responsible for DNA recognition. Phosholipase D superfamily members catalyze many different kinds of reactions, and include several endonucleases. The structure of the N-terminal domain of BfiI probably resembles that of the related Nuc endonuclease from *Salmonella typhimurium* [Stuckey JA, Dixon JE (1999) Crystal structure of a phospholipase D family member. Nat Struct Biol 6:278–284]. Nuc contains an eight-stranded mixed β-sheet flanked by five α-helices (PDB code 1byr), somewhat reminiscent of the common REase catalytic core. However, its catalytic residues include a His, Lys, and Asp positioned on loops between β-strands, emphasizing the mechanistic differences between this family of enzymes and most REases. The phospholipase D-related BfiI, along with the topoisomerase-related NaeI, underlines the diversity and flexible requirements for REase activity.

References

Aggarwal AK (1995) Structure and function of restriction endonucleases. Curr Opin Struct Biol 5:11–19

Athanasiadis A, Vlassi M, Kotsifaki D, Tucker PA, Wilson KS, Kokkinidis M (1994) Crystal structure of PvuII endonuclease reveals extensive structural homologies to EcoRV. Nat Struct Biol 1:469–475

Baldwin GS, Sessions RB, Erskine SG, Halford SE (1999) DNA cleavage by the EcoRV restriction endonuclease: roles of divalent metal ions in specificity and catalysis. J Mol Biol 288:87–103

Ban C, Yang W (1998) Structural basis for MutH activation in *E. coli* mismatch repair and relationship of MutH to restriction endonucleases. EMBO J 17:1526–1534

Beernink PT, Segelke BW, Hadi MZ, Erzberger JP, Wilson DM 3rd, Rupp B (2001) Two divalent metal ions in the active site of a new crystal form of human apurinic/apyrimidinic endonuclease, Ape1: implications for the catalytic mechanism. J Mol Biol 307:1023–1034

Beese LS, Steitz TA (1991) Structural basis for the 3′-5′ exonuclease activity of *Escherichia coli* DNA polymerase I: a two metal ion mechanism. EMBO J 10:25–33

Bhagwat AS, Lieb M (2002) Cooperation and competition in mismatch repair: very short-patch repair and methyl-directed mismatch repair in *Escherichia coli*. Mol Microbiol 44:1421–1428

Bond CS, Kvaratskhelia M, Richard D, White MF, Hunter WN (2001) Structure of Hjc, a Holliday junction resolvase, from *Sulfolobus solfataricus*. Proc Natl Acad Sci USA 98:5509–5514

Bozic D, Grazulis S, Siksnys V, Huber R (1996) Crystal structure of *Citrobacter freundii* restriction endonuclease Cfr10I at 2.15 Å resolution. J Mol Biol 255:176–186

Brautigam CA, Steitz TA (1998) Structural and functional insights provided by crystal structures of DNA polymerases and their substrate complexes. Curr Opin Struct Biol 8:54–63

Bujnicki JM (2000) Phylogeny of the restriction endonuclease-like superfamily inferred from comparison of protein structures. J Mol Evol 50:39–44

Bujnicki JM, Rychlewski L (2001a) Grouping together highly diverged PD-(D/E)XK nucleases and identification of novel superfamily members using structure-guided alignment of sequence profiles. J Mol Microbiol Biotechnol 3:69–72

Bujnicki JM, Rychlewski L (2001b) Identification of a PD-(D/E)XK-like domain with a novel configuration of the endonuclease active site in the methyl-directed restriction enzyme Mrr and its homologs. Gene 267:183–191

Bunting KA, Roe SM, Headley A, Brown T, Savva R, Pearl LH (2003) Crystal structure of the *Escherichia coli* dcm very-short-patch DNA repair endonuclease bound to its reaction product-site in a DNA superhelix. Nucleic Acids Res 31:1633–1639

Carrick KL, Topal MD (2003) Amino acid substitutions at position 43 of NaeI endonuclease – evidence for changes in NaeI structure. J Biol Chem 278:9733–9739

Chandrasegaran S, Smith HO, Amzel ML, Ysern X (1986) Preliminary X-ray diffraction analysis of HhaII endonuclease–DNA cocrystals. Proteins 1:263–266

Cheetham GM, Steitz TA (2000) Insights into transcription: structure and function of single-subunit DNA-dependent RNA polymerases. Curr Opin Struct Biol 10:117–123

Cheng X, Balendiran K, Schildkraut I, Anderson JE (1994) Structure of PvuII endonuclease with cognate DNA. EMBO J 13:3927–3935

Conlan LH, Dupureur CM (2002a) Dissecting the metal ion dependence of DNA binding by PvuII endonuclease. Biochemistry 41:1335–1342

Conlan LH, Dupureur CM (2002b) Multiple metal ions drive DNA association by PvuII endonuclease. Biochemistry 41:14848–14855

Cowan JA (1998) Metal activation of enzymes in nucleic acid biochemistry. Chem Rev 98:1067–1088

Declais AC, Hadden J, Phillips SE, Lilley DM (2001) The active site of the junction-resolving enzyme T7 endonuclease I. J Mol Biol 307:1145–1158

Deibert M, Grazulis S, Janulaitis A, Siksnys V, Huber R (1999) Crystal structure of MunI restriction endonuclease in complex with cognate DNA at 1.7 Å resolution. EMBO J 18:5805–5816

Deibert M, Grazulis S, Sasnauskas G, Siksnys V, Huber R (2000) Structure of the tetrameric restriction endonuclease NgoMIV in complex with cleaved DNA. Nat Struct Biol 7:792–799

Deva T, Krishnaswamy S (2001) Structure-based sequence alignment of type-II restriction endonucleases. Biochim Biophys Acta 1544:217–228

Dupureur CM, Conlan LH (2000) A catalytically deficient active site variant of PvuII endonuclease binds Mg(II) ions. Biochemistry 39:10921–10927

Engler LE, Sapienza P, Dorner LF, Kucera R, Schildkraut I, Jen-Jacobson L (2001) The energetics of the interaction of BamHI endonuclease with its recognition site GGATCC. J Mol Biol 307:619–636

Erskine SG, Halford SE (1998) Reactions of the EcoRV restriction endonuclease with fluorescent oligodeoxynucleotides: identical equilibrium constants for binding to specific and non-specific DNA. J Mol Biol 275:759–772

Fedor MJ (2000) Structure and function of the hairpin ribozyme. J Mol Biol 297:269–291

Freemont PS, Friedman JM, Beese LS, Sanderson MR, Steitz TA (1988) Cocrystal structure of an editing complex of Klenow fragment with DNA. Proc Natl Acad Sci USA 85:8924–8928

Friedhoff P, Sheybani B, Thomas E, Merz C, Pingoud A (2002) *Haemophilus influenzae* and *Vibrio cholerae* genes for mutH are able to fully complement a mutH defect in *Escherichia coli*. FEMS Microbiol Lett 208:121–126

Friedhoff P, Thomas E, Pingoud A (2003) Tyr-212: A key residue involved in strand discrimination by the DNA mismatch repair endonuclease MutH. J Mol Biol 325:285–297

Fuxreiter M, Osman R (2001) Probing the general base catalysis in the first step of BamHI action by computer simulations. Biochemistry 40:15017–023

Galburt EA, Stoddard BL (2002) Catalytic mechanisms of restriction and homing endonucleases. Biochemistry 41:13851–13860

Grabowski G, Jeltsch A, Wolfes H, Maass G, Alves J (1995) Site-directed mutagenesis in the catalytic center of the restriction endonuclease EcoRI. Gene 157:113–118

Grazulis S, Deibert M, Rimseliene R, Skirgaila R, Sasnauskas G, Lagunavicius A, Repin V, Urbanke C, Huber R, Siksnys V (2002) Crystal structure of the Bse634I restriction endonuclease: comparison of two enzymes recognizing the same DNA sequence. Nucleic Acids Res 30:876–885

Groll DH, Jeltsch A, Selent U, Pingoud A (1997) Does the restriction endonuclease EcoRV employ a two-metal-Ion mechanism for DNA cleavage? Biochemistry 36:11389–401

Hadden JM, Convery MA, Declais AC, Lilley DM, Phillips SE (2001) Crystal structure of the Holliday junction resolving enzyme T7 endonuclease I. Nat Struct Biol 8: 62–67

Hadden JM, Declais AC, Phillips SE, Lilley DM (2002) Metal ions bound at the active site of the junction-resolving enzyme T7 endonuclease I. EMBO J 21:3505–3515

Haq I, O'Brien R, Lagunavicius A, Siksnys V, Ladbury JE (2001) Specific DNA recognition by the Type II restriction endonuclease MunI: the effect of pH. Biochemistry 40:14960–14967

Heitman J, Ivanenko T, Kiss A (1999) DNA nicks inflicted by restriction endonucleases are repaired by a RecA- and RecB-dependent pathway in *Escherichia coli*. Mol Microbiol 33:1141–1151

Hickman AB, Li Y, Mathew SV, May EW, Craig NL, Dyda F (2000) Unexpected structural diversity in DNA recombination: the restriction endonuclease connection. Mol Cell 5:1025–1034

Horton JR, Cheng X (2000) PvuII endonuclease contains two calcium ions in active sites. J Mol Biol 300:1049–1056

Horton NC, Dorner LF, Perona JJ (2002a) Sequence selectivity and degeneracy of a restriction endonuclease mediated by DNA intercalation. Nat Struct Biol 9:42–47

Horton NC, Newberry KJ, Perona JJ (1998) Metal ion-mediated substrate-assisted catalysis in Type II restriction endonucleases. Proc Natl Acad Sci USA 95:13489–13494

Horton NC, Otey C, Lusetti S, Sam MD, Kohn J, Martin AM, Ananthnarayan V, Perona JJ (2002b) Electrostatic contributions to site specific DNA cleavage by EcoRV endonuclease. Biochemistry 41:10754–10763

Huai Q, Colandene JD, Chen Y, Luo F, Zhao Y, Topal MD, Ke H (2000) Crystal structure of NaeI-an evolutionary bridge between DNA endonuclease and topoisomerase. EMBO J 19:3110–3118

Huai Q, Colandene JD, Topal MD, Ke H (2001) Structure of NaeI–DNA complex reveals dual-mode DNA recognition and complete dimer rearrangement. Nat Struct Biol 8:665–669

Jeltsch A, Alves J, Maass G, Pingoud A (1992) On the catalytic mechanism of EcoRI and EcoRV. A detailed proposal based on biochemical results, structural data and molecular modelling. FEBS Lett 304:4–8

Jeltsch A, Alves J, Wolfes H, Maass G, Pingoud A (1993) Substrate-assisted catalysis in the cleavage of DNA by the EcoRI and EcoRV restriction enzymes. Proc Natl Acad Sci USA 90:8499–8503

Jeltsch A, Kroger M, Pingoud A (1995a) Evidence for an evolutionary relationship among type-II restriction endonucleases. Gene 160:7–16

Jeltsch A, Maschke H, Selent U, Wenz C, Kohler E, Connolly BA, Thorogood H, Pingoud A (1995b) DNA binding specificity of the EcoRV restriction endonuclease is increased by Mg2+ binding to a metal ion binding site distinct from the catalytic center of the enzyme. Biochemistry 34:6239–6246

Jeltsch A, Pleckaityte M, Selent U, Wolfes H, Siksnys V, Pingoud A (1995 c) Evidence for substrate-assisted catalysis in the DNA cleavage of several restriction endonucleases. Gene 157:157–162

Jen-Jacobson L (1997) Protein-DNA recognition complexes: conservation of structure and binding energy in the transition state. Biopolymers 44:153–180

Jo K, Topal MD (1995) DNA topoisomerase and recombinase activities in NaeI restriction endonuclease. Science 267:1817–1820

Jo K, Topal MD (1996a) Changing a leucine to a lysine residue makes NaeI endonuclease hypersensitive to DNA intercalative drugs. Biochemistry 35:10014–10018

Jo K, Topal MD (1996b) Effects on NaeI–DNA recognition of the leucine to lysine substitution that transforms restriction endonuclease NaeI to a topoisomerase: a model for restriction endonuclease evolution. Nucleic Acids Res 24:4171–4175

Jose TJ, Conlan LH, Dupureur CM (1999) Quantitative evaluation of metal ion binding to PvuII restriction endonuclease. J Biol Inorg Chem 4:814–823

Junop MS, Yang W, Funchain P, Clendenin W, Miller JH (2003) In vitro and in vivo studies of MutS, MutL and MutH mutants: correlation of mismatch repair and DNA recombination. DNA Repair (Amst) 2:387–405

Kim YC, Grable JC, Love R, Greene PJ, Rosenberg JM (1990) Refinement of EcoRI endonuclease crystal structure: a revised protein chain tracing. Science 249:1307–1309

Komori K, Sakae S, Daiyasu H, Toh H, Morikawa K, Shinagawa H, Ishino Y (2000) Mutational analysis of the *Pyrococcus furiosus* Holliday junction resolvase hjc revealed functionally important residues for dimer formation, junction DNA binding, and cleavage activities. J Biol Chem 275:40385–40391

Kostrewa D, Winkler FK (1995) Mg^{2+} binding to the active site of EcoRV endonuclease: a crystallographic study of complexes with substrate and product DNA at 2 Å resolution. Biochemistry 34:683–696

Kovall R, Matthews BW (1997) Toroidal structure of lambda-exonuclease. Science 277:1824–1827

Kovall RA, Matthews BW (1998) Structural, functional, and evolutionary relationships between lambda-exonuclease and the Type II restriction endonucleases. Proc Natl Acad Sci USA 95:7893–7897

Kovall RA, Matthews BW (1999) Type II restriction endonucleases: structural, functional and evolutionary relationships. Curr Opin Chem Biol 3:578–583

Lagunavicius A, Grazulis S, Balciunaite E, Vainius D, Siksnys V (1997) DNA binding specificity of MunI restriction endonuclease is controlled by pH and calcium ions: involvement of active site carboxylate residues. Biochemistry 36:11093–11099

Lagunavicius A, Siksnys V (1997) Site-directed mutagenesis of putative active site residues of MunI restriction endonuclease: replacement of catalytically essential carboxylate residues triggers DNA binding specificity. Biochemistry 36:11086–11092

Lins RD, Straatsma TP, Briggs JM (2000) Similarities in the HIV-1 and ASV integrase active sites upon metal cofactor binding. Biopolymers 53:308–315

Loh T, Murphy KC, Marinus MG (2001) Mutational analysis of the MutH protein from *Escherichia coli*. J Biol Chem 276:12113–12119

Lovell S, Goryshin IY, Reznikoff WR, Rayment I (2002) Two-metal active site binding of a Tn5 transposase synaptic complex. Nat Struct Biol 9:278–281

Lukacs CM, Kucera R, Schildkraut I, Aggarwal AK (2000) Understanding the immutability of restriction enzymes: crystal structure of BglII and its DNA substrate at 1.5 Å resolution. Nat Struct Biol 7:134–140

Martin AM, Horton NC, Lusetti S, Reich NO, Perona JJ (1999a) Divalent metal dependence of site-specific DNA binding by EcoRV endonuclease. Biochemistry 38:8430–8439

Martin AM, Sam MD, Reich NO, Perona JJ (1999b) Structural and energetic origins of indirect readout in site-specific DNA cleavage by a restriction endonuclease. Nat Struct Biol 6:269–277

Mordasini T, Curioni A, Andreoni W (2003) Why do divalent metal ions either promote or inhibit enzymatic reactions? The case of BamHI restriction endonuclease from combined quantum-classical simulations. J Biol Chem 278:4381–4384

Newman M, Lunnen K, Wilson G, Greci J, Schildkraut I, Phillips SE (1998) Crystal structure of restriction endonuclease BglI bound to its interrupted DNA recognition sequence. EMBO J 17:5466–5476

Newman M, Strzelecka T, Dorner LF, Schildkraut I, Aggarwal AK (1994). Structure of restriction endonuclease BamHI phased at 1.95 Å resolution by MAD analysis. Structure 2:439–452

Nishino T, Komori K, Ishino Y, Morikawa K (2001a) Dissection of the regional roles of the archaeal Holliday junction resolvase Hjc by structural and mutational analyses. J Biol Chem 276:35735–35740

Nishino T, Komori K, Tsuchiya D, Ishino Y, Morikawa K (2001b) Crystal structure of the archaeal holliday junction resolvase Hjc and implications for DNA recognition. Structure (Camb) 9:197–204

Nishino T, Morikawa K (2002) Structure and function of nucleases in DNA repair: shape, grip and blade of the DNA scissors. Oncogene 21:9022–9032

O'Loughlin TJ, Xu Q, Kucera RB, Dorner LF, Sweeney S, Schildkraut I, Guo HC (2000) Crystallization and preliminary X-ray diffraction analysis of *Msp*I restriction endonuclease in complex with its cognate DNA. Acta Crystallogr D Biol Crystallogr 56 Pt 12:1652–1655

Perona JJ (2002) Type II restriction endonucleases. Methods 28:353–364

Perona JJ, Martin AM (1997) Conformational transitions and structural deformability of EcoRV endonuclease revealed by crystallographic analysis. J Mol Biol 273:207–225

Pingoud A, Jeltsch A (1997) Recognition and cleavage of DNA by type-II restriction endonucleases. Eur J Biochem 246:1–22

Pingoud A, Jeltsch A (2001) Structure and function of Type II restriction endonucleases. Nucleic Acids Res 29:3705–3727

Pingoud V, Kubareva E, Stengel G, Friedhoff P, Bujnicki JM, Urbanke C, Sudina A, Pingoud A (2002) Evolutionary relationship between different subgroups of restriction endonucleases. J Biol Chem 277:14306–14314

Rice P, Craigie R, Davies DR (1996) Retroviral integrases and their cousins. Curr Opin Struct Biol 6:76–83

Roberts RJ, Vincze T, Posfai J, Macelis D (2003) REBASE–restriction enzymes and methyltransferases. Nucleic Acids Res 31:418–420

Rosenberg JM (1991) Structure and function of restriction endonucleases. Curr Opin Struct Biol 1:104–113

Sam MD, Perona JJ (1999) Catalytic roles of divalent metal ions in phosphoryl transfer by EcoRV endonuclease. Biochemistry 38:6576–6586

Sapranauskas R, Sasnauskas G, Lagunavicius A, Vilkaitis G, Lubys A, Siksnys V (2000) Novel subtype of Type IIs restriction enzymes. *Bfi*I endonuclease exhibits similarities to the EDTA-resistant nuclease Nuc of *Salmonella typhimurium*. J Biol Chem 275:30878–30885

Selent U, Ruter T, Kohler E, Liedtke M, Thielking V, Alves J, Oelgeschlager T, Wolfes H, Peters F, Pingoud A (1992) A site-directed mutagenesis study to identify amino acid residues involved in the catalytic function of the restriction endonuclease EcoRV. Biochemistry 31:4808–4815

Skirgaila R, Grazulis S, Bozic D, Huber R, Siksnys V (1998) Structure-based redesign of the catalytic/metal binding site of Cfr10I restriction endonuclease reveals importance of spatial rather than sequence conservation of active centre residues. J Mol Biol 279:473–481

Skirgaila R, Siksnys V (1998) Ca^{2+}-ions stimulate DNA binding specificity of Cfr10I restriction enzyme. Biol Chem 379:595–598

Soundararajan M, Chang Z, Morgan RD, Heslop P, Connolly BA (2002) DNA binding and recognition by the IIs restriction endonuclease MboII. J Biol Chem 277:887–895

Stanford NP, Halford SE, Baldwin GS (1999) DNA cleavage by the EcoRV restriction endonuclease: pH dependence and proton transfers in catalysis. J Mol Biol 288:105–116

Steitz TA (1998) A mechanism for all polymerases. Nature 391:231–232

Steitz TA, Steitz JA (1993) A general two-metal-ion mechanism for catalytic RNA. Proc Natl Acad Sci USA 90:6498–6502

Stoddard BL (1998) New results using Laue diffraction and time-resolved crystallography. Curr Opin Struct Biol 8:612–618

Tamulaitis G, Solonin AS, Siksnys V (2002) Alternative arrangements of catalytic residues at the active sites of restriction enzymes. FEBS Lett 518:17–22

Tsutakawa SE, Jingami H, Morikawa K (1999a) Recognition of a TG mismatch: the crystal structure of very short patch repair endonuclease in complex with a DNA duplex. Cell 99:615–623

Tsutakawa SE, Morikawa K (2001) The structural basis of damaged DNA recognition and endonucleolytic cleavage for very short patch repair endonuclease. Nucleic Acids Res 29:3775–3783

Tsutakawa SE, Muto T, Kawate T, Jingami H, Kunishima N, Ariyoshi M, Kohda D, Nakagawa M, Morikawa K (1999b) Crystallographic and functional studies of very short
patch repair endonuclease. Mol Cell 3:621–628

van der Woerd MJ, Pelletier JJ, Xu S, Friedman AM (2001) Restriction enzyme BsoBI-
DNA complex: a tunnel for recognition of degenerate DNA sequences and potential
histidine catalysis. Structure (Camb) 9:133–144

Venclovas C, Timinskas A, Siksnys V (1994) Five-stranded beta-sheet sandwiched with
two alpha-helices: a structural link between restriction endonucleases EcoRI and
EcoRV. Proteins 20:279–282

Viadiu H, Aggarwal AK (1998) The role of metals in catalysis by the restriction endonuclease BamHI. Nat Struct Biol 5:910–916

Vipond IB, Baldwin GS, Halford SE (1995) Divalent metal ions at the active sites of the
EcoRV and EcoRI restriction endonucleases. Biochemistry 34:697–704

Vipond IB, Halford SE (1995). Specific DNA recognition by EcoRV restriction endonuclease induced by calcium ions. Biochemistry 34:1113–1119

Wah DA, Bitinaite J, Schildkraut I, Aggarwal AK (1998) Structure of FokI has implications for DNA cleavage. Proc Natl Acad Sci USA 95:10564–10569

Wah DA, Hirsch JA, Dorner LF, Schildkraut I, Aggarwal AK (1997) Structure of the multimodular endonuclease FokI bound to DNA. Nature 388:97–100

Winkler FK (1992) Structure and function of restriction endonucleases. Curr Opin
Struct Biol 2:93–99

Winkler FK, Banner DW, Oefner C, Tsernoglou D, Brown RS, Heathman SP, Bryan RK,
Martin PD, Petratos K, Wilson KS (1993) The crystal structure of EcoRV endonuclease and of its complexes with cognate and non-cognate DNA fragments. EMBO J
12:1781–1795

Wlodawer A (1999) Crystal structures of catalytic core domains of retroviral integrases
and role of divalent cations in enzymatic activity. Adv Virus Res 52:335–350

Xu Q, O'Loughlin TJ, Kucera RB, Schildkraut I, Guo H-C (2001) Structural study of
restriction enzyme MspI/DNA complex by x-ray crystallography. Biophys J 80:296a

Xu SY, Schildkraut I (1991) Cofactor requirements of BamHI mutant endonuclease E77 K
and its suppressor mutants. J Bacteriol 173:5030–5035

Zhou EX, Reuter M, Meehan EJ, Chen L (2002) A new crystal form of restriction endonuclease EcoRII that diffracts to 2.8 Å resolution. Acta Crystallogr D Biol Crystallogr
58:1343–1345

Protein Engineering of Restriction Enzymes

J. ALVES, P. VENNEKOHL

1 Introduction

Restriction endonucleases are among the most accurate enzymes known (Pingoud and Jeltsch 2001). Consequently, changing their very high sequence specificity is one of the scientific goals in studying these enzymes (Jeltsch et al. 1996). This review focuses on protein engineering regarding Type II restriction enzymes. They comprise the vast majority of the about 3600 entities listed in REBASE today (Roberts et al. 2003), although the number of those studied in detail is far smaller.

First, we will discuss different approaches to changing the recognition of a given sequence starting with rational design using site-directed mutagenesis and progressing to random mutagenesis combined with in vivo selection. Then, we will describe attempts to lengthen the recognition sequence by designing additional enzyme contacts to DNA. As the dimeric nature of all restriction enzymes doubles the effect of each amino acid exchange, we will also discuss engineering of subunit composition. Finally, we will speculate on possible future directions of protein engineering of restriction enzymes.

Table 1 gives an overview on enzymes, their mutants and properties that are mentioned in this review.

2 Modifying Contacts of Restriction Enzymes to Their Recognition Sequence

The first cocrystal structure of a DNA-binding protein in complex with its binding sequence to be solved was the one of the EcoRI restriction endonuclease (McClarin et al. 1986). It featured only six amino acid residues contact-

J. Alves, P. Vennekohl
Department of Biophysical Chemistry, Medical School Hanover, Carl-Neuberg-Str. 1, 30625 Hanover, Germany

Nucleic Acids and Molecular Biology, Vol. 14
Alfred Pingoud (Ed.)
Restriction Endonucleases
© Springer-Verlag Berlin Heidelberg 2004

Table 1. Overview on enzymes, their mutants and properties

Enzyme	Mutant	Methodology[a]	Properties
Enhanced cleavage activity (Sect. 2)			
PvuII	D34G	rand. mut.	Loss of DNA binding for central GC base pair, still only cleaves its recognition site CAGCTG
EcoRI	I197A	site dir. mut.	Specific activity five times higher than wt-enzyme, although not involved in DNA recognition
BamHI	C54A	rat. des.	Higher specific activity but no direct DNA contact
HindIII	E86K	rat. des.	Higher specific activity but no direct DNA contact
Recognition of modified sequences (Sect. 2)			
EcoRV	T94V	rat. des.	Cleaves Sp-methylphosphonate four orders of magnitude faster than unmodified substrate
EcoRI	Q115A or A142G	site dir. mut.	Cleavage of GAAUTC as fast as GAATTC
EcoRV	N188Q	rat. des.	Prefers GATAUC over GATATC
EcoRI	G140A	site dir. mut.	Prefers GAAUTC or GAATUC over GAATTC about ten times at low activity level
EcoRI	M137Q	site dir. mut.	GAATT5 mC is preferred ten times over GAATTC but not in 5 mCGAATT5 mCG
BamHI	N116H/ S118G	in vivo select.	GGmATCC is strongly preferred over GGATCC
Recognition of altered sequences (Sect. 2)			
EcoRI	G140S/ N141S/ R145K	site dir. mut.	Attempted change of recognition from GAATTC to GGATCC, almost no cleavage activity or binding activity, unstable proteins
EcoRI	M137Q/ Arg200/ Arg203	site dir. mut.	Attempted change of recognition from GAATTC to CAATTG, cleavage activity but no change in specificity
Enhancement of relaxed specificity (Sect. 2)			
BsoBI	D246A	sat. mut.	Prefers CCCGGG (70-fold in binding and 100-fold reduced cleavage) over CTCGAG of the former CYCGRG recognition
BsoBI	A32T/ T40P/ Q140L/ D246A	rand. mut.	Tenfold higher cleavage activity compared to D246A but still with enhanced specificity
Eco57I	T862N	in vivo select.	Generation of a relaxed specificity CTGRAG from CTGAAG (selection for methylase)
BstYI	K133N/ S172N	in vivo select. rand. mut.	Some mutants with preference for AGATCY instead of RGATCY; combination produces mutants with only sixfold less cleavage activity for AGATCT and no cleavage for GGATCC
Lengthening of recognition sequence (Sect. 3)			
EcoRV	A181E	sat. mut.	Preference for TGATATCA tenfold in oligodeoxynucleotide and plasmid
EcoRV	K104R	rat. des.	Preference for TGATATCA tenfold in oligodeoxynucleotide and plasmid

Table 1. (*Continued*)

Enzyme	Mutant	Methodology[a]	Properties
EcoRV	A181E/ K104R	rat. des.	Preference for TGATATCA of the single mutants did not result in more specificity for the double mutant
EcoRV	N92T/ S183A/ T222S	dir. ev.	25-fold preference for AT flanked GATATC sited
EcoRV	R226AorV	Ala scan., sat. mut.	Preference for nonflexible surrounding sequences of GATATC
EcoRI	I197Q	site dir. mut.	Possible additional contact outside of the recognition sequence GAATTC led to no enhancement of specificity
EcoRI	A138N/ M137Q/ I197Q	site dir. mut.	More active than I197Q but still no enhancement in specificity
EcoRI	GlyAB/ I197Q	rat. des.	Possible additional contact outside of the recognition sequence GAATTC led to no enhancement of specificity
EcoRI	A138N/ GlyAB/ I197Q	site dir. mut.	Cleaves only TGAATTCA on λ DNA but no real contact made to the outer base pair; instable protein, stabilized with Ca²⁺ it cleaves the canonical site again
EcoRI	K130E	site dir. mut.	Higher selectivity due to flexible or nonflexible neighboring sequences
EcoRI	Myb133	rat. des.	Needs Ca²⁺ for stabilization and no longer cleaves GAATTC on λ DNA, cleavage of AAG-GAATTCCTA identified with a random substrate pool, protein only partially folded

Changed subunit composition (Sect. 4)

Enzyme	Mutant	Methodology	Properties
EcoRI	L158D-I230K	site dir. mut.	Heterodimeric protein L158D-I230K shows increase in specific activity compared to either homodimer, still very unstable enzyme
AlwI	N.AlwI	rat. des.	Monomeric enzyme, recognizes its specific site but nicks a single strand with high specificity

Single-chain nuclease (Sect. 4)

Enzyme	Mutant	Methodology	Properties
PvuII	Single-chain PvuII	rat. des.	Nearly as active as the homodimeric wild-type enzyme
EcoRI	EcoFus	rat. des.	Deleted 65 C-terminal amino acids and 72 N-terminal amino acids, detectable cleavage activity but not to be purified
EcoRI	EcoRI NCNC	rat. des.	Bridging linker between C- and N-terminus, not to be purified due to linker cleavage

[a] Ala scan.: alanine scanning; dir. ev.: directed evolution scheme; in vivo select.: in vivo selection; rand. mut.: random mutagenesis; rat. des.: rational design; site dir. mut.: site-directed mutagenesis; sat. mut.: saturation mutagenesis; site dir. mut.: site-directed mutagenesis

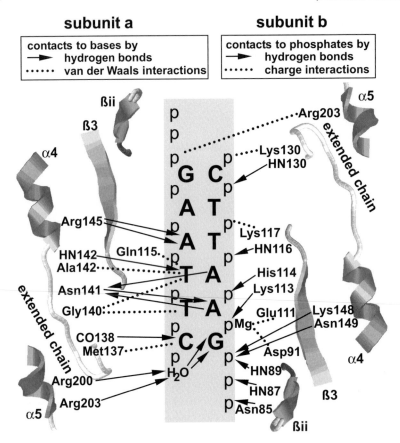

Fig. 1. Interaction of restriction endonuclease EcoRI with its recognition sequence. Each subunit establishes the same contacts but to get a clearer picture in subunit *a* the phosphate and in subunit b the base contacts are omitted. Furthermore, all contacts between amino acid residues which are of course important for coupling of restriction to catalysis and for subunit interactions are not shown

ing one base pair each which led to proposals of how the specificity of this enzyme could be changed (Rosenberg 1987). The revised structure (Kim et al. 1990) shows a much more complicated network of interacting residues as outlined in Fig. 1. We will use the EcoRI endonuclease as the prime example in this review.

Each subunit of the homodimeric enzyme forms nine hydrogen bonds and five hydrophobic interactions to the bases of the recognition sequence (on the left side of Fig. 1) and 12 phosphate contacts (on the right side of Fig. 1). Because of the intrinsic symmetry of both the enzyme and palindromic recognition sequence, 52 specific interactions are made overall. To establish so many contacts in the small space of six base pairs the major groove of the DNA has to be widened by kinking the phosphodiester backbone. Nevertheless, all these contacting residues are close to each other and interwoven by direct and water-mediated interactions which are omitted from Fig. 1 for clarity. Similar complexity is found in other cocrystal structures of restriction endonucleases (for review see Pingoud and Jeltsch 2001). Therefore, this redundant and over-determined recognition that seems to be a prerequisite for the very high specificity of these enzymes, has to be overcome to perform successful protein engineering. At least, every effort has to be embedded in this network of contacts.

2.1 Site-Directed Mutagenesis

Site-directed mutagenesis is used primarily to verify the contacts seen in the cocrystal structure. However, changing specificity was also always intended, because not only conservative changes or substitutions by alanine were made, but side chains were introduced which should have the potential of establishing different contacts. We and others have mutated all residues depicted in Fig. 1 (Wolfes et al. 1986; Geiger et al. 1989; Alves et al. 1989; Needels et al. 1989; King et al. 1989; Wright et al. 1989; Hager et al. 1990; Heitman and Model 1990; Oelgeschläger et al. 1990; Jeltsch et al. 1993; Flores et al. 1995; Grabowski et al. 1995; Muir et al. 1997; Windolph and Alves 1997; Fritz et al. 1998; Ivanenko et al. 1998; Küster 1998; Rosati 1999).

The results of all these mutagenesis experiments suggest a general rule: The change of a contacting residue results in a sometimes very large drop in catalytic activity but not in a change of specificity.

This has to be expected because mutant enzymes are immediately toxic to their host by cleaving sequences which are not protected by the accompanying methylase. In the Type II systems, protection is achieved usually by a separate enzyme specific for the same sequence. Therefore, during evolution a strong selection pressure reduced all potentialities to cleave unprotected sequences even when misincorporation during translation creates mutated enzymes. This explains both parts of the rule: The large drop in catalytic activity is due to a strict coupling of recognition and catalysis and the redundancy results in a tolerance to the loss of some specific contacts.

A good example is the PvuII endonuclease in which Asp34 interacts with the guanine of the central base pair in the recognition sequence CAGCTG (Nastri et al. 1997). The D34G mutant was isolated as a nearly inactive enzyme with strong DNA binding. Closer inspection of its DNA-binding properties showed that it has completely lost its specificity for the central two base pairs but still cleaves only the canonical sequence with four orders of magnitude reduced activity.

However, there are also deviations from the general rule. Firstly, not all mutations lead to a reduction of cleavage activity. The EcoRI single mutant I197A has a five times higher specific activity than the wild-type enzyme

(Küster and Alves 1995). It is not involved in DNA recognition (Ivanenko et al. 1998) and must have some structural role although this is not obvious in the crystal structure. The same is true for the C54A mutant of BamHI (Mukhopadhyay and Roy 1998) and possibly for the E86K mutant of HindIII (Tang et al. 2000), an enzyme whose structure is not yet known. Therefore, the low cleavage rate of restriction enzymes can be slightly improved.

A larger enhancement should have consequences on accuracy as is found in another class of mutants mostly at non-contacting residues. They cleave faster at sequences deviating by one base pair from the recognition sequence (Heitman and Model 1990; Flores et al. 1995). This behavior of so-called promiscuous mutants is better described as interference with the strict coupling of recognition and catalysis as the inherent low inaccuracy of the enzymes is strongly enhanced while the canonical cleavage is not necessarily faster.

Secondly, the few cases of changed specificity have to be seen as deviations from the rule as well. The finding that the EcoRV endonuclease cleaves oligodeoxynucleotide substrates with a Sp-methylphosphonate at GATATpC as fast as the unmodified sequence suggests that the contacting residue Thr94 might contact both substrates by simply rotating its methyl group to the Sp-methylphosphonate or its hydroxyl group to the normal phosphate oxygen. This led to the design of the T94V mutant which cleaves the Sp-methylphosphonate four orders of magnitude faster than the unmodified substrate (Lanio et al. 1996). This is the best change in specificity found for restriction enzymes, although for a completely artificial recognition sequence.

In several instances, the recognition of thymine bases could be modified such that a discrimination against uracil is no longer possible or even uracil is the preferred substrate. In EcoRI the mutants Q115A (Jeltsch et al. 1993) and A142G (Küster 1998) cleave at GAAUTC as fast as at GAATTC while the wild type enzyme shows a tenfold difference in favor of the canonical sequence. The EcoRV mutant N188Q with a similar behavior at GATAUC was generated (Wenz et al. 1994).

The EcoRI mutant G140A even prefers the GAAUTC or GAATUC sequence about ten times although at a low activity level (Küster 1998). The added methyl group on the protein not only needs space but also can hydrophobically interact with the C5 position of a pyrimidine. We use this mutant as a starting point to establish recognition of a different base pair by mutating also Asn141 and Arg145 which contact the complementary adenine (see Fig. 1; Gerschon 2000; Vennekohl, unpubl.). The main difficulty is that each additional mutation reduces the catalytic activity further. At such a low level, the incorporation of mutations with a promiscuous effect only slightly enhances the cleavage rate. Additionally, these triple or quadruple mutants tend to be more unstable which not only hampers purification but may also be responsible for a low level of unspecific nuclease activity interfering with specific DNA cleavage. One mutant produces a transient band pattern, but to date it could not be identified which sites are cleaved.

Another hydrophobic contact in EcoRI recognition led also to a specificity change. Among the mutants for Met137 contacting the cytosine residue (Fig. 1), only M137Q shows no reduced cleavage activity. The impaired interaction is possibly compensated by a repositioning of the neighboring Ala138 for which mutations with a promiscuous effect are already known. Cleavage experiments with oligodeoxynucleotide substrates containing modified bases established a preference of this mutant for a methyl group at C5 of the pyrimidine (Küster 1998). The best substrate was GAATT5 mC which is preferred ten times over GAATTC by the mutant. This is exactly opposite for the wild type enzyme which is inhibited by the cytosine methylation. As 5-methyl cytosine is a natural base in eukaryotic DNA in the sequence context CpG, we also tested the palindromic extension of the EcoRI recognition sequence 5 mCGAATT5 mCG. However, the effect of the additional methyl group next to the 5'-site of the recognition sequence counteracts the preference of the mutant (Peters, in prep.).

However, we also use the M137Q mutant as a starting point to change the outer base pair recognition of the EcoRI sequence. This is especially difficult because the guanine contacts are mediated by a water molecule positioned by the long amino acid residues Arg200 and Arg203 (Fig. 1). Lysine residues at these positions have the potential to reorient the water in context with the M137Q mutation. Unfortunately, the double and triple mutants show no change in specificity (Peters, in prep.). We are left with the fact that in the case of restriction enzymes with a well-defined recognition sequence we only were able to change specificity to bases not normally found in DNA.

Obviously, it would be advantageous to test DNA recognition without a large and rigid enzyme framework. For EcoRI we showed that an oligopeptide with the sequence of the extended chain which harbors most of the base-contacting residues (Fig. 1) is able at high concentrations to protect the EcoRI sites in plasmid DNA against cleavage (Jeltsch et al. 1995 b). One of the reasons for low affinity binding of the peptide was argued to be the necessity of peptide dimerization on the DNA to mimic an EcoRI dimer bound to the recognition sequence. Therefore, we synthesized a bidentate peptide with two covalently joined extended chain sequences and variants of it (Vennekohl 2000), which still have very low binding constants to GAATTC containing oligodeoxynucleotide. Peptide libraries comprising variants of the extended chain motif show an increase in unspecific binding especially if the positive charge of the peptides was raised. Furthermore, we showed in cleavage protection assays, which allow for much higher peptide concentrations, that the sequence specificity of the bidentate peptide is relaxed to $GAN_{(1or2)}TC$ and, therefore, cannot mimic an EcoRI recognition exactly.

For enzymes with degenerate recognition sequences it may be easier to enhance specificity to a more defined sequence. The BsoBI restriction endonuclease recognizes CYCGRG where Y stands for both pyrimidines and R for both purines. This actually leads to the recognition of the palindromic sequences CCCGGG and CTCGAG as well as the asymmetric sequence CTCGGG (which is identical to CCCGAG because of the antiparallel DNA double strand). In the cocrystal structure, Asp246 contacts the purine via a water molecule (van der Woerd et al. 2001). Saturation mutagenesis at this residue led to the mutant D246A which prefers CCCGGG (Zhu et al, 2003). This is due to a 70-fold stronger binding of this sequence in the context of a 100-fold reduced cleavage activity overall. Thus CTCGAG is nearly not cut at all. For improvement of the catalytic activity of the mutant the authors used random mutagenesis and the SOS induction assay (Heitman and Model 1991) in which a reporter gene shows the induction of the SOS response by multiple DNA cleavage events. Most of the secondary mutations were outside the DNA–protein interface as we described before for other restriction enzymes. Some of the more active secondary mutants lost the new sequence preference but one quadruple mutant retained it in the context of a tenfold higher cleavage activity. This work is promising and also demonstrates the power of random mutagenesis followed by in vivo selection.

2.2 In Vivo Selection Systems

The first in vivo systems designed to select out specificity changes inactivated the catalytic center and used the resulting mutant as a transcriptional repressor. The highest selection pressure is achieved if cells die without repressor binding. Fisher et al. (1995) used cells lysogenic for the phage P22 in which the gene of an antirepressor of lytic genes was controlled by a catalytically inactive EcoRI mutant. They tried to isolate a mutant binding to the methylated EcoRI sequence which also would be a very dangerous mutant for a restriction system. They only found activity mutants of the EcoRI methylase. It may well be that the drop in binding strength which normally accompanies mutations of contacting residues prevents good repression and even if binding to the new sequence does occur it cannot protect the cells.

For selection of BamHI specificity mutants, cells with a spectinomycin resistance gene controlled by an antisense RNA were used. Repressor binding to the antisense promoter allows the cells to become resistant against the antibiotic. This system worked fine for the identification of catalytic site residues by omitting the methylase gene in the cells (Dorner and Schildkraut 1994) as well as for the identification of mutations of contacting residues which still allow binding to the canonical sequence (Dorner et al. 1999). Reactivation of the catalytic center was possible for 14 out of 17 of these mutants. Some peculiarities in the cleavage reaction led to a closer inspection of the N116H/S118G double mutant which in contrast to the wild type enzyme strongly prefers GGmATCC over GGATCC in binding and cleavage (Whitaker et al. 1999). For selection of new sequence specificities the binding sequence has to be changed. Using this approach, some BamHI mutants have been selected to bind to the sequences AGATCT, GCATGC, as well as CAGCTG instead of GGATCC. But the reactivation of the catalytic center proved to be unsuccessful or nothing interesting could be selected (L.F. Dorner, R.B. Kucera, I. Schildkraut, pers. comm.). This raises the question whether a selected mutant which binds a different sequence can be activated again to cleave at this sequence. The D34G mutant of PvuII described above argues against this, too.

However, there is also an example in favor of the feasibility of this approach. A subclass of the Type II restriction enzymes combines the catalytic centers of endonuclease and methylase activity in one polypeptide chain. Sequence specificity is controlled by the same DNA-binding domain. Rimseliene et al. (2003) have used Eco57I to establish a general scheme for their specificity change. After inactivation of the catalytic center of the endonuclease the DNA-binding domain is mutagenized and the resulting plasmid ensemble amplified in vivo where they are target for methylation by the mutant they code for. The selection is performed in vitro by cleaving the purified plasmid pool with a restriction endonuclease of a different specificity. GsuI was used in this case. It cleaves at CTGGAG which deviates in one base pair from the recognition sequence CTGAAG of Eco57I. Only those plasmids stay uncut which are methylated at the new sequence. These are used for reactivation of the endonuclease. The resulting single mutant T862N exhibits a relaxed specificity CTGRAG recognizing both the sequences of Eco57I as well as of GsuI. Actually this scheme is a selection for a methylase with new sequence specificity but it shows that reactivation of endonucleolytic activity after a specificity change is possible.

The selection scheme becomes more elaborate if one tries to directly select for an active restriction enzyme. A prerequisite for this is to protect the cells against the newly generated cleavage activity. Therefore, selection is only possible for those specificities for which a methylase gene is already at hand. The counterselection against the wild-type specificity as well as all undesired ones is obvious, as these activities kill the cell. The positive selection for the desired specificity needs sorting out of all those cells harboring the vast majority of inactive enzymes produced by random mutagenesis. We and others used challenging phages to kill the cells which cannot restrict. In principle, this should work fine, but it needs a mutant enzyme with high cleavage activity as well as the ability for efficient facilitated diffusion to find its cleavage sites quickly (see Jeltsch and Urbanke, this Vol.). If the mutant is about 100-fold less active than the wild-type enzyme restriction drops significantly. Taking into account the results from the site-directed mutagenesis experiments described above, it is very unlikely to find specificity mutants with high enough cleavage rates. We do not know of any positive results from such selection experiments.

Again, the situation seems to be better for specificity enhancement of restriction enzymes with degenerate sequences. BstYI recognizes RGATCY. Its gene was randomly mutagenized and introduced into cells harboring solely the BglI methylase which is specific for one of the three BstYI sequences: AGATCT (Samuelson and Xu 2002). Mutant genes were isolated from the surviving clones and reintroduced in cells with a SOS reporter gene to select for active endonucleases. Some of the mutants showed a preference for AGATCT. By combining their mutations, an enzyme was created which is only sixfold less active than the wild type but no longer cleaves GGATCC.

3 Lengthening of Recognition Sequences

Considering the results described so far, it seems to be easier to enhance sequence specificity than to design a new one. For a restriction enzyme with nondegenerate recognition this means to leave the evolved contacts intact and to add some specificity outside. Actually restriction enzymes with longer recognition sequences are very useful, as the DNAs manipulated in vitro become longer. At the moment, only 13 restriction enzymes with eight basepair recognition sequences are listed in REBASE (Roberts et al. 2003).

From cocrystal structures of EcoRV it was predicted that the A181E mutant should prefer CGATATCN and the K104R mutant NGATATC(G/T) (Horton and Perona 1998). Again, site-directed mutagenesis was used to verify such predictions (Schöttler et al. 1998). Saturation mutagenesis of Ala181 gave several mutants with slight preferences for outside base pairs. A181E in contrast to the predictions, preferred TGATATCA tenfold in oligodeoxynucleotide as well as in plasmid substrates. Lanio et al. (2000) extended these studies by mutations of Lys104. Again, they did not attain the expected specificities. Combining the A181E mutation with K104R, which exhibits a slight preference for the same sequence, did not result in a more specific double mutant.

It is not strictly proven that these mutants establish a base contact. It also may be that they are influenced by the sequence context of the cleavage site and by that become more selective. Selectivity mutants can not be seldom for EcoRV. A directed evolution scheme of three successive rounds with about 500 manually inspected mutants yielded several mutants with preferences for AT-flanked cleavage sites including one triple mutant with a 25-fold preference (Lanio et al. 1998). While from this study it is not clear which property of the recognition sequence leads to the preferential cleavage, Wenz et al. (1998) have directly tested oligodeoxynucleotides with nonflexible oligo-dA•oligo-dT or flexible oligo-dAT flanking sequences. They started with the R226A mutant obtained in an alanine scanning for possible phosphate contacts. This mutant shows some preference for a nonflexible surrounding which is enhanced by a factor of 100 in the R226 V mutant.

For the creation of a lengthened recognition sequence based on the EcoRI endonuclease, we chose Ile197 (Rosati and Alves 1999). This residue is ideally

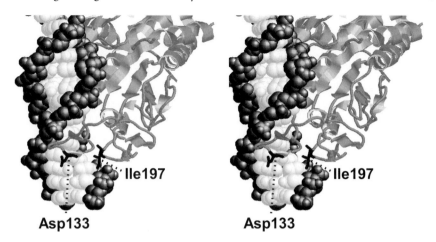

Fig. 2. Stereo-view of Asp133 and Ile197 on top of the major groove which were used to lengthen the recognition sequence of restriction endonuclease EcoRI

positioned above the major groove outside the recognition sequence (Fig. 2). However, not any mutation led to an enhancement of specificity. We considered three possible reasons for this.

First, the enzyme might not need an additional contact as long as all interactions to the normal recognition sequence are made. Therefore, we interfered with canonical recognition by secondary mutations, but all mutants behaved as if Ile197 was not changed.

Second, a longer side chain such as glutamine or arginine may be so flexible that equivalent interactions may be made with varying base pairs. Therefore, we tried to integrate Ile197 mutations into the existing network of interacting residues by introducing M137Q and A138N mutation which points toward Ile197. Some of the triple mutants are even more active than the I197A mutant described above but showed no indication of enhanced specificity. In mutants combining these two approaches, the properties of the mutations interfering with normal recognition dominated.

The third reason may be that Ile197 is too far away from the bases to establish contacts even with long side chains. As it is located at the tip of a short loop, we introduced glycine residues before (position A) and/or behind (position B) Ile197 to move it deeper into the major groove. Cleavage activity was good for the construct with the two glycine residues (GlyAB/I197Q), but again only at the canonical sites. As the former reasons also apply for this mutant, we added mutations to interfere with normal interaction or to connect the glutamine to the other interacting residues. This last mutant A138N/GlyAB/I197Q cleaves only two of the five EcoRI sites in bacteriophage λ DNA. These two are the only EcoRI sites which are palindromic up to eight base pairs TGAATTCA.

This looks promising but does not prove a real contact to the outer adenine. We already have found several mutants which cleave slowly at one or more sites of λ DNA. We studied this phenomenon by using a plasmid pool harboring all possible sequence surroundings at an EcoRI site varying three base pairs before and behind it (Windolph et al. 1997). The mutants tested exhibit a higher selectivity due to flexible or nonflexible neighboring sequences whose influence is ignored by the wild-type enzyme. Several rounds of selection were needed to find the selectivity preferences. In contrast to this behavior, a mutant establishing a real base contact should give a clear result in a single selection round. This is not the case with the mutant described above. Successive rounds of selection failed because the mutant is only active for a few days after purification. Stabilization of enzyme structure with Ca^{2+} ions is possible (Windolph and Alves 1997), but this stabilizes canonical cleavage, too.

Although an enhanced selectivity may be a good starting point for lengthening the recognition sequence we have to stabilize the new enzyme conformation and not the canonical one. For that purpose we will need additional mutations probably outside of the protein–DNA interface. A rational design of these is not possible yet.

Given the difficulties to design and incorporate new base contacts into the framework of the restriction enzyme we also tried to use preformed modules like a zinc finger or one repeat of the Myb transcription factor (von Witzendorff 2002). These DNA-binding modules were selected because they are small (26 and 50 amino acid residues, respectively) and should fold independently. Besides, it is known that they need further modules to interact strongly with DNA. Therefore, this approach is different from the one described in Kandavelou et al., this Vol. about chimeric nucleases as we use the part of the restriction enzyme for DNA recognition and cleavage and not as a sole nuclease without sequence specificity.

Choosing a suitable place to incorporate such a module is of great importance. As both the N- and C-termini of the EcoRI endonuclease are far away from the DNA, we fused these modules into the tip of the inner (before position 133) or outer arm (before position 197) which are located above the major groove (Fig. 2). Hopefully, they would not interfere with the EcoRI structure but find their interaction partner in the DNA. The Myb133 fusion needs Ca^{2+} ions for stabilization but did not cleave the EcoRI sites on λ DNA. Using the plasmid pool described above, we identified the partially palindromic sequence AAGGAATTCCTA as recognition sequence in a single selection round. With an oligodeoxynucleotide substrate this preference could be verified under conditions where we used vast excess of unspecific DNA. Therefore, the fusion mutant is still able to cleave GAATTC if the sequence is isolated in an oligodeoxynucleotide substrate but not in the presence of many other sequences. Circular dichroism spectroscopy revealed that not all mutant molecules are properly folded as the amount of secondary structure is less than anticipated for complete folding of both EcoRI and Myb part. Again, additional changes are necessary to get a stable mutant with a new specificity.

4 Changing the Subunit Composition

All orthodox Type II restriction enzymes work as dimers of identical subunits. Therefore, two single strands can be cleaved in a single binding event. Recognition of a palindromic sequence makes evolution of specific contacts more economical as only half of them have to be evolved. Thus, study of these enzymes becomes more difficult because each mutation has double effect as it automatically occurs twice in the homodimer. Furthermore, understanding of communication between the two subunits is a complicated task.

Several attempts to generate heterodimeric enzymes were made. The simplest way is to use two genes with fusions to different affinity tags and to co-express them in one cell. Purification using both affinity materials consecutively sorts out homodimeric proteins and yields heterodimeric enzymes (Stahl et al. 1996; Wende et al. 1996). In principle, this opens the way to asymmetric recognition which may be easier to engineer than a symmetrical one. However, due to the use of largely identical sequences of the two genes, the danger of recombinational modification arises. This led to high instability of two EcoRI genes cloned into a polycistronic message (Vennekohl, unpubl. results).

We tried to destabilize the homodimeric interface and simultaneously stabilize a heterodimeric protein. As in most enzymes, the monomer–monomer interface of EcoRI is mainly hydrophobic. We mutated two of these residues (Leu158 and/or Ile230) which contact each other from both subunits to aspartate or lysine. The L158D mutation destabilizes the homodimer to a large extent, while the other single mutants turned out to be dimeric (Vennekohl and Alves 1999). In vitro formation of heterodimers is complicated by the fact that the I230K mutant is a rather stable dimer by itself. Thus, longer incubation times are necessary and the amount of heterodimeric enzyme is hard to determine. One indication of heterodimer generation is an increase in specific activity as the L158D monomer is a nickase and sequence recognition is completely lost. Some preferences for cleavage of distinct phosphodiester bonds seem to reflect the flexibility of the sugar–phosphate backbone and the sequence dependent fine structure of the double helix, which leads to preferential cleavage of many so-called unspecific nucleases.

The Type IIS enzyme AlwI gave a completely different result (Xu et al. 2001). It is homologous to the monomeric but sequence-specific nickase N.BstNBI. Swapping of the dimerization domain resulted in a monomeric N.AlwI which still recognizes its sequence but nicks a single strand with high activity. Obviously the composition of Type IIS enzymes out of a DNA-binding domain specific for the whole recognition sequence and a separated nuclease domain is helpful in these experiments. The second nuclease domain is needed only for cleavage of the other single strand.

Another way to attain monomeric restriction enzymes is to fuse the subunits into a single chain. This was done successfully for the PvuII endonuclease where the C-terminus of one subunit is close to the N-terminus of the other (Simoncsits et al. 2001). The single chain PvuII is nearly as active as the wild type enzyme.

In case of EcoRI, fusion is more complicated as C- and N-terminus of both subunits are far away from each other. Therefore, we fused the two subunits where they come close, deleting 65 residues from the C-terminal end of the first part and 72 residues from the N-terminal end of the other (von Pall de Tolna 2001). Although cleavage activity is detectable in crude extracts, the deletions impeded purification in large scale. We also tried to bridge the distance between C- and N-terminus with a linker comprising 21 amino acid residues (Ha-Thi 2002). This linker is sensitive to cellular proteases which again impedes purification of the fusion protein.

This single-chain EcoRI is very active as we experienced unexpected difficulties during its construction. Normally, for all cloning of EcoRI mutants, we use a gene with a good but strongly repressible promoter (p_L) and a bad ribosomal binding sequence which allows for cloning in cells without the methylase gene. Even though this works for wild-type EcoRI, it was not possible for the fusion construct. Obviously, the low transcription rate which always occurs is tolerated by the cells as long as the few mRNA molecules are translated slowly to inactive monomers, as those have to find each other to form an active enzyme. The single-chain EcoRI does not need to dimerize and immediately cleaves DNA of its host cell. Even one active restriction enzyme may kill the cell. This could be the reason that restriction enzymes evolved to be dimeric. During the phase of establishing a new restriction-modification system, the endonuclease is poorly expressed while the methylase prevails. In this critical phase the need to bring together two monomers to assemble an active endonuclease may be helpful.

5 Future Directions (in Vitro Evolution)

We have described that despite very large efforts, engineering of restriction endonucleases was rewarded by little success. It appears that these enzymes are fine-tuned conformational machines which tolerate only few structural deviations. On the other hand, the structural homology of some of these enzymes even in the small sample of structures solved today shows that they are evolutionarily related (Bujnicki 2001). This can also be deduced from their primary sequences although with much less confidence (Jeltsch et al. 1995a; Anton et al. 1997; Pingoud et al. 2002; Bitinaite et al. 2002), which indicates that larger changes are necessary to evolve an enzyme with a new specificity.

Nowadays, we have tools at hand to perform evolution in vitro (Dower and Mattheakis 2002). In principle, one has to produce a library of mutant genes to obtain a large variety of proteins and to select afterwards those enzymes with the desired specificity and activity. Because the amount of all possible mutants for a typical protein is enormous, these steps have to be repeated several times.

However, in vitro two main difficulties arise. Firstly, phenotype and genotype have to be coupled. As we want to select the properties of proteins and reverse translation is not possible, we have to link each enzyme physically to its mRNA which can be reverse transcribed to the gene. This is one of the reasons for creating single-chain restriction endonucleases. The most popular methods only work with monomeric proteins. These are phage display (Hoess 1993), ribosome display (Hanes and Plückthun 1997) and mRNA–protein fusions (Roberts and Szostak 1997). The first has the disadvantage that the phages are produced by cells. It is better to perform all steps in vitro because transformation into cells reduces the diversity from 10^{13} to about 10^8.

The second difficulty is the selection step itself. The properties of many restriction endonucleases suggest the following procedure: In the absence of Mg^{2+}, many of these enzymes bind strongly to their recognition sequence. Variants with other specificity or with low binding strength can be washed away. Then catalytically efficient variants can be eluted by adding Mg^{2+}. Therefore, one can easily select for an efficient restriction enzyme. However, one will start with very inefficient enzymes which have to be optimized in later steps. Not to lose promising candidates in the first selection rounds may be difficult. Furthermore, this selection scheme is not compatible with ribosome display, as high Mg^{2+} concentrations are used to fix protein and mRNA on the ribosome. Thus, the enzyme will dissociate from the ribosome during the first selection phase.

Despite all difficulties, we are convinced that in vitro evolution experiments will tell us a lot more about restriction enzymes as was the case with other systems. Especially the contribution of residues outside the protein–DNA interface will show us their importance for activity and specificity (Shimotohno et al. 2001).

References

Alves J, Rüter T, Geiger R, Fliess A, Maass G, Pingoud A (1989) Changing the hydrogen bonding potential in the DNA binding site of EcoRI by site-directed mutagenesis drastically reduces the enzymatic activity, not however, the preference of this restriction endonuclease for cleavage within the site -GAATTC-. Biochemistry 28:2678–2684

Anton BP, Heiter DF, Benner JS, Hess EJ, Greenough L, Moran LS, Slatko BE, Brooks JE (1997) Cloning and characterization of the BglII restriction-modification system reveals a possible evolutionary footprint. Gene 187:19–27

Bitinaite J, Mitkaite G, Dauksaite V, Jakubauskas A, Timinskas A, Vaisvila R, Lubys A, Janulaitis A (2002) Evolutionary relationship of Alw26I, Eco31I and Esp3I, restriction endonucleases that recognise overlapping sequences. Mol Genet Genomics 267:664–672

Bujnicki JM (2001) Understanding the evolution of restriction-modification systems: clues from sequence and structure comparisons. Acta Biochim Pol 48:935–967

Dorner LF, Schildkraut I (1994) Direct selection of binding proficient/catalytic deficient variants of BamHI endonuclease. Nucleic Acids Res 22:1068–174

Dorner LF, Bitinaite J, Whitaker RD, Schildkraut I (1999) Genetic analysis of the base-specific contacts of the BamHI restriction endonuclease. J Mol Biol 285:1515–1523

Dower WJ, Mattheakis LC (2002) In vitro selection as a powerful tool for the applied evolution of proteins and peptides. Curr Opin Chem Biol 6:390–398

Fisher EW, Yang MT, Jeng S, Gardner JF, Gumport RI (1995) Selection of mutations altering specificity in restriction-modification enzymes using the bacteriophage P22 challenge-phage system. Gene 157:119–121

Flores H, Osuna J, Heitman J, Soberon X (1995) Saturation mutagenesis of His114 of EcoRI reveals relaxed-specificity mutants. Gene 157:295–301

Fritz A, Küster W, Alves J (1998) Asn141 is essential for DNA recognition by EcoRI restriction endonuclease. FEBS Lett 438:66–70

Geiger R, Rüter T, Alves J, Fliess A, Wolfes H, Pingoud V, Urbanke C, Maass G, Pingoud A, Duesterhoeft A, Kroeger M (1989) Genetic engineering of EcoRI mutants with altered amino acid residues in the DNA binding site. Physico-chemical investigations give evidence for an altered monomer/dimer equilibrium for the Gln144Lys145 and Gln144Lys145Lys200 mutants. Biochemistry 28:2667–2677

Gerschon D (2000) Proteindesign zur Veränderung der Sequenzspezifität der Restriktionsendonuklease EcoRI. Doctoral Thesis, School of Veterinary Medicine, Hanover

Grabowski G, Jeltsch A, Wolfes H, Maass G, Alves J (1995) Site-directed mutagenesis in the catalytic center of the restriction endonuclease EcoRI. Gene 157:113–118

Ha-Thi MC (2002) Konstruktion und Expression von Fusionsdimeren der Restriktionsendonuklease EcoRI. Diploma Thesis, University of Hanover

Hager PW, Reich NO, Day JP, Coche TG, Boyer HW, Rosenberg JM, Greene PJ (1990) Probing the role of glutamic acid 144 in the EcoRI endonuclease using aspartic acid and glutamine replacements. J Biol Chem 265:21520–21526

Hanes J, Plückthun A (1997) In vitro selection and evolution of functional proteins by using ribosome display. Proc Natl Acad Sci USA 94:4937–4942

Heitman J, Model P (1990) Mutants of the EcoRI endonuclease with promiscuous substrate specificity implicate residues involved in substrate recognition. EMBO J 9:3369–3378

Heitman J, Model P (1991) SOS induction as an in vivo assay of enzyme-DNA interactions. Gene 103:1–9

Hoess RH (1993) Phage display of peptides and protein domains. Curr Opin Struct Biol 3:572–579

Horton NC, Perona JJ (1998) Role of protein-induced bending in the specificity of DNA recognition: crystal structure of EcoRV endonuclease complexed with d(AAAGAT) + d(ATCTT). J Mol Biol 227:779–787

Ivanenko T, Heitman J, Kiss A (1998) Mutational analysis of the function of Met137 and Ile197, two amino acids implicated in sequence-specific DNA recognition by the EcoRI endonuclease. Biol Chem 379:459–465

Jeltsch A, Alves J, Oelgeschläger T, Wolfes H, Maass G, Pingoud A (1993) Mutational analysis of the function of Gln115 in the EcoRI restriction endonuclease, a critical amino acid for recognition of the inner thymidine residue in the sequence -GAATTC-and for coupling specific DNA binding to catalysis. J Mol Biol 229:221–234

Jeltsch A, Kröger M, Pingoud A (1995 a) Evidence for an evolutionary relationship among Type II restriction endonucleases. Gene 160:7–16

Jeltsch A, Alves J, Urbanke C, Maass G, Eckstein H, Zhang L, Bayer E, Pingoud A (1995 b) A dodecapeptide comprising the extended chain-α4 region of the restriction endonuclease EcoRI specifically binds to the EcoRI recognition site. J Biol Chem 270:5122–5129

Jeltsch A, Wenz C, Wende W, Selent U, Pingoud A (1996) Engineering novel restriction endonucleases: principles and applications. Trends Biotechnol 14:235–238

Kim Y, Grable JC, Love R, Greene PJ, Rosenberg JM (1990) Refinement of EcoRI endonuclease crystal structure: a revised protein chain tracing. Science 249:1307–1309

King K, Benkovic SJ, Modrich P (1989) Glu-111 is required for activation of the DNA cleavage center of EcoRI endonuclease. J Biol Chem 264:11807–11815

Küster W (1998) Bedeutung hydrophober Kontakte für die sequenzspezifische DNA-Erkennung der Restriktionsendonuklease EcoRI. Doctoral Thesis, University of Hanover

Küster W, Alves J (1995) Mutational analysis of the function of Ile197 in the EcoRI restriction endonuclease. Biol Chem 376:S122

Lanio T, Jeltsch A, Pingoud A (1998) Towards the design of rare cutting restriction endonucleases: Using directed evolution to generate variants of EcoRV differing in their substrate specificity by two orders of magnitude. J Mol Biol 283:59–69

Lanio T, Jeltsch A, Pingoud A (2000) On the possibilities and limitations of rational protein design to expand the specificity of restriction enzymes: a case study employing EcoRV as the target. Protein Eng 13:275–281

Lanio T, Selent U, Wenz C, Wende W, Schulz A, Adiraj M, Katti SB, Pingoud A (1996) EcoRV-T94 V: A mutant restriction endonuclease with an altered substrate specificity towards modified oligodeoxynucleotides. Protein Eng 9:1005–1010

McClarin JA, Frederick CA, Wang B-C, Boyer HW, Grable J, Rosenberg JM (1986) Structure of DNA-EcoRI endonuclease recognition complex at 3 Å resolution. Science 234:1526–1541

Muir RS, Flores H, Zinder ND, Model P, Soberon X, Heitman J (1997) Temperature-sensitive mutants of the EcoRI endonuclease. J Mol Biol 274:722–737

Mukhopadhyay P, Roy KB (1998) Protein engineering of BamHI restriction endonuclease: replacement of Cys54 by Ala enhances catalytic activity. Protein Eng 11:931–935

Nastri HG, Evans PD, Walker IH, Riggs PD (1997) Catalytic and DNA binding properties of PvuII restriction endonuclease mutants. J Biol Chem 272:25761–25767

Needels MC, Fried SR, Love R, Rosenberg JM, Boyer HW, Greene PJ (1989) Determinants of EcoRI endonuclease sequence discrimination. Proc Natl Acad Sci USA 86:3579–3583

Oelgeschläger T, Geiger R, Rüter T, Alves J, Fliess A, Pingoud A (1990) Probing the function of individual amino acid residues in the DNA binding site of the EcoRI restriction endonuclease by analysing the toxicity of genetically engineered mutants. Gene 89:19–27

Pingoud A, Jeltsch A (2001) Structure and function of Type II restriction endonucleases. Nucleic Acids Res 29:3705–27

Pingoud V, Kubareva E, Stengel G, Friedhoff P, Bujnicki JM, Urbanke C, Sudina A, Pingoud A (2002) Evolutionary relationship between different subgroups of restriction endonucleases. J Biol Chem 277:14306–14314

Rimseline R, Maneliene Z, Lubys A, Janulaitis A (2003) Engineering of restriction endonucleases: using methylation activity of the bifunctional endonuclease Eco57I to select the mutant with novel sequence specificity. J Mol Biol 327:383–391

Roberts RJ, Vincze T, Posfai J, Macelis D (2003) REBASE: restriction enzymes and methyltransferases. Nucleic Acids Res 31:418–420

Roberts RW, Szostak JW (1997) RNA-peptide fusions for the in vitro selection of peptides and proteins. Proc Natl Acad Sci USA 94:12297–12302

Rosati O (1999) Untersuchung und Design von DNA-Kontakten der Restriktions-endonuklease EcoRI inner- und außerhalb der Erkennungssequenz. Doctoral Thesis, University of Hanover

Rosati O, Alves J (1999) Towards the design of a 8 base pair cutter based on the restriction endonuclease EcoRI. Protein Sci 8:54

Rosenberg JM, McClarin JA, Frederick CA, Wang B-C, Grable J, Boyer HW, Greene P (1987) Structure and recognition mechanism of EcoRI endonuclease. Trends Biochem 12:395

Samuelson JC, Xu SY (2002) Directed evolution of restriction endonuclease *Bst*YI to achieve increased substrate specificity. J Mol Biol 319:673–683

Schöttler S, Wenz C, Lanio T, Jeltsch A, Pingoud A (1998) Protein engineering of the restriction endonuclease EcoRV – structure-guided design of enzyme variants that recognize the base pairs flanking the recognition site. Eur J Biochem 258:184–191

Shimotohno A, Oue S, Yano T, Kuramitsu S, Kagamiyama H (2001) Demonstration of the importance and usefulness of manipulating non-active-site residues in protein design. J Biochem (Tokyo) 129:943–948

Simoncsits A, Tjörnhammar ML, Rasko T, Kiss A, Pongor S (2001) Covalent joining of the subunits of a homodimeric Type II restriction endonuclease: single-chain PvuII endonuclease. J Mol Biol 309:89–97

Stahl F, Wende W, Jeltsch A, Pingoud A (1996) Introduction of asymmetry in the naturally symmetric restriction endonuclease EcoRV to investigate intersubunit communication in the homodimeric protein. Proc Natl Acad Sci USA 93:6175–6180

Tang D, Ando S, Takasaki Y and Tadano J (2000) Mutational analyses of restriction endonuclease-HindIII mutant E86 K with higher activity and altered specificity. Protein Eng 13:283–289

van der Woerd MJ, Pelletier JJ, Xu S, Friedman AM (2001) Restriction enzyme BsoBI-DNA complex: a tunnel for recognition of degenerate DNA sequences and potential histidine catalysis. Structure-(Camb) 9:133–144

Vennekohl P (2000) Bedeutung der Dimerisierung für Spezifität und Katalyse der Restriktionsendonuklease EcoRI. Doctoral Thesis, University of Hanover

Vennekohl P, Alves J (1999) EcoRI restriction endonuclease goes heterodimeric. Protein Sci 8:55

von Pall de Tolna S (2001) Mutagenesestudien zur Löslichkeitsverbesserung eines fusionierten Dimers der Restriktionsendonuklease EcoRI. Diploma Thesis, University of Hannover

von Witzendorff D (2002) Fusion von DNA-Bindungsdomänen mit der Restriktions-endonuklease EcoRI zur Erweiterung der Spezifität. Diploma Thesis, University of Hanover

Wende W, Stahl F, Pingoud A (1996) The production and characterization of artificial heterodimers of the restriction endonuclease EcoRV. Biol Chem 377:625–632

Wenz C, Selent U, Wende W, Jeltsch A, Wolfes H, Pingoud A (1994) Protein engineering of the restriction endonuclease EcoRV: Replacement of an amino acid residue in the DNA binding site leads to an altered selectivity towards unmodified and modified subtrates. Biochem Biophys Acta 1219:73–80

Wenz C, Hahn M, Pingoud A (1998) Engineering of variants of the restriction endonuclease EcoRV that depend in their cleavage activity on the flexibility of sequences flanking the recognition site. Biochemistry 37:2234–2242

Windolph S, Alves J (1997) Influence of divalent cations on inner-arm mutants of restriction endonuclease EcoRI. Eur J Biochem 244:134–139

Windolph S, Fritz A, Oelgeschläger T, Wolfes H, Alves J (1997) Sequence context influencing cleavage activity of the K130E mutant of the restriction endonuclease EcoRI identified by a site selection assay. Biochemistry 36:9478–9485

Whitaker RD, Dorner LF, Schildkraut I (1999) A mutant of BamHI restriction endonuclease which requires N6-methyladenine for cleavage. J Mol Biol 285:1525–1536

Wolfes H, Alves J, Fliess A, Geiger R, Pingoud A (1986) Site directed mutagenesis experiments suggest that Glu 111, Glu 144 and Arg 145 are essential for endonucleolytic activity of EcoRI. Nucleic Acids Res 14:9063–9080

Wright DJ, King K, Modrich P (1989) The negative charge of Glu-111 is required to activate the cleavage center of EcoRI endonuclease. J Biol Chem 264:11816–11821

Xu Y, Lunnen KD, Kong H (2001) Engineering a nicking endonuclease N.AlwI by domain swapping. Proc Natl Acad Sci USA 98:12990–12995

Zhu Z, Zhou J, Friedman AM, Xu SY (2003) Isolation of BsoBI restriction endonuclease variants with altered substrate specificity. J Mol Biol 330:359–372

Engineering and Applications of Chimeric Nucleases

K. Kandavelou, M. Mani, S. Durai, S. Chandrasegaran

1 Introduction

Each human cell contains about 3×10^9 base pairs (bp) within its genome. With the first draft sequence of the human genome now available, biologists estimate that there are about 30,000–40,000 different genes within the genome (IHGSC 2001, Venter et al. 2001). This is fewer than originally anticipated, but still a huge number. These genes code for all of the human body's proteins. Simple mutations within the coding region of critical genes can lead to the formation of abnormal proteins, resulting in disease phenotypes, premature death, or failure of an embryo to develop. Furthermore, mutations that affect the regulatory region of genes can result in aberrant gene expression within cells, and give rise to cancer phenotypes. The 'Holy Grail' of the Human Genome Project is 'Gene Therapy', that is, how genes might someday be used, modified, or even changed to correct human disease.

Gene therapy is based on a simple concept: Mutations within the coding or regulatory regions of certain critical genes give rise to disease phenotypes. Correction of these mutations to normal alleles will result in the reversion of disease phenotypes to normal phenotypes. This is particularly true of monogenic diseases. Thus, gene therapy provides a new paradigm for treating human disease by correcting the causative genetic defect. However, the implementation of this novel concept into clinical practice and therapeutic reality has proven to be extremely difficult; and, the progress has been very slow. This is because targeting a particular defective gene for correction among 30,000 or more human genes within the genome requires exquisite sequence-specificity of the gene therapy vectors, which are the vehicles used to deliver the therapeutic genes into cells. This problem is further compounded by the fact

K. Kandavelou, M. Mani, S. Durai, S. Chandrasegaran
Department of Environmental Health Sciences, The Johns Hopkins University
Bloomberg School of Public Health, 615 North Wolfe Street, Baltimore,
Maryland 21205–2179, USA

Nucleic Acids and Molecular Biology, Vol. 14
Alfred Pingoud (Ed.)
Restriction Endonucleases
© Springer-Verlag Berlin Heidelberg 2004

that these genes and their regulatory sequences, which account for about 5 % of the human genome, are buried within the context of a 3x10⁹-bp DNA sequence. Current gene therapy vectors lack the required sequence specificity that is necessary for the targeted correction of a defective gene within the genome. As a result, the therapeutic genes are delivered and inserted randomly within the genome that is at many sites other than the location of the genetic defect. Such random insertions of the therapeutic gene at unwanted locations are mutagenic, especially when the insertions occur in critical genes (Connolly 2002). Over time, these insertions can and do give rise to cancer or other disease phenotypes. Many of the difficulties associated with gene therapy are likely to be overcome if one could insert the corrected version of the mutation at the precise location of the genetic defect within the genome.

How then does one achieve targeted correction of a gene defect within cells? Biologists have known for a while that when a defined chromosomal break is introduced at a unique site within a genome, homologous recombination is induced at that site in a large fraction of cells in a population (Bibikova et al. 2001; Vasquez et al. 2001). The challenge then, so far unfulfilled, is to develop a general means of introducing a double strand break (DSB) uniquely at a given locus in the genome to induce homologous recombination. In this chapter, we discuss the progress towards targeted correction of a genetic defect. This involves two steps: (1) engineering of chimeric nucleases, the molecular tools necessary to make a chromosomal DSB (double-strand break) at a chosen site within the human genome; (2) application of these molecular tools for targeted chromosomal cleavage and correction of a genetic defect by inducing homologous recombination at that locus in cells.

2 Engineering of Chimeric Nucleases

In order to make a unique chromosomal DSB within a mammalian genome, we need restriction enzymes that recognize DNA sequences of 16 bp or more in length. Such restriction enzyme sites will occur once every 4^{16} (=4.3x10⁹) bp on average, which is about once per human genome. Most bacterial enzymes typically recognize short palindromes that are 4 to 8 bp in length (Mani et al. 2003a, b; Kandavelou et al. 2003). This means that they will cut DNA on average once every 4^4 to 4^8 (=256 to 65,536) bases depending on the restriction enzyme used (Mani et al. 2003a; Chandrasegaran and Smith 1999). For instance, one can expect an 8-bp site to occur about 45,776 times within the human genome. Obviously, bacterial restriction enzymes are not useful for the task at hand, which is to introduce a targeted chromosomal break at a site of our own choosing within the human genome.

Over a decade ago, our laboratory set out to engineer chimeric nucleases, the molecular tools necessary for introducing a targeted chromosomal DSB within a mammalian genome. The research started with the study of FokI restriction endonuclease, a bacterial Type IIS restriction enzyme. FokI recognizes the nonpalindromic pentadeoxyribonucleotide, 5′-GGATG-3′:5′-CATCC-3′, in duplex DNA and cleaves 9/13 nucleotides downstream of the recognition site (Sugisaki and Kanazawa 1981). It does not recognize any specific sequence at the site of cleavage. This property implies the presence of two separate protein domains within FokI: one for sequence-specific recognition of DNA and the other for the endonuclease activity (Fig. 1A). Once the DNA-binding domain is anchored at the recognition site, a signal is transmitted to

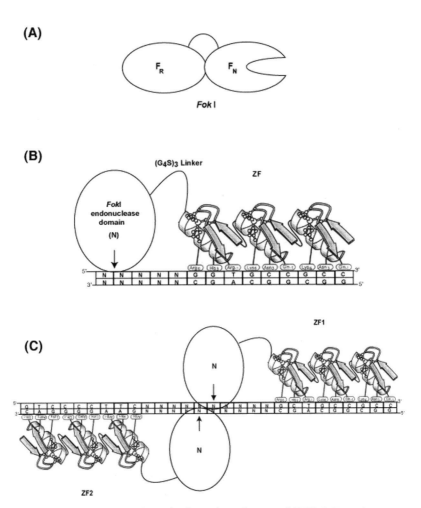

Fig. 1. A Schematic representation of FokI endonuclease and ZFN. **A** Domain structure of FokI; **B** ZFN bound to its 9-bp cognate site; and **C** A pair of ZFN with different specificities bound to their target sequence. Since ZFN requires two copies of the 9-bp recognition sites in an inverted orientation in order to dimerize and produce a DSB, it effectively has an 18-bp recognition site. F_R FokI recognition domain; F_N or N FokI cleavage domain; *ZF1* and *ZF2* zinc finger proteins

the endonuclease domain, probably through allosteric interactions, and the cleavage occurs. We reasoned that one may be able to swap the FokI recognition domain with other naturally occurring DNA-binding proteins that recognize longer DNA sequences or other designed DNA-binding motifs to create chimeric nucleases. In this way, Type IIS enzymes like FokI are probably ideal candidates for engineering novel sequence specificities.

2.1 Functional Domains in FokI Restriction Endonuclease

Several laboratories including ours have independently cloned the FokI restriction-modification system from *Flavobacterium okeanokoites* (Kita et al. 1989; Looney et al. 1989; Wu and Chandrasegaran 1989, unpubl. work). These labs have also reported overproduction and purification of FokI endonuclease. As a first step, we probed the domain structure of FokI by limited proteolysis using trypsin. Our studies on proteolytic fragments of FokI endonuclease revealed an N-terminal DNA-binding domain and a C-terminal domain with nonspecific DNA cleavage activity (Li et al. 1992; Li and Chandrasegaran 1993; Fig. 1A). Waugh and Sauer (1993, 1994) showed that single amino acid substitutions decouple the DNA-binding and strand scission activities of FokI endonuclease. The mode of DNA-binding by FokI was analyzed by DNA foot-printing methods (Li et al. 1993; Yonezawa and Sugiura 1994). These studies showed a lack of protection at the cleavage site of DNA by FokI. Wah et al. (1997, 1998) reported the crystal structures of FokI and FokI - bound to its cognate site, which confirmed the modular nature of FokI endonuclease. The endonuclease domain appears to be sequestered in a "piggyback" fashion by the recognition domain. This is consistent with the DNA foot-printing analysis. Thus, the crystal structure of FokI endonuclease is in complete agreement with the model derived from rigorous biochemical studies.

2.2 Chimeric Nucleases

The modular nature of FokI endonuclease suggested that it might be feasible to engineer chimeric nucleases by fusing other DNA-binding proteins to the cleavage domain of FokI. The obvious next step was to create novel chimeric nucleases by simply swapping the DNA recognition domain of FokI with other DNA-binding proteins that recognize longer DNA sequences. This indeed proved to be the case. Several novel fusion nucleases were generated in our lab by linking the nonspecific FokI cleavage domain to other DNA-binding proteins. The latter include the three common eukaryotic DNA-binding motifs namely the helix-turn-helix motif, the zinc finger motif and the basic helix-loop-helix protein containing a leucine zipper motif (bzip). Our lab reported the creation of the first chimeric nuclease by fusing the *Drosophila Ubx* homeodomain to the FokI cleavage domain (Kim and Chandrasegaran 1994; Kim et al. 1996). This was followed by the fusion of zinc finger motifs and the N-terminal 147 amino acids of the yeast Gal4 protein respectively, to the cleavage domain of FokI (Kim et al. 1998). Such engineered nucleases were shown to make specific cuts very close to their expected recognition sequences in vitro. Since then other labs have reported making FokI nuclease domain fusions with various other DNA-binding proteins (Huang et al. 1996a; Kim et al. 1997a; Lee et al. 1998). FokI fusions have been used to identify high- and low-affinity binding sites of transcription factors in vitro (Ruminy et al. 2001), to study recruitment of various factors to the promoter sites in vivo using the method called protein position identification with a nuclease tail (PIN*POINT) assay (Kim et al 1999a; Lee et al. 1998, 1999a, b, 2001) and to analyze Z-DNA conformation-specific proteins (Kim et al. 1997a, 1999b). Chimeric nucleases now form a novel class of engineered nucleases in which the non-specific cleavage domain of FokI is fused to other DNA-binding motifs. The most important chimeric nucleases are those based on zinc-finger DNA-binding motifs (Kim et al. 1996; Fig. 1B).

2.3 Zinc Finger Binding and Specificity

The modular structure of zinc finger domains (ZF) and modular recognition by zinc finger proteins make them the most versatile of DNA recognition motifs for designing artificial DNA-binding proteins (reviews Pabo et al. 2001; Beerli and Barbas 2002). Each zinc finger consists of about 30 amino acids and folds into a $\beta\beta\alpha$-structure, which is stabilized by the chelation of a zinc ion by the conserved Cys_2His_2 residues. Each finger typically recognizes a 3-bp DNA sequence by inserting the α-helix into the major groove of DNA (Pavletich and Pabo 1991). Binding of longer DNA sequences is achieved by linking several of these zinc finger motifs in tandem. Each finger, because of variations of certain key amino acids in the α-helix of one-zinc finger to the next, makes its own unique contribution to DNA-binding affinity and specificity (Fig. 2). Because they appear to bind as independent modules, zinc fingers can be linked together in a peptide designed to bind a predetermined DNA site. Although more recent studies suggest that there might be a synergistic interaction between adjacent zinc fingers (Isalan et al. 1998) and that the zinc finger–DNA recognition is more complex than originally perceived, it still appears that zinc finger motifs will provide an excellent framework for designing DNA-binding proteins with a variety of new sequence-specificities. In theory, one can design a zinc finger for each of the 64 possible triplet codons, and using a combination of these fingers, one could design a protein for sequence-specific recognition of any segment of DNA. At present, the rules relating to zinc finger sequences/DNA-binding preferences and redesigning

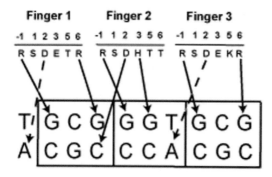

of DNA-binding specificities of zinc finger proteins are well understood only for the 5'-GNN-3' triplets. Although numerous structural studies of zinc finger protein complexes with DNA are available, the information is not yet sufficient for rational design of zinc finger domains that bind to any given triplet (Pabo et al. 2001).

The creation of zinc finger chimeric nucleases (ZFN) that recognize and cleave any target sequence depends on the reliable creation of zinc finger proteins (ZF) that can specifically recognize a target sequence. A phage display selection method could be used to identify zinc finger proteins that bind to a particular DNA site (Griesman and Pabo 1997; Isalan et al. 1998). This is the method of choice for designer zinc finger proteins. Briefly, a library of zinc fingers with randomized amino acids at specific positions (–1 to 6) within the α-helix is fused to a viral coat protein of a phage that is used to infect bacteria. The virus is then allowed to replicate and reproduce in culture to make more copies. Since the fusion proteins are displayed on the surface of the virions, they can be enriched for highly specific and functional zinc finger proteins that bind a chosen target substrate with high affinity. Three different selection methods based on the phage display – parallel selection, sequential selection, and bipartite selection – have been reported in literature using Zif268-derived phage libraries for selection of designer zinc fingers (Greisman and Pabo 1997; Segal et al. 1999; Isalan et al. 2001). These libraries are constrained by the target site overlap problem because of the presence of Asp[2] at position 2 of the α-helix in each of the three-zinc finger domains of the Zif268 (Fig. 2). The Asp[2] of all three fingers make a cross-strand contact to a base outside the canonical triplet site (Pavletich and Pabo 1991, 1993). This precludes the presence of a simple general recognition code that will facilitate the rational design of zinc finger-based DNA-binding domains. This also makes selection of zinc finger proteins by using phage display libraries more difficult. Three strategies have since been developed to overcome this problem: (1) zinc fingers 1 to 3 can be sequentially selected in the context of neighboring zinc fingers (Griesman and Pabo 1997); (2) targets can be selected from libraries with simultaneous randomization of amino acid residues in two adjacent fingers (Segal et al. 1999); and (3) bipartite selection which combines the sequential selection with the parallel selection technique (Isalan et al. 2001). However, these approaches do not provide for the selection of zinc finger domains that function as independent, modular recognition motifs. They select for targets that are dependent on their inter-domain interaction for their sequence specificity.

An alternate approach based on a bacterial two-hybrid system has been described for the isolation of a candidate zinc finger protein in an in vivo context from a randomized Zif268-derived mutant library of >10^8 members in size (Joung et al. 2000). This method is analogous to the yeast two-hybrid system previously described by Hochschild and coworkers to study protein–protein interactions (Dove et al. 1997). In the bacterial two-hybrid system, zinc finger–DNA interactions are required for cell growth and survival. However, this approach also suffers from the same limitation as the phage display methods in that it does not allow for selection of zinc finger motifs that act as independent, modular recognition domains since it also utilizes a Zif268-derived mutant library.

In collaboration with Dr. Marc Ostermeier's lab, we are developing a double-reporter, one-hybrid system for rapidly selecting zinc finger proteins and improving their sequence specificities. This system will also allow for identification of zinc finger motifs that act as independent modular units. This will be done by using a mutant zinc finger library that is based on consensus backbone framework for each and every finger within the protein; and by limiting the amino acid at position +2 of the α-helix of each finger to a glycine residue thus eliminating the cross-strand base contact that occurs outside the 3-bp site in the Zif268 derived libraries due to the presence of Asp^2.

The one hybrid, double reporter system that is being developed in our lab is based on the one hybrid system originally described by Hochschild and coworkers that is used to detect protein–DNA interactions (Hu et al. 2000). In their system, the gene for a DNA binding domain is fused to a subunit of E. coli RNA polymerase. The fusion is then used to activate transcription from a lac-derived promoter provided the binding site for the DNA-binding domain is present and centered at the –63 position. The reporter gene under the control of the lac-derived promoter is β-galactosidase. Thus, this method provides a way to screen and assay for interaction between the DNA-binding domain and various DNA-binding sites positioned at –63. Our system is designed to create one hybrid, double reporter system for evaluating and evolving DNA-binding specificity of zinc finger proteins. The gene coding for the zinc finger is fused to a subunit of E. coli RNA polymerase. The fusion protein is then used to activate transcription of a reporter gene under the control of a lac-derived promoter provided the zinc finger binding site is placed at an appropriate distance upstream of the promoter. In our system, we incorporate two separate operons each containing one reporter gene under the control of a lac-derived promoter. The only difference between the two operons is the nature of the reporter gene and the target zinc finger binding sites, which are placed upstream of the promoter.

In this way, binding of a zinc finger protein to two different sites can be evaluated simultaneously. We plan to employ two different reporter systems as well. In the first, the antibiotic resistance genes coding for chloramphenicol and tetracycline, respectively, will be placed under the control of lac-derived promoters on separate operons. In the second, we have chosen two reporter genes namely the green fluorescent protein (GFP) and the red fluorescence protein (dsRED) to be placed under the control of lac-derived promoters on separate operons. The fluorescent system offers selection of zinc finger mutants with improved sequence specificity using quantitative flow cytometry and fluorescence activated cell sorting.

Recent advances suggest that a combination of design and selection is best suited to identify custom zinc finger DNA-binding proteins for known target sites. So far, only zinc finger modules that specifically recognize 5'-GNN-3' and 5'-ANN-3' triplets have been identified through design and selection strategies (Dreier et al. 2000, 2001, Liu et al. 2002). Studies are underway to determine the zinc finger recognition preferences for the other triplets, 5'-CNN-3' and 5'-TNN-3', respectively. In many instances, the current knowledge of zinc finger preferences for 5'-GNN-3' and 5'-ANN-3' is more than sufficient for designing and/or selecting a zinc finger protein to target a specific site within a gene of the human genome.

The ability to design or select zinc fingers with desired specificity implies that DNA-binding proteins containing zinc fingers will be made to order. Therefore, we reasoned that one could design "artificial" nucleases (ZFN) that will cut DNA at any preferred site by making fusions of zinc finger proteins (ZF) to the cleavage domain (N) of FokI endonuclease. We have been successful in engineering several novel ZFN by fusing three-zinc finger proteins to the cleavage domain of FokI (Kim et al. 1996, 1997b). In these fusions, the three Cys_2His_2 zinc finger DNA-binding domain is at the N-terminus and the FokI nuclease domain is at the C-terminus; they are connected by a flexible 15 amino acid $(Gly_4Ser)_3$ linker to achieve efficient DSB at the predicted 9-base pair recognition site. We have shown that the fusion of FokI cleavage domain to the zinc finger motif does not alter the sequence specificity of the zinc finger protein and it does not change its binding affinity significantly (Smith et al. 1999). It must be made clear that sequence specificity of the engineered ZFN is only as good as the zinc finger motif that was used to make it. If a zinc finger shows affinity for degenerate sites, then, the engineered ZFN will cut at those degenerate binding sites albeit with lower affinity.

Although, our lab was the first to fuse another functional moiety (FokI cleavage domain) to zinc finger proteins, our long-term focus and efforts have remained on the development of chimeric nucleases. Other functional domains, like activator domains and repressor domains have been fused to the designed zinc finger motifs to form hybrid proteins that act as transcrip-

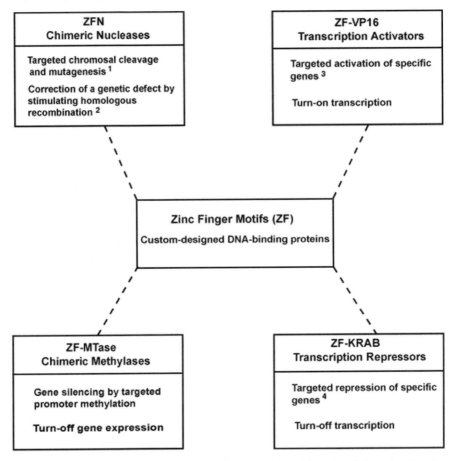

Fig. 3. Zinc finger protein platform technology. Additional functional domains like the FokI cleavage domain (N), activator domain (VP16), repressor domain (KRAB) and methylases (Mtase) can be fused to the designed zinc finger motifs (ZF) to form chimeric proteins that act as chimeric nucleases (ZFN), transcriptional activators (ZF-VP16), transcriptional repressors (ZF-KRAB) and targeted methylases (ZF-Mtase), respectively. The biological applications for these chimeric proteins, reported in literature, are indicated in their respective boxes. [1]Bibikova et al. (2002); [2]Bibikova et al. (2001), Porteus and Baltimore (2003); [3]Rebar et al. (2002); [4]Beerli et al. (1998)

tion activators and transcription repressors within cells (Kim et al. 1997b; Beerli et al. 1998; Ordiz et al. 2002; Stege et al. 2002; Fig. 3). More recently, engineered transcription factors have been shown to stimulate blood vessel growth and wound healing by the induction of the gene coding for VEGF-A (vascular endothelial growth factor A) in mice (Rebar et al. 2002). Cytosine methylation has been targeted preferentially to pre-determined sequences by attaching CpG-specific DNA methyltransferase to Zif 268 three-zinc finger proteins (Xu and Bestor 1997; McNarma et al. 2002; Fig. 3). It appears that important applications in medicine and biological research arising from zinc finger platform technology are almost certain to follow.

2.4 Mechanism of Cleavage by Zinc Finger Chimeric Nucleases (ZFN)

A three-zinc finger protein recognizes nine base pairs. An issue relevant to gene targeting is the ability of the cleavage reagent to distinguish the chosen target from all other sequences in a complex genome. In theory, any sequence shorter than 16-bp would be present in multiple copies in the human genome. In particular, any given 9-bp sequence is predicted to occur 10^4 times within the human genome. Therefore, the three-zinc finger nucleases are expected to make about 10,000 DSB within the human genome. How then does one achieve a unique, targeted chromosomal DSB within a genome?

Our study on the mechanism of cleavage by the three-zinc finger nucleases (Smith et al. 2000) indicates that the cleavage domain must dimerize in order to cut DNA. Therefore, efficient cleavage requires two zinc finger-binding sites situated in close proximity in an inverted orientation (Fig. 1C). Such a configuration of binding sites orients the nuclease domains towards each other promoting dimerization. As expected, the spacer length between the inverted repeats is critical for the dimerization of the nuclease domains to occur. This in turn is determined by the $(Gly_4Ser)_3$ linker, which connects the zinc fingers to the FokI cleavage domain. Thus, ZFN prefer inverted repeats as their substrate to make a DSB.

However, at high enzyme concentrations, single binding sites are cut; one molecule binds at its cognate site, while the second probably binds nonspecifically nearby on the DNA facilitating the dimerization of the FokI catalytic domain. The first clue for FokI dimerization came from the crystal structure of the FokI endonuclease, which revealed the interacting interface between the catalytic domains (Wah et al. 1997, 1998). In addition, kinetic studies on the mechanism of cleavage by FokI further supported that dimerization of FokI enzyme was required to produce a DSB (Bitinaite et al. 1998).

Since ZFN requires two copies of the 9-bp recognition sites in a tail-to-tail orientation in order to dimerize and produce a DSB, it effectively has an 18-bp recognition site (Fig. 1C). In principle, the binding sites need not be identical provided the ZFN that bind both sites are present. We have shown that two ZFN with different sequence specificities collaborate to produce a DSB when their binding sites are appropriately placed and oriented with respect to each other (Smith et al. 2000). Since the recognition specificity of zinc fingers can be manipulated experimentally, ZFN could be engineered so as to target a unique site within a genome.

3 Application of Chimeric Nucleases

The modular structure of FokI endonuclease and zinc finger proteins has made it possible to create artificial nucleases that will cut DNA at a pre-determined site. This approach opens the way to generate many novel ZFN with tailor-made sequence specificities desirable for various applications in biotechnology and medicine.

3.1 Stimulation of Homologous Recombination Through Targeted Cleavage in Frog Oocytes Using ZFN: Recombinogenic Repair

How might these ZFN be used in the correction of a genetic defect? One approach would be to recruit the pre-existing cellular machinery to achieve the goal. Cells of many different organisms use recombination to repair DNA damage, especially DSB. If left un-repaired such DSB in a cellular chromosome would be lethal. Carroll and coworkers (1996) have shown that recombination in frog oocytes proceeds by exonuclease resection and then annealing of complementary strands. This is called single-strand annealing and this same mechanism is thought to operate in mammalian cells. One could utilize this process for gene-targeting experiments. The essence of this approach is to target the chromosomal DNA for sequence-specific cleavage at or near the site where one wants the recombination to occur. Exonuclease would then resect the DNA at that chromosomal site, generating single-strand tails (Lin et al. 1984). Recombination would proceed by a single-strand annealing mechanism with the exogenous DNA that is present in the cell. The homologous foreign DNA would thus be incorporated at this chromosomal site. If the chromosomal target site is not cleaved, then it would not be accessible for exonucleolytic resection. In this case, the exogenous DNA will be degraded while the target remains unchanged. Thus, making a targeted DSB would greatly stimulate homologous recombination between the exogenous DNA and a chromosomal sequence. Such experiments have been performed using Group I intron-encoded homing endonucleases in yeast, cultured mammalian and plant cells (Carroll 1996; Jasin 1996). Although the target sites for these enzymes range from 15–40-bp in length, they exhibit a broad range of sequence specificities and cutting frequencies because of their binding to degenerate sites. In addition, their sequence specificities are not amenable to easy manipulation. As a result they have only a limited use in gene-targeting experiments.

Recently, we have shown in collaboration with Dana Carroll's lab that the three-zinc finger chimeric nuclease, Zif-QQR-F_N, finds and cleaves its target in vivo (Bibikova et al. 2001). This was tested by microinjection of DNA substrates and the enzyme into frog eggs. Double-strand cleavage required an inverted repeat of the 9-bp target (Fig. 1C). When the appropriate sites were placed in the recombination substrate, this DNA was cleaved in frog oocytes by the injected enzyme and homologous recombination ensued. These microinjection experiments have shown that greater than 90% of the substrate was cleaved in vivo and almost all cleaved substrates underwent homologous recombination. Although, in this proof-of-principle experiment, an extrachromosomal target was used to test the idea, two important conclusions can be drawn: (1) chimeric nucleases find and cleave their targets within cells; and (2) the targeted DSB stimulates homologous recombination at the site of cleavage, provided an undamaged template of the homolog is available for recombinogenic repair within cells. This is what normally occurs in cells: when one copy of a gene on one chromosome is damaged, it is repaired using the other undamaged allele on the paired chromosome as a template.

3.2 Targeted Chromosomal Cleavage and Mutagenesis in Fruit Flies Using ZFN: Mutagenic Repair

The obvious next step was to show specific cleavage of a chromosomal target within cells by chimeric nucleases. Recently, Dana Carroll's lab reported specific chromosomal cleavage and mutagenesis in *Drosopila melanogaster* (Bibikova et al. 2002). They designed a pair of ZFN that targeted the yellow (Y) gene on the X-chromosome of *Drosophila*. The nucleases were cloned under the control of a heat shock promoter and then introduced into *Drosophila* genome via P element-mediated transformation. Only when both ZFN were expressed in the developing larvae did somatic mutations occur specifically in the Y gene (yellow gene of *Drosophila* X chromosome). The yellow mosaics were observed in 46% of males expressing both ZFN. No yellow mosaics were observed in control larvae that expressed a single ZFN or without heat shock. Germline Y mutations were recovered from 5.7% of males but none from the females tested. This was as expected in female fruit flies because the targeted DSB would be repaired by recombination with the uncut Y homolog. In males, only simple ligation via nonhomologous end joining (NHEJ) is available to repair the chromosomal damage. DNA sequence analysis of the mutants revealed that all mutations were small deletions and/or insertions localized to the targeted site, which was typical of repair by NHEJ (Gloor et al. 2000; Bibikova et al. 2002). Many of these mutations led to frame shifts or deletion of essential codons of the Y gene. As a result, the male larvae emerging from the heat shock experiment showed yellow patches in the otherwise dark posterior abdomen. In subsequent experiments, Carroll's lab has recovered *y* mutations in female fruit flies, but always at a much lower frequency than that in male flies. This suggests that if the cleavage efficiency of ZFN was high enough to cut both homologues of the Y gene simultaneously, one can produce female germline mutations. Since accurate repair from the homolog appears not to be fully efficient, it might be possible to induce germline mutations in autosomal recessive genes using ZFN (Bibikova et al. 2002). Furthermore, since DSB stimulates mutagenic repair that essentially operates in all organisms, targeted cleavage by ZFN could facilitate generation of directed mutations in a variety of cells and organisms. In a corollary experiment, Bibikova et al. (2003) have shown in *Drosophila*, that gene targeting is enhanced using designed zinc finger nucleases. The ZFN- induced gene-targeting frequencies are about ten fold greater than those observed without target cleavage.

Several important conclusions can thus be drawn from this study: (1) chimeric nucleases can be used to target and cleave a specific chromosomal site within cells; (2) organisms that express ZFN, like the *Drosophila* embryos in this case, are viable and they go on to mature into adults; (3) chimeric nucleases can be used to generate directed mutations in organisms and cells, including targeted germline mutations in autosomal recessive genes – provided the ZFN sequence-specificity and cleavage is high enough to target both copies of the alleles within the genome; and (4) gene targeting – the process of gene replacement by homologous recombination – is enhanced through targeted DSB using designed ZFN.

Similar experiments are underway in our lab using mammalian cells. We plan to correct a tyrosinase gene mutation in albino mouse melanocytes using ZFN. The long-term goal is to correct a tyrosinase gene mutation in albino mouse embryonic stem cells. Melanocytes are melanin pigment-producing cells of the skin. Tyrosinase is a key enzyme for melanin synthesis and pigmentation and it catalyzes the conversion of L-tyrosine to melanin. Melanocytes from albino mice contain a homozygous point mutation (TGT to TCT) in the tyrosinase gene. As a result, the amino acid cysteine is changed to a serine in the tyrosinase enzyme (Shibahara et al. 1990; Alexeev and Yoon 1998). Correction of this point mutation should restore the tyrosinase enzyme activity; and hence confer pigmentation to the albino melanocytes, resulting in dark cells. We have engineered ZFN to target a specific site within the tyrosinase gene for cleavage. In vitro studies indicate that as expected ZFN bind and cleave the target site contained in a plasmid. These constructs are currently being tested in cell culture studies. Continued expression of ZFN is toxic to the melanocytes, which results in cell death. This warrants controlled expression of ZFN in these cells as in the case of experiments done using *Drosophila* larvae. After completion of this study, we plan to perform similar experiments using mouse embryonic stem cells with the eventual goal of generating dark progeny from albino mouse parents.

3.3 Gene Targeting in Human Cells Using ZFN

Gene targeting is a powerful technique to incorporate genetic changes in mammalian genomes. However, this process is very inefficient even in murine embryonic stem cells where it has been successfully applied to produce transgenic mouse "knockouts". Only one cell among 10^6 cells treated with donor DNA has the desired mutant phenotype (Capecchi 1989). Powerful selection techniques are then needed to enrich for the cells carrying the directed mutation. Recently, Porteus and Baltimore (2003) have described a system based on the correction of a mutated GFP gene to study gene targeting in somatic cells. They have shown that gene targeting using ZFN is stimulated over 2000-fold by targeted chromosomal cleavage. This corresponds to a targeted correction rate of 3–5 % of the cells. This is very encouraging indeed since these targeting rates are in the range of being therapeutically useful.

4 Future Experiments

The next logical step is to consider gene targeting using chimeric nucleases as a form of gene therapy for the correction of genetic defects that give rise to human diseases, particularly, in the treatment of monogenic diseases, which arise from deleterious mutations in a single gene.

It would entail the following steps:

1. Identify a target site within the gene of interest. Inverted sequences of the form $(NNC)_3...(GNN)_3$ separated anywhere between 6 to 16 bases make for excellent targets. The target sequence could be within several hundred base pairs from the mutation site for gene conversion.
2. Design and/or select zinc finger proteins (ZF) that recognize the target.
3. Convert the engineered ZF to zinc finger chimeric nucleases (ZFN).
4. Deliver the ZFN and normal gene directly into the nucleus either by microinjection or through viral vectors; ZFN direct targeted chromosomal DSB and stimulate homologous recombination (through recombinogenic repair) with the exogenously provided normal gene. To achieve targeted chromosomal cleavage and mutagenesis by NHEJ (through mutagenic repair) only ZFN need to be delivered into the cells.
5. Monitor for homologous recombination at the target site.

Direct and rigorous experimentation is required to unequivocally establish the induction of homologous recombination at the targeted chromosomal site using human cell lines. Furthermore, it is important to show that unwanted recombination or insertions of the therapeutic gene do not occur at nonhomologous sites at other parts of the genome due to residual cleavage by ZFN. The genomic integrity and stability of treated cells need to be intact at all other loci except at the targeted chromosomal DSB after the ZFN treatment. These studies should be followed by an application to correct a genetic defect in a transgenic animal model.

The focus of our lab is now on two such research projects. In the first, the goal is to achieve targeted chromosomal cleavage and mutagenesis of the

CCR5 (β chemokine receptor 5) gene in human cell lines. In the second, our aim is to attempt targeted correction of a gene mutation in the CFTR (cystic fibrosis transmembrane conductance regulator) gene associated with cystic fibrosis in humans. Both of these projects are described briefly below.

4.1 Targeted Chromosomal Cleavage and Mutagenesis of the CCR5 Gene in a Human Cell Line

A 32-nucleotide deletion (Δ32) within the β-chemokine receptor 5 (CCR5) gene has been identified in individuals who remain uninfected despite extensive exposure to HIV-1 (Huang et al. 1996b; Liu et al. 1996). CCR5 is the major co-receptor on CD4+ cells through which HIV gains entry into the cells. The virus is unable to infect CD4+ lymphocytes and macrophages that are homozygous for the Δ32 mutant CCR5 allele. The goal here is to induce targeted mutagenesis of the CCR5 gene in primate or human cell lines by using engineered ZFN (Fig. 4A).

We have identified two zinc finger target sites near the Δ32 locus of the CCR5 gene. We have engineered ZFN to target and cleave one of these sites. In

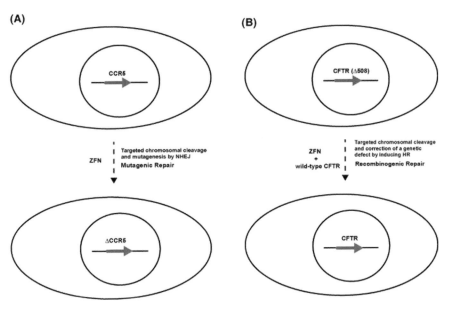

Fig. 4. Gene targeting in a human cell line using ZFN. **A** Targeted chromosomal cleavage and mutagenesis by nonhomologous end joining (NHEJ) through mutagenic repair using engineered chimeric nucleases. In this experiment, only ZFN are introduced into cells. **B** Targeted chromosomal cleavage and correction of a genetic defect by inducing homologous recombination (*HR*) through recombinogenic repair. In this experiment, both ZFN and the normal gene are introduced into cells

vitro studies indicate that the engineered ZFN bind and cleave the target site encoded in a plasmid as expected. Experiments are underway to test this construct in cell culture studies. Our long-term goal is to induce targeted Δ32 deletion at the chromosomal locus encoding the CCR5 gene in hematopoietic stem cells (CD34+ cells) of individuals who are at high risk for HIV infection. The HIV-1 resistant autologous cells will then be amplified and expanded in cell culture. These cells will then be used to reperfuse the bone marrow of these individuals, thereby making their CD4+ lymphocytes and macrophages resistant to HIV-1 infection.

4.2 Targeted Correction of a CFTR Genetic Defect in a Human Cell Line Using ZFN

Cystic fibrosis (CF) is a lethal autosomal recessive hereditary disorder characterized by chronic obstruction and infection of the airway, pancreatic insufficiency, and abnormally high electrolyte concentration in sweat. Although CF is a disease that affects multiple organs, respiratory disease is the major cause of morbidity and mortality: 95 % of CF patients die of respiratory failure. Normal efflux of chloride ions occurs across respiratory epithelial cell membranes in response to elevated cAMP (cyclic adenosine monophosphate). This is lacking in cells derived from CF individuals, which in turn limits water secretion from epithelial cells. As a result, dehydration and excessively thick airway mucus ultimately results in mucous obstruction, infection and inflammation, followed by damage to the airway epithelium. CF is caused by mutations of the cystic fibrosis transmembrane conductance regulator gene (CFTR gene). The most common mutation is a 3-bp deletion leading to the loss of phenylalanine at position 508 in the CFTR protein, known as ΔF508 mutation (Korst et al. 1995; Steen 1997). Our goal is to attempt correction of ΔF508 mutation using ZFN in a human cell line (Fig. 4B).

We have identified target sites within the CFTR gene near the ΔF508 mutation. We have designed ZFN to bind and cleave these targets. These are being tested in vitro for DNA binding and sequence-specific cleavage of the target. Targeted correction of ΔF508 mutation involves somatic cells unlike the CCR5 experiment. The latter is to be done using hematopoietic stem cells that can be expanded in culture and reperfused into patients. The delivery of the CFTR gene to the target site and correction of the gene defect needs to occur in a sufficient number of cells in order to be helpful to the patients. One might have to use lenti-viral vectors for efficient delivery of the therapeutic gene.

5 Potential Limitations of the Chimeric Nuclease Technology

Several things need to be considered when applying this technology in gene therapy.

These include the following: (1) ZF modules that recognize specifically each and every one of all possible triplets have not yet been identified. Only ZF modules for 5'-GNN-3' and 5'-ANN-3' triplets are known. In principle, this is likely to limit the number of human genes that could be targeted by ZFN approach. However, because the target sequence could be within several hundred bp from the mutation site for gene conversion, this may not pose a problem. (2) Targeting a single site within a gene of the human genome may not be very efficient. The identified target, if present in multiple copies within the gene of interest, will be an ideal locus for efficient cleavage. (3) Controlled and regulated expression of ZFN is necessary and highly preferable since continued expression of ZFN within cells is toxic and cell death by apoptosis ensues. Ideally, one would like to express ZFN for a short while within cells, where hopefully they will find their target and induce a DSB, and then disappear via protein degradation pathways. Therefore, methods are needed for controlled and regulated expression of ZFN within mammalian cells. (4) In most cases, both alleles of autosomal recessive genes need to be targeted and mutated to observe a phenotypic effect. This is true of CCR5 gene targeting experiments as well. Heterozygous cells for CCR5Δ32 mutant allele confer only partial protection for about 1–2 years against HIV-1 infection before all cells eventually get infected (Liu et al. 1996). Homozygous cells for CCR5Δ32 mutant allele are resistant to HIV-1 infection. While it may be feasible to target both copies of the CCR5 gene within cells using ZFN, it may not be a very robust process and may prove to be laborious and cumbersome because multiple rounds of selection may be needed.

As an alternate route, one could envision targeted insertion of an inhibitor sequence like for a small interfering RNA (siRNA) within one allele of CCR5, and thereby, control the gene expression from the other homolog. Incidentally, one can use a similar strategy to simultaneously control HIV-1 infection by both M-tropic and T-tropic viruses, which use CCR5 and CXCR4 (α-chemokine receptor 4) co-receptors, respectively.

Furthermore, it appears that both copies of CCR5 gene can be mutated without any adverse side effects to individuals (Proudfoot 2002). The simplest explanation for this is that there are alternate mechanisms available within cells to compensate for the loss of CCR5 function (Gu et al. 2003). This makes CCR5 an ideal locus within the human genome for targeted insertion of other genes for controlled and regulated expression within cells.

6 Future Outlook

Gene targeting using chimeric nucleases is an emerging new technology. Recent advances now make it possible to design and/or select zinc finger proteins capable of recognizing virtually any 18-bp target sequence. This is long enough to specify a unique address within mammalian genomes. In principle, chimeric nucleases (ZFN) that combine the nonspecific cleavage domain of FokI endonuclease with zinc finger proteins (ZF) offer a general way to deliver site-specific DSB to the genome, and stimulate homologous recombination at that site. In the case of chimeric nucleases, binding of two ZFN each recognizing a 9-bp inverted site is necessary, since dimerization of the FokI cleavage domain is required to produce a DSB. Therefore, ZFN effectively have an 18-bp recognition site. Recent experiments from several labs indicate that ZFN find and cleave their chromosomal target; and as expected, they induce local homologous recombination at the site of cleavage (Bibikova et al. 2001, 2002; Porteus and Baltimore 2003). These studies provide a glimpse of potential future therapeutic applications of ZFN in modifying and rewiring the human genome itself. This is a great time to be a biologist. There is an excited anticipation that this novel emerging technology will make targeted correction of a genetic defect feasible, especially in treating monogenic diseases. Several areas of research appear to be converging and coalescing to make gene therapy a reality. These include the availability of the first draft sequence of the human genome. The *complete nucleotide sequence of the genome* is soon to follow this technical feat. Also, great strides and progress are being made in *stem cell* research. The first applications of chimeric nuclease technology using ZFN to treat human diseases are likely to occur in *ex-vivo gene therapy using stem cells*. Here, the desired cells can be identified through selection, expanded in culture and replenished into patients. Thus, the availability of chimeric nucleases, a new type of engineered molecular scissors that target a unique site within the human genome, will contribute and greatly aid the feasibility of genome engineering that is targeted rewiring of the genome. Ethical issues aside, it is not unreasonable to expect that in a decade or two, all the technical problems associated with gene delivery will be overcome; and that gene therapy will be routinely used in a clinical setting, finally signifying a paradigm shift in the treatment of human diseases. Only then, the full impact of the Human Genome Project will be felt, its full potential and goals realized, and its promises fulfilled.

Acknowledgements. The work on chimeric nuclease technology is funded by a grant from National Institutes of Health (GM 53923). We thank Dr. Jean Chin at NIH for her constant support and encouragement of this research. We thank Drs. Alfred Pingoud, Marc Ostermeier, Linzhao Cheng, P.C. Huang, and Ms. Fei Jamie Dy for helpful comments and suggestions. We also thank Drs. Matthew Porteus and David Baltimore for providing their manuscript prior to publication.

References

Alexeev V, Yoon K (1998) Stable and inheritable changes in genotype and phenotype of albino melanocytes induced by an RNA-DNA oligonucleotide. Nat Biotech 16:1343–1346

Beerli RR and Barbas CF III (2002) Engineering polydactyl zinc-finger transcription factors. Nat Biotech 20:135–141

Beerli RR, Segal DJ, Dreier B, Barbas CF III (1998) Toward controlling gene expression at will: specific regulation of the erbB-2/HER-2 promoter by using polydactyl zinc finger proteins constructed from modular building blocks. Proc Natl Acad Sci USA 98:14628–14633

Bibikova M, Golic M, Golic KG, Carroll D (2002) Targeted chromosomal cleavage and mutagenesis in *Drosophila* using zinc finger nucleases. Genetics 161:1169–1175

Bibikova M, Carroll D, Segal, DJ, Trautman JK, Smith J, Kim YG, Chandrasegaran S (2001) Stimulation of homologous recombination through targeted cleavage by chimeric nucleases. Mol Cell Biol 21:289–297

Bibikova M, Beumer K, Trautman JK, Carroll D (2003) Enhancing gene targeting with designed zinc finger nucleases. Science 300:764

Bitinaite J, Wah DA, Aggarwal AK, Schildkraut I (1998) *Fok*I dimerization is required for DNA cleavage. Proc Natl Acad Sci USA 95:10570–10575

Capecchi MR (1989) Altering the genome by homologous recombination. Science 244:1288–1292

Carroll D (1996) Homologous genetic recombination in *Xenopous*: Mechanisms and implications for gene manipulation. Prog Nucleic Acid Res Mol Biol 54:101–125

Chandrasegaran S, Smith J (1999) Chimeric restriction enzymes: what is next? Biol Chem 380:841–848

Connolly JB (2002) Lentiviruses in gene therapy clinical research. Gene Ther 24:1730–1740

Dove SL, Joung JK, Hochschild A (1997) Activation of prokaryotic transcription through arbitrary protein–protein contacts. Nature 386:627–630

Dreier B, Segal DJ, Barbas CF III (2000) Insights into the molecular recognition of the 5′-GNN-3′ family of DNA sequences by zinc finger domains. J Mol Biol 303:489–502

Dreier B, Beerli RR, Segal DJ, Flippin JD, Barbas III, CF (2001) Development of Zinc finger domains for recognition of the 5′-ANN-3′ family of DNA sequences and their use in the construction of artificial transcription factors. J Biol Chem 276:29466–29478

Gloor GB, Moretti J, Mouyal J, Keeler KJ (2000) Distinct P-element excision products in somatic and germline cells of *Drosophila melanogaster*. Genetics 155:1821–1830

Greisman HA, Pabo CO (1997) A general strategy for selecting high-affinity zinc finger proteins for diverse DNA target sites. Science 275:657–661

Gu Z, Steinmetz LM, Gu X, Scharfe C, Davis RW, Li WH (2003) Role of duplicate genes in genetic robustness against null mutations. Nature 421:63–66

Hu JC, Kornacker MG, Hochschild A (2000) *Escherichia coli* one- and Two-hybrid systems for the analysis and identification of protein–protein interactions. Methods 20:80–94

Huang B, Shaeffer CJ, Li Q, Tsai MW (1996a) Splase: a new class IIS zinc finger restriction endonuclease with specificity for Sp1 binding sites. J Prot Chem 15:481–489

Huang Y, Paxton WA, Wolinsky SM, Neumann AU, Zhang L, He T, Kang S, Ceradini D, Jin Z, Yazdanbakhsh K, Kunstman K, Erickson D, Dragon E, Landau NR, Phair J, Ho DD, Koup RA (1996b) The role of a mutant CCR5 allele in HIV-1 transmission and disease progression. Nat Med 11:1240–1243

International Human Genome Sequencing Consortium (2001) Initial sequencing and analysis of the human genome. Nature 409:860–921

Isalan M, Klug A, Choo Y (1998) Comprehensive DNA recognition through concerted interactions from adjacent zinc fingers. Biochemistry 37:12026–12033

Isalan M, Klug A, Choo Y (2001) A rapid, generally applicable method to engineer zinc fingers illustrated by targeting the HIV-1 promoter. Nat Biotech 19:656–660

Jasin M (1996) Genetic manipulation of genomes with rare-cutting endonucleases. Trends Genet 12:224–228

Joung JK, Ramm EI, Pabo CO (2000) A bacterial two-hybrid selection system for studying protein–DNA and protein–protein interactions. Proc Natl Acad Sci USA 97:7382–7387

Kandavelou K, Mani M, Reddy SPM, Chandrasegaran S (update 2003) Ligation: theory and practice. In: Encyclopedia of life sciences. Nature Publishing Group, London

Kim YG, Chandrasegaran S (1994) Chimeric restriction endonuclease. Proc Natl Acad Sci USA 91:883–887

Kim YG, Cha J, Chandrasegaran S (1996) Hybrid restriction enzymes: zinc finger fusions to Fok I cleavage domain. Proc Natl Acad Sci USA 93:1156–1160

Kim YG, Kim PS, Herbert A, Rich A (1997a) Construction of Z-DNA-specific restriction endonuclease. Proc Natl Acad Sci USA 94:12875–12879

Kim YG, Shi Y, Berg JM, Chandrasegaran S (1997b) Site-specific cleavage of DNA–RNA hybrids by zinc finger/*Fok*I cleavage domain fusions. Gene 203:43–49

Kim YG, Smith J, Durgesha M, Chandrasegaran S (1998) Chimeric restriction enzyme: GAL4 fusion to *Fok*I cleavage domain. Biol Chem 379:489–495

Kim MK, Lee JS, Chung JH (1999a) In vivo transcription factor recruitment during thyroid hormone receptor-mediated activation. Proc Natl Acad Sci USA 96:10092–10097

Kim YG, Lowenhaupt K, Schwartz T, Rich A (1999b) The interaction between Z-DNA and Zab domain of double-stranded RNA adenosine deaminase characterized using fusion nucleases. J Biol Chem 274:19081–19086

Kita K, Kotani H, Sugisaki H, Takanami M (1989) The *Fok*I restriction-modification system: organization and nucleotide sequences of the restriction-modification system. J Biol Chem 264:5751–5756

Korst RJ, McElvaney NG, Chu CS, Rosenfeld MA, Mastrangeli A, Hay J, Brody SL, Eissa NT, Danel C, Jaffe HA, Crystal RG (1995) Gene therapy for the respiratory manifestations of cystic fibrosis. Am J Respir Crit Care Med 151:S75-S87

Lee JS, Lee CH, Chung JH (1998) Studying the recruitment of Sp1 to the β-globin promoter with an in vivo method: Protein position identification with nuclease tail (PIN*POINT). Proc Natl Acad Sci USA 95:969–974

Lee JS, Lee CH, Chung JH (1999a) The beta-globin promoter is important for recruitment of erythroid kruppel-like factor to the locus control region in erythroid cells. Proc Natl Acad Sci USA 96:10051–10055

Lee CH, Murphy MR, Lee JS, Chung JH (1999b) Targeting a SWI/SNF-related chromatin remodeling complex to the β-globin promoter in erythroid cells. Proc Natl Acad Sci USA 96:12311–12315

Lee JS, Ngo H, Kim D, Chung JH (2001) Erythroid Krüppel-like factor is recruited to the CACCC box in the β-globin promoter but not to the CACCC box in the γ-globin promoter: The role of the neighboring promoter elements. Proc Natl Acad Sci USA 97:2468–2473

Li L, Chandrasegaran S (1993) Alteration of the cleavage distance of *Fok*I restriction endonuclease by insertion mutagenesis. Proc Natl Acad Sci USA 90:2764–2768

Li L, Wu LP, Chandrasegaran S (1992) Functional domains in *Fok*I restriction endonuclease. Proc Natl Acad Sci USA 89:4275–4279

Li L, Wu LP, Clarke R, Chandrasegaran S (1993) Insertion and deletion mutants of Fok I restriction endonuclease by insertion mutagenesis. Proc Natl Acad Sci USA 90:2764–2768

Lin FL, Sperle K, Sternberg N (1984) Model for homologous recombination during transfer of DNA into mouse L cells: role for the ends in the recombination process. Mol Cell Biol 4:1020–1034

Liu Q, Xia Z, Zhong X, Case CC (2002) Validated zinc finger protein designs for all 16 GNN DNA triplet targets. J Biol Chem 277:3850–3856

Liu R, Paxton WA, Choe S, Ceradini D, Martin SR, Horuk R, MacDonald ME, Stuhlmann H, Koup RA, Landau NR (1996) Homozygous defect in HIV-1 coreceptor accounts for resistance of some multiply-exposed individuals to HIV-1 infection. Cell 86:367–77

Looney MC, Moran LS, Jack WE, Feehery GR, Benner JS, Slatko BE, Wilson GG (1989) Nucleotide sequence of the FokI restriction-modification system: separate strand specificity domain in the methyltransferase. Gene 80:209–216

Mani M, Kandavelou K, Chandrasegaran S (update 2003a) Restriction enzymes. In: Encyclopedia of life sciences. Nature Publishing Group, London

Mani M, Kandavelou K, Wu J, Chandrasegaran S (2003b) Restriction enzymes: essential tools for analyzing and manipulating deoxyribonucleic acid. In: Encyclopedia of human genome. Nature Publishing Group, London

McNarma AR, Hurd PJ, Smith AEF, Ford KG (2002) Characterisation of site-biased DNA methyltransferases: specificity, affinity and subsite relationships. Nucleic Acids Res 30:3818–3830

Ordiz MI, Barbas CF III, Beachy RN (2002) Regulation of transgene expression in plants with polydactyl zinc finger transcription factors. Proc Natl Acad Sci USA 99:13290–13299

Pabo CO, Peisach E, Grant RA (2001) Design and selection of novel Cys_2His_2 zinc finger proteins. Annu Rev Biochem 70:313–340

Pavletich NP, Pabo CO (1991) Zinc finger–DNA recognition: crystal structure of a Zif268-DNA complex at 2.1 Å. Science 252:809–817

Pavletich NP, Pabo CO (1993) Crystal structure of a five-finger GLI-DNA complex: new perspectives on zinc fingers. Science 261:1701–1707

Porteus MH, Baltimore D (2003) Chimeric nucleases stimulate gene targeting in human cells. Science 300:763

Proudfoot AEI (2002) Chemokine receptors: Multifaceted therapeutic targets. Nature Rev Immun 2:106–115

Rebar EJ, Huang Y, Hickey R, Nath AK, Meoli D, Nath S, Chen B, Xu L, Liang Y, Jamieson AC, ZhangL Spratt SK, Case CC, Wolffe A and Giordano FJ (2002) Induction of angiogenesis in a mouse model using engineered transcription factors. Nat Med 12:1427–1432

Ruminy P, Derambure C, Chandrasegaran S, Salier JP (2001) Long-range identification of hepatocyte nuclear factor-3 (Fox A) high and low-affinity binding sites with a chimeric nuclease. J Mol Biol 310:523–535

Segal DJ, Dreier B, Beerli RR, Barbas CF III (1999) Toward controlling gene expression at will: selection and design of zinc finger domains recognizing each of the 5′-GNN-3′ DNA target sequences. Proc Natl Acad Sci USA 96:2758–2763

Smith J, Berg JM, Chandrasegaran S (1999) A detailed study of the substrate specificity of a chimeric restriction enzyme. Nucleic Acids Res 27:674–681

Smith J, Bibikova M, Whitby FG, Reddy AR, Chandrasegaran S, Carroll D (2000) Requirements for double-strand cleavage by chimeric restriction enzymes with zinc finger DNA-recognition domains. Nucleic Acids Res 28:3361–3369

Shibahara S, Okinaga S, Tomkta Y, Takeda A, Yamamoto H, Sato M, Takeuchi T (1990) A point mutation in the tyrosinase gene of BALB/c albino mouse causing the cysteine–serine substitution at position 85. Eur J Biochem 189:455–461

Steen CD (1997) Cystic fibrosis: inheritance, genetics and treatment. Brit J Nursing 6:192–199

Stege JT, Guan X, Ho T, Beachy RN, Barbas CF III (2002) Controlling gene expression in plants using synthetic zinc finger transcription factors. Plant J 32:1077–1086

Sugisaki H, Kanazawa S (1981) New restriction endonucleases from *Flavobacterium okeanokoites* (*Fok*I) and *Micrococcus luteus* (MluI). Gene 16:73–78

Vasquez MK, Marburger K, Intody S, Wilson JH (2001) Manipulating the mammalian genome by homologous recombination. Proc Natl Acad Sci USA 98:8403–8410

Venter JC, Adams MD, Myers EW et al. (2001) The sequence of the human genome. Science 291:1304–1351

Wah DA, Hirsch JA, Dormer LF, Schildkraut I, Aggarwal AK (1997) Structure of the multimodular endonuclease *Fok*I bound to DNA. Nature 399:97–100

Wah DA, Bitinaite J, Schildkraut I, Aggarwal AK (1998) Structure of FokI has implications for DNA cleavage. Proc Natl Acad Sci USA 95:10564–10569

Waugh DS, Sauer RT (1993) Single amino acid substitutions uncouple the DNA-binding and strand scission activities of *Fok*1 endonuclease. Proc Natl Acad Sci USA 90:9596–9600

Waugh DS, Sauer RT (1994) A novel class of *Fok*1 restriction endonuclease mutant that cleave hemi-methylated substrates. J Biol Chem 269:369–379

Xu GL, Bestor T (1997) Cytosine methylation targeted to pre-determined sequences. Nat Genet 17:376–378

Yonezawa A, Sugiura Y (1994) DNA-binding mode of class-IIS restriction endonuclease *Fok*I revealed by DNA footprinting analysis. Biochem Biophys Acta 1219:369–379

Subject Index

DATE DUE